Timber Engineering

Timber Engineering

Edited by

Sven Thelandersson
Lund University, Sweden

and

Hans J. Larsen
Technical University of Denmark, Lyngby, Denmark

JOHN WILEY & SONS, LTD

Telephone (+44) 1243 779777

Email (for orders and customer service enquiries): cs-books@wiley.co.uk
Visit our Home Page on www.wileyeurope.com or www.wiley.com

Other Wiley Editorial Offices

John Wiley & Sons Inc., 111 River Street, Hoboken, NJ 07030, USA

Jossey-Bass, 989 Market Street, San Francisco, CA 94103-1741, USA

Wiley-VCH Verlag GmbH, Boschstr. 12, D-69469 Weinheim, Germany

John Wiley & Sons Australia Ltd, 33 Park Road, Milton, Queensland 4064, Australia

John Wiley & Sons (Asia) Pte Ltd, 2 Clementi Loop #02-01, Jin Xing Distripark, Singapore 129809

John Wiley & Sons Canada Ltd, 22 Worcester Road, Etobicoke, Ontario, Canada M9W 1L1

Wiley also publishes its books in a variety of electronic formats. Some content that appears in print may not be available in electronic books.

Library of Congress Cataloging-in-Publication Data

Timber engineering/edited by Sven Thelandersson and Hans J. Larsen.
p. cm.
Includes bibliographical references and index.
ISBN 0-470-84469-8 (alk. paper)
1. Building, Wooden. 2. Timber. I. Thelandersson, Sven, 1944- II. Larsen, Hans J.

TA666 .T495 2003
624.1′84 – dc21

2002027217

British Library Cataloguing in Publication Data

A catalogue record for this book is available from the British Library

ISBN 0-470-84469-8

Typeset in 10/12pt Times by Laserwords Private Limited, Chennai, India
Printed and bound by CPI Antony Rowe, Eastbourne, East Sussex
This book is printed on acid-free paper responsibly manufactured from sustainable forestry in which at least two trees are planted for each one used for paper production.

Contents

List of Contributors

Simon Aicher	Forschungs- und Materialprufungsanstalt fur das Bauwesen Otto-Graf-Institut, Universität Stuttgart, Pfaffenwaldring 4 D-70569 Stuttgart (Vaihingen), Germany
Hans J. Blass	Universität Karlsruhe, Ingenieurholzbau und Baukonstruktion Eschenweg 2, 76297 Stutensee, Germany
Ario Ceccotti	Trees and Timber Institute – National Research Council of Italy, NRC-TTI, via Barazzuoli 23, 50136 Florence, Italy
Ricardo O. Foschi	The University of British Columbia, Department of Civil Engineering, 2324 Main Mall, Vancouver, B.C., V6T 1Z4 Canada
Per Johan Gustafsson	Lund University, Structural Mechanics, Box 118, SE-221 00 Lund, Sweden
Preben Hoffmeyer	Technical University of Denmark, DTU, Department of Civil Engineering, Brovej – Building 118, DK-2800 Kgs. Lyngby, Denmark
Tord Isaksson	Lund University, Structural Engineering Box 118, SE-221 00 Lund, Sweden
Carl-Johan Johansson	SP, Swedish National Testing and Research Institute, Box 857, SE-501 15 Borås, Sweden
Erol Karacabeyli	Forintek Canada Corp., Western Laboratory, 2665 East Mall, Vancouver B.C., V6T 1W5 Canada
Frank Lam	Dept. of Wood Science, Univ. of British Columbia, 2329 West Mall, Vancouver, B.C., V6T 1Z4 Canada
Hans J. Larsen	Technical University of Denmark, DTU, Department of Civil Engineering, Brovej – Building 118, DK-2800 Kgs. Lyngby, Denmark
Annika Mårtensson	Lund University, Structural Engineering, Box 118, SE-221 00 Lund, Sweden
Alpo Ranta-Maunus	VTT Building Technology, Structures, P.O. Box 1805, FIN-02044 VTT, Finland

Jacob Nielsen	Aalborg University, Building Technology and Structural Engineering, Sohngaardsholmsvej 57, DK-9000 Aalborg, Denmark
Marjan Popovski	Forintek Canada Corp., Western Laboratory, 2665 East Mall, Vancouver B.C., V6T 1W5 Canada
Helmut Prion	Depts. of Civil Engineering and Wood Science, Univ. of British Columbia, 2329 West Mall, Vancouver, B.C., V6T 1Z4 Canada
Erik Serrano	Lund University, Structural Mechanics, Box 118, SE-221 00 Lund, Sweden
Ian Smith	University of New Brunswick, Structural/Timber Engineering, P.O. Box 44 555, Fredericton, NB, E3B 6C2 Canada
Sven Thelandersson	Lund University, Structural Engineering Box 118, SE-221 00 Lund, Sweden

Preface

Timber engineering deals mainly with the structural aspects of timber construction, including material behaviour and properties on wood element level. The field is closely related to wood science, which provides an important basis for the structural behaviour on a larger scale. For basic elements of wood science, the reader is referred to Dinwoodie (2000). The relations between clear wood properties and the behaviour of timber in structural sizes is described in a very instructive way by Madsen (1992). Another book by Madsen (2000) gives a comprehensive overview of the structural behaviour of timber connections.

The purpose of the present book is to summarise the state-of-the-art in timber engineering, with a focus on key issues for the understanding of

- timber as a modern engineered construction material with controlled and documented properties;
- the background for the design of structural systems based on timber and engineered wood products; and
- the background for the structural design of joints in structural timber systems.

To read this book it is necessary to have a prior knowledge of the elementary design of timber structures, as well as a general background in structural engineering at a basic level. The book could be thought of as a teaching resource at masters level, or correspondingly. For more basic elements of timber engineering, including methods of design, the reader is referred to STEP (1995), for example.

The book has been written by a number of international renowned experts, who were invited to write high quality state-of-the-art papers on their specific subject with a selection of key references. The result is a collection of advanced teaching materials, with a good chance of being usable for some future time period.

So that the content of the book has a high actuality in time and space, it is mainly concerned with research-based knowledge, and is independent of codes and standards. Codified rules developed from research are given on some occasions, but they should then be seen as examples. The goal has been to present the best knowledge available at the time of writing about the phenomena treated in the book. The intention is that readers world wide should be able to use this book as a basis for understanding the background for any particular code.

The book starts with a general introduction to timber engineering. Then, the contributions from different authors are grouped into the following sections:

Part One: Basic properties of wood-based structural elements (seven chapters).

Part Two: Design aspects on timber structures (five chapters).

Part Three: Joints and structural assemblies (five chapters).

A general overview is presented as an introduction for each of the sections.

Lund and Copenhagen 2002
Sven Thelandersson, Hans J. Larsen

REFERENCES

Dinwoodie J.M. (2000) *Timber: Its nature and behaviour*, 2nd ed., E&FN Spon, London.

Madsen B. (1992) *Structural Behaviour of Timber*. Timber Engineering Ltd., Vancouver, Canada.

Madsen B. (2000) *Behaviour of Timber Connections*. Timber Engineering Ltd., Vancouver, Canada.

STEP (1995) *Timber Engineering, STEP*, Volumes 1 and 2, Centrum Hout, Netherlands.

1

Timber Engineering – General Introduction

Sven Thelandersson

1.1 TIMBER – OUR OLDEST BUILDING MATERIAL

Protection and shelter against wind, rain and cold is a very basic need for mankind. Since ancient times, wood has been the most important material used for this purpose. In developed cultures, the art of house construction was already quite advanced several thousand years ago. Figure 1.1 shows an archaeological reconstruction of a so-called long house in Central Europe from 3000 BC (Kuklik, 2000). The width of this type of house was in the range of 5.5–7 m and the length varied from 20–45 m. The main structural elements were made from round timber. This can be seen as an early example of a timber framed house, which in various forms has been used ever since, especially in forested regions.

Some parts of Asia have a very long history of timber construction. In Japan the oldest timber structure still in existence is from the seventh century (Yasumura, 2000). A typical historical timber building is the three storey pagoda, shown in Figure 1.2. This building, which still exists, was constructed in 730 AD. It has a double roof in each storey supported by a central wooden pole.

Another region with a long tradition of timber construction is Scandinavia, where wood is a resource that has always been easily available. The oldest existing timber building in Scandinavia is the Borgund church in Norway, built in the twelfth century (see Figure 1.3). The load-bearing

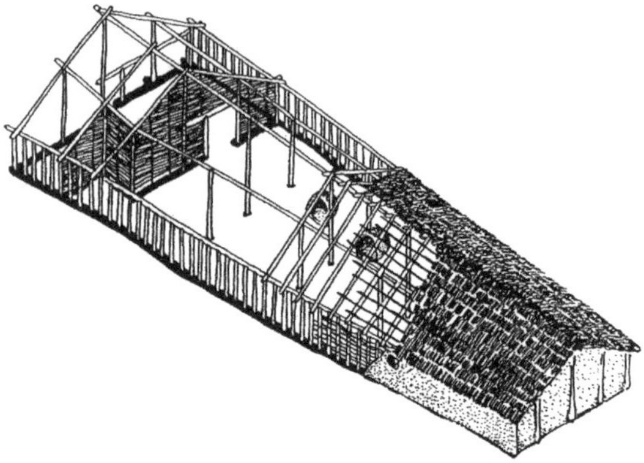

Figure 1.1 Archaeological reconstruction of long house from 3000 BC (Reproduced from Kuklik, 2000 by permission of Petr Kuklik, Prague University)

Figure 1.2 Three storey pagoda, Yakusiji Toto, built 730 (Yasemura, 2000). (Reproduced by permission of M. Yasemura)

structure is a three-dimensional frame with roundwood poles and horizontal timber trusses connected by semi-rigid arch-shaped joints.

Figure 1.3 Borgund church, Norway, twelfth century (Photo: Lund University)

Another important application for timber construction is bridges. Before the emergence of modern structural materials such as steel and concrete, timber was the dominating structural material in bridge construction. In 55 BC, the emperor Julius Caesar had a 140 m long temporary timber bridge built across the Rhine (Figure 1.4). The bridge was 5–6 m wide, and allowed two lane traffic. Only 10 days were needed to complete the bridge (Timber Bridges, 1996).

One of the oldest existing timber bridges is the 222 m long Chapel Bridge in Luzern, Switzerland, which was built in 1333 (Stadelmann, 1990). This bridge is a well-known tourist attraction. As with many other bridges in Switzerland, it is covered by a roof, which effectively protects the wood from biological deterioration. Unfortunately, the bridge was partly destroyed by a fire in 1993, but has been rebuilt in its original form (see Figure 1.5).

Although historic timber structures have disappeared to a greater degree than, for example, structures made of stone, these examples show that timber has excellent durability provided that the structures are properly designed and maintained. One aspect of long-term durability of timber structures is that they are often designed in such a way that damaged elements can easily be replaced.

The fact that timber has been used extensively as a building material for a very long time does not mean that we have a deep scientific understanding

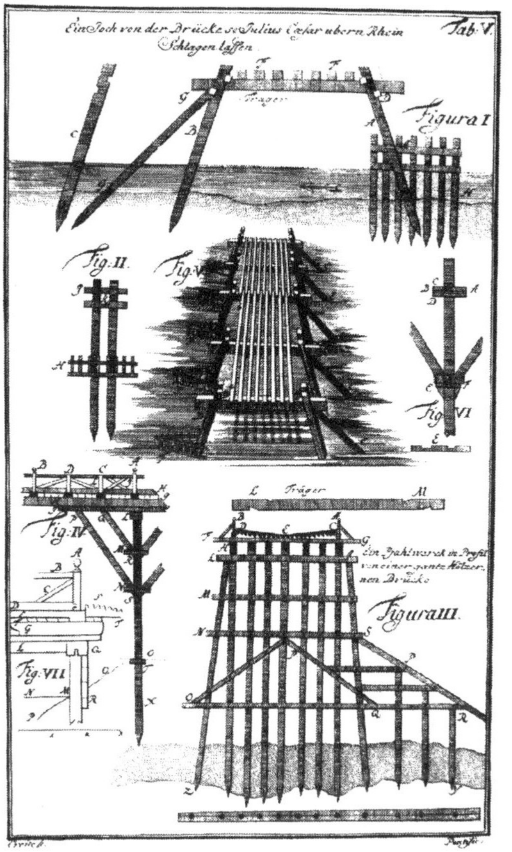

Figure 1.4 Caesar's bridge across Rhine 55 BC (Jacob, 1726)

Figure 1.5 Chapel bridge in Luzern, Switzerland originally built 1333, restored after fire damage 1993 (Photo: Sven Thelandersson)

of the behaviour of the material. On the contrary, timber construction has to a large extent been based on empirical experience and craftsmanship. Wood is therefore often seen as a material with inadequate control and documentation of its properties and behaviour. The purpose with this book is to present recent advances in research to improve our understanding of the structural performance of timber.

1.2 MODERN TIMBER CONSTRUCTION

Today, the growing stock volume of wood worldwide is estimated to 490 billion m^3 (FAO, 2000a). The total world production of timber in 1999 was 3275 million m^3 (FAO, 2000b). It is estimated that around 55% of this volume is used as fuel. A substantial part of the raw material is used for pulp and paper, and 317 million tons of paper was produced worldwide in 1999. The total volume of sawn timber and panel products produced in the same year was 592 million m^3, which is 18% of the total raw material production. The relative amounts of different wood products are shown in Table 1.1.

Timber is used as a major structural material in a great variety of building and civil engineering applications. *Lightweight timber frame systems* (based on structural timber, engineered wood products and panels) may be used for single family houses, multi-storey residential buildings and commercial buildings. Similar elements are used as walls and roofs in industrial buildings. Timber is often used for roof construction in buildings, even

Table 1.1 Production of sawn goods and wood-based panels in the world, 1999 (FAO, 2000b)

Product	Volume, 10^6 m^3	Volume, %
Sawn goods, softwood	323.2	55
Sawn goods, hardwood	108.5	18
Fibreboard	30.2	5
Particle board	75.2	13
Veneer sheets	6.4	1
Plywood	48.1	8
Total	591.6	100

if the rest of the structure is made from concrete or steel. *Large-scale timber systems* (based on glulam, LVL and other engineered wood products) may be used for industrial and commercial buildings with long spans, as well as for bridges, parking decks, etc. Worldwide, the potential to increase the utilisation of timber for these applications in the future is large.

There are many general advantages in using timber for building purposes. It is an environmentally friendly, easily recyclable material. The energy consumption during production is very low compared to that of other building materials. Timber has a low weight in relation to strength, which is advantageous for transport, erection and production. The foundation can be simplified and low inertia forces make the building less sensitive to earthquakes. Furthermore, wood has aesthetic qualities, which give great possibilities in architectural design.

Building systems based on wood has a great potential to be rational and cost-effective. Experiences from North America, where timber frame building systems have a dominating position in the market, indicate that it is possible to reduce the cost of low-to-medium rise buildings significantly by using lightweight building systems based on timber and panel products. These systems have the advantage of simple construction techniques at the building site, and very short construction times can be achieved.

1.3 THE TIMBER FRAME BUILDING CONCEPT

Timber frame buildings are built up by a skeleton of timber joists and studs, covered with panels fastened to the wood elements. Wood-based panels, such as plywood, OSB, fibre-board or chipboard, with structural quality, are commonly used in timber frame buildings. Gypsum panels or similar products are also widely used in combination with timber, mainly to provide fire resistance. A typical timber frame house is shown in Figure 1.6.

The timber frame concept is also very competitive for multi-storey, multi-residential buildings up to 5–6 storeys (see Figure 1.7).

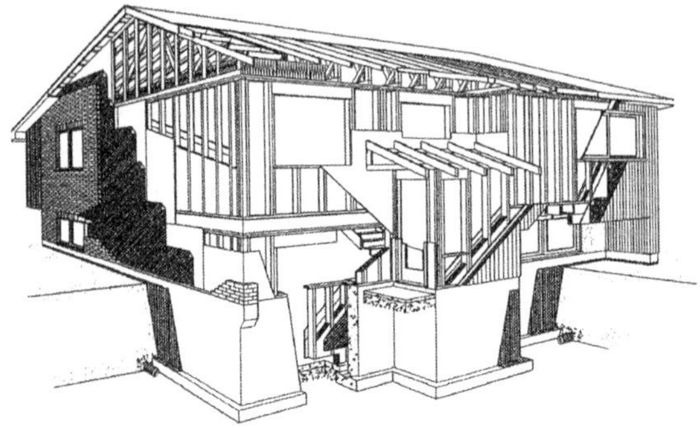

Figure 1.6 The anatomy of a typical timber frame small house

Timber frame systems can be conceived as composite wall and floor units built up from timber framing, panel products, insulation, cladding, etc., with good possibilities to adapt the design to various requirements. One and the same composite unit in a timber frame system can be utilised for the

- transfer of vertical loads,
- stabilization of wind and earthquake loads,
- physical separation,
- fire separation,
- sound insulation, and
- thermal insulation.

It is important that the design is made so that all the relevant requirements are met in an optimal way. In the design of walls and floors, different aspects can be identified as critical. The factors governing the design of walls are, in order of priority:

- fire resistance,
- horizontal stabilisation,
- sound insulation, and
- vertical loading.

A typical design of a load-bearing, stabilising wall used for separation between flats is shown in Figure 1.8.

For the design of floors, the most important factors are, in order of priority:

- impact sound insulation,
- vibration control,

Figure 1.7 Four storey platform frame timber house under construction (Reproduced by permission of Holger Staffansson, Skanska AB)

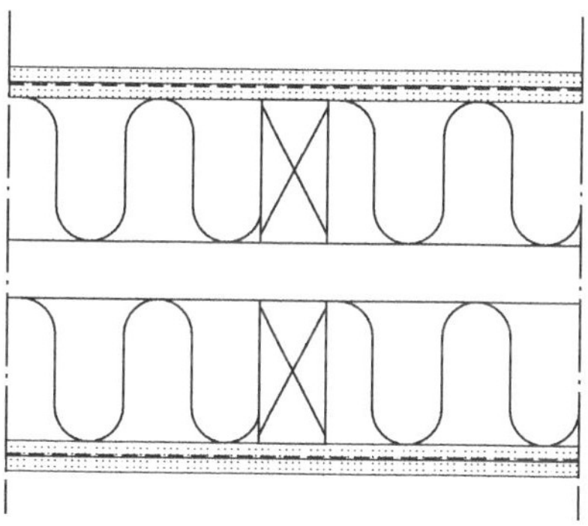

Figure 1.8 Horizontal section of double wall separation between flats. Double fire guard gypsum panels are used

- simplicity in production, and
- possibility for the installation of services.

In many countries, customers expect acoustic performance of a high standard. It is possible to meet these requirements with lightweight floor structures, but the solutions often become complicated and expensive. For this reason, there is a need to develop better floor solutions. One alternative could be to use floors based on laminated timber decking, where the material cost is higher, but the floor is still competitive due to a simpler production process. Composite floors with concrete on top of timber decking, or in combination with timber joists are to some extent used in Central Europe.

A very crucial issue for the efficiency of a timber frame system is the solution of wall-floor joints. In the design of such joints, a number of aspects must be taken into account:

- sound performance,
- load transfer vertically and in shear,
- fire separation,
- shrinkage and settlement of the joint (perpendicular to the grain),
- thermal insulation,
- air tightness,
- ease of erection,
- degree of prefabrication of walls and floors, and
- economy.

In the platform frame system commonly used in North America, standard solutions are available for wall/floor joints (see Figure 1.9a). Since the vertical forces are transferred in compression perpendicular to the grain in the whole joint, substantial shrinkage and vertical settlements will occur. The deformations created by this can be quite difficult to handle in a multi-storey building. An example of a wall-floor joint solution, designed to minimise shrinkage settlements in the joint, is shown in Figure 1.9b.

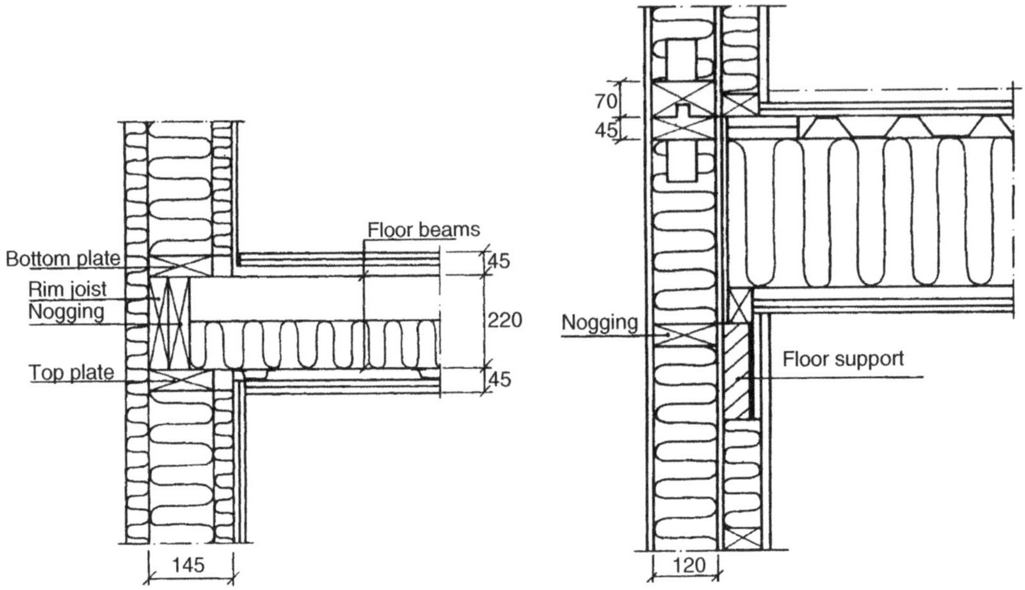

Figure 1.9 Wall-floor joints. (a) Platform frame design, (b) design to minimise shrinkage

In concrete and masonry buildings, staircase and elevator shafts as well as cross walls can be used to stabilise the building. In timber frame construction, timber frame shear walls are used for lateral stabilisation against wind and earthquakes (see Figure 1.10). For multi-storey timber framed buildings, the issue of stabilisation is not trivial. A trend towards narrow houses to provide good daylight, as well as requirements of high acoustic performance, makes it more difficult to stabilise the buildings in an economical manner. To ensure good acoustic performance, double walls are used between flats (see Figure 1.8), and to prevent flanking transmission, the horizontal floor diaphragms should be disconnected at the double walls. However, continuity in the floor diaphragms is needed for efficient horizontal stabilisation. This conflict must be resolved by the structural engineer. Recent research has shown that a better understanding of the force transfer in timber frame systems can contribute towards achieving a more economical and flexible design of the stabilising system in timber frame buildings (e.g. see Andreasson (2000) and Paevere (2001)).

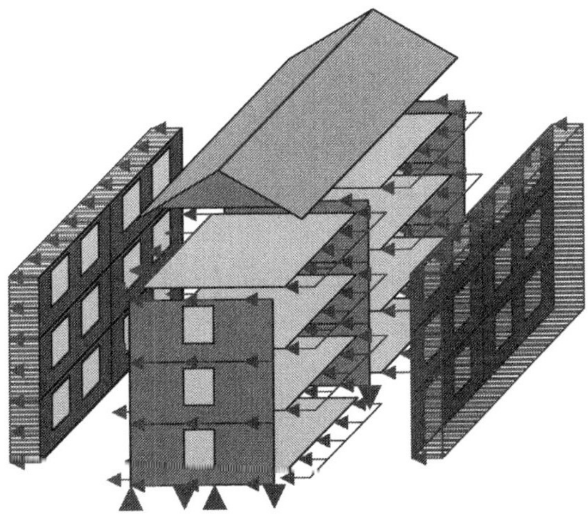

Figure 1.10 Force transfer in multistorey timber frame building under lateral loading. (Drawing: Sverker Andreasson. Reproduced by permission of Lund University)

The economy of multi-storey timber houses depends very much upon whether a simple bracing system is sufficient, i.e. if the forces can be transferred through the wall boards required for fire protection, or whether it is necessary to use more expensive wood-based panels and extra studs and anchorages.

1.4 LARGE-SCALE TIMBER CONSTRUCTION

The maximum dimension of solid timber sawn from logs is of the order of 300 mm or even less, depending on species and growth region. This means that the maximum possible span of structural timber beams in practice is limited to 5–7 m. Before the appearance of engineered wood products such as glulam, timber trusses were therefore commonly used to achieve larger spans, which is often needed in roof and bridge construction. Timber trusses produced from structural timber are still the most common solution for roof structures in small residential houses (Figure 1.11). This type of truss is today almost exclusively designed with punched metal plate fasteners, creating stiff and strong truss joints at a low cost. Timber roof trusses are frequently used for spans of up to 12 m, but can be designed for spans up to 30–40 m.

Another possibility to extend the spans for timber structures is to use laminated beams, i.e timber laminations stacked on top of each other and structurally connected to form members with large cross-sections. Early applications used mechanical fasteners, such as bolts, dowels and rods, to connect the laminations. The potential of the lamination technique was, however, not fully exploited until synthetic glues became generally available in the early twentieth century (see Figure 1.12). Glued laminated timber or *glulam* became one of the first engineered wood products, and is still very competitive in modern construction. By bending the laminations before gluing, it can be produced in curved shapes. The cross-section depth is in principle unlimited, but for practical reasons maximum depths are of the order of 2 m. This makes glulam an ideal material to create structures for large spans. A variety of structural systems based on straight and curved glulam members has been developed for roofs with spans of up to 100 m (see Figure 1.13). Today, several other wood-based products are available for large-scale timber structures, such as Laminated Veneer Lumber (LVL) and Parallel Strand Lumber (PSL) (see Chapter 4 in Part One). These products are suitable for larger spans in a similar way as straight glulam members.

Figure 1.12 Glulam arch roof for Stockholm central railway station, built 1925. Reproduced by permission of Svenskt Limträ AB

For straight and tapered beams, spans of 30 m and more can be achieved. Figure 1.14 shows a typical glulam roof with straight tapered beams for hall buildings.

Spans of up to 50 m can be realised by three hinge frames built up by two curved glulam elements, as shown in Figures 1.13 and 1.15. Frames can also be made from straight members, with

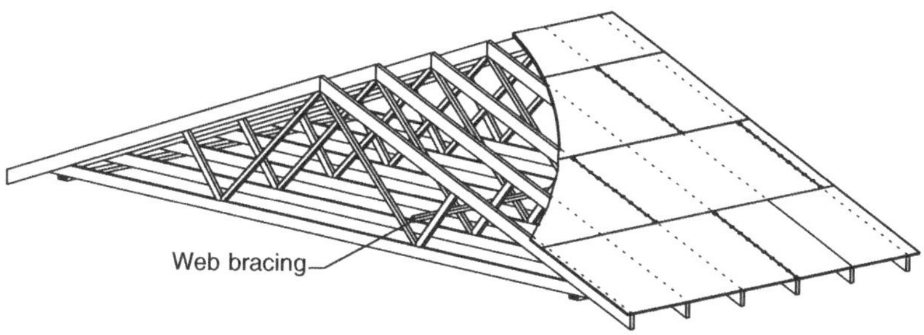

Figure 1.11 Typical roof truss for a single family house (Drawing: Jacob Nielsen, Aalborg University, Denmark)

SYSTEM	DESCRIPTION	SPAN m	SECTION DEPTHS
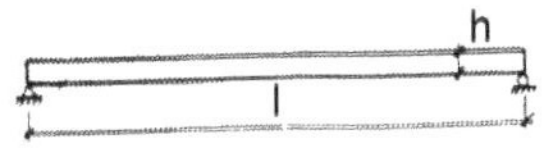	Straight beam	< 30	***h* ≈ *l*/17**
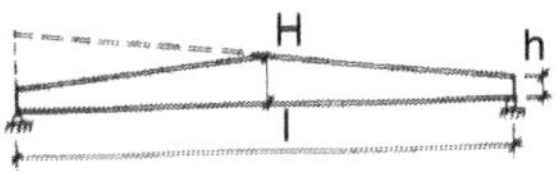	Tapered beam	10–30	***h* ≈ *l*/30** ***H* ≈ *l*/16**
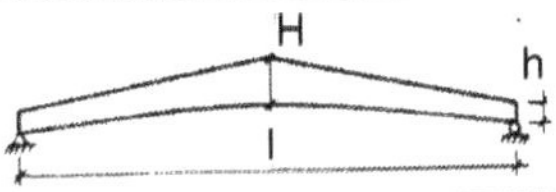	Pitched cambered beam	10–20	***h* ≈ *l*/30** ***H* ≈ *l*/16**
	Three-hinged truss with tie-rod	15–50	***h* ≈ *l*/30**
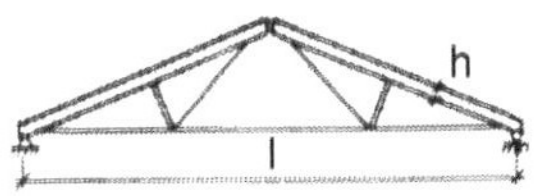	Three-hinged truss with tie-rod and trussed beams	20–100	***h* ≈ *l*/40**
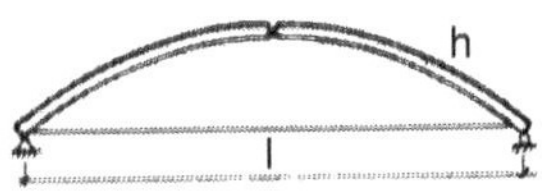	Three-hinged arch	20–100	***h* ≈ *l*/50**
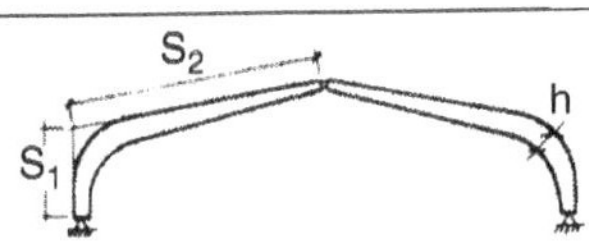	Three-hinged curved frame	15–50	$h \approx (s_1+s_2)/15$
	Three-hinged frame with finger-jointed knee	10–35	$h \approx (s_1+s_2)/13$
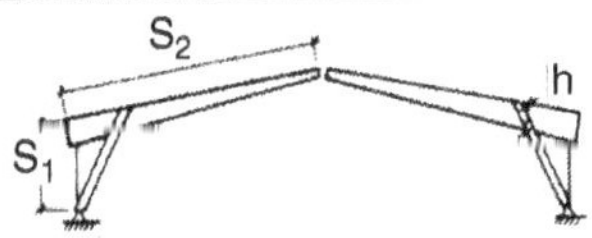	Three-hinged frame with trussed knee	10–35	$h \approx (s_1+s_2)/15$

Figure 1.13 Structural systems for glulam (Glulam Manual, 1995. Reproduced by permission of Svenskt Limträ AB)

the moment resisting frame corners manufactured using a finger jointing technique. Several types of mechanical joints are also commonly used for such frame corners, but they are generally less effective in resisting moments.

Similarly, plane arches can be constructed by pre-manufactured curved glulam elements for spans of up to 100 m. An example is shown in Figure 1.16. For spans larger than 60 m, the difficulty associated with transport of the curved elements is usually a restricting factor. Glulam arches are normally designed as three-hinged, which gives simpler joints to be arranged at the building site. Also, a statically determinate system is usually preferred to avoid restraint forces from moisture movements in the wood.

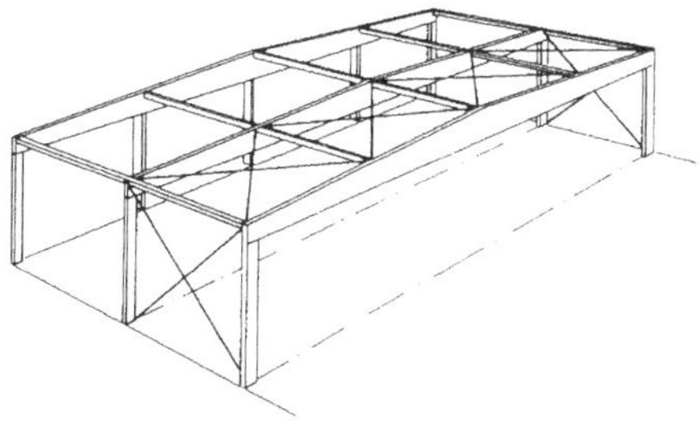

Figure 1.14 Roof structures with straight glulam beams can be used for spans of up to 30–40 m. Steel wires are used for diagonal wind bracing

Figure 1.15 Three-hinged glulam curved frame. Reproduced by permission of Svenskt Limträ AB

Figure 1.16 Glulam arch structure made from two curved glulam elements. Reproduced by permission of Svenskt Limträ AB

For very large spans, glulam trusses may be used. Very efficient truss joints can be made by slotted-in steel plates combined with dowel fasteners. Rational methods for manufacturing such joints have been developed, making truss systems of this type competitive. Examples of trussed glulam structures are shown in Figures 1.17 and 1.18.

Glulam or other engineered wood products may also be used efficiently for spatial frames and dome structures. Special detailing solutions are usually

Figure 1.17 Glulam arch truss with a free span of 86 m for a sports facility in Lillehammer, Norway (Photo: Sven Thelandersson)

Figure 1.18 Roof structure for Gardemoen air terminal, Oslo, Norway. The main girder is a slightly curved glulam truss, covered by plywood for the sake of appearance. Reproduced by permission of Svenskt Limträ AB

Figure 1.19 Interior from the Tacoma Dome, Washington, USA with a 162 m span (Photo: Sven Thelandersson)

Figure 1.20 Glulam roof for a swimming hall in Bad Durrheim, Germany, built 1987 (Reproduced by permission of Arge Holz, Dusseldorf)

developed for the three-dimensional joints in such systems. Figure 1.19 shows one of the largest timber structures in the world, with a diameter of 162 m.

Many good architects around the world prefer to work with wood as a major structural material. There are numerous examples worldwide of large span timber buildings with excellent architecture and innovative design. An example is shown in Figure 1.20.

Another important application for large-scale timber construction is bridges. For small spans, straight beams of solid wood, glulam or other engineered wood products can be used as the primary load-bearing elements. Trusses, arches or framed structures can be used as primary structures for larger spans. A very common solution for the bridge deck is to use laterally prestressed timber plates of the type shown in Figure 1.21. As an alternative to lateral prestress, the planks can be nailed to each other, although this gives lower transverse stiffness of the deck.

In modern bridge construction, timber is growing in popularity for foot and bicycle bridges as well as road bridges with moderate spans, especially in the USA, Central Europe and Scandinavia. One reason for this is environmental awareness and the trend towards the use of ecologically sound materials in construction. New efficient jointing techniques developed in recent years are also very important for competitiveness in timber bridge construction. An excellent example of a modern timber bridge for road traffic is shown in Figure 1.22.

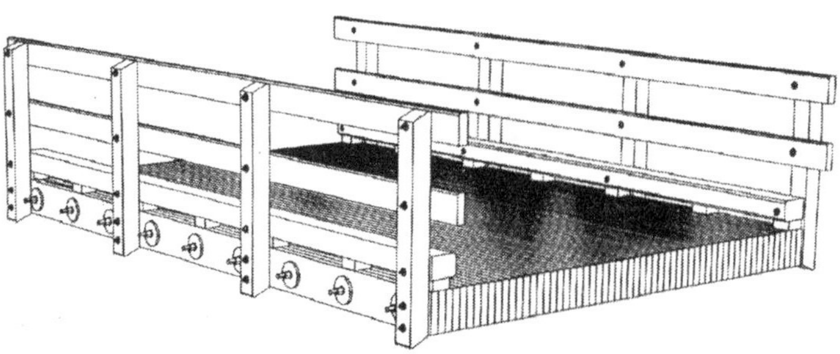

Figure 1.21 Bridge with laminated timber decking (Reproduced by permission of Svenskt Trä)

Figure 1.22 Europabrücke, Murau, Austria, built 1993. Glulam structure with concrete bridge deck (Reproduced by permission of Holger Gross)

A key factor for timber bridge design is durability. Preservative chemical treatment is not an attractive alternative considering environmental policies of today. However, by careful design and detailing, the wood material in a timber bridge can be kept more or less constantly dry, so that biological decay is avoided and long lifetimes can be achieved with or without very limited use of preservative treatment.

Laminated decks of the type shown in Figure 3.16 have also become popular for floors in house construction. With such floor structures, combined with concrete or other materials, good solutions for sound insulation and fire can be achieved at reasonable costs. Massive timber constructions are sometimes also used for the whole structural building system, including wall units.

REFERENCES

Andreasson S. (2000) Three-Dimensional Interaction in Stabilisation of Multi-Storey Timber Frame Building Systems. Report TVBK-1017, Division of Structural Engineering, Lund University, Sweden.

FAO (2000a) *The Global Forest Resource Assessment 2000*, Summary report.

FAO (2000b) *Yearbook of Forest Products.*

Glulam Manual (1995) Svenskt Limträ, Stockholm (in Swedish).

Jacob L. (1776) Brücken und Brückenbau (Theatrum Pontificale, oder Schauplatz der Brückenbaues). In Th Schäfer GmbH Hannover, 1982.

Kuklik P. (2000) Development of timber framed houses in Central Europe. *Proc. of COST E5 Workshop, Timber Frame Building Systems*, Ed. A. Cecotti, S. Thelandersson, Venice, Italy.

Paevere P. (2001) Full-Scale Testing, Modelling and Analysis of Light-Framed Structures Under Earthquake Loading. PhD thesis, Civil and Environmental Engineering, University of Melbourne.

Stadelmann W. (1990) *Holzbrücken der Schweiz – ein Inventar.* Verlag Bündner Monatsblatt, Chur, Switzerland.

Timber Bridges (1996) Träinformation AB, Stockholm, Sweden (in Swedish).

Yasemura M. (2000) Seismic performance of timber structures in Japan. *Proc. of COST E5 Workshop, Timber Frame Building Systems*, Ed. A. Cecotti, S. Thelandersson, Venice, Italy.

PART ONE

Basic Properties of Wood-Based Structural Elements

2

Introduction: Wood as a Construction Material

Sven Thelandersson

2.1 WOOD – A HIGH STRENGTH FIBRE COMPOSITE

Wood material is biologically 'produced' in the growing tree to meet the needs of the tree itself. In this respect, it is a high quality fibre composite, optimally designed to resist loads acting on the tree, but also to provide transport of water and nutritional agents. The stem and branches of the tree are designed to resist gravity loads and wind loads. The wood structure is adapted to create maximum strength in stressed directions, whereas in other directions the strength is quite low. The wood cells in conifers are oblong in shape, with length in the range 2–4 mm and a length/diameter ratio of about 100:1. They are predominantly oriented in one direction, called the fibre direction or grain direction. This is the strong direction, generally parallel with the longitudinal axis of the stem. In the vicinity of branches, the fibres are redirected to form very efficient 'structural' joints between the stem and the branches (see Figure 2.1). For conifers, the fibre structure in zones predominantly under compression (e.g. in the lower part of branches) is different from 'normal' wood and optimised to resist compression stresses. Under special growth conditions, a similar fibre structure will be formed

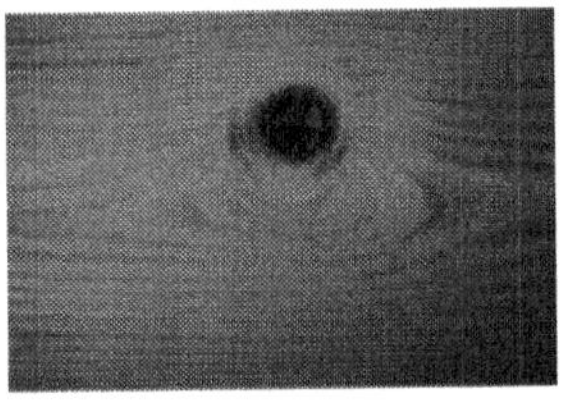

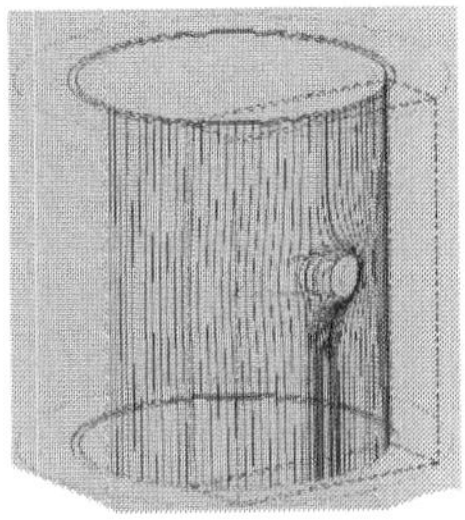

Figure 2.1 (a) The fibre orientation around a knot creates a structurally efficient joint between branch and stem, (b) the fibre structure resembles flow lines around the knot (Foley, 2000)

in the stem, especially in the lower part of the tree. This type of wood is called 'compression wood'.

The strength and stiffness in the fibre direction are very large in relation to the weight of the material, especially in tension. Table 2.1 shows a comparison of strength/density ratios for some structural materials.

For clear wood this ratio is significantly higher than for other building materials. Also, in the case of structural timber, where natural defects reduce strength, the strength in relation to weight is of the same order of magnitude as ordinary structural steel. The strength/density ratio is more advantageous for fibre reinforced synthetic polymers, but the price of such products is also significantly higher.

The chain from micro level to macro level for wood is illustrated in Figure 2.2.

Timber engineering deals with the properties and performance on the engineering levels above clear wood. However, clear wood properties and more detailed levels are very important for a basic understanding of the key problems in timber engineering. In this book, the term 'wood' is usually reserved for small clear wood elements more or less free from defects and irregularities on the macro level. The term 'timber', or sometimes 'lumber', is used to describe elements on a structural scale, where the influence of macro defects on the performance is important. Wood being a biological material means that there is a considerable variability at all levels.

2.2 STRUCTURAL TIMBER

Sawn timber in structural dimensions is a non-homogeneous material, which contains growth defects in the form of knots, zones with compression wood, oblique fibre orientation, etc. Such growth characteristics, which were once created to serve the needs of the tree, will usually reduce the strength significantly when the timber is sawn and used for other purposes. This is described in the following excellent way by Borg-Madsen (1992):

> *Mother nature spent millions of years devising an arrangement whereby the branches grew perpendicular to the stem of the tree without impairing the strength of the branch and retaining the full strength of the stem. ... It is well known that poles made from the stem are very strong but*

Table 2.1 Strength/density ratios for some structural materials

Material	Density kg/m^3	Strength, MPa	Strength/Density 10^{-3} MPa·m^3/kg
Structural steel	7800	400–1000	50–130
Aluminium	2700	100–300	40–110
Concrete, compression	2300	30–120	13–50
Clear softwood, tension	400–600	40–200	100–300
Clear softwood, compression	400–600	30–90	70–150
Structural timber, tension	400–600	15–40	30–80
Glass fibre in epoxi, typical value, tension			500
Carbon fibre in epoxi, typical value, tension			1000

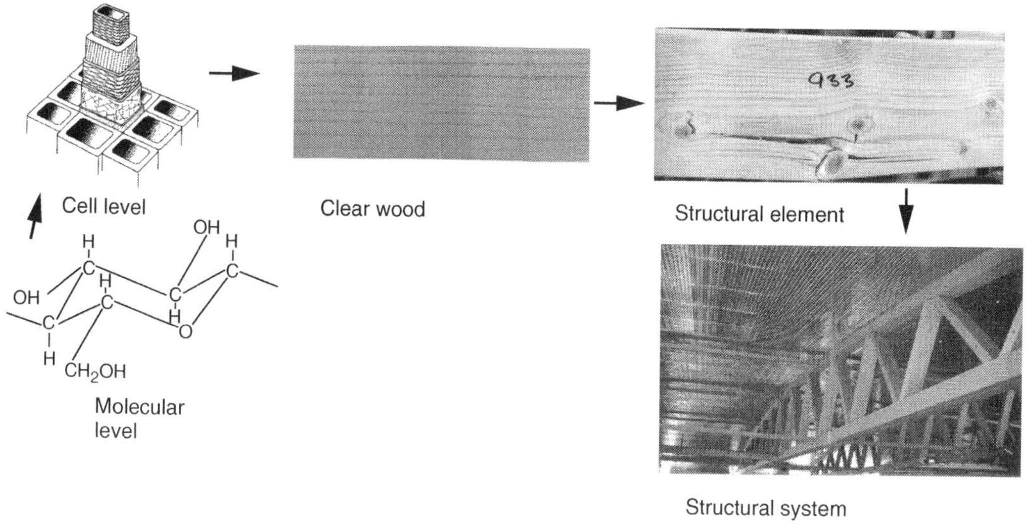

Figure 2.2 The wood chain from micro level to macro level

when we cut through the knots representing the intricate handiwork of mother nature, what was a proficient pattern to carry the branch becomes a severe weakness. The interlocking of the fibres is destroyed, leaving the unsupported fibres where tension perpendicular to grain stresses can initiate crack growth.

Therefore, the mechanical behaviour of timber cannot be derived with any reliability from the properties of clear wood. Furthermore, the presence and character of knots and other defects vary from one timber board to another, which means that the structural properties of sawn timber exhibits a significant variability. Strength properties of structural timber are therefore usually determined by direct testing of timber elements according to a standardised methodology, and strength is defined on the element level rather than on the material level. Strength data associated with structural timber therefore reflects moment, tension, compression and shear capacity of a timber element, even if the data are expressed in stress units, obtained assuming that the theory of elasticity is valid. The bending strength, for instance, is defined as the moment capacity of a timber beam determined by testing, divided by the elastic section modulus. The influence of defects is implicitly included in the strength values specified, and they can only be applied if the stresses are determined by elastic theory.

This is the reason why different strength values are valid for different loading modes such as bending, tension and compression. Figure 2.3 shows typical cumulative distribution functions for strength of timber in tension, compression and bending.

The coefficient of variation in bending for structural timber is in the range 20–40%, depending

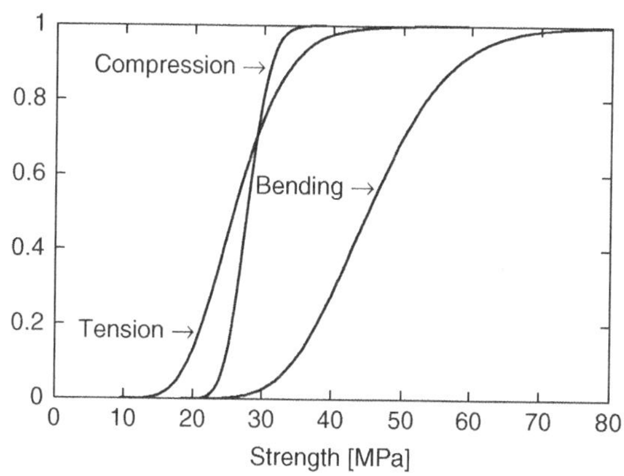

Figure 2.3 Cumulative distribution of the strength of structural timber (Spruce, Picea Abies, grown in Scandinavia) for different loading modes (Hansson & Thelandersson, 2001)

upon species and growth conditions, as well as on the method and strategy used to grade the timber.

When timber is to be used for structural purposes, some type of non-destructive evaluation of strength (or grading) is required to guarantee the safety of the structure. Historically, the most common method was *visual grading* of each timber element, following specified rules. Nowadays, a number of different *machine grading* methods are also used, although visual grading is still very common in practice. Typical for all types of grading methods is that one or several indicating properties are observed in each timber element, and the grading is based on the expected correlation between the indicative properties and strength. This correlation is generally not very high, which means that there is an uncertainty associated with the grading procedure which has to be considered. The state-of-the-art concerning strength grading of structural timber is presented by Johansson in Chapter 3.

The fact that the load-bearing capacity of timber is governed by the presence and characteristics of random growth defects means that the strength of timber elements also depends upon the size of the structural element itself, and the way in which it is loaded. This is described in more detail by Isaksson in Chapter 4.

2.3 ENGINEERED WOOD PRODUCTS

Engineered Wood Products (EWP) represent a broad class of materials intended for structural use. They are typically produced from wood which has been processed to smaller fractions by sawing, peeling, chipping, slicing or defibration. The wood constituents used for EWP can be sawn laminations, veneers, strands, flakes or sawdust. The constituents are bonded together to form panel products, timber-like elements in different sizes, or shaped structural products. The bonding between constituents is most often made with adhesives mixed with the constituents or sprayed on their surfaces, with application of heat and pressure in the production process. One type of products are made by gluing solid sawn timber or laminations

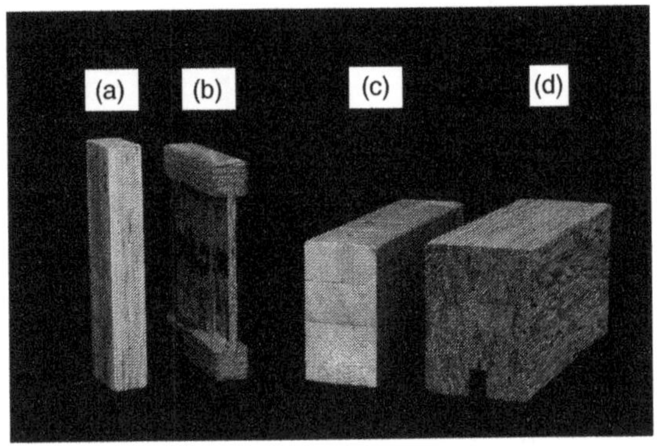

Figure 2.4 Some examples of Engineered Wood Products. (a) LVL, (b) Composite I-beam, (c) Glulam, (d) Parallel strand lumber

into larger structural members. Some examples are shown in Figure 2.4.

EWPs have several advantages:

- They can be produced in sizes which are not limited by the tree dimensions.
- They allow a more efficient utilisation of the raw wood material.
- The strength reducing effects of defects present in solid wood will be more or less neutralised, depending on the type of product.
- Their dimensional stability and tolerances are generally significantly better than for sawn timber.
- They can more easily be adapted to market requirements.

A disadvantage is that EWPs, as a rule, are more expensive. Moreover, the ecological image of wood as a natural material is impaired by incorporating adhesives into the products.

The growth defects of solid wood are either removed in the production process or distributed in the finished product so that the strength is affected less and the variability of the product becomes smaller. This means a more efficient utilisation of the material. As an example, the probability density functions for glulam and structural timber are compared in Figure 2.5. Although glulam is made from laminations of structural timber, the decisive strength, defined as the fifth percentile,

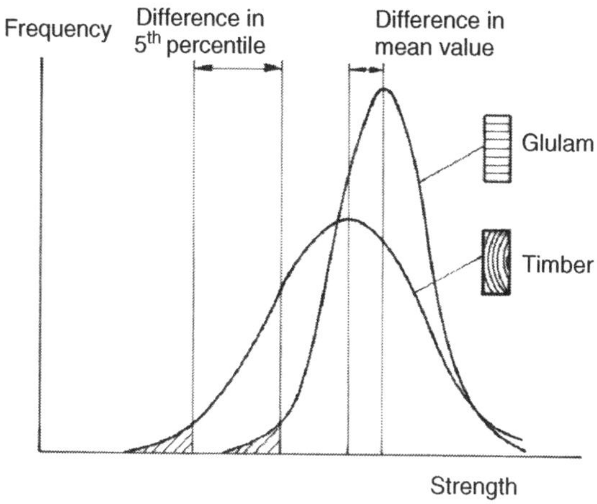

Figure 2.5 Probability density functions for solid wood and glulam

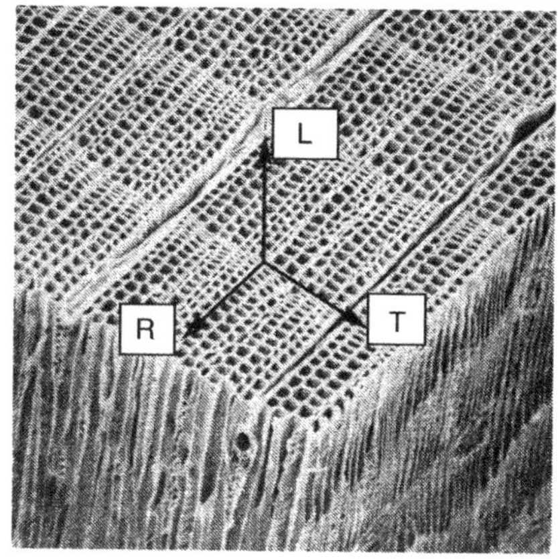

Figure 2.6 Microscopic picture of softwood. The cells have a typical tubular shape

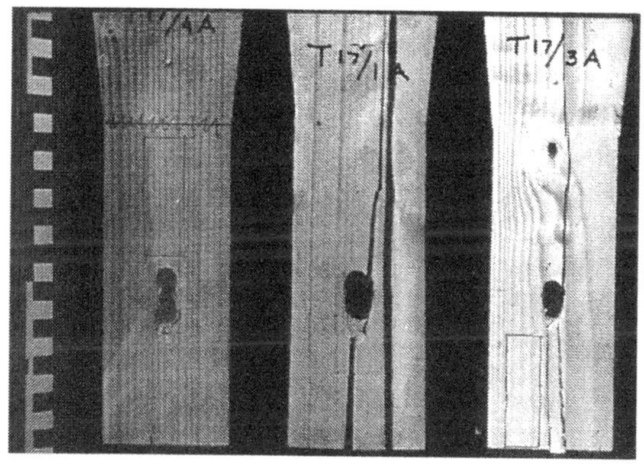

Figure 2.7 Failure in tension perpendicular to the grain in a timber joint.

is significantly higher for glulam than for timber. The reason for this is the load sharing between laminations in the glulam, which allows locally weak zones to redistribute stress to adjacent stronger regions.

Similar mechanisms will be active in other types of EWP.

A detailed presentation of the mechanical behaviour and properties of glulam can be found in Chapter 5 by Serrano, and an overview of currently used Engineered Wood Products and their characteristics is given by Lam and Prion in Chapter 6.

2.4 THE ANISOTROPIC NATURE OF WOOD

Wood is anisotropic, i.e. its physical properties depend upon direction. The tension strength is 30–50 times smaller perpendicular to the grain than in the fibre direction, and most people know that it is quite easy to split wood along planes parallel to the fibres. The reason for this is quite obvious in view of the anatomy and fibre structure of wood (see Figure 2.6). The weakness exhibited by wood in cross grain directions must always be carefully considered in the design and detailing of timber structures.

Failures of timber structures observed in practice are often due to tension perpendicular to the grain, which commonly occurs in curved elements, in joints in timber structures (see Figure 2.7) and in discontinuities created by wholes and notches in timber members. The effective load-bearing capacity of timber members under tension perpendicular to the grain depends upon the magnitude of the stressed volume (e.g. see Barrett, 1974).

Our present understanding of fracture processes related to perpendicular to grain failure is limited. In engineering design, empirical rules for detailing, such as minimum recommended distances between fasteners as well as edge distances in mechanical joints, are commonly employed to avoid the risk for splitting failure. The reliability of such design methods is limited, however, especially when applied in large-scale real world situations, different from the laboratory tests from which they are derived.

A promising line of research during recent years is the application of rational fracture mechanics, which has contributed significantly to a better understanding and more reliable design methods concerning perpendicular to grain failure. A comprehensive summary of the findings from this research is given by Gustafsson in Chapter 7.

The strong anisotrophy in wood is also valid for stiffness properties. The elastic modulus perpendicular to grain is typically 50–80 times smaller than parallel to grain. When wood is subjected to compression perpendicular to grain, significant deformations will occur even at low load levels. Another manifestation of wood anisotrophy is moisture-induced strain, i.e. shrinkage and swelling. Large moisture-induced strains occur in directions perpendicular to grain, typically 5–10 times larger than parallel to grain. Likewise, strains induced by moisture change are 1.5–2 times larger in the tangential than in the radial direction. For this reason, cross-sections of sawn timber change shape when they are initially dried and subsequently when moisture changes occur (see Figure 2.8).

2.5 WOOD AND MOISTURE

Wood in green condition contains large amounts of water. The moisture content (weight of water in relation to weight of dry wood) can be of the order 100% or more in sapwood. Part of the water is free and stored in cavities of the wood cell, the rest is bound in cell walls. When wood is dried, the free water disappears first, and at around 30% moisture content almost all of the remaining water is bound in the cell wall. This moisture state is called the 'fibre saturation point'. When wood is dried below fibre saturation, mechanical properties will depend upon moisture content, so that both strength and stiffness will increase when the wood becomes drier.

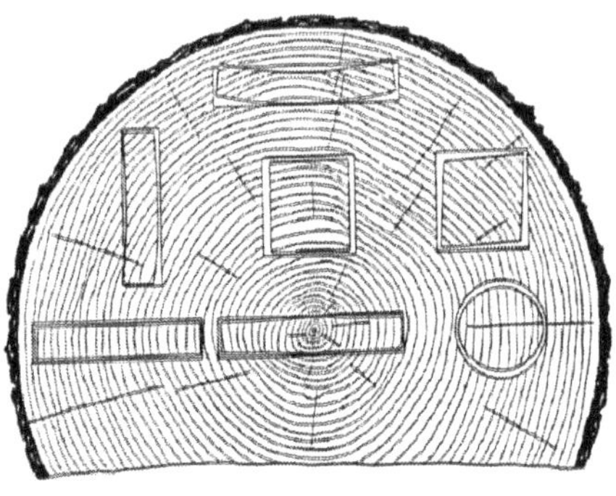

Figure 2.8 Distortion of cross-sections after drying of sawn timber taken from different positions in the log

Wood is hygroscopic and interacts with the humidity of the ambient air. The moisture content in a wood element will always change towards an equilibrium corresponding to the relative humidity of the surroundings. Building components in practice are usually exposed to changing climate conditions of a more or less random nature. The response of a wood element is quite slow, so that the average moisture content of cross-sections in structural sizes will respond only to changes in weekly or monthly averages of humidity. This slowness in response implies that the moisture distribution in timber elements becomes non-uniform. Figure 2.9 shows the continuous change in moisture distribution over the width of a glulam cross-section, exposed to outdoor conditions under shelter during 45 weeks starting from November.

The presence of moisture gradients of the type shown in Figure 2.9 means that moisture-induced stresses will develop in timber elements, due to constrained shrinkage or swelling. Such stresses may be present in all material directions, but they are negligible in the longitudinal direction, where the strength is also high. In cross grain directions, however, moisture-induced stresses are much larger, and can easily become significant in relation to the low strength in these directions. The internal stresses in glulam cross-sections induced by the moisture gradients shown in Figure 2.9 were determined by testing (Jönsson, 2001). Tensile stresses larger than the characteristic tensile strength (0.5 MPa) were present under a period of 12 weeks during the winter season.

Surface cracking of timber elements due to non-uniform moisture-induced stresses is very

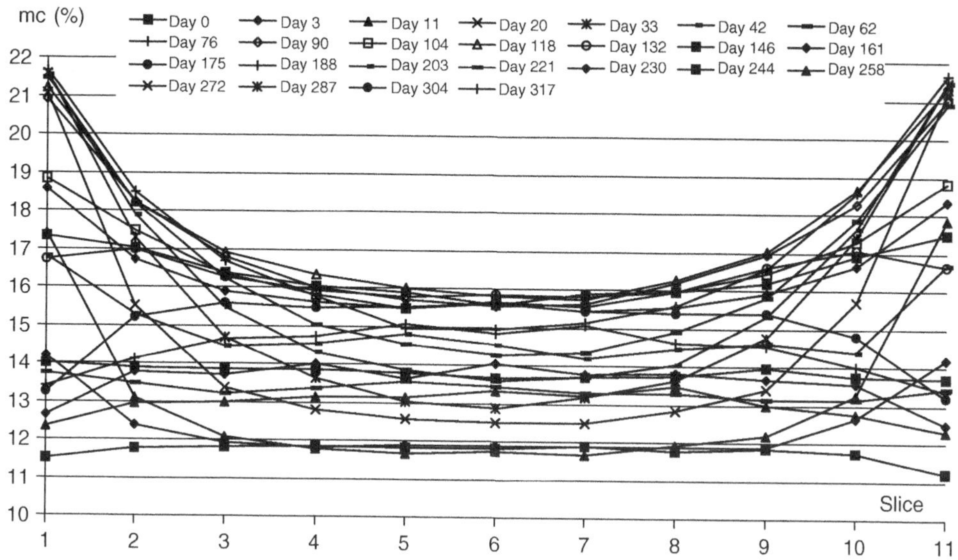

Figure 2.9 Measured moisture content distributions in glulam cross-section (90×270 mm^2) exposed to outdoor conditions under shelter during a winter and summer period (Jönsson, 2001). The specimens were conditioned in RH 60% before the test

common. The presence of moisture-induced internal stresses has a strong influence on load-bearing capacity associated with tension perpendicular to grain failure modes. The fact that internal moisture-induced stresses are always more or less present in timber structures is not normally considered in engineering design, implying that predictions of failure in tension perpendicular to grain are associated with large uncertainties. This is probably an important reason for the large frequency of accidental failures of this type in practice.

Further research is clearly needed to understand the complex interaction between moisture and mechanical performance of timber structures. The present state-of-the-art in this field is presented by Ranta-Maunus in Chapter 9.

2.6 DURATION OF LOAD EFFECTS IN TIMBER

The strength depends upon the rate of loading and duration of the load for all structural materials. This is of little practical importance for most materials, but for timber the loss of strength under sustained load is so large that it has to be considered in everyday engineering design. The design strength values are adjusted by a reduction factor, depending upon the expected load duration. For solid timber the strength reduction under ten years load duration can be as high as 40%.

The mechanisms behind the strong duration of load effects exhibited by wood is still not fully understood, although a large amount of research has been carried out. There are several circumstances which makes research about duration of load effects quite difficult, expensive and time consuming. Experimental investigations take a long time if results of relevance for the timescales of interest in practice are to be found. Quantification of duration of load effects involves a comparison between the strengths of specimens under long-term loading and standard short-term tests. These two different tests have to be performed on two different sets of specimens, since the same piece of wood cannot be broken twice. The large variability of wood, especially for solid timber, makes any observations of the relative effect of duration of load uncertain. For structural timber the most common method is to compare matched samples, but to do this in a reliable way, a large number of specimens is needed.

Furthermore, duration of load tests are commonly performed under constant load acting until failure occurs, or alternatively with some type of ramp loading applied at different rates. For a timber structure in practice, a significant part of the load will vary in a random manner over the lifetime of the structure. It is still an open question as to how results from duration of load tests should be interpreted and used for arbitrary time variations of loads. Several theoretical damage models have been proposed, but due to the experimental difficulties mentioned above, the validation of such models is not satisfactory. An excellent summary of the state-of-the-art concerning duration of load effects for wood, timber and wood-based products is given by Hoffmeyer in Chapter 8.

REFERENCES

Barrett J.D. (1974) Effect of size on tension perpendicular to grain strength of Douglas Fir. *Wood and Fiber* **6**(2): 126–143.

Foley C. (2001) A three-dimensional paradigm of fiber orientation in timber. *Wood Science and Technology*, **35**, 435–465.

Hansson M. and Thelandersson S. (2001) Capacity of timber roof trusses considering statistical system effects. *Holz als Roh und Werkstoff.* In press.

Jönsson J. (2001) Moisture induced stresses in glulam cross sections. *Proc. CIB W18*, Paper 34-12-4, Venice, Italy.

Madsen B. (1992) *Structural Behaviour of Timber*. Timber Engineering Ltd, Vancouver, Canada.

3

Grading of Timber with Respect to Mechanical Properties

Carl-Johan Johansson

3.1 INTRODUCTION

A pre-requisite for the use of timber in load-bearing constructions is that the strength and stiffness properties are known and can be controlled to stay within desirable limits. This cannot be achieved in the same way as for man-made products such as steel, concrete, plastics and wood fibre board, where a certain material quality is obtained by changing the composition of the raw materials or by changing some of the environmental conditions (temperature, pressure, etc.). There are a few examples where the properties of solid wood can be influenced in a similar way. It is possible to improve the durability of wood to different levels by chemical or thermal treatment. With chemical modification the dimensional stability as well as the durability can be controlled. However, as far as the mechanical properties of wood are concerned, the only realistic way of obtaining quality within desired limits is grading. There are presently two types of strength grading systems:

- *Visual strength grading*, which is based on visual inspection to ensure that the pieces do not have visible defects in excess of the limits specified in the relevant grading rule.
- *Machine strength grading*, where the pieces are passed through a machine which measures one or several parameters non-destructively. Based on these measurements, strength and stiffness are predicted.

Some more or less primitive forms of visual strength grading have been used ever since man began using wood in load-bearing constructions. It was, however, not until the beginning of the twentieth century that formal grading rules were established. In the USA ASTM Standard D245, originally published in 1927, contained guidelines for developing grading rules (Madsen, 1992). From the 1930s onwards, grading rules were introduced in various European countries (Glos, 1995). The grading rules developed over the last 60 years differ widely with respect to grading criteria, number of grades and grade limits. This is due to the large number of wood species, timber qualities and building traditions. All visual grading rules have in common the general deficiency that, for practical reasons, only visually recognisable characteristics can be taken into consideration, and only simple combination rules are possible. Important strength determining characteristics such as density cannot be assessed satisfactorily.

The idea of using non-destructive measurements to predict strength and use that as a basis for grading timber was proposed concurrently in the USA and Australia in the late 1950s (Madsen, 1992). The main purpose was to improve the accuracy of strength grading processes with the aim of achieving a better utilisation of the available timber qualities.

Strength grading is not only a matter of strength. To use timber as a structural material, its stiffness properties need to be known and controlled. The density is also of importance as the strength of mechanical connections in timber is linked to it. When the terms 'strength grading' and 'strength class' are used in the further text, they normally refer to grading with respect to strength as well as stiffness and density.

The purpose here is to describe the basis for strength grading, with the main focus on machine strength grading. Timber properties of importance will be described as well as the relationship between properties and how they vary within the timber.

Different techniques for non-destructive testing applicable to timber will be presented, as well as most of the machine grading systems that are available on the market. In connection with this, an explanation will be given of how the accuracy of the grading method affects the yield in a certain strength class.

3.2 STRENGTH AND STIFFNESS OF TIMBER AND RELATED PROPERTIES

3.2.1 General

Development of accurate grading systems requires a profound knowledge and understanding of how the strength and stiffness properties of timber depend upon the clear wood properties, and of the effect of different kinds of defects, such as knots, slope of grain and top failure. The main emphasis here is on the properties of European species used in construction.

3.2.2 Clear Wood Properties – Interrelation and Variation

Looking at the microstructure of wood with its cells, it is reasonable to regard it as being built up of small tubes bonded together (see Figure 3.1). The tube width as well as the wall thickness can vary, but the wall material properties are very similar in all wood species. This in particular applies to the wall material density, which is about 1500 kg/m^3 (Kollmann *et al.*, 1968). This means that wood material properties such as Modulus

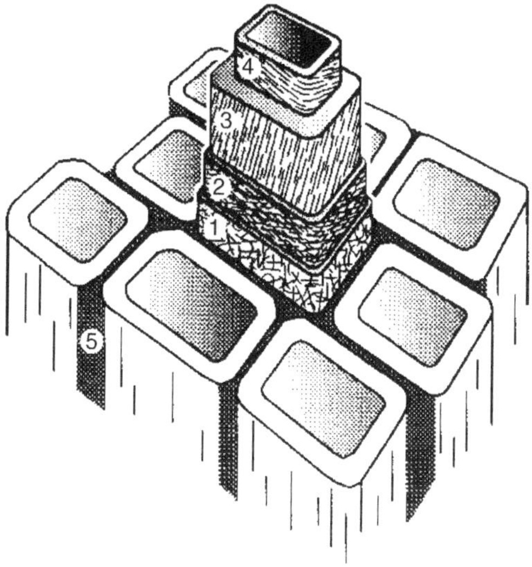

Figure 3.1 Schematic drawing of the microstructure of wood (Ormarsson, 1999)

Of Elasticity (MOE) and strength along the grain and density are strongly linked to the cell or tube wall thickness. This explains why the correlation between clear wood properties in general is very good. For small, defect-free specimens of Norway spruce (*Picea abies*), Foslie (1971) found the following coefficient of determination values (R^2):

- Bending strength vs. modulus of elasticity $R^2 = 0.76$
- Bending strength vs. density $R^2 = 0.66$
- Modulus of elasticity vs. density $R^2 = 0.64$

In timber the wood properties are not uniformly distributed, as can be seen in Figure 3.2 (Steffen *et al.*, 1997). In this particular case, the density varies from about 400 to almost 600 kg/m^3, and the MOE from 10 000–22 000 N/mm^2. As far as the MOE is concerned, Dahlbom *et al.* (2000) and Bengtsson (2000) obtained similar results. They measured altogether 3400 small defect-free specimens of Norway spruce, and found that the MOE increased by a factor of about 2 from the pith and outwards. Figure 3.2 and results from Wormuth (1993) indicate that the density varies less. Wormuth also presents examples where the density is almost constant over the whole timber cross-section, but where the MOE increases from about 6000 N/mm^2 at the pith to 12 000 N/mm^2 at an outer corner of a 67 × 195 mm cross-section of Norway spruce timber. This, in a way, contradicts what was stated earlier about the good correlation between MOE and density, and indicates that there are other important factors that control the relationship. One is the microfibril angle in the S_2 layer (Persson, 2000) (see Figure 3.3). The density is rather uniformly distributed, but the MOE drops from the twelfth growth ring to the pith due to the strong increase in microfibril angle. Another factor is spiral grain (Ormarsson, 1999), which normally seems to affect the MOE less than the microfibril angle.

3.2.3 Defects in Timber

Timber normally contains defects of various kinds, such as knots, top rupture, compression wood, slope of grain, decay, bark pockets, wane and resin pockets. The seriousness of these defects is not only a matter of how much each individual defect reduces the strength, but also how often they occur. Knots are regarded as very serious defects as they can greatly reduce strength and are almost always present in large numbers in a piece of timber. Top ruptures can have a dramatic effect on strength, but they occur less frequently and normally as a

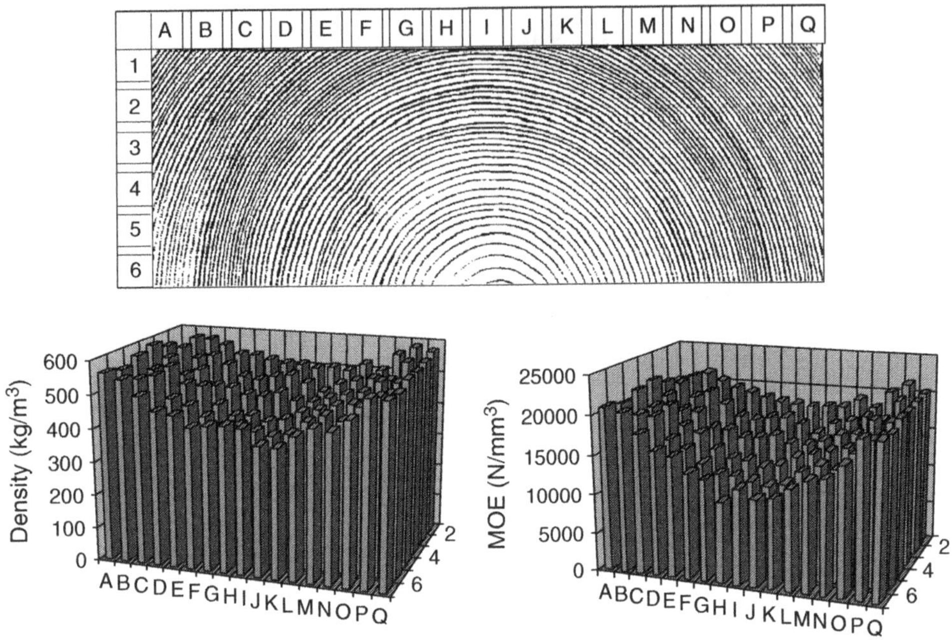

Figure 3.2 Distribution of density (at 12% m. c.) and MOE along the grain determined on 9 × 9 × 200 mm specimens cut from a 67 × 195 mm defect-free board section of Norway spruce (*Picea abies*) (Steffen *et al.*, 1997)

single defect in a piece of timber. Johansson *et al.* (1998) give an indication of the importance of different defects (see Figure 3.4). Norway spruce timber was graded by two independent graders at two different institutes using the following criteria:

- Wide face knots ≤1/5 of the width
- Slope of grain ≤1/14
- Narrow face knots ≤1/3 of the thickness
- Annual ring width <4 mm
- Knot clusters ≤1/5 of the width +1/3 of the thickness
- Top failure – not allowed
- Hard decay – not allowed
- Wane ≤1/10 of the width and thickness, respectively
- Compression wood – not allowed
- Resin pockets – minor ones allowed
- Bark pockets – length ≤2 times timber width, depth ≤1/4 of thickness and width ≤1/6 of timber width.

Knots, and in particular narrow face knots, are by far the most important downgrading defects. They account for 66% (Grader 1) and 78% (Grader 2) of the downgrading. The other defects are far less influential. Compression wood is one example. In addition to being moderately influential, it is also difficult to assess. This most likely explains why Grader 1 has a downgrading percentage of 17%, whereas Grader 2 only has 1%.

The importance of knots is also underlined by the fact that in destructive bending or tension tests, the failure is almost exclusively caused by knots. Schniewind *et al.* (1971) tested 2 × 6 inch redwood timber (*Sequoia sempervirens*), and in 95% of the cases failure occurred at knots or local slope of the grain, which in most cases was related to knots. Further evaluation of the results from Johansson *et al.* (1998) shows that 91% of the failures were caused by knots, and only 1% by slope of the grain. In the remaining 8% it was not possible to clearly identify what caused the failure.

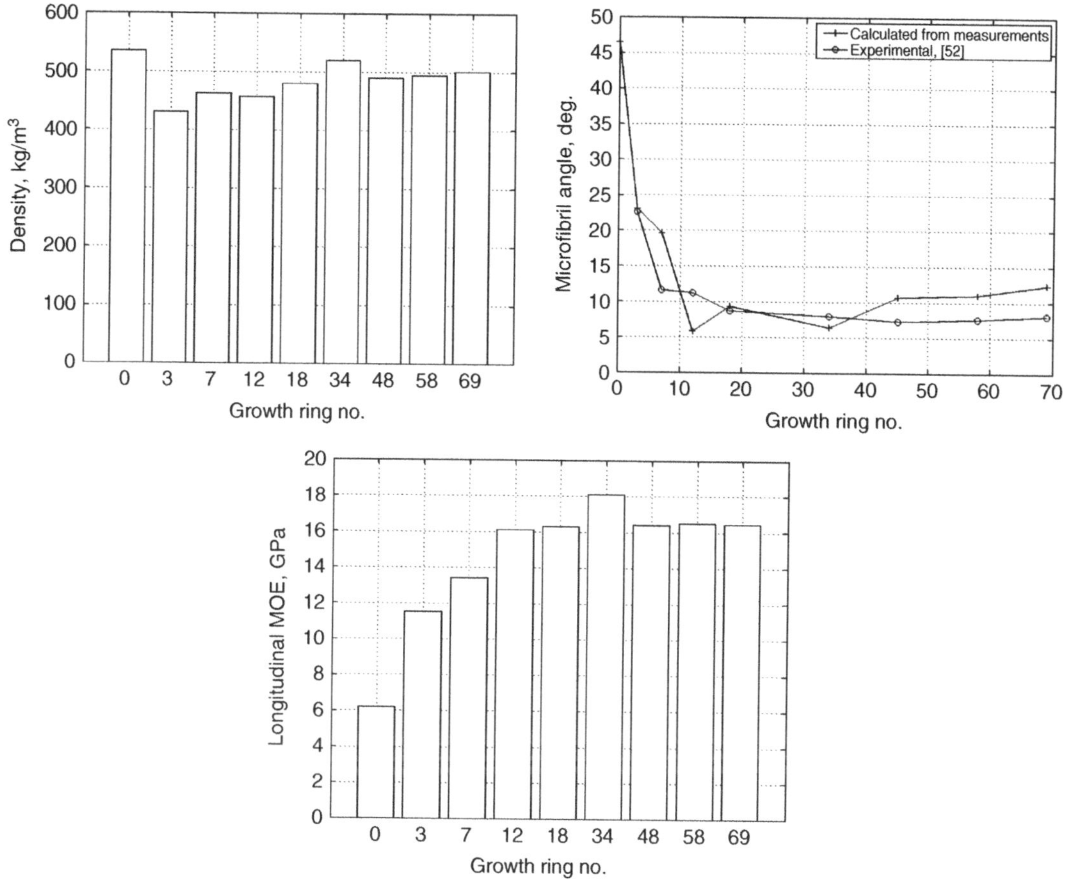

Figure 3.3 Effect of the microfibril angle on the longitudinal MOE in Norway spruce (Persson, 2000)

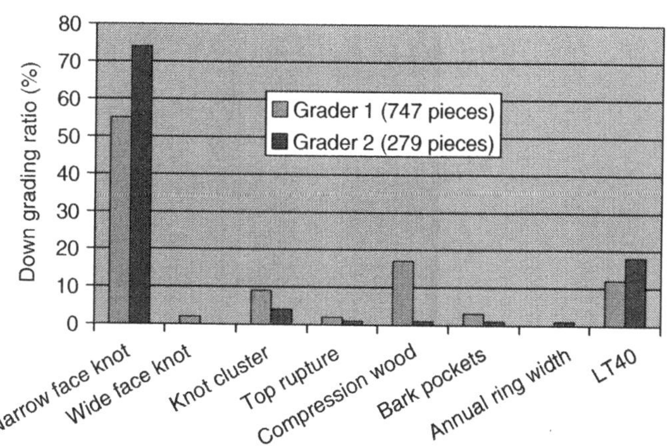

Figure 3.4 Distribution of downgrading defects for grading of Norway spruce performed at two different institutes. LT40 denotes pieces that fulfil the grading criteria. Annual ring width is not to be regarded as a defect, but is included as it is an important visual grading parameter (Johansson *et al.*, 1998)

Knots can occur in clusters, which is normally the case in species like Norway spruce, or they are solitary, as in for example Douglas fir (*Pseudotsuga menziesii*). Isaksson (1999) demonstrates the presence of knots in Norway spruce timber by using the parameter CWAR (Clear Wood Area), which is the knot free portion of the timber cross-section (see Figure 3.5). Minima corresponding to knot clusters occur at every 0.5–0.6 m. The distance normally increases with the growth rate.

Knots affect the strength of timber both by the cross-sectional reduction they cause and by the effect of the related fibre distortion. The effect on the stiffness is primarily linked to the fibre distortion.

The use of slope of grain as a strength reducing parameter does not normally refer to knot-related fibre deviations, but to the general slope of grain that occurs for instance in boards cut from curved logs or from butt logs due to the conical shape of the stem at the butt (Perstorper, 1999).

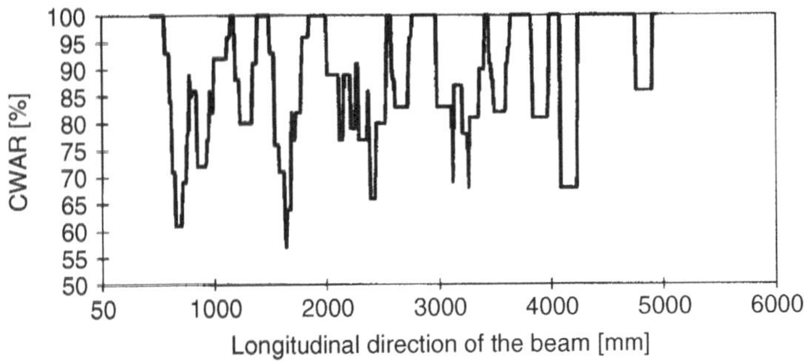

Figure 3.5 CWAR (Clear Wood Area) along a piece of Norway spruce timber. CWAR = 100(1-TKAR)%, where TKAR is the sum of the projected knot areas located within 150 mm of the timber length divided by the cross-sectional area (Isaksson, 1999)

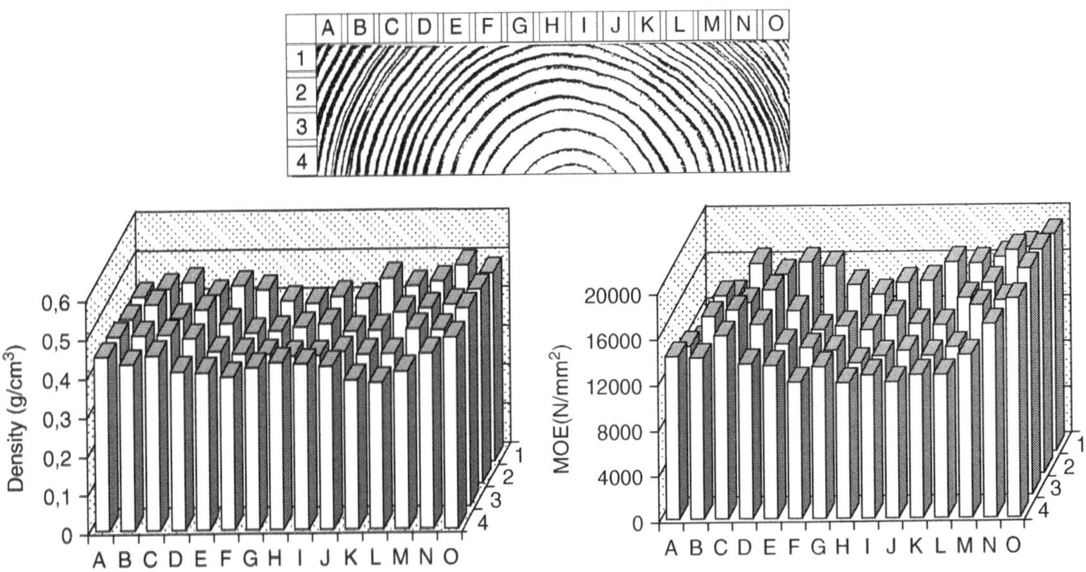

Figure 3.6 Distribution of density and longitudinal MOE in a section of Norway spruce timber containing compression wood, see upper left corner (Wormuth, 1993)

Compression wood develops in areas of the stem exposed to large compressive stresses during growth, for instance due to wind, and can be observed as annual rings with a darker colour and slightly eccentric shape. Compression wood can only be found in conifers. It is developed to counteract the effect of the compressive stresses. At the microscopic level, compression wood can be distinguished from normal wood by the thicker cell walls of the late wood tracheids. Of great importance is also that the S_2 layer is modified and has a much larger microfibril angle than in normal wood. The effect of this is a strongly reduced longitudinal MOE. This can be seen in Figure 3.6, where there is compression wood in the upper left corner of the cross-section. The density is rather evenly distributed, but a pronounced drop in MOE can be noted.

3.3 RELATION BETWEEN STRENGTH/STIFFNESS AND TIMBER CHARACTERISTICS THAT CAN BE MEASURED NON-DESTRUCTIVELY

3.3.1 General Overview

Table 3.1 presents an overview of relationships between strength on the one hand and a number of characteristics that can be measured non-destructively on the other. The strength of these relationships is indicated by the coefficient of determination, R^2. As can be seen, the R^2 values differ considerably between different investigations. This is most likely due to the fact that the material and test methods differ.

3.3.2 Knots

Knots alone are poor strength predictors. The R^2 values for Norway spruce in Table 3.1 vary between 0.16 and 0.42. If the location of the knots in the timber cross-section is also taken into account, the correlation can be improved quite considerably. By only using the knot area ratio (see definition for TKAR in Figure 3.5) Schniewind *et al.* (1971) obtained an R^2 value of 0.32 for redwood (*Sequoia sempervirens*) tested in tension. By also using information about the position of the knots in the timber cross-section, they managed to improve the correlation to $R^2 = 0.50$. Johansson (1976), who studied Norway spruce in tension, however, only noted an improvement of the R^2 value from 0.30 to 0.35.

Kunesh *et al.* (1972) studied Douglas fir (*Pseudotsuga menziesii*) with knot diameters ranging from 3/4–$3\frac{1}{4}$ inches. They separated edge knots and centre knots and obtained R^2 values of 0.67 and 0.69, respectively, for the relationship between tensile strength and knot size.

Table 3.1 Correlation coefficients from various investigations of the relationship between strength and non-destructively obtained characteristics for Norway spruce (*Picea abies*) timber (Hoffmeyer, 1995). The numbers in the table head refer to the following investigations: 1: Johansson *et al.* (1992), 2: Hoffmeyer (1984), 3: Hoffmeyer (1990), 4: Lackner *et al.* (1988), 5: Glos *et al.* (1982), 6: Johansson (1976)

Characteristics that can be measured non-destructively	Coefficient of determination R^2						
	Bending strength				Tensile strength		
Source	1	2	3	4	1	5	6
Knots	0.27	0.20	0.16	0.25	0.36	0.42	0.30
Annual ring width	0.21	0.27	0.20	0.44	0.36	0.33	0.28
Density	0.16	0.30	0.16	0.40	0.38	0.29	0.38
MOE, bending or tension	0.72	0.53	0.55	0.56	0.70	0.69	0.58
MOE, flatwise, short span							0.74
Knots and annual ring width combined	0.37	0.42	0.39		0.49		
Knots and density combined	0.38		0.38		0.55	0.61	0.64
Knots and MOE	0.73	0.58	0.64		0.70	0.76	0.78

3.3.3 Density and Annual Ring Width

Both density and annual ring width are regarded as measures of the clear wood strength and stiffness. The correlation with the strength of timber, however, is poor, with R^2 values ranging from 0.16 to 0.40 for the density and from 0.20 to 0.44 for the annual ring width, according to Table 3.1.

Table 3.2 Correlation (R^2) between density, annual ring width and latewood ratio. About 150 pieces in each sample, except for sample G that contains 75 pieces (Johansson *et al.*, 1998)

Sawmill	Density vs. annual ring width	Density vs. latewood ratio	Latewood ratio vs. annual ring width
E	0.38	0.51	0.64
G	0.20	0.12	0.06
H	0.28	0.39	0.48
I	0.21	0.15	0.08
K	0.18	0.40	0.47
L	0.38	0.09	0.15
M	0.38	0.42	0.64
S	0.30	0.16	0.24
All	0.25	0.12	0.26

Table 3.3 Relation between strength and MOE of 40 × 145 mm Norway spruce expressed as a coefficient of determination (R^2) (Johansson *et al.*, 1998)

Sawmill	Tension		Bending	
	N	R^2	N	R^2
E	149	0.65	50	0.73
G	150	0.70		
H	150	0.62		
I	149	0.51	50	0.60
K	152	0.69	50	0.54
L	153	0.68		
M	139	0.66		
S	154	0.63	56	0.63

Johansson *et al.* (1998) studied the relationships between density, annual ring width and also latewood ratio in Norway spruce from eight different sawmills in Finland, Norway and Sweden. Looking at the whole material, they concluded that the correlations were modest, if they existed at all. However, when analysing properties from each individual sawmill, it was verified that relationships exist, but depending on the origin of the timber (see Table 3.2).

3.3.4 Modulus of Elasticity

The stiffness, which is normally expressed as a MOE, is by far the best individual predictor of strength. The coefficient of determination R^2 in Table 3.1 and Table 3.3 ranges from 0.51 to 0.73. The highest value was obtained for a relatively short test span in flatwise bending. In contrast to what was observed for the density and the annual ring width, the coefficient of correlation for strength vs. MOE seems to be independent of the origin of the timber (see Table 3.3). The regression function, however, can differ considerably from one sawmill to another (see Table 3.4). The constant A in the simple regression function ranges from 2.78 for sawmill L to 3.14 for sawmill I, so for a certain MOE timber from I is about 13% stronger than the timber from L. Similar results were obtained by Chrestin *et al.* (1997), who studied the bending strength of timber from seven sawmills in Sweden. The main reason why the relation between strength and MOE can differ from one location to another is the presence of knots, and that the knot sizes differ. Timber from sawmill I had a mean knot area ratio which was only 70% of that of timber from L.

The relationship between bending strength and MOE can also be affected by the way in which the timber is dried. Bengtsson *et al.*

Table 3.4 Regression function for 40 × 145 mm Norway spruce timber from eight different locations in Finland, Norway and Sweden (Johansson *et al.*, 1998). Tensile strength (f_t) as a function of flatwise MOE ($E_{m,flat}$)

$f_t = A \cdot 10^{-3} E_{m,flat}$	Origin							
	E	G	H	I	K	L	M	S
A	2.89	3.03	3.04	3.14	3.01	2.78	2.80	2.87

(2000) tested Norway spruce timber dried in a common low temperature process and at high temperature (115–120 °C). The edgewise MOE was not affected, but the mean bending strength was roughly 10% lower in the high temperature dried material.

The MOE for timber, the way it is normally determined, contains not only information about the clear wood strength properties, but also to a large extent about knots and slope of grain. The knots affect the stiffness by disturbing the surrounding grain, which decreases the longitudinal stiffness. MOE measurements are also likely to pick up compression wood and spiral grain, at least if they are severe. Looking at the strength versus MOE plot in Figure 3.7, one can assume that points above the regression line have smaller than average knots, and points below the line have larger.

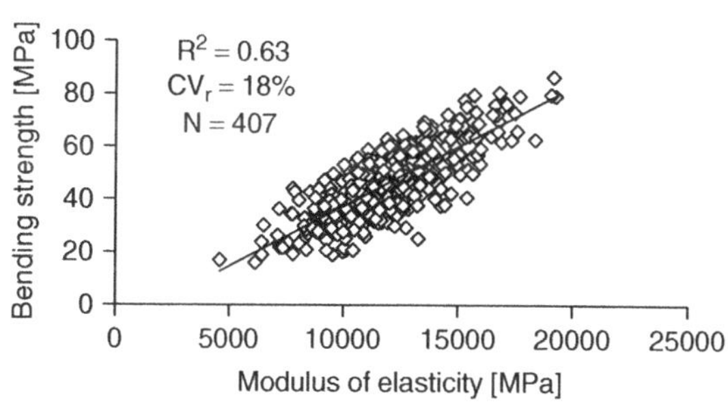

Figure 3.7 Relationship between edgewise bending strength and modulus of elasticity for Norway spruce (Hoffmeyer *et al.*, 1999). Function of linear regression: $f_m = 0.0045\ MOE\text{-}8.14$

Concerning the relationship between the MOE and other non-destructive parameters, Hoffmeyer (1995) concludes that only the density of the timber shows a reasonably good correlation. The coefficients of determination in investigations he studied were between 0.25 and 0.50, and additional information about knots does not seem to improve the prediction of MOE.

It is surprising, in a way, that the correlation between MOE and density for timber is rather poor considering that the correlation for clear wood is good ($R^2 = 0.64$; see Section 3.2.2), and that the effect of knots seems to be small. One explanation can be found in Figure 3.2 and 3.6, where it is shown that the MOE and the density vary over the timber cross-section. The density used in most studies is determined on a test piece cut across the timber, which means that the density obtained is a mean value. The MOE, on the other hand, can be determined either in edgewise or flatwise bending or in tension. The edgewise MOE values will, to a large extent, be controlled by the MOE at the edges, which, at least for Norway spruce, seems to be higher than the average MOE in the cross-section. MOE determined in tension tests is in principle equal to the mean MOE in the cross-section, and the correlation with the mean density should be better than between edgewise MOE and mean density. Coefficients of determination found in the literature seem to indicate that the correlation between tensile MOE and density is better (see Table 3.5).

Table 3.5 Correlation coefficients from various investigations of the relationship between MOE and density for Norway spruce (*Picea abies*) timber. The numbers in the table head refer to the following investigations: 1:Johansson (1976.), 2:Johansson *et al.* (1998), 3:Chrestin (1997), 4:Boström (1994)

Source	1		2		3		4	
	Dim. (mm)	R^2	Dim. (mm)	R^2	Dim. (mm)	R^2	Dim. (mm)	R^2
Tensile MOE vs. density	33 × 155	0.69	40 × 95	0.50				
			40 × 145	0.44				
			40 × 195	0.25				
Edgewise MOE vs. density			40 × 145	0.38	45 × 145	0.49	45 × 120	0.18
							45 × 195	0.39
							70 × 170	0.26

3.3.5 Combination of Non-Destructive Parameters

In the previous sections it was demonstrated how individual parameters relate to the strength and stiffness properties. By combining two or several parameters the prediction can be improved. By how much is a matter of whether or not the parameters are interrelated. Hoffmeyer (1995) used multiple regression analysis and obtained the following results:

- Bending strength vs. knots, annual ring width and density: $R^2 \approx 0.4$.
- Tensile strength vs. knots, annual ring width and density: $R^2 \approx 0.5$.

Combining information about knots and annual ring width as well as combining knots and density gives considerably better correlation than when the parameters are used individually. This is also demonstrated by Hoffmeyer *et al.* (1999) (see Figure 3.8). Note that for bending strength, the narrow face knot contains most of the knot information, and that adding wide face knot and knot cluster data only increases the R^2 value marginally.

Modulus of elasticity alone gives a better prediction than density, annual ring width and knot data combined (Hoffmeyer, 1995). This is, as has already been pointed out, due to the fact that the MOE for timber also contains information on the clear wood properties (density/annual ring width), knots, slope of grain and possibly also other characteristics. It is therefore not surprising that rather little is gained by combining MOE with other data. This is demonstrated in Table 3.6.

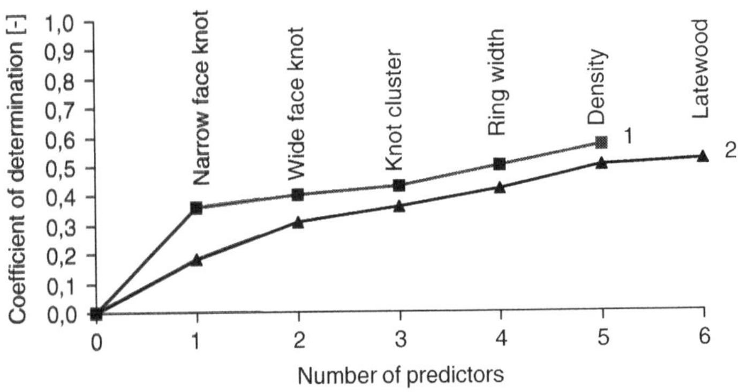

Figure 3.8 Multiple regression analysis for bending (1) and tensile (2) strength (Hoffmeyer *et al.*, 1999). Knots are measured according to INSTA 142 (Anon. 1997)

Table 3.6 Effect of adding knot data to MOE on the coefficient of determination. MOE determined in flatwise bending

Source	Coefficient of determination (R^2)	
	Strength vs. MOE	Strength vs. MOE and knot data
Johansson, (1992)		
• 58 × 120 mm Norway spruce, bending	0.49	0.59
• 34 × 145 mm Norway spruce, tension	0.66	0.71
Johansson *et al.* (1998)		
• 40 × 145 mm Norway spruce, tension	0.55	0.65
Orosz, 1969		
• 50 × 100 mm Southern pine, tension	0.59	0.76

Johansson *et al.* (1992) and Boström (1994) both showed that for edgewise bending it is enough to add edge knot data to the MOE to improve the correlation. Other knot information seems to be redundant.

3.4 GRADING OF TIMBER – EFFECT OF GRADING ACCURACY ON THE YIELD

The basis of grading is an estimation of the strength properties based on measurement of one or several characteristics which are related to the strength properties. The ability of a grading system to estimate strength accurately depends upon how well the measured characteristic(s) can predict the true strength of the timber, and how well the characteristics can be measured. The coefficient of determination, R^2, can be used as a measure of how strong the relationship between the grading characteristic and the strength is. The measurement error can be represented by a coefficient of variation, CoV.

Take, for instance, visual grading. The R^2 values presented in the previous chapter are based on knot and annual ring width data measured carefully by a ruler. In the practical visual grading situation they are estimated at some distance with the naked eye. The measurement error is likely to be large.

A statistical analysis of machine strength grading has been carried out by Hoffmeyer (1995), based on a model developed by Glos *et al.* (1982), to demonstrate the importance of grading accuracy. A stochastic model is used, where the relationship between strength and grading parameter is described by means of a two-dimensional normal distribution. Further, it is assumed that a linear relationship between strength and grading parameter exists. A typical result is shown in Figure 3.9. In this example, the timber is assumed to have a mean bending strength of 45 N/mm^2 and a coefficient of variation of 25%. These values are typical for Norway spruce from the Nordic countries. The graph describes conditions relating to grading of class C30 and better (lower fifth percentile bending strength $\geq$30 N/mm^2). For a machine with a coefficient of variation, CoV, lower than 0.1, the grading yield increases considerably with the coefficient of determination for the relationship between strength and measured characteristic up to 0.4. For values above that the increase in yield is moderate. Most existing machines have a CoV-value of less than 0.1. However, the coefficient of determination is

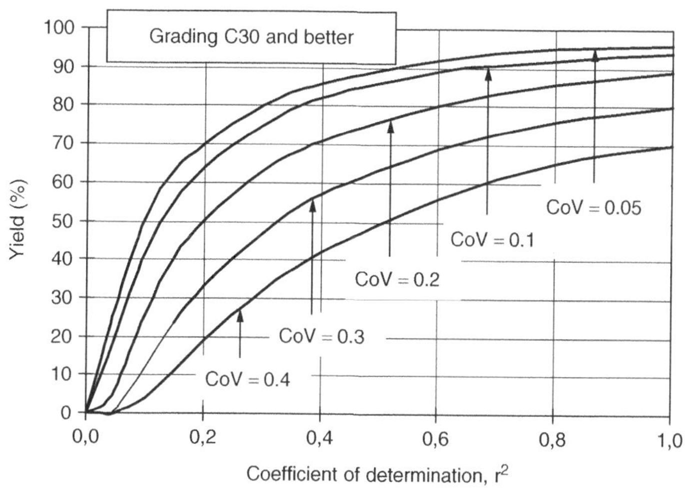

Figure 3.9 Yield as a function of the coefficient of determination, R^2, between the grading characteristic and the bending strength and the coefficient of variation (CoV) of the machine when determining the grading characteristic (Hoffmeyer, 1995)

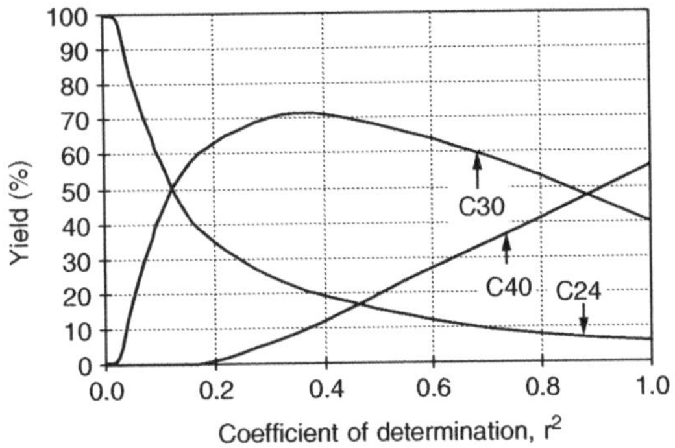

Figure 3.10 Yield as a function of the coefficient of determination, R^2, between the grading characteristic and the bending strength (Hoffmeyer, 1995). The coefficient of variation of the measurement error of the machine when determining the grading characteristic is set to CoV = 0.1, which is a realistic value

important for the yield in high strength classes and for classes close together. In Figure 3.10, which is based on the same assumptions on the timber characteristics as those used in Figure 3.9, the yield for the C24/C30/C40 class combination is presented. There is an almost linear relationship between the coefficient of determination and the yield in the C40 class.

3.5 MACHINE GRADING PRINCIPLES

3.5.1 Flatwise Bending

The most widely used machine grading principle is to measure the flatwise bending stiffness. In list of commercial grading machines in Hoffmeyer (1995), seven out of ten types in the world were so-called bending machines. Since then, new machine types have been introduced, but it seems that less than 5% of the total number of machines are based on principles other than flatwise bending.

The flatwise bending stiffness/MOE has the advantage of being very simple to measure. In principle a three-point loading test is carried out continuously. The span lengths are between 600 and 1200 mm. The most common is around 900 mm. The timber is passed through the machine between load and support rollers (see the example in Figure 3.11). Two systems for measuring the flatwise stiffness/MOE exist: (1) the load is kept constant and the deflection measured; (2) the timber is forced to a certain pre-set deflection, and the load required to maintain this deflection is measured. The bending stiffness EI can be determined according to

$$EI = \frac{F \cdot l^3}{48 \cdot \delta}$$

where F is the applied force, l the span length and δ the deflection. The flatwise MOE can be obtained by dividing the bending stiffness by the moment of inertia I. Figure 3.12 presents MOE distributions along three different pieces of timber.

As was emphasised earlier, the stiffness, EI and MOE contain information not only about the clear wood properties, but also about knots, slope of grain, etc. Furthermore, the grading machine contributes with shear deformations, indentations

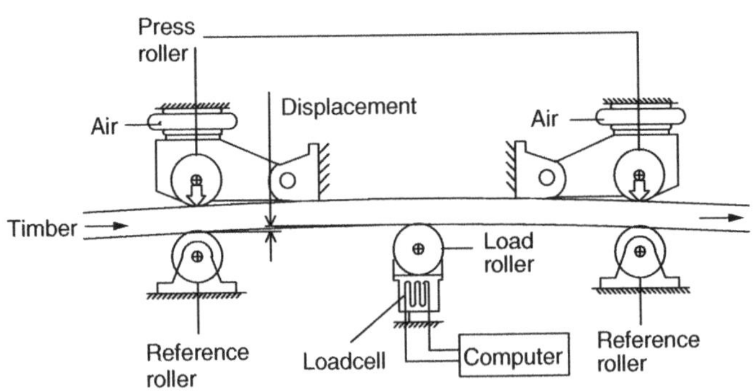

Figure 3.11 Typical bending type machine

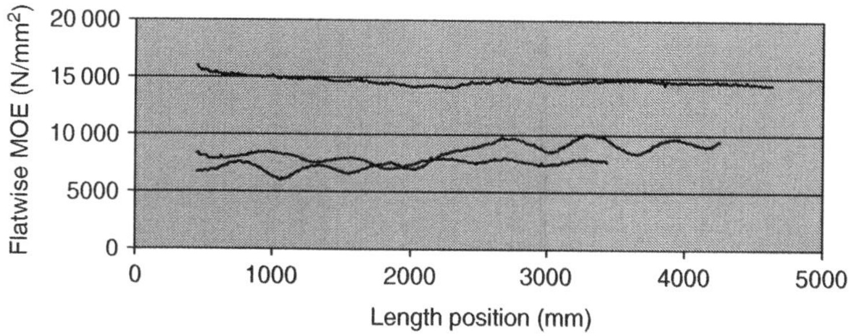

Figure 3.12 Flatwise MOE distribution along different pieces of 40 × 145 mm spruce timber. Data from Johansson *et al.* (1998). In the two bottom pieces minima corresponding to knot clusters are clearly visible. The top piece with high MOE has no or only very small knots

in the timber at the load and support rollers, as well as deformations in the machine itself. This means that MOEs can be machine-specific and also thickness-specific.

Timber is graded in sawn or planed condition. If the timber is planed it is either to its final dimension or just to obtain a uniform thickness for the grading operation. In sawn condition, thickness variations can be considerable and can cause considerable measurement errors. A deviation of 5% from a nominal thickness will affect the flatwise MOE by a factor of 1.16.

Vibrations in the timber during passage through the machine can cause considerable measurement errors, which increase with the throughput speed. Moreover, the bending principle has the disadvantage of not allowing assessment of the timber ends.

3.5.2 Measurement of Density by Means of Radiation Techniques

With x-ray or gamma rays, the density of the material can be determined. By using an array of senors, it is possible to determine the density distribution both across and along the timber. Knots have a higher density than the surrounding material, and therefore a measure of the size, location and possibly even the shape of knots can be obtained (Boström, 1999). The accuracy is strongly related to the number of sensors.

Schajer (2001) used a multiple-detector density x-ray scanner on 2 × 4 southern yellow pine. The scanner gave a measure of the clear wood properties as well as the size and location of knots. Both a one-detector and a three-detector scanner were used to estimate bending strength. The one-detector system gave an R^2 value of 0.65 and the three-detector system a value of 0.73. Bending strength was also estimated by means of flatwise bending stiffness, and gave an R^2 value of only 0.53.

3.5.3 Modulus of Elasticity Determined from Resonant Vibrations

The possibility to determine the MOE of timber by using dynamic testing was already being discussed in the early 1930s (Hoffmeyer, 1995). In recent years, the method has lead to practical applications thanks to the strong development of measurement techniques associated with dynamic testing and the use of the Fast Fourier Transform (FFT).

Both longitudinal and bending modes can be used. The resonance frequency of a longitudinal vibration in a beam can be described by the following expression:

$$f_{A-n} = \gamma_{A-n} \cdot n \cdot \sqrt{\frac{E_A}{\rho \cdot L^2}} \quad \text{(Hz)}$$

where the constant γ_{A-n} depends upon the support conditions, n is the mode number, ρ is the density and L the length of the beam. If both ends of the beam are free (i.e. the beam is floating) and the resonance frequency of the first axial mode is measured, then the MOE can be expressed by

$$E_A = 4 \cdot \rho \cdot L^2 \cdot f_{A-1}^2$$

The 'dynamic' bending MOE can be determined in a corresponding way (Hoffmeyer, 1995).

For defect-free wood the 'dynamic' MOE correlates well with the true value. (Wormuth, 1993) obtained an R^2 value of 0.85 between dynamic and static tensile MOE for 10 × 10 × 200 mm Norway spruce specimens. Also, for timber containing defects such as knots the correlation is fairly good. Larsson (1997) tested 38 × 89, 38 × 140 and 38 × 184 mm Norway spruce, and obtained R^2 values between 0.68 and 0.91. He also studied the relationship between dynamic MOE and edgewise bending strength, and found R^2 values ranging from 0.39 to 0.61. Sauter (1992) studied 60 × 120 mm Douglas fir *(Pseudotsuga menziesii (Mirb.) Franco)* and obtained a coefficient of determination of 0.71 for dynamic flatwise MOE versus edgewise bending strength.

One disadvantage of using the measurement of resonant vibrations to predict the strength and stiffness properties of timber seems to be that the method only delivers a weighted mean MOE, whereas the so-called bending machines are in principle capable of detecting the minimum MOE along a piece of timber. On the other hand, tests in a laboratory environment indicate that the dynamic MOE predicts the strength as well as or even better than the bending machines (Hoffmeyer, 1995).

3.5.4 Optical Detection of Knots and Other Defects

Several camera systems for the optical detection of timber characteristics have been developed for appearance grading. The most common camera type is the CCD (Charged Coupled Device). Pictures from the cameras are processed in image analysis programs which extract defects and other characteristics appearing on the timber surface. A further development is the use of the so-called tracheid effect (Åstrand, 1996), i.e. the spread of light in the wood surface. The denser the material, the less spread of light. This enables a rough estimate of the wood density at the surface, and since knots normally have a higher density than the surrounding wood, they will be easier to detect using the tracheid effect (see Figure 3.13). There are some systems in use or being developed where cameras are integrated parts of the strength grading system (Boström, 1999).

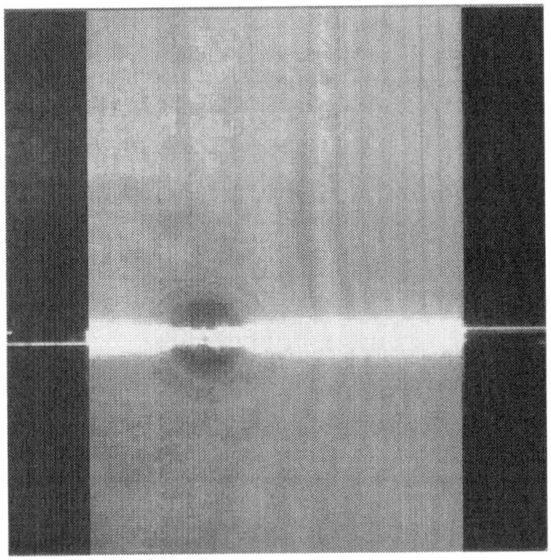

Figure 3.13 Illustration of the tracheid effect. The timber surface is illuminated by two line lasers. The spread of light is less in the dense knot material

3.5.5 Other Measurement Principles

Other principles worth mentioning are:

- Measurements of longitudinal stress wave speed. A pulse can be generated using ultrasonic sound or by the impact from a hammer at the timber end. For a slender rod the stress wave speed can be expressed as $c_{rod} = \sqrt{E/\rho}$ according to Clough *et al.* (1975). Sandoz (1989) obtained an R^2 value of 0.50 between stress wave velocity and bending strength of spruce.
- The microwave technique for the detection of knots has been applied in timber grading (Finnograder; see Section 3.6.1). A new system has been developed in Australia (Leicster *et al.*, 1994).

3.5.6 Combined Measurements

As was indicated in Section 3.3.5, combining the measurements of two or several characteristics leads to better prediction accuracy. There are, however, only a few applications. One is a combination of x-ray and flatwise bending. Another is cameras for the detection of knots and other defects in combination with flatwise bending. Boström (1998) studied the latter combination, in which the previously mentioned tracheid effect is used. By adding edge knot data obtained by camera and image processing to the flatwise MOE, he improved the strength prediction accuracy considerably, with an R^2-value from 0.59 to 0.65 for Norway spruce and for Maritime pine (*Pinus pinaster. Sol.*) from 0.49 to 0.56.

Cameras or x-ray can also be used in connection with flatwise MOE just to overcome the disadvantage that the timber ends cannot be assessed in bending. In this case, the prime purpose is to check that the ends do not differ significantly from the middle parts in terms of defects and density.

3.6 GRADING MACHINES ON THE MARKET

3.6.1 Presentation of Machine Types

Grading machines have been used in North America, UK, Australia, New Zealand and Scandinavia since the 1960s. In Germany grading machines were not used until the early 1990s. The most important grading machines are presented in Table 3.7. They were all developed for use in production lines, and in all except for the Dynagrade system, the timber is fed lengthwise through the

Table 3.7 Existing grading machines worldwide. (Based on information from Hoffmeyer (1995) and Stenman (2000))

Name	Number of machines world wide	Measurement principle	Grading speed
Computermatic/ Micromatic	137	Flatwise bending. Measurement of deflection. Span 914 mm. Constant load. Shape arm to measure bow.	≤105 m/min
CookBolinder/ Tecmach	66	Flatwise bending. Measurement of load. Span 900 mm. Constant deflection. Bow is considered by either running the timber through twice after turning 180 degrees or using two machines, one after the other.	≤200 m/min
EuroGrecomat (Mainly Germany)	6	Flatwise bending. Measurement of load and deflection. Span either 700 or 900 depending on timber thickness. X-ray for determination of density and detection of knots.	≤120 m/min
Raute Timgrader (Mainly Finland)	19	Flatwise bending. Measurement of load. Span 510 mm. Constant deflection. Timber forced to s-shape.	≤135 m/min
CLT and 7200LS (mainly North America)	59	Flatwise bending. Measurement of load. Constant deflection. Timber forced to s-shape.	≤600 m/min
Dart (Australia-New Zealand)	–	Flatwise bending. Measurement of deflection. Constant load. Shape arm for measurement of bow.	≤250 m/min
Ersson ESG-240	8	Flatwise bending. Measurement of load. Constant deflection. Bow measured by laser sensors.	≤200 m/min
Newness XLG	25	X-ray with 11 detectors across.	≤600 m/min
GradeMaster 403	5	Measurement of resonant longitudinal vibrations and mass. Scanning on all four sides.	≤20 pieces/min
Dynagrade	30	Measurement of resonant longitudinal vibrations. Vibrations induced by steel ball on a lever arm. Response picked up by microphone.	≤100 pieces/min

Figure 3.14 Dynagrade grading machine in a trimming line with a length measuring device to the far left and measurement of resonant vibrations and computer to the left

machine. Dynagrade is normally placed in the trimming line and the timber is transported transversely (see Figure 3.14).

Other machines or equipment worth mentioning, according to Hoffmeyer (1995) and Boström (1999), are:

- TRU Timber Grader (125 machines in South Africa).
- Finnograder (a machine developed in Finland at the beginning of the 1980s. A microwave technique is used to detect knots and slope of grain and γ-rays determine the density. No machines in operation).
- Sylvatest (equipment based on ultrasonic pulse measurement).
- Metriguard E-computer (measures the frequency response).

3.6.2 Accuracy of the Machines

Only a few investigations exist where machines have been compared on equal terms, i.e. using the same timber and the same grading speed. Boström (1994) studied CookBolinder, Computermatic, Finnograder and Sylvatest. Three dimensions of Norway spruce, 45 × 195 mm (121 pieces), 70×170 mm (124 pieces) and 45 × 120 mm (170 pieces), were passed through the machines. In the summary in Table 3.8, the machines are compared on the basis of the coefficient of determination, R^2, which in this case includes both the uncertainty in relationship between the grading characteristics (strength vs. grading parameter) and the machine's measurement error. The 45 × 195 mm timber appears to give better grading accuracy than the other dimensions. The fact, however, is that the 45 × 195 mm material had a much larger variation in strength and stiffness properties than the other dimensions, and this affects the R^2 value. This underlines the importance of using the same material when comparing machines.

Betzold (1999) studied CookBolinder, Computermatic, Ersson ESG240 and Dynagrade. The same 400 pieces of 45 × 145 mm Norway spruce were passed through all four machines. The test material had been divided into four equal samples, with 100 pieces in each, which had been dried in different ways.

As can be seen in Table 3.9, Dynagrade predicted strength with roughly the same accuracy as the other machines, despite the fact that it was

Table 3.8 Comparison of grading accuracy of different grading machines (Boström, 1994). The coefficient of determination R^2 is used as a measure of the accuracy

Grading machine type	Grading speed	Grading accuracy (R^2)					
		Machine flat wise MOE vs. edgewise bending strength			Machine flat wise MOE vs. edgewise MOE		
		45 × 120	70 × 170	45 × 195	45 × 120	70 × 170	45 × 195
CookBolinder	50 m/min	0.51	0.48	0.73	0.71	0.71	0.87
Computermatic	50 m/min	0.41	0.60	0.62	0.57	0.84	0.68
Finnograder	150 m/min	0.34	0.38	0.55	–	–	–
Sylvatest	–	0.34	0.33	0.62	–	–	–

Table 3.9 Comparison of grading accuracy of different grading machines (Betzold, 1999). The coefficient of determination R^2 is used as a measure of the accuracy

Grading machine type	Grading speed	Grading accuracy (R^2)							
		Machine flatwise MOE vs. edgewise bending strength				Machine flatwise MOE vs. edgewise MOE			
		Sample 1	Sample 2	Sample 3	Sample 4	Sample 1	Sample 2	Sample 3	Sample 4
CookBolinders	40 m/min	0.65	0.56	0.46	0.54	0.76	0.80	0.69	0.80
Computermatic	40 m/min	0.56	0.47	0.43	0.51	0.70	0.66	0.63	0.75
Ersson ESG240	100 m/min	0.61	0.57	0.54	0.53	0.76	0.70	0.76	0.80
Dynagrade	100 pieces/min	0.56	0.40	0.49	0.53	0.54	0.34	0.48	0.42

Table 3.10 Comparison of four grading machines with respect to accuracy and yield. The coefficient of determination R^2 is used as a measure of accuracy. The grades are special ones for glued laminated timber laminations, and the class designation refers to the characteristic tensile strength

Grading machine type Speed	R^2 Machine MOE vs. tensile strength	Yield (%) for combinations of grades						
		23/15		18	26/14.5		22/11	
		High	Low		High	Low	High	Low
Computermatic 40 m/min	0.44	47	53	99	27	73	67	33
CookBolinder 40 m/min	0.50	53	46	98	31	69	68	32
Dynagrade 40 pieces/min	0.47	52	47	98	37	63	65	35
Ersson ESG 240 80 m/min	0.41	52	48	100	15	85	66	34

run at normal production speed and the others at a lower speed. In the case of CookBolinder and Computermatic, the speed was much lower. Dynagrade is more or less insensitive to grading speed, whereas the bending type machines tend to induce vibrations in the timber which can severely reduce the grading accuracy. The vibrations increase with speed.

Johansson *et al.* (1998) also included the yield in the comparison (see Table 3.10). The yield is of decisive importance as it affects the economy of grading. The yield is of course strongly related to the accuracy of the machine, but also to how the settings for each grade or grade combination have been derived. Johansson *et al.* therefore calculated settings using the same principle, EN 519 (Anon, 1995a), for all four machines. A total of around 1200 pieces of Norway spruce, mainly 40 × 145 mm, were used. Four machine types were studied.

It appears that the yields obtained in the machines are fairly similar except for grade 26. In this grade the lowest yield is obtained by the machines with the lowest grading accuracy.

3.7 OPERATION AND CONTROL OF GRADING MACHINES

Machines can be operated in either machine or output control systems (Anon, 1995a). Machine control means that the quality of the graded timber is controlled only by setting the machine and continuously checking the performance. This requires an assessment of each machine type based on a very extensive experimental study of the machine type, including testing of large quantities of timber to obtain setting values for each species, strength class, dimension, etc. Machine control systems dominate in Europe. Derivation of settings values plays a very central role in this system. So far, at least three different procedures have been used in Europe. A new approach was introduced by Rouger (1997), in which the machine's performance is compared with that of a perfect machine capable of grading each piece of timber to its optimum grade. In the comparison a cost analysis is used where weighting factors are applied to pieces that are upgraded or downgraded. For pieces which are wrongly upgraded, the weighting factors have an adverse effect on the acceptance of the machine and its settings because the use of such pieces can be unsafe. For pieces wrongly downgraded the effect of the weighting factors is less severe, because the penalty is the cost of using larger cross-sections or more pieces of timber.

Output control is mainly used in North America, and in this system a limited number of graded pieces are proof-loaded to check that the quality of the timber is within desirable limits. In output control systems the demand on the grading machine is less strict.

In the output control systems initial settings are established using much less testing than in machine control systems. Instead, the settings are checked regularly by means of a proof-loading procedure and adjusted if necessary.

One output control system is described in EN 519, and the standards that will supersede it. Here it is required that 10 specimens from each strength class are sampled per working shift. The pieces are tested in edgewise bending, and a proof-load corresponding to 96% of the characteristic strength for the class in question. The MOE is also controlled. Cusum charts are used to control the process. If the process is out of control, adjustments of the settings may be necessary.

3.8 PROPERTIES AND YIELD OF MACHINE STRESS GRADED TIMBER IN RELATION TO VISUALLY GRADED TIMBER

Due to the higher grading accuracy, machine strength grading gives a higher yield in relation to what is obtained by visual grading. This is demonstrated in Table 3.11 which presents results from investigations where the same pieces of

Table 3.11 Comparison of visual and machine strength grading. Strength values and density are characteristic fifth percentiles and MOE is a mean value. Visual grades S10 and S8 are according to ECE-rules (1982), and LT30 and LT20 are grades in INSTA 142 (Anon, 1997)

Visual grading					Machine strength grading				
Visual grade	Yield (%)	Strength (N/mm^2)	MOE (N/mm^2)	Density (kg/m^3)	Machine grade	Yield (%)	Strength (N/mm^2)	MOE (N/mm^2)	Density (kg/m^3)
Johansson et al. (1992) – Grading of 58 × 120 *mm planed Norway spruce – Strength and MOE determined in bending*									
S10	12	35.1	14650	–	T30M	50	35.1	14540	–
S8	63	28.5	12920	–	T24M	42	28.2	11460	–
Johansson et al. (1998) – Grading 46 to 52 mm thick Norway spruce – Strength and MOE determined in tension									
LT30	34	21.8	13000	390	22	74	22.9	13400	397
LT20	41	16.6	11500	385	11	26	14.1	9900	355

timber have been graded both visually and by machine. In the investigation by Johansson *et al.* (1992), the S10 (visual) and T30M (machine) have the same strength, which is a coincidence, and almost the same MOE. The yield in S10 is, however, only 12%, to be compared with the 50% in T30M. In practice, grades corresponding to S8 and T24M are the most common and looking at S8 and T24M, the difference in yield is less, 75% compared to 92%. Note that the MOE for T24M is less than for S8, but both grades correspond to the European strength class C24 with a mean MOE of 11 000 N/mm^2.

3.9 FUTURE DEVELOPMENT

Important driving forces are the demands for better grading yield in general, the ability to grade high strength material and high grading speed.

Higher yield requires more advanced and accurate non-destructive methods and combinations of methods. None of the methods applied in machine strength grading today is capable of correctly mapping the variability of the timber, for instance the density and MOE variation from the pith outwards and the size, shape, orientation and location of the knots. As knots are the most important strength reducing characteristics, it is a great challenge to be able to better predict the strength reducing effect of knots.

Another approach is to carry out grading earlier in the production process. One way is to grade the logs to obtain an enrichment of logs that, after sawing, is likely to give higher yields of structural timber. This is already done to some extent by using visual criteria. Perstorper (1999) shows that measurement of resonant vibrations is an option worth developing further. He measured Norway spruce logs with a length of 4–5 meters and diameter of 320–390 mm. Afterwards, 45 × 145 mm boards were produced from the logs and tested in bending. The coefficient of determination for the relation between edgewise bending strength of individual boards and the MOE based on axial vibrations was 0.47. Oja (1999) used an x-ray Logscanner, and with data from it he created a model based on log diameter, taper, percentage of heartwood, density and density variations to predict the MOE of individual boards from the logs. He managed to predict the MOE with an R^2 value of 0.48. Oja's model was further developed to predict the strength of individual boards (Bengtsson, 2001). Based on measurement of Scots pine (*Pinus silvestris*) logs, the bending strength of individual boards cut from the logs was predicted with an R^2 of 0.55.

Yet another approach is strength grading by means of surface scanning techniques. There are several systems for surface scanning on the market, but these are exclusively used for appearance grading. It is likely that these systems will be further developed for grading with respect to strength.

Proof loading was mentioned in connection with output control systems as a means for checking the machine settings. This technique is also used as a grading method in Australia (Leicester, 2001), and has been developed further (Ziethén, 2001). One aspect to consider is to what extent proof loading affects the pieces that survive (Lam *et al.*, 2001).

REFERENCES

Anon (1982) ECE-rules. ECE recommended standards for stress grading and finger jointing of structural coniferous sawn timber. Timber Bulletin for Europe. Volume XXXIV. Supplement 16. United Nations Commission for Europe, Geneva.

Anon (1995a) EN 519. Structural timber – Grading – Requirements for machine strength graded timber and grading machines.

Anon (1995b) EN 338. Structural timber – Strength classes.

Anon (1997) INSTA 142. Nordic visual strength grading rules.

Bengtsson C. (2000) Stiffness of spruce wood – Influence of moisture conditions. *Holz als Roh- und Werkstoff*, **58**, 344–352.

Bengtsson C., Betzold D. (2000) Bending strength and stiffness of Norway spruce timber – influence of high temperature drying. *Proc. International Symposium on Wood Machining. Properties of wood and wood composites related to wood machining*, Vienna, Austria.

Bengtsson C. (2001) Swedish National Testing and Research Institute. Personal communication.

Betzhold D. (1999) Maschinelle Festigkeitssortierung. Einfluss der Hochtemperatur-trocknung auf die elastomechanischen Eigenschaften des Snittholzes. Diplomarbeit, Fachhochschule Eberswalde.

Boström L. (1994) Machine strength grading. Comparison of four different grading systems. Swedish National Testing and Research Institute, SP REPORT 1994:49.

Boström L. (1998) Machine strength grading using smart image sensors for defect detection. *Proc. 5th World Conference on Timber Engineering*, Montreux, Switzerland.

Boström L. (1999) State-of-the art on machine strength grading. *Proc. 1st RILEM Symposium on Timber Engineering*, Stockholm, Sweden.

Clough R. Penzien J. (1975) *Dynamics of Structures*. MacGraw-Hill, New York.

Chrestin H., Boström L. (1997) The influence if timber origin on machine strength grading yield. *Proc. International IUFRO S 5.02 Timber Engineering Conference*, Copenhagen, Denmark.

Dahlbom O., Petersson H., Ormarsson S. (2000) Stiffness and shape stability grading analysis of sawn timber based on experimentally found growth characteristics. *Proc. World Conference on Timber Engineering*, Whistler Resort, British Columbia, Canada.

Foslie M. (1971) Strength properties of Norway Spruce. Part 3 – Strength properties of small defect free specimens. Report No 42. Norwegian Institute of Wood Technology. (In Norwegian.)

Glos P., Heimeshoff B. (1982) Möglichkeiten und Grenzen der Festigkeitssortierung von Brettlamellen für den Holzleimbau. In: *Ingenieurholzbau in Foschung und Praxis* (Ehlbeck und Steck). Bruderverlag, Karlsruhe. (In German.)

Glos P. (1995) Strength grading. STEP lecture A6, Centrum Hout, The Netherlands.

Hoffmeyer P. (1984) Om konstruktionstraes styrke och styrkesortering. I Skovteknologi. Et historiskt og perspektivisk strejtog. Dansk Skovforening. (In Danish.)

Hoffmeyer P. (1990) Failure of wood as influenced by moisture and duration of load. Doctoral Thesis, State University of New York, College of Environmental Science and Forestry, Syracuse, New York.

Hoffmeyer P. (1995) Styrkesortering ger mervärde. Del 2 – Tillgaenglig teknik. Danmarks Tekniske Universitet, Laboratoriet for Bygningsmaterialer, Teknisk Rapport 335.

Hoffmeyer P., Bräuner L., Boström L. and Solli K.H. (1999) Tensile strength of glulam laminations of Nordic spruce. *Proc. Pacific Timber Engineering Conference*, Rotorua, New Zealand.

Isaksson T. (1999) Modelling the variability of bending strength in structural timber. Length and load configuration effects. Doctoral thesis, Division of Structural Engineering. Lund Institute of Technology, Lund.

Johansson C.-J. (1976) Tensile strength of glulam laminations. Chalmers University of Technology, Steel and Timber Structures, Internal report no. S76:18. (In Swedish.)

Johansson C.-J., Brundin J. and Gruber R. (1992) Stress grading of Swedish and German timber. A comparison of machine stress grading and three visual grading systems. Swedish National Testing and Research Institute, SP REPORT 1998:38.

Johansson C.-J., Boström L., Bräuner L., Hoffmeyer P., Holmquist C. and Solli K.H. (1998) Laminations for glued laminated timber – Establishment of strength classes for visual strength grades and machine settings for glulam laminations of Nordic origin. Swedish National Testing and Research Institute, SP REPORT 1998:38.

Kollmann F.P. and Côte Jr, W.A. (1968) *Principle of Wood Science and Technology. Part I – Solid Wood.* Springer-Verlag, New York.

Kunesh R.H. and Johnsson J.W. (1972) Effect of single knots on tensile strength of 2 by 8 inch Douglas fir dimension lumber. *Forest Products Journal*, **22**(1), 32–36.

Lackner R. and Foslie M. (1988) Gran fra Vestlandet – Styrke och sortering. Norwegian Institute of Wood technology, Report. 74. (In Norwegian.)

Larsson D. (1997) Mechanical characterization of engineering materials by modal testing. Doctoral thesis, Chalmers University of Technology, Gothenburg, Sweden.

Lam F., Abayakoon S., Svensson S. and Gyamfi C. (2001) Influence of proof loading on the reliability of members. *Proc. International Council for Building Research Studies and Documentation*, Working Commission W18 – Timber Structures, Meeting Thirty four, Venice, Italy.

Leicester R.H. (1988) Rogue factors in proof grading. *Proc. 1988 International Conference on Timber Engineering*, Seattle, USA.

Leicster R.H., Seath C.A., Breitinger H.O. and Shakiba S.B. (1994) Development of a commercial microwave stresser, CSIRO, Division of Building and Engineering. Melbourne, Australia.

Madsen B. (1992) *Structural Behaviour of Timber.* Timber Engineering Ltd.

Oja J. (1999) X-ray measurements of properties of saw logs. Doctoral thesis, Luleå University of Technology, Luleå, Sweden.

Ormarsson S. (1999) Numerical analysis of the moisture-related distortion in sawn timber. Doctoral thesis, Department of Structural Mechanics, Chalmers University of Technology, Gothenburg, Sweden.

Orosz I. (1969) Modulus of elasticity and bending strength ration as indicators of tensile strength of lumber. *Journal of Materials*, **4**(4), 842–864.

Persson K. (2000) Micro mechanical modelling of wood and fibre properties. Doctoral thesis, Lund University, Department of Mechanics and Materials.

Perstorper M. (1999). Dynamic testing of logs for prediction of timber strength. *Proc. Pacific Timber Engineering Conference*, Rotorua, New Zealand.

Rouger F. (1997) A new statistical method for the establishment of machine settings. *Proc. International Council for Building Research Studies and Documentation*, Working Commission W18 – Timber Structures, Meeting Thirty, Vancouver, Canada.

Sandoz J. (1989) Grading of construction timber by ultrasound. *Wood Science and Technology*, **23**, 95–108.

Sauter U. (1992) Technologische Holzeigenschaften der Douglasie (*Pseudotsuga menziesii (Mirb.) Franco*) als ausprägung unterschiedlicher Wachstumsbedingungen. Doctoral thesis, Fakultät der Albert-Ludwigs-Universität zu Freiburg.

Schajer G.S. (2001) Lumber strength grading using x-ray scanning. *Forest Products Journal*, **51**(1), 43–50.

Schniewind A.P. and Lyon P.E. (1971) Tensile strength of redwood dimension lumber – II. Prediction of strength values. *Forest Products Journal*, **21**(8), 18–27.

Steffen A., Johansson C.-J. and Wormuth E.-W. (1997) Study of the relationship between flatwise and edgewise moduli of elasticity of sawn timber as a means to improve mechanical strength grading technology. *Holz als Roh- und Werkstoff*, **55**, 245–253.

Stenman B. (2000) Swedish National Testing and Research Institute. Personal communication.

Wormuth E.-W. (1993) Untersuchung des Verhältnisses von flachkant zu hochkant ermitteltem Elastizitätsmodul von Schnittholz zur Verbesserung der maschinellen Festigkeitssortierung. Master's thesis, Department of Wood Technology, University of Hamburg.

Ziethén R. (2001) Swedish National Testing and Research Institute. Personal communication.

Åstrand E. (1996) Automatic inspection of sawn wood. Doctoral thesis, Linköping University, Sweden.

4

Structural Timber – Variability and Statistical Modelling

Tord Isaksson

4.1 OBJECTIVES

This section includes a general review of research on the topic of variability and statistical modelling of structural timber. The reasons for the variability within pieces of timber are presented. Further, different theories for describing the variability are reviewed. The variability is the reason why timber has a length- and load-dependent load-bearing capacity. Theories for describing these effects are reviewed.

4.2 INTRODUCTION

Structural timber displays considerable strength variability between and within members, which makes it difficult to evaluate its reliability and to design timber elements and systems in a rational way. Most of today's engineering design methods for timber are based on elementary theory of structures assuming homogeneous material. In reality, due to the presence of various defects in the wood, the variation in strength along a timber beam results in a strength which is dependent on the length of the element and the type of loading, i.e. the moment distribution of the beam. These effects are only to a limited extent accounted for in today's codes. Apart from length and type of loading, the strength is also dependent on climate conditions and the duration of load. These effects are not included in the discussion here.

One way of describing the variability phenomena and its effects on, for example, the strength of a timber beam is proposed in a model introduced by Riberholt *et al.* (1979). It is assumed that timber is composed of localised weak zones connected by segments of clear wood, and that failure is primarily initiated in these weak zones (see Figure 4.1). The weak zones correspond to knots or groups of knots, which are distributed along the length at random.

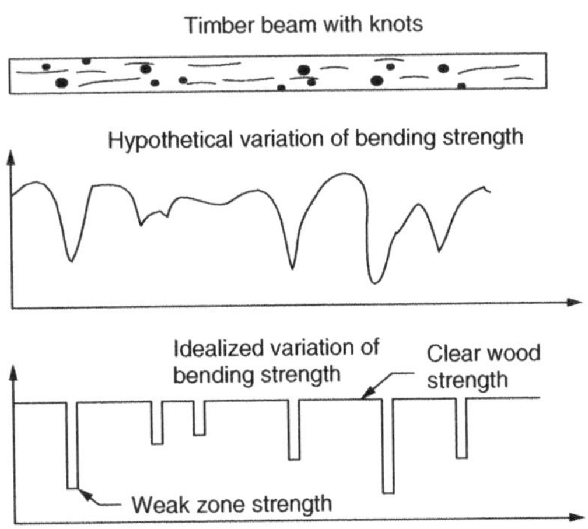

Figure 4.1 Modelling of lengthwise variation of bending strength in timber beams according to Riberholt *et al.* (1979)

The length of a timber beam affects the apparent strength, since there is a higher probability of having a weaker section within a longer beam than in a shorter one. This has been shown in many investigations, e.g. see reviews in Madsen (1992) and Barrett and Fewell (1990).

Furthermore, the apparent strength of a timber beam will also depend upon the nature of the moment distribution along the beam. For instance, a beam with a point load at mid-span can be expected to have an apparent strength that is higher than a beam of the same length loaded with a constant bending moment along the entire length. This has also been confirmed experimentally by several investigations (e.g. Madsen, 1992).

Figure 4.2 shows an example of how the length and type of loading influence the load carrying capacity of a beam. The strength given in some codes represents more or less the strength of the weakest section. This corresponds to a constant bending moment along the beam.

Figure 4.3 shows a trussed rafter. The upper chord is an example of a member, which is over-designed using the principles in today's code. The peaks of the varying moment distribution have a low probability of coinciding with the lowest strength of the member.

4.3 PROPERTIES OF WOOD

Timber, regardless of species and origin, consists of defects distributed throughout more or less

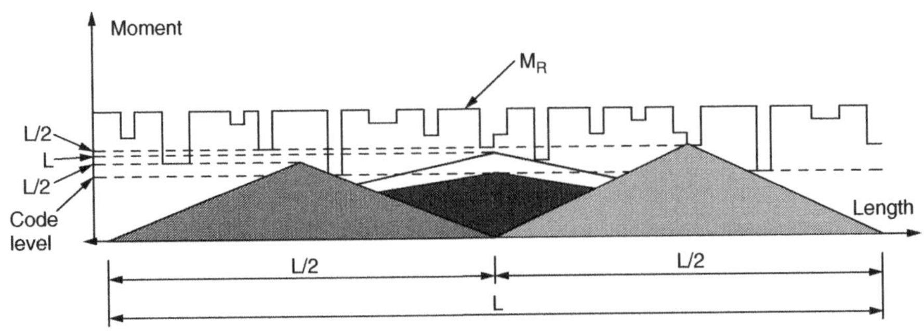

Figure 4.2 Example of a lengthwise variation in bending strength and maximum bending moment due to loading for different lengths of a beam

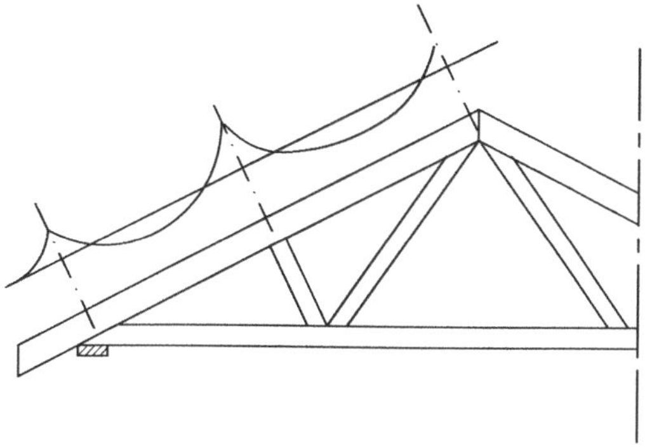

Figure 4.3 The upper chord in the trussed rafter is a practical example of where the probability is low that the maximum bending moment will coincide with the weakest section. This means that the load carrying capacity is not fully utilised

defect-free wood. The structure and properties of timber are the results of millions of years of evolution aimed at providing the most efficient system to optimise growth, and to support the crown under occurring loads, such as wind and snow. In most cases, man's use of timber does not coincide with the natural conditions it is 'designed' for. Unlike other structural materials, such as steel and concrete, timber cannot be produced to a particular specification. Instead, timber has to be divided into groups with a certain desired range of properties. The end use of the material decides which properties are most important. For example, appearance may be important for use in furniture, while strength is important for use in structures (studs, joists, roof trusses).

Due to the structure of trees, the material properties are significantly different in the longitudinal (stem) direction and transversal (cross) direction (see Figure 4.8). That is, timber is a highly anisotropic material. Some macroscopic characteristics valid for softwood are discussed below. The microstructure and ultrastructure are not considered. This is, however, not to say they are unimportant to the behaviour of timber.

In the transversal direction the *growth rings* are an influential parameter for strength. Large annual growth rings mean a low density and thus a lower strength. A growth ring consists of a lighter and a darker part, i.e. the earlywood evolved during spring and early summer and the latewood evolved during summer. During the first 10–20 years of the tree's life, the wood is characterised as juvenile wood (e.g. see Dinwoodie, 2000), with a lower strength and stiffness. The growth rings near to the pith are called 'heartwood' and the outer rings 'sapwood'. The heartwood is characterised by the absence of living cells, while the sapwood is where liquid transport takes place. These two types of wood have no significant effect on strength.

The strength is strongly dependent on the occurrence of natural defects such as *knots*, compression wood and grain deviation. A knot is a branch that is embedded in the main stem of a tree, and it is undoubtedly the most common and influential defect. As long as the branch is growing the knot is known as 'tight'. When the branch dies the knot is called an 'encased knot'. For softwoods, the branches occur as whorls at more or less regular intervals producing knot clusters (see Figures 4.4–4.6). The sawing pattern and location in the log influences how the knots will affect the strength of a wood member. The reduction in strength due to knots is due to the deviation in grain angle that occurs around knots (see Figure 4.7). The main reason for the strength reducing effect of a knot is that it produces tension perpendicular to the grain, which is the weakest direction in wood.

Compression wood is a result of the tree's reaction to external forces. Softwood produces

Figure 4.4 Norway Spruce trees, around 25–30 years old. The branches are dead up to about halfway up the trunk

Figure 4.5 Norway Spruce trees, aged around 50–60 years. The lower dead branches have fallen off

Figure 4.6 Norway Spruce trees, aged about 80 years, ready for harvesting. Note the regularity at which the branches occur

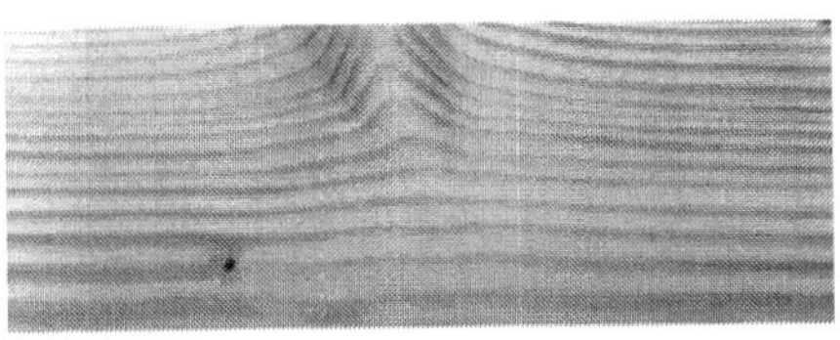

Figure 4.7 Slope of grain at a knot

compression wood in areas of high compression. These areas are darker than normal due to a higher latewood proportion in the growth ring. Although compression wood has a higher density, the stiffness is low. The failure in beams with compression wood is often more brittle.

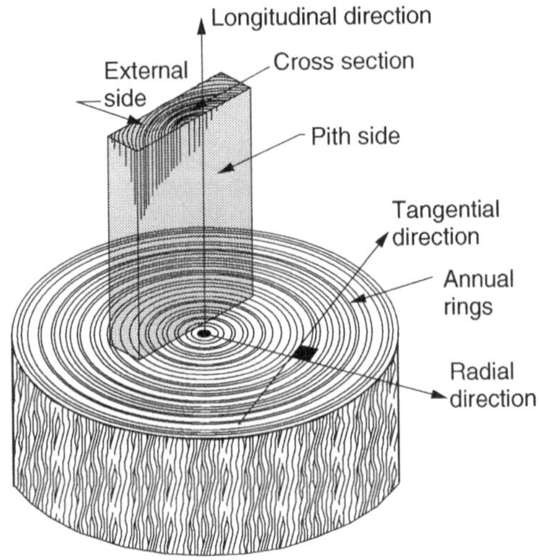

Figure 4.8 Definition of some geometrical parameters

Grain deviation is defined as the angle between the grain and the longitudinal direction of the wooden element. Grain deviation can be due to spiral grain, pronounced tapering of the stem, crook and sawing of the log. Again, as with knots, grain deviation leads to tension perpendicular to the grain, and thus a reduction in strength.

Wood is a hygroscopic material and there is a continuous exchange of moisture with the surroundings. The moisture content influences both strength and stiffness parameters. Generally, the higher the moisture content, the lower the strength and stiffness, at least below the fibre saturation point.

4.4 MODELLING OF VARIABILITY WITHIN AND BETWEEN MEMBERS

Most of the work that has been done in the field of modelling the within-member variability of timber properties has been on tension parallel to the grain. This work started in the mid-1980s with publications by Kline, Woeste and Bendtsen (1986), Showalter, Woeste and Bendtsen (1987) and Taylor and Bender (1989). These studies were focused on the tension mode, probably because it is a decisive parameter in the design of the lower chord in timber trusses, and because it is easier to test. An important benefit of having a model of the within member variation of various properties is that it dramatically reduces the amount of testing necessary to determine, for example, length and load configuration factors. Furthermore, regarding the design of systems, i.e. taking into account the favourable behaviour of members when placed in a system, the model can be quite useful.

The following sections present some of the research that has been done on within-member variability. The modulus of elasticity in bending is an essential parameter because it correlates well with both tension and bending strength. It should be noted that studies of within-member variation of the modulus of elasticity were done as early as the 1960s. This parameter is of course easier to measure as the test is non-destructive.

There are several references on the variation of the Modulus Of Elasticity (MOE) within members from the mid-1960s to the early 1980s. Examples are Corder (1965), Kass (1975), Suddarth and Woeste (1977) and Foschi and Barrett (1980). With increasing computer capacity, modelling of variability became more accessible and easy to execute. Kline *et al.* (1986) probably took the initial step of modelling and simulating within-member properties using a stochastic model. They were looking for a better representation of timber stiffness than just an average stiffness for the whole piece of timber, i.e. including the lengthwise variation. Their MOE measurements were performed as illustrated in Figure 4.9. The beams were 4.9 m long and the flatwise MOE was determined in four 760 mm segments.

Before continuing to discuss their findings, the term *Lag-k* serial correlation needs an introduction. This is a measure of the statistical dependency between, as in this case, the MOEs from different segments, i.e. it indicates whether there is a stronger dependency between closely located segments than between segments far apart, but still within the same beam. The *Lag-k* serial correlation is the correlation between an observation from a segment i and an observation from segment $i + k$. If $k = 2$, *Lag-k* means the correlation between strengths of segments 1 and 3 and segments 2 and 4, and so on. For a data set $x_1, x_2, \ldots, x_n$ the *Lag-k* serial correlation r_k is given by Equation (4.1). $\overline{x}$ is the mean of all x_i.

$$r_k = \frac{\sum_{i=1}^{n-k} (x_i - \overline{x})(x_{i+k} - \overline{x})}{\sum_{i=1}^{n} (x_i - \overline{x})^2} \tag{4.1}$$

Kline *et al.* found that the serial correlation decreased with increasing k. The lengthwise variation in MOE was modelled by a second order Markov model according to Equation (4.2):

$$X_{i+1} = \beta_1 X_i + \beta_2 X_{i-1} + \varepsilon_{i+1} \tag{4.2}$$

where X_i are the MOE-values, β_i are multiple regression coefficients and ε is a random error. Necessary measures are taken to eliminate bias in the generated MOE values.

Showalter *et al.* (1987) focused on the fact that tensile strength is dependent on the size of the member. They developed a model of the lengthwise variation in tensile strength to be able to, for example, analyse the strength of the lower chord in trusses. Tension tests were performed on members such as that shown in Figure 4.9. They were only able to test sections 1 and 4, as sections 2 and 3 were held in clamping devices. The results showed a significant length-dependent tensile strength for the species they tested, Southern Pine. Their model consists of one Markov process generating correlated MOE values and one process generating correlated tension strength values. The MOE and tension strengths are correlated. The *Lag-3* serial correlation was found by the tests, and the *Lag-1* and *Lag-2* serial correlations were estimated from the *Lag-3* serial correlation. They found the *Lag-k* correlation to decrease with increasing k, i.e. two

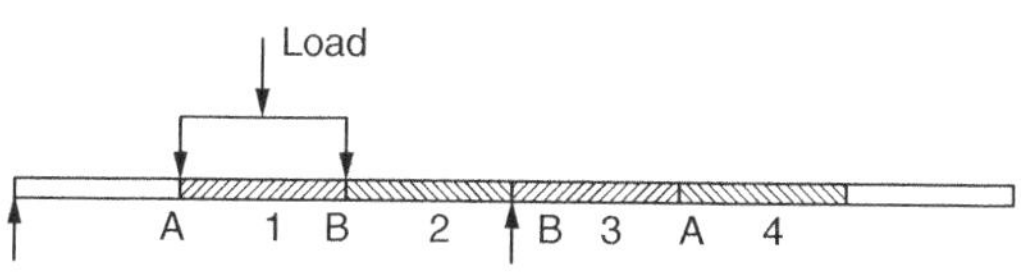

Figure 4.9 Four 760 mm segments, 1–4, tested for flatwise MOE (Kline *et al.*, 1986)

sections next to each other showed a better correlation than sections located further away. The model showed satisfactory agreement with the tension tests on beams of different lengths.

Lam and Varoglu (1991a) carried out an experimental study on the variation of tensile strength along the length of timber. The Spruce–Pine–Fir lumber members were tested over a length of 610 mm enabling four to six sections to be tested along a beam with a total length of 6.1 m. The results showed a decreasing correlation between sections the further apart they were. Sections separated by more than 1.83 m could be considered to be statistically independent. Lam and Varoglu (1991b) followed up this work with an article on the development of the model. The minimum tensile strength was represented by a three-parameter Weibull distribution, and the mean of tensile strengths, which is correlated to the minimum strength, was considered to be normally distributed and expressed as a function of the minimum tensile strength. Further, the tensile strength along the length of a member is assumed to be log-normally distributed. The Monte Carlo simulation technique was then used to generate the tensile strength profile of a desired number of beams. The decreasing *Lag-k* correlation, or the autocorrelation as Lam and Varoglu call it, was still not taken into account. This was done by introducing a moving average process. The tensile strength is given by Equation (4.3). Using this model, they found good agreement with their experimental results:

$$T(x_i) = \mu + t(x_i) - a_1 t(x_{i-1}) - a_2 t(x_{i-2}) - a_3 t(x_{i-3}) \qquad (4.3)$$

where

$T(x_i)$ is the tensile strength of section x_i
μ is the mean of the tensile strengths within a member
t is the moving average process
a_i are weighting factors.

Lam and Barrett (1991) presented a technique to remove trends from the within-member variation of compressive strength. Experimental data showed a clear trend in the variation of compressive strength along the length of a member. This must be removed before the parameters of the model can be estimated, after which the trend can be added again. This technique, together with Kriging analysis of the compressive strength along a member, proved to be successful, and they concluded that these two techniques could be applied to models of within-member bending and tension strength.

The lengthwise variability of bending stiffness of timber beams was investigated by Czmoch in 1991. In his tests, the beams were loaded in five steps and the correlation functions (*Lag-k* serial correlations) were determined for each load level. He used two models, one explaining a random stiffness around a global mean value calculated for the whole set of beams, and the other using a random variable for the mean of an individual beam together with another random process representing the local fluctuations within a beam. The variance of local fluctuations within a member was found to be greater than the variance of the individual means for all but the first load level. The correlation between sections was found to decrease relatively quickly as the distance between the sections increased.

Taylor and Bender (1991) continued refinement of the model of within-member variation of tensile strength and modulus of elasticity in bending. The purpose was to obtain a model capable of reliability analysis of glue-laminated beams where the tensile strength and modulus of elasticity are governing parameters in the design. By using Monte Carlo simulations, the strength and stiffness of glulam beams could be predicted. The experimental tests were performed according to the procedures used by Kline *et al.* (1986) and Showalter *et al.* (1987). That is, they determined the bending stiffness in four sections of the beam and tested two of the sections in tension. The *Lag-1*, *2* and *3* serial correlations were determined for the modulus of elasticity while only the *Lag-3* serial correlation could be determined for the tensile strength. The other *Lag-k* serial correlations were estimated. To be able to take into account the dependency between tensile strength and modulus of elasticity in the model, the *Lag-k* cross-correlation was determined, i.e. the correlation between the tensile strength of one section and the modulus of elasticity of a section k sections away.

The tensile strengths and the modulus of elasticity were fitted to the appropriate statistical distribution. Since the strength and stiffness of different sections are assumed to have the same kind of distribution, only one distribution was used for each property. Furthermore, the correlations that were discussed above were determined directly from test results where possible, or estimated using a Markov process.

The research hitherto reviewed on the variability of properties within timber members has been on tensile strength and stiffness. However, in 1979 Riberholt and Madsen introduced a model for the lengthwise variation of bending strength. The strength along a beam (here called clear wood strength) is assumed to be constant, except where defects occur (see Figure 4.1).

The clear wood strength and the weak zone strengths are described by stochastic variables, while a Poisson process represents the occurrence of weak sections. The parameter in the Poisson process is estimated from direct measurements of the distance between defects. At that time they did not have any results from direct measurement of the strength of weak sections. Instead, two methods were used to predict the strength of weak sections. Method 1 was based on the assumption that the strengths of the weak sections are independent. Using results from tests on beams in their weakest section, as prescribed in the Eurocode, it is possible to estimate a distribution representing the strength of weak sections. Method 2 employed data from machine stress grading of timber together with a regression relation between the output from the machine and bending strength. They found no reason to reject the theory of a constant *Lag-k* serial correlation, i.e. the dependency between the strength of two sections is independent of the distance between them. A correlation coefficient between 0.50 and 0.67 was found to be reasonable.

In 1991 the first step was taken towards using a model of the variation in bending strength within members in a study of the effect of length and load configuration on the load-carrying capacity of a beam. Czmoch, Thelandersson and Larsen (1991) used the model proposed by Riberholt *et al.* (1979). What they found was a significant lack of important experimental information on which to base and verify a model. The following assumptions were made:

- Timber is composed of short weak zones connected by sections of clear wood.
- The weak zones correspond to knots or groups of knots and are randomly distributed.
- Failure occurs only in the middle of the weak zones.
- The strength of the weak zones is randomly distributed.

The model should enable studies of a number of important issues in the design of timber members and systems, such as:

- effect of length and load configuration on the load carrying capacity,
- influence of test procedures in determining the characteristic bending strength,
- use of reliability theory in design.

The following data were identified as necessary input to the model:

- Distribution function of the distance to the first weak zone.
- Distribution function of the distance between weak zones.
- Correlation function for distances between weak zones.
- Distribution function of the strength of weak zones.
- Correlation function for strengths of weak zones.

Very limited information was found in the literature, so they performed a sensitivity analysis using the Riberholt model. Reasonable assumptions could be made for most of the input. The results showed quite high length and load configuration effects. The effect of the test standard for determining the characteristic strength was also significant. The need for experimental tests of the variability within members was pointed out.

Leicester *et al.* (1991) presented a discrete co-relationship with Weibull theory to model the variability in bending strength between and within members. The classical Weibull theory does not take into account the correlations in strength within members and between members cut from the same piece of timber. Leicester introduced the correlation within members to the Weibull model, and gave expressions for the strength in clear wood and at defects.

In Isaksson (1999) a statistical model of the variability in bending strength within and between pieces of timber is presented. The model is based on the ideas of Riberholt *et al.* (1979), i.e. timber is composed of localised weak sections connected by segments of clear wood (see Figure 4.1). The input to the model is based on an experimental investigation. In about 130 timber beams (length around 5 meters), the bending strength was determined in 5–6 weak sections. The test set-up is shown in Figure 4.10. All knots appearing on the surface of the beam were recorded. The knot data is used to determine various knot measures enabling to identify weak sections.

The random variables of the model are:

- distance between weak sections,
- length of weak sections,
- bending strength of weak sections, and
- bending strength between weak sections.

A weak section is primarily identified by knot measures, and the distance between weak sections and the width of a weak section are based on knot measurements.

It was found that the variability in bending strength within and between members could be modelled according to Equation (4.4). This is based on the observation that the correlation coefficient between weak sections is independent of the distance between the sections within the same member. A model based on logarithmic values is used. The model can be visualised as in Figure 4.11.

$$\ln(f_{ij}) = \mu + \tau_i + \varepsilon_{ij} \tag{4.4}$$

where

f_{ij} is the strength of beam i in section j
μ is the logarithm mean of all weak sections
τ_i is the difference between the logarithm mean of weak sections within one member i and μ
ε_{ij} is the difference between weak section j in member i and the value $\mu + \tau_i$. The mean equals zero and the standard deviation is σ_j.

The standard deviations are related according to Equation (4.5).

$$\sigma_{ij} = \sqrt{\sigma_i^2 + \sigma_j^2} \tag{4.5}$$

where

σ_i is the standard deviation between members
σ_j is the standard deviation within a member
σ_{ij} is the standard deviation of the whole sample of beams and weak sections.

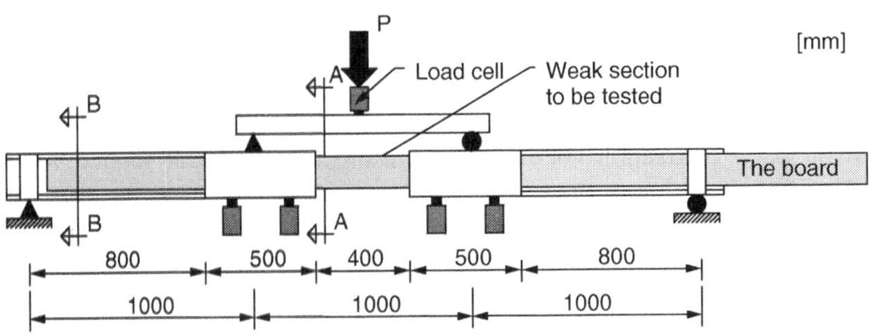

Figure 4.10 Test arrangement for testing the bending strength in several sections within a member (Isaksson, 1999)

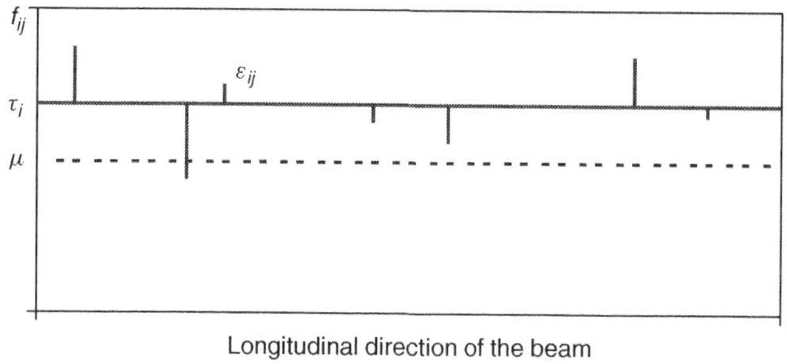

Figure 4.11 Statistical model of strength in weak sections (Isaksson, 1999)

The correlation coefficient becomes

$$r = \frac{\sigma_i^2}{\sigma_i^2 + \sigma_j^2} \tag{4.6}$$

In addition to the experimental data for Norway Spruce, some data for Radiata Pine was found in the literature (Leicester *et al.*, 1996). Radiata Pine shows a higher variability in strength than Norway Spruce.

Table 4.1 summarises the input to the statistical model. A sensitivity study of the different parameters of the model was performed. The distance between weak sections is represented by a Gamma distribution and the width of a weak section is set to be 150 mm. The strength of the clear wood between weak sections was given the strength of the strongest of the weak sections.

A similar study on the lengthwise variation of bending strength was started in Sweden at Trätek (Swedish Institute for Wood Technology Research). Källsner and Ditlevsen (1994, 1997) performed an experimental study on 26 beams, cross-section 45 by 120 mm^2. The logs came from a specific area, i.e. the growth conditions were as identical as possible to provide as homogeneous a sample as possible. Weak sections were tested by cutting them out (knot clusters), finger jointing them to pieces of higher strength and testing them according to the normal procedure (see Figure 4.12). Out of the 197 tests of weak section strengths, 42% of the failures occurred at the finger joints, 1% in the stronger timber outside the weak section, 24% were of mixed type, and only 34% failed in the weak sections. Due to the many failures outside the weak sections, the evaluation of the test results had to be refined using statistical tools to include the failures in the joints.

They used a so-called 'hierarchical model' with two levels to represent the variation in bending strength between and within members. The *Lag-k* correlation was found to be constant, i.e. an equi-correlation. The bending strength was modelled using the same principles as in Isaksson (1999), e.g. Equation (4.4).

Källsner *et al.* (1997) assigned numbers to the above parameters as follows:

Table 4.1 Summary of input to the model of variability in bending strength

Parameter	Model	
	Norway Spruce	Radiata Pine
Distance [mm]	Gamma (2.5445, 194.12)	Exponential (1, 1300)
Width [mm]	150	150
Weak section strength [MPa]	LogN(4.01977, 0.247136)	LogN(3.75443, 0.38523)
Coeff. of corr. between weak section strengths	0.565	0.55
Strong section strength	greatest weak section strength	greatest weak section strength

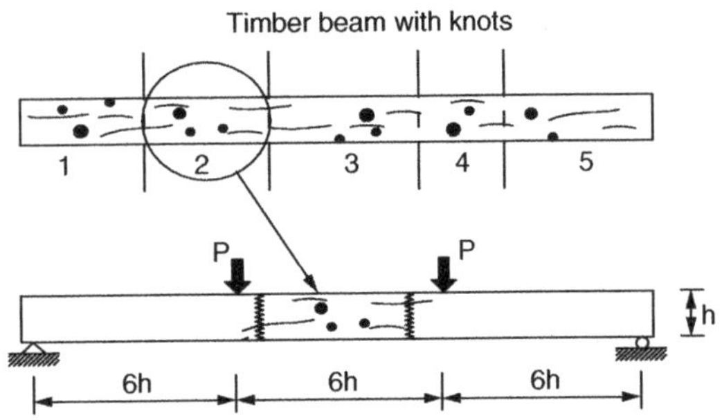

Figure 4.12 Selection and testing of weak sections (Källsner *et al.*, 1994)

$\mu = 49.77$ MPa
$\sigma_i = 7.03$ MPa
$\sigma_j = 6.33$ MPa
$r = 0.55$

Using the model together with the above numbers, the influence on the load-carrying capacity of the number of weak sections in a beam was studied. As could be expected, the longer the beam, the lower the strength. However, the decrease in load-carrying capacity predicted by the model was not as high as that found by testing of longer beams. They present two possible reasons for this. First, the testing of single weak sections does not include the possibility of neighbouring weak sections interacting. That is, cracks could not propagate from one section to another. This is possible when testing longer beams with several weak sections. Secondly, testing of longer beams means that more elastic energy is stored and released at failure. The testing of longer beams revealed 5–15% lower load-carrying capacity compared with the prediction of the statistical model.

Williamson (1994) carried out a study on the correlation coefficient between strengths of timber elements of a single beam. He found several reasons for a dependency between sections within a beam:

- Sawing variations tend to apply to an entire beam.
- Pith in a beam tends to occur all the way along.
- The wood in a beam originates from the same distance from the pith, and thus has a uniform density.
- The wood will come from a similar height up the tree, and therefore branches (knots) will have similar characteristics.

He collected experimental results from various studies, most of them referred to earlier in this section. The correlation coefficient was determined and the corresponding 95% confidence interval. The overall mean of the coefficient was 0.58, and out of 15 data sets, one mean fell outside the confidence interval, which is what one would expect. No evidence was found against the hypothesis that timber, independent of species, has the same correlation coefficient and that it does not decrease with separation distance.

4.5 THEORIES FOR DESCRIBING SIZE AND LOAD CONFIGURATION EFFECTS

4.5.1 Size Effects According to the Weibull Theory

The most widely adopted way of handling size and load configuration effects is to use the so-called weakest link theory. This theory says that "when subjected to tension, a chain is as strong as its weakest link", and was introduced by Pierce (1926), Tucker (1927) and Weibull (1939). They studied brittle materials. To explain this theory let us consider a material with a reference volume subjected to tension. The probability P_f that this volume will fail when subjected to a given load s is given by Equation (4.7):

$$P_f = P(R < s) = F(s) \tag{4.7}$$

where F is the statistical probability distribution of the strength of the material. If the strength fits the Weibull distribution it is written as shown in Equation (4.8). Since the calculations are performed using stresses, the notation s is changed to σ:

$$P(R < \sigma) = F(\sigma) = 1 - \exp\left(-\left(\frac{\sigma - b}{a}\right)^{1/k}\right) \tag{4.8}$$

This is a three-parameter Weibull distribution with scale parameter a, shape parameter k and location parameter b. A nice feature of the Weibull distribution is that the minimum of statistically independent and identically distributed Weibull variables is also Weibull distributed. That is, for a series of N reference volumes the probability of failure for the system can be written as Equation (4.9).

$$\begin{aligned} P_f &= P(R_i < \sigma \text{ for any } i = 1, \ldots, N) \\ &= 1 - P(\sigma \leq R_i \text{ for all } i = 1, \ldots, N) \\ &= 1 - \prod_{i=1}^{N} P(\sigma \leq R_i) \\ &= 1 - \prod_{i=1}^{N}(1 - P(R_i < \sigma)) \end{aligned} \tag{4.9}$$

Introducing Equation (4.8) into Equation (4.9) gives Equation (4.10).

$$\begin{aligned} P_f &= 1 - \prod_{i=1}^{N}(1 - P(R_i < \sigma)) \\ &= 1 - \prod_{i=1}^{N}\left(1 - \left(1 - \exp\left(-\left(\frac{\sigma - b}{a}\right)^{1/k}\right)\right)\right) \\ &= 1 - \left(\exp\left(-\left(\frac{\sigma - b}{a}\right)^{1/k}\right)\right)^N \\ &= 1 - \exp\left(-N\left(\frac{\sigma - b}{a}\right)^{1/k}\right) \end{aligned} \tag{4.10}$$

The number of reference volumes N is directly connected to the total volume V, that is, N can be replaced by V in Equation (4.10). When comparing two volumes V_1 and V_2, and keeping the probability of failure P_f constant, we get the following relation between the two 'load capacities' σ_1 and σ_2, Equation (4.11).

$$\begin{aligned} & P_f(\sigma_1) = P_f(\sigma_2) \\ & \Rightarrow 1 - \exp\left(-V_1\left(\frac{\sigma_1 - b}{a}\right)^{1/k}\right) \\ & = 1 - \exp\left(-V_2\left(\frac{\sigma_2 - b}{a}\right)^{1/k}\right) \\ & \Rightarrow V_1\left(\frac{\sigma_1 - b}{a}\right)^{1/k} = V_2\left(\frac{\sigma_2 - b}{a}\right)^{1/k} \\ & \Rightarrow \frac{V_1}{V_2} = \left(\frac{\sigma_2 - b}{\sigma_1 - b}\right)^{1/k} \\ & \Rightarrow \frac{\sigma_2 - b}{\sigma_1 - b} = \left(\frac{V_1}{V_2}\right)^k \end{aligned} \tag{4.11}$$

If b is zero we have the two-parameter Weibull distribution, and Equation (4.11) is reduced to Equation (4.12):

$$\frac{\sigma_2}{\sigma_1} = \left(\frac{V_1}{V_2}\right)^k \tag{4.12}$$

This is the equation that explains the size effect according to the weakest link theory for the pure tension mode. For the case of bending, the equation must be modified to take into account the stress variation due to the type of loading. This will be discussed in more detail in Section 4.5.2 on load configuration effects.

For timber the volume is typically a product of the cross-sectional area, height (H) multiplied by width (W), and the length (L), i.e. the volume V equals HWL. Equation (4.12) can then be rewritten according to Equation (4.13):

$$\left(\frac{\sigma_2}{\sigma_1}\right) = \left(\frac{H_1}{H_2}\right)^{k_H}\left(\frac{W_1}{W_2}\right)^{k_W}\left(\frac{L_1}{L_2}\right)^{k_L} \tag{4.13}$$

where k_H, k_W and k_L are the shape factors related to each dimension. If the tension strength of timber would be Weibull distributed, then these shape factors would be the same and independent of which dimension was used for the evaluation. This is not the case for structural timber. The main reason for this is that the variation in strength does not result in independent identically distributed Weibull variables. Furthermore, the variation in strength is different in the three directions.

The width of members of structural timber is often kept constant, i.e. the width effect is generally not of interest. Equation (4.13) is then reduced to Equation (4.14):

$$\frac{\sigma_2}{\sigma_1} = \left(\frac{H_1}{H_2}\right)^{k_H} \left(\frac{L_1}{L_2}\right)^{k_L} \quad (4.14)$$

Since strength tests of structural timber are performed at a constant span-to-depth ratio (in Eurocode the ratio is set to 18), it is possible to substitute one of the parameters in Equation (4.14) and rewrite it as Equation (4.15).

$$\frac{\sigma_2}{\sigma_1} = \left(\frac{H_1}{H_2}\right)^{k_H} \left(\frac{H_1}{H_2}\right)^{k_L} = \left(\frac{H_1}{H_2}\right)^{k_H+k_L} = \left(\frac{H_1}{H_2}\right)^{k_{HL}} \quad (4.15)$$

k_{HL} is a factor taking into account both length and height effects, while k_H only accounts for the height effect. The exponent k_{HL} is for example given as 0.2 in the Eurocode 5 (see Equation 4.16):

$$k_h = \begin{cases} (150/h)^{0.2} \\ 1.3 \end{cases} \quad (4.16)$$

where h is the depth of the beam in mm.

4.5.2 Load Configuration Effects According to the Weibull Theory

If the volume V as was defined in the previous section, is submitted to a varying stress distribution $\sigma(x, y, z)$, the total probability of failure follows Equation (4.17):

$$P_f = 1 - \exp\left(-\int_{x,y,z\in V} \left(\frac{\tilde{\sigma}(x, y, z)}{a}\right)^{1/k} dV\right) \quad (4.17)$$

where

$$\tilde{\sigma}(x, y, z) = \begin{cases} 0 & \text{if } \sigma(x, y, z) \le b \\ \sigma(x, y, z) - b & \text{if } \sigma(x, y, z) > b \end{cases}$$

The integrand in Equation (4.17) is often referred to as the failure intensity $\lambda(x, y, z)$, defined by

$$\lambda(x, y, z) = \left(\frac{\tilde{\sigma}(x, y, z)}{a}\right)^{1/k} \quad (4.18)$$

Let us consider a beam with dimensions according to Figure 4.13. According to the Weibull theory, the load-carrying capacity will be dependent on type of loading, i.e. the stressed volume. The load-carrying capacity of a beam subjected to a constant bending moment and a point load at mid-span will be compared below. The location parameter b is set to zero.

For the beam with constant moment the stress is only dependent on the y-variable, i.e. the height of the beam, Equation (4.19):

$$\tilde{\sigma}(x, y, z) = \sigma_{\max,1}\frac{y}{h} \quad 0 \le y \le h \quad (4.19)$$

where $\sigma_{\max,1}$ is the maximum stress in the cross-section. The subscript 1 denotes constant moment

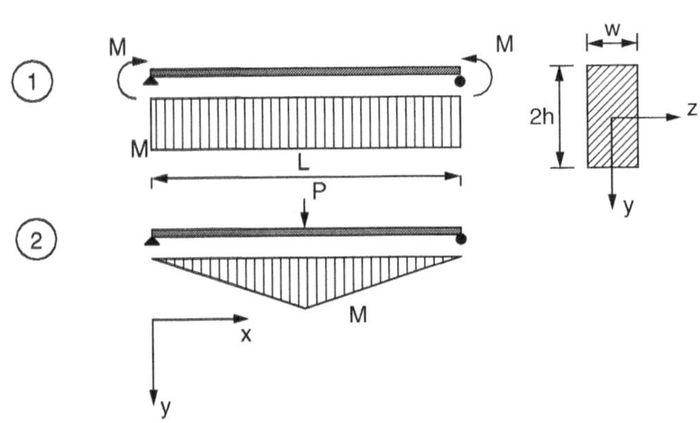

Figure 4.13 Two loading conditions, a given span and cross-section

loading. From Equation (4.17), the probability of failure then becomes

$$P_{f1} = 1 - \exp\left(-\int_V \left(\frac{\tilde{\sigma}(x,y,z)}{a}\right)^{1/k} dV\right)$$

$$= 1 - \exp\left(-\left(\frac{\sigma_{\max,1}}{a}\right)^{1/k} \frac{kLwh}{k+1}\right) \quad (4.20)$$

Let us follow the same procedure for a beam subjected to a point load at mid-span. The stress is now dependent on both x and y and is given by Equation (4.21):

$$\tilde{\sigma}(x,y,z) = \sigma_{\max,2}\frac{x}{L/2}\frac{y}{h}$$

$$0 \le x \le L/2; 0 \le y \le h$$

$$\tilde{\sigma}(x,y,z) = \sigma_{\max,2}\frac{L-x}{L/2}\frac{y}{h}$$

$$L/2 \le x \le L; 0 \le y \le h \quad (4.21)$$

where $\sigma_{\max,2}$ is the maximum stress in the beam and 2 denotes the point load case. The probability of failure is then given by Equation (4.22).

$$P_{f2} = 1 - \exp\left(-\int_V \left(\frac{\tilde{\sigma}(x,y,z)}{a}\right)^{1/k} dV\right)$$

which gives

$$P_{f2} = 1 - \exp\left(-\left(\frac{\sigma_{\max,2}}{a}\right)^{1/k} \frac{Lwh}{\left(\frac{1}{k}+1\right)^2}\right) \quad (4.22)$$

The intention is to maintain a constant probability of failure, independent of the type of loading, i.e. $P_{f1} = P_{f2}$. From Equations (4.20) and (4.22)

$$\left(\frac{\sigma_{\max,1}}{a}\right)^{1/k} \frac{kLwh}{(k+1)} = \left(\frac{\sigma_{\max,2}}{a}\right)^{1/k} \frac{k^2Lwh}{(k+1)^2}$$

$$\Rightarrow \frac{\sigma_{\max,2}}{\sigma_{\max,1}} = \left(\frac{k}{k+1}\right)^k \quad (4.23)$$

For the case of a constant bending moment, the term $\frac{kLwh}{(k+1)}$ can be referred to as the reference stress volume V_{ref}, i.e. the corresponding volume if the beam were subjected to a constant maximum stress along its length. The beam with a point load has an equivalent volume under constant stress equal to $\frac{k^2Lwh}{(k+1)^2}$, or $\frac{kV_{ref}}{(k+1)}$. The ratio between the stress volumes for the two load cases shown in Figure 4.13, $\frac{k}{(k+1)}$, is called the fullness factor λ, i.e. the fullness of the stress distribution compared with the constant bending mode.

In Figure 4.14, seven different types of loading conditions of a beam are illustrated. These cover most of the possible variations in the bending moment. Table 4.2 summarises the load configuration effect based on the Weibull theory by giving the fullness parameter λ (Colling, 1986; Johnson, 1953). That is, the relation between two different loading conditions is given by Equation (4.24), where the subscripts refer to the load cases as given in Figure 4.14:

$$\frac{\sigma_j}{\sigma_i} = \left(\frac{\lambda_i}{\lambda_j}\right)^k \quad (4.24)$$

Also given in the table is the load configuration factor using two different reference loadings for

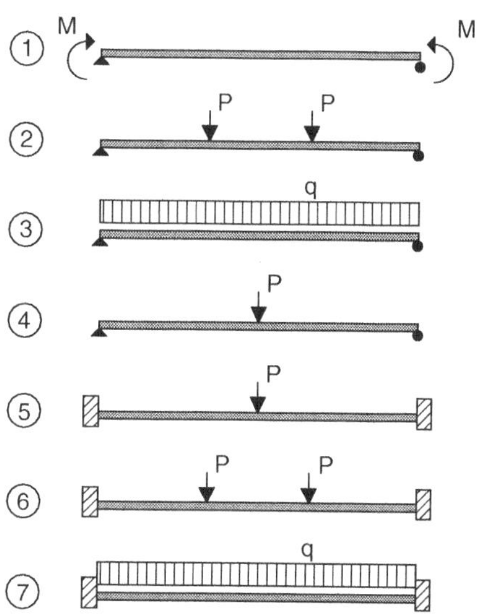

Figure 4.14 Illustration of seven different types of load

Table 4.2 Fullness parameter λ for the seven load cases given in Figure 4.14. The two left columns show the load configuration factor for $k = 0.2$

Load Case (LC) according to Figure 4.14	Fullness parameter λ	Load configuration factors ($k = 0.2$) No 1 as base	No 2 as base
1	1	1	0.850
2	$\frac{(3k+1)}{3(k+1)}$	1.176	1
3	$\frac{(k^3+0.345k^2-0.027k+0.0013)}{k^2(k+1)}$	1.221	1.038
4	$\frac{k}{(k+1)}$	1.431	1.217
5	$\frac{k}{(k+1)}$	1.431	1.217
6	$\frac{4}{9(k+1)}+\frac{2}{9(k+1)}\left(\frac{1}{2}\right)^{1/k}+\frac{1}{3}\left(\frac{1}{2}\right)^{1/k}$	1.635	1.390
7	for $k = 0.2$ $\lambda^5 = 0.58$ for $k = 0.125$ $\lambda^8 = 0.67$	1.724	1.466

$k = 0.2$. The only parameter which must be known is the shape parameter k. This is what makes the Weibull theory so attractive to use. The relation between the different loading conditions is derived mathematically.

4.5.3 Size and Load Configuration Effects According to Reliability Methods

In this section the effect of length and load configuration on the load carrying capacity of a beam is analysed using a reliability method. In most building design codes, e.g. the Eurocode, indicative target reliability indices in the ultimate limit state are given. Eurocode 1 prescribes, with a return period of one year, that the probability of failure must be less than 10^{-6}, which corresponds to a reliability index of 4.7.

Based on these target reliability indices, it is possible to evaluate the length and load configuration effects using level II probabilistic methods (see Chapter 11). A reference length and load configuration is selected. The maximum possible load on a given beam configuration for a given reliability index is determined. This is now the reference load. For other length and load configurations, the maximum possible load is also determined for the same reliability index. The ratio between the expected value of the actions is a measure of how much the load can be increased or decreased compared with the reference beam. The expected value of the action is, in other words, scaled up or down relative to the reference action to preserve the reliability index. The principles of using reliability methods are presented in Chapter 11. See also, for example, Thoft-Christensen *et al.* (1982).

The problem of using this method, and the probable reason for it not being used more frequently, is that it requires a great deal of experimental test results for different lengths and load configurations. In Isaksson (1999) a statistical model of a beam with knowledge of the variation in bending strength between and within beams is presented. This makes it possible to simulate a beam of any length, to apply any kind of load to it and to repeat this many times. For each length and load configuration, it is possible to determine the mean and variance of the load carrying capacity.

4.6 SIZE AND LOAD EFFECTS REPORTED IN THE LITERATURE

The topic of size and load configuration effect for structural timber is quite frequently occurring in the literature. Mostly it has been various forms of fitting to the Weibull distribution. In the 1960s, Bohannan (1966) was probably the first to apply the Weibull theory to wood, in this case clear defect-free wood laminations. Later the theory was applied for structural timber. The basis for determining material properties of timber is the bending test. The grading of structural timber is based on the bending strength and therefore most experimental results are on bending tests. Table 4.3 shows the shape parameter k of the Weibull distribution (see Section 4.5.1) for a few investigations on the size effect in the bending mode. Since the bending moment in most practical cases varies along the length of the member, it is not just the probability of having a major defect within the span, but also the probability of this defect being highly stressed. Failure in bending can be divided into two types, namely tension and compression failures (see Figure 4.15). Sometimes a third category is used, mixed failure. The Weibull theory would show the best agreement for brittle tension failures.

Contrary to the bending mode where the stresses often vary along the length of a member, a member in tension is usually loaded by a force that is constant along the length. Thus failure will occur in the weakest section. The longer the member, the more likely to have a larger defect, and consequently a lower strength. While bending results in a variable stress over the volume, the tension mode results in a high stress over the whole volume. This explains why the tension strength is generally lower than the bending strength. Table 4.4 shows some references for the shape parameter in tension. The bottom chord of a truss is an example where the tension stresses may vary along the length of the chord. The strength of the flange of a wooden I-beam is also mainly governed by the tension strength.

Finally, the compression mode is similar to the tension mode, since in most cases members have a constant stress along the length. There is, however, a difference in the failure mode. If buckling is prevented the fibres will wrinkle under the compression stress. This will of course result in a ductile failure compared to the brittle tension failure. As a consequence of the more ductile behaviour, the size effect will be less pronounced. There are few compression tests reported in the literature. In Madsen (1992), the length effect is given as 0.10.

The width effect is generally negligible, since most structural timber has the same width. For the bending mode the effect on strength of changing

Table 4.3 Length effect in the bending mode. Test results from different references

Reference	Material	k (shape factor Weibull distr.)	
		mean	fifth percentile
Barrett and Fewell (1990)	S-P-F, H-F[a]	0.17	0.17
Ehlbeck and Colling (1990)[b]	–	0.15	–
Madsen (1992)[c]	S-P-F, H-F[a]	0.18	0.23
Rouger and Fewell (1994)[d]	Spruce, Fir	0.21	–
Källsner *et al.* (1994, 1997)	Norway Spruce	0.10	–
Isaksson (1999)			
– simply supported	Norway Spruce	0.11	0.07
– fixed support	Norway Spruce	0.07	0.03
– simply supported	Radiata Pine[e]	0.15	0.13
– fixed support	Radiata Pine[e]	0.08	0.06

[a]S-P-F: spruce-pine-fir, H-F: Hemlock-fir [b]Material and at what fractile the evaluation was made, is unknown [c]Mean value of several investigations [d]Probably evaluated at the mean [e]Based on Leicester *et al.* (1996)

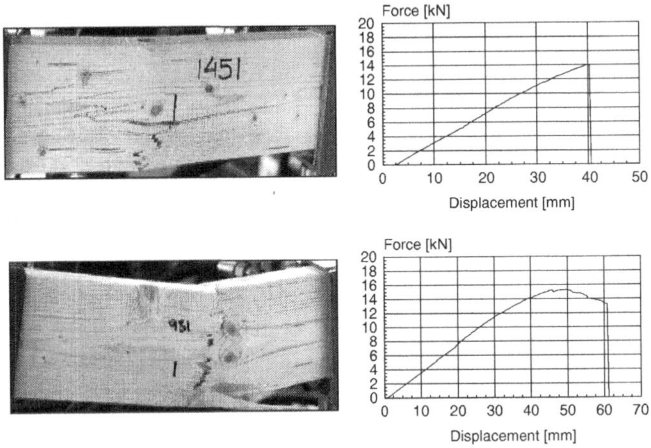

Figure 4.15 Failure types in the bending mode. The upper picture and diagram showing a brittle tension failure. The lower showing a ductile compression failure (Isaksson, 1999)

Table 4.4 Length effect in the tension mode. Test results from different references

Reference	Material	*k* (shape factor Weibull distr.) mean	*k* (shape factor Weibull distr.) fifth percentile
Barrett and Fewell (1990)	S-P-F, H-F [a]	0.17	0.17
Madsen (1992)	S-P-F, H-F [a]	0.14	0.19

[a]S-P-F: spruce-pine-fir, H-F: Hemlock-fir

depth is very small compared to the effect of changing the length. For the tension mode, however, the depth effect is more pronounced. Madsen (1992) indicates a depth effect of the same order as the length effect in the tension mode.

4.7 SIZE AND LOAD CONFIGURATION EFFECTS BASED ON MODELS OF VARIABILITY IN PROPERTIES

This section will combine the theories for describing size and load configuration effects (see Section 4.5), and the models of the variability of material properties within and between pieces of timber (see Section 4.4). The simulated 'test' results used to quantify length load configuration effects is in this case a sample of 10 000 beams for each species, i.e. Norway Spruce and Radiata Pine. Only the bending mode is considered. See Isaksson (1999) for a more thorough presentation.

Two different ways of evaluating the effects are used, direct comparison of mean strengths and preserving the reliability index.

4.7.1 Length Effect

Direct Comparison of Strengths – Weibull Theory

Using the theories in Section 4.5.1, the length effect can be evaluated on the samples of simulated beams. Figure 4.16 shows the length effect for seven different Load Configurations (LC) defined in Figure 4.14. The results obtained from pure Weibull theory with the shape factor equal to 0.1 and 0.2 are also shown for comparison. The material is Norway Spruce. As can be seen,

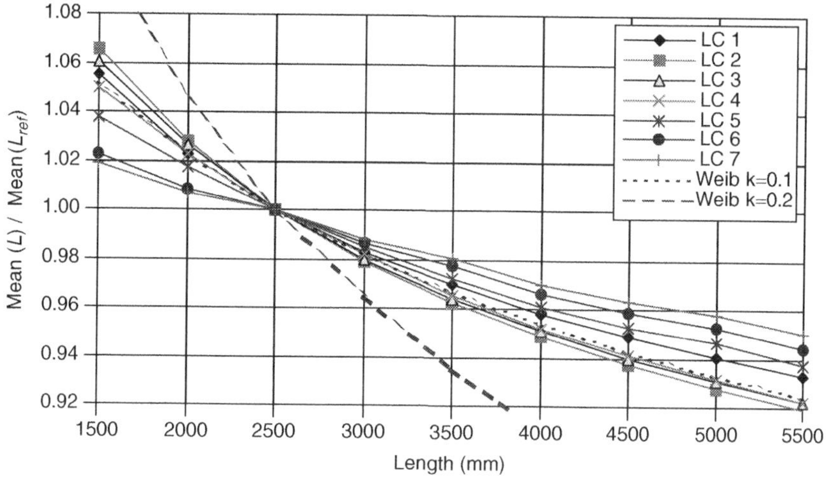

Figure 4.16 Ratio between mean strength of a beam with length L and mean strength of a reference beam with length L_{ref} (2500 mm), Norway Spruce (Table 4.1). Also shown is the length effect according to the Weibull theory using $k = 0.1$ and 0.2

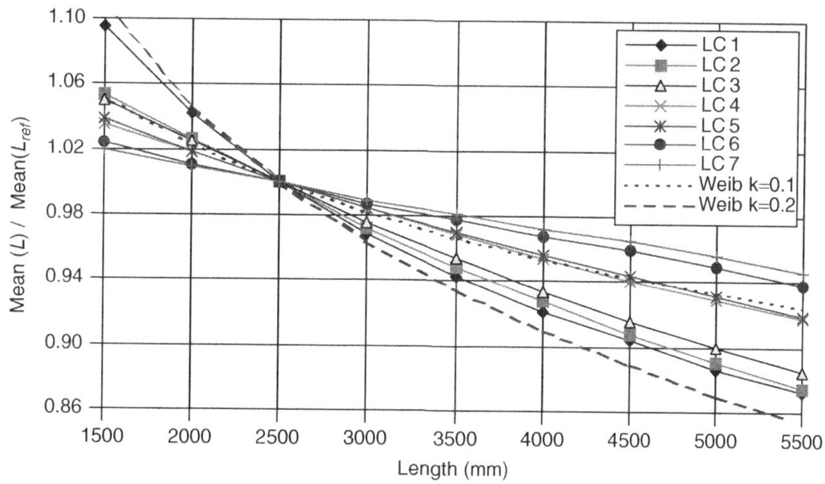

Figure 4.17 Ratio between mean strength of a beam with length L and mean strength of a reference beam length L_{ref} (2500 mm), Radiata Pine (Table 4.1). Also shown is the length effect according to the Weibull theory using $k = 0.1$ and 0.2

the effect is less pronounced for beams with fixed supports, i.e. the more uniform the bending moment is the smaller the length effect. The results for simply supported beams are close to Weibull theory with the shape parameter equal to 0.1, while the beams with fixed supports shows even smaller effects. Figure 4.17 shows the results for Radiata Pine, which is a material with higher variability. Consequently, the effect is stronger. For simply supported beams, the results are close to using a shape factor of 0.15, and for fixed supports 0.10.

The results from the simulations show that it is difficult to represent the behaviour for timber using Weibull theory with one single shape factor.

Preserving the Reliability Index

The samples of different lengths with different loading conditions were also evaluated using

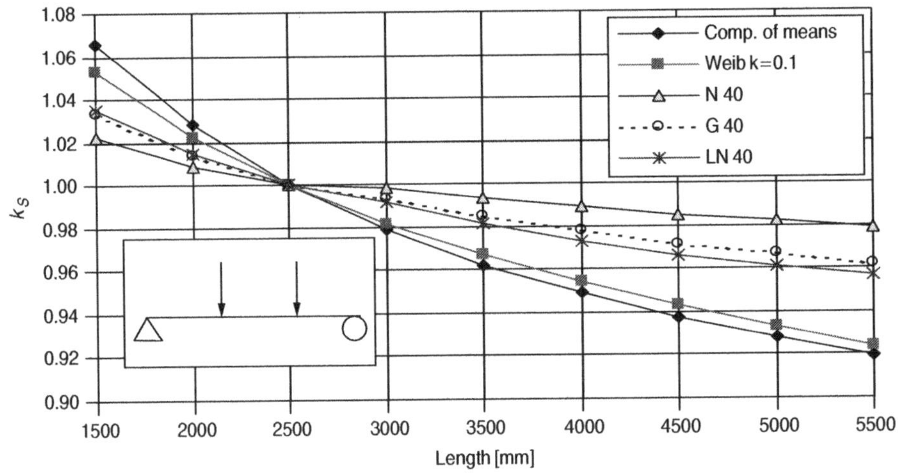

Figure 4.18 The correction of strength due to length for the reference load configuration (third point loading). N (normal), G (Gumbel) and LN (log-normal) refer to the distribution of the action. *COV* of the action is 40%

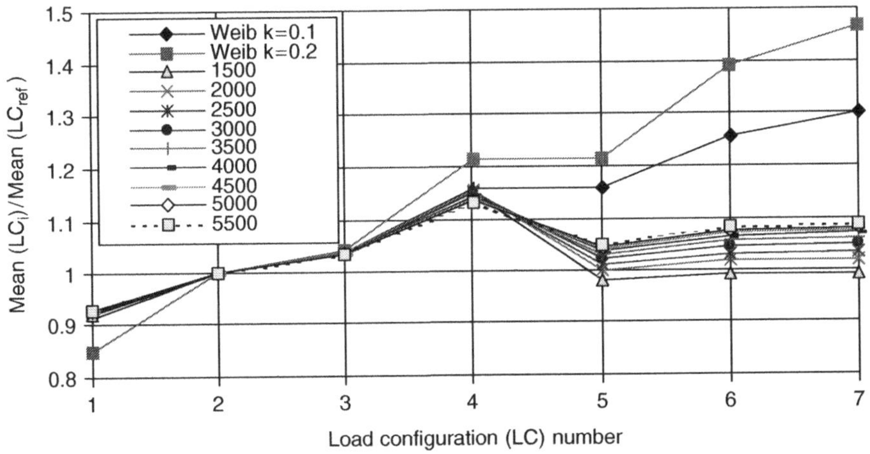

Figure 4.19 Ratio between mean strength of Load Configuration (LC) i and the mean strength of the reference load configuration (LC 2, see Figure 4.14), Norway Spruce. Also shown is the load configuration effect according to the Weibull theory using $k = 0.1$ and $k = 0.2$

reliability methods. The results were found not to be sensitive to the selected level of safety β. Three different assumptions were made concerning the statistical distribution of the action: normal, log-normal or extreme (Gumbel). The resistance was always described by a log-normal distribution. The coefficients of variation of the action were set to both 25% and 40%. The load-carrying capacities of different lengths and load configurations were compared with the capacity of a reference beam, 2.5 m long and loaded at the third points. The comparison was carried out at constant reliability index, β. A higher coefficient of variation for the action (COV_S) resulted in a very small increase in effect. The three assumptions on the distribution of the action gave more or less the same length effect. Generally, the length effect is less pronounced and almost negligible compared with the Weibull results. Figure 4.18 shows the correction of strength for the different assumptions described above. k_s is the ratio between the mean strength of the reference beam and the mean strength of a beam with a different length and/or load configuration. The results are valid for the Norway Spruce material.

4.7.2 Load Configuration Effect

Direct Comparison of Strengths – Weibull Theory

Using the theories in Section 4.5.2, the load configuration effect can be evaluated on the samples of simulated beams. Figure 4.19 shows the load effect for seven different Load Configurations (LC). The material is Norway Spruce. As can be seen, the effect is most pronounced for load configuration 4 (LC 4), which is the simply supported beam with a single point load at mid-span. While the results show less effect for beams with fixed supports, the Weibull theory using k equal to 0.1 or 0.2 shows increasing effect. The shape parameter for simply supported beams is close to 0.1, while it is less for beams with fixed supports. Figure 4.20 shows the results for Radiata Pine. The effect is stronger. For simply supported beams the shape factor is close to 0.15, and for fixed supports 0.10.

Preserving the Reliability Index

The effect of load configuration is shown in Figure 4.21 for the reference length of 2.5 m. For beams with fixed supports, the Weibull theory overestimates the effect due to the fact that the shape factor k changes with load configuration, i.e. evaluating the shape factor on test results using different load configurations will produce different values. k decreases with increasing load configuration number. The reliability methods show small effects. It is important to note that the effects shown for load are purely statistical, and no redistribution of stresses above the proportional limit is included. For the indeterminate beams, for load configurations 5–7, a higher strength can be expected (approximately 10% according to Östlund (1971).

4.7.3 Length and Load Configuration Effects in Code Format

The code format for deriving the design strength can generally be given as in Equation (4.15):

$$f_{d1} = \frac{k_{\text{mod}} f_k}{\gamma_m} \tag{4.25}$$

where

k_{mod} is a modification factor taking into account the effect on the strength of duration of load and moisture content,

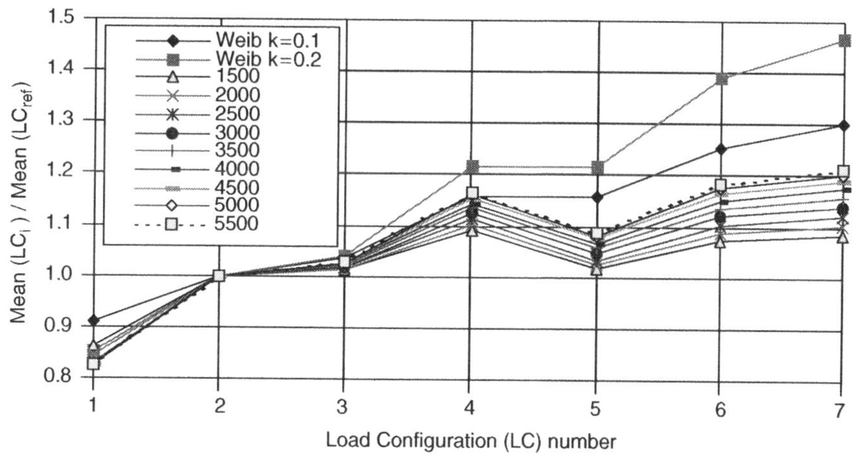

Figure 4.20 Ratio between mean strength of Load Configuration (LC) i and the mean strength of the reference load configuration (LC 2, see Figure 4.14), Radiata Pine. Also shown is the load configuration effect according to the Weibull theory using $k = 0.1$ and $k = 0.2$

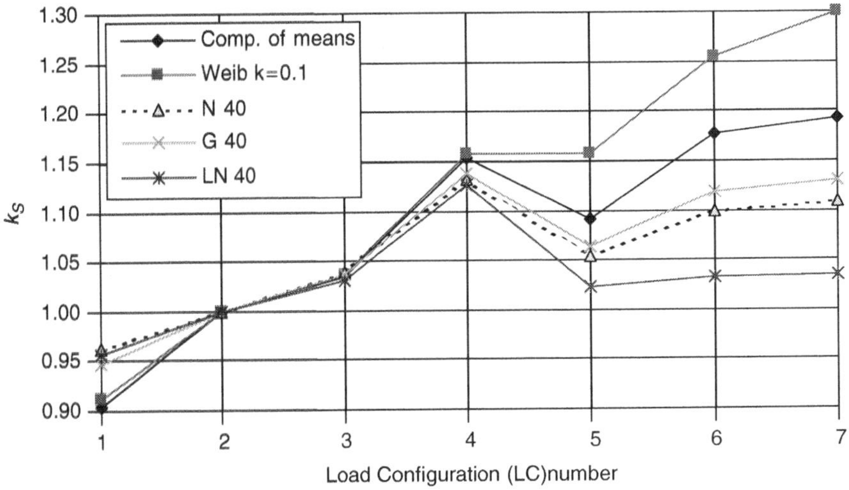

Figure 4.21 The correction of strength due to load configuration for the reference length (2.5 m). N (normal), G (Gumbel) and LN (Lognormal) refer to the distribution of the action. COV of the action is 40%

f_k is the characteristic strength, and
γ_m is a partial safety factor for material property.

The characteristic strength is determined according to a test standard under a specific load configuration and length. Let us now consider the possibility of including a factor α, which is a function of the length L and load configuration ξ. This function is selected to minimise the variability in strength. Isaksson (1999) found the following function:

$$\alpha = 1.4 - 0.45\xi - 0.03L \tag{4.26}$$

(L in mm). ξ is given in Table 4.5.

Table 4.5 Load configuration coefficient, ξ, for the different loading conditions studied

Load Configuration (LC) (see Figure 4.14)	ξ
1	1
2	0.667
3	0.667
4	0.5
5	0.5
6	0.444
7	0.385

The modified equation for the design strength is given by Equation (4.17):

$$f_{d2} = \alpha \frac{k_{\text{mod}} f_k}{\gamma_m} \tag{4.27}$$

Figure 4.22 shows the ratio between the design strength using α according to the above, and the design strength as given in the Swedish code (BKR99) using $\alpha = 1$. As can be seen, the design strength using this function increases for most lengths and load configurations. The only exception is the constant bending moment.

4.8 STRUCTURAL SYSTEMS

So far, the discussion has been on how the variability of material properties within members affects the load carrying capacity of timber. Single members can of course be assembled in a structural system. Analysing such systems is outside the scope of this section. Two examples of systems are shown in Figure 4.23, a floor system and a roof truss. These two represents a parallel and serial system, respectively. For a truss the failure is likely to occur when the weakest member fails, while failure of a beam in the floor structure not necessarily means failure of the whole system.

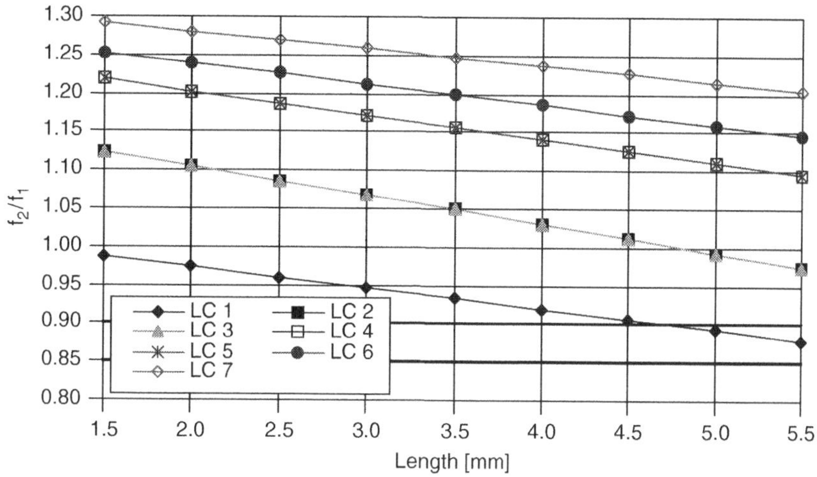

Figure 4.22 The ratio between f_2 and f_1 as a function of length and load configuration (Norway Spruce)

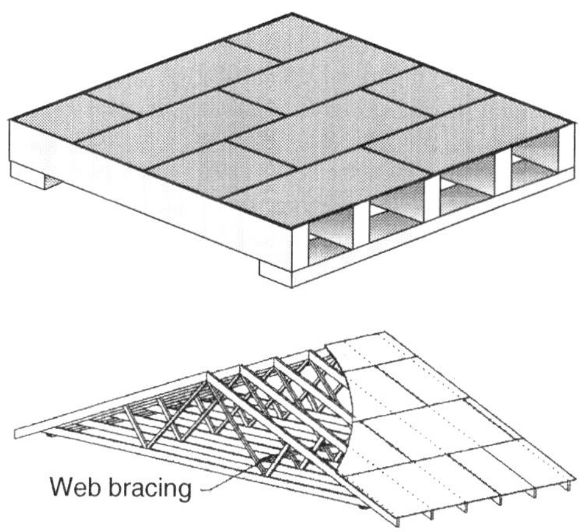

Figure 4.23 Two examples of structural systems. The truss system is taken from Chapter 19

Load sharing (i.e. weaker beams are less loaded than stronger ones) results in a higher load carrying capacity of the system – a system effect. A roof built up by several trusses can also be regarded as a parallel system.

REFERENCES

Anon (1982) *Sortering av sågat virke av furu och gran. 'Gröna boken'. (Instructions for grading of sawn timber products of pine and spruce)* (in Swedish). AB Svensk Trävarutidning. Stockholm, Sweden.

Anon (1981) *Instruktion för sortering och märkning av T-virke. (Instructions for grading and marking of T-timber)* (in Swedish). T-virkesföreingen. Stockholm, Sweden.

Barrett J.D. and Fewell A.R. (1990) *Size factors for the bending and tension strength of structural timber.* CIB/W18A – Timber structures. Paper 23-10-3, Lisbon, Portugal.

Bohannan B. (1966) *Effect of size on bending strength of wood members.* United States Forest Service Research paper FPL 56.

Colling F. (1986) Einfluβ des Volumens und der Spannungsverteilung auf die Festigkeit eines Rechteckträgers. *Holz als Roh- und Werkstoff,* **44**, 179–183.

Corder S.E. (1965) Localised deflection related to bending strength of lumber. *Proc. Second Symposium on Nondestructive Testing of Wood,* Ed. W.L. Galligan, Washington State university, Pullman, WA.

Czmoch I. (1991) Lengthwise variability of bending stiffness of timber beams. *Int. Timber Eng. Conf.*, **2**, 2.158–2.165, London, UK.

Czmoch I. Thelandersson S. and Larsen H.J. (1991) *Effect of within member variability on bending strength of structural timber.* CIB/W18A – Timber structures. Paper 24-6-3, Oxford, UK.

Dinwoodie J.M. (2000) *Timber: Ist Nature and Behaviour,* 2nd Ed. E&FN Spon, London, UK.

Ehlbeck J. and Colling F. (1990) *Bending strength of glulam beams. A design proposal.* CIB/W18A – Timber structures. Paper 23-12-1, Lisbon, Portugal.

Eurocode 1 (1994) *Basis of design and actions on structures. Part 1: Basis of design.* European Committee for Standardization, Bruxelles, Belgium.

Eurocode 5 (1993) *Design of timber structures – Part 1-1: General rules and rules for buildings*. European Committee for Standardization, Bruxelles, Belgium.

Foschi R.O. (2002) Reliability of Structures with timber and wood based products. In: *Timber Engineering*, Ed. Thelandersson, S. and Larsen, M.J., Wiley, Chichester.

Foschi R.O. and Barrett J.D. (1980) *Glue-laminated beam strength: A model. ASCE J. Struct. Div.* **106**(ST8).

Isaksson T. (1999) *Modelling the variability of bending strength in structural timber – Length and load configuration effects*. Report TVBK-1015, Department of Structural engineering, Lund University, Sweden.

Johnson A.I. (1953) *Strength, safety and economical dimensions of structures*. Bulletin no 12, Division of building statics and structural eng., Royal Institute of Technology, Stockholm, Sweden.

Kline D.E. Woeste F.E. and Bendtsen B.A. (1986) Stochastic model for modulus of elasticity of lumber. *Wood and Fiber Science*, **18**(2), 228–238.

Kass A.J. (1975) Middle ordinate method measures stiffness variation within pieces of lumber. *Forest Prod. J.*, **25**(3), 33–41.

Källsner B. and Ditlevsen O. (1994) *Lengthwise bending strength variation of structural timber*. IUFRO, Sydney, Australia.

Källsner B. Ditlevsen O. and Salmela K. (1997) *Experimental verification of a weak zone model for timber in bending*. IUFRO S5.02 Timber engineering, pp. 389–404, Copenhagen, Denmark.

Lam F. and Barrett J.D. (1991) Modelling compression strength spatial variation properties of lumber using trend removal and kriging analyses. *Intern. Timber Eng. Conf.*, **2**, 2.135–2.150, London, UK.

Lam F. and Varoglu E. (1991a) Variation of tensile strength along the length of lumber. Part 1. Experimental. *Wood Science and Technology*, **25**, 351–360.

Lam F. and Varoglu E. (1991b) Variation of tensile strength along the length of lumber. Part 2. Model development and verification. *Wood Science and Technology*, **25**, 449–458.

Leicester R. and Breitinger H.O. (1991) A discrete corelationship with Weibull. *Intern. Timber Eng. Conf.*, **2**, 2.151–2.157, London, UK.

Leicester R.H. Breitinger H.O. and Fordham H.F. (1996) *Equivalence of in-grade testing standards*. CIB/W18A – Timber structures. Paper 29-6-2, Bordeaux, France.

Madsen B. (1992) *Structural behaviour of timber*. Timber engineering Ltd, Vancouver Canada.

Nielsen (2002) Trusses and joints with punched metal plate fasteners. In: *Timber Engineering*, Ed. Thelandersson, S. and Larsen, H.J., Wiley, Chichester.

Pierce F.T. (1926) Tension tests for cotton yarn. *J. Textile Institute*, T155–T368.

Riberholt H. and Madsen P.H. (1979) *Strength of timber structures, measured variation of the cross sectional strength of structural lumber*. Report R 114, Struct. Research Lab., Technical University of Denmark.

Rouger F. and Fewell T. (1994) *Size effects in timber: Novelty never ends*. CIB/W18A – Timber structures, Paper 27-6-2, Sydney, Australia.

Showalter K.L. Woeste F.E. and Bendtsen B.A. (1987) *Effect of length on tensile strength in structural lumber*. Forest Prod. Lab., Research paper FPL-RP-482, Madison, WI.

Suddarth S.K. and Woetse F.E. (1977) Influences of variability in loads and modulus of elasticity on wood column strength. *Wood Science*, **10**(2), 62–67.

Taylor S.E. and Bender D.A. (1989) A method for simulating multiple correlated lumber properties. *Forest Products J.*, **39**(7/8), 71–74.

Taylor S.E. and Bender D.A. (1991) Stochastic model for localized tensile strength and modulus of elasticity in timber. *Wood and Fiber Science*, **23**(4), 501–519.

Thoft-Christensen P. and Baker M.J. (1982) *Structural Reliability Theory and its Applications*. Springer-Verlag, Berlin.

Tucker J. (1927) A study of compressive strength dispersion of material with applications. *J. Franklin Institute*, **204**, 751–781.

Weibull W. (1939) *A statistical Theory of Strength of Materials*. Ing. Vet. Akad. Proc. 151, Stockholm.

Williamson J. (1994) *Statistical dependence of timber strength*. IUFRO Timber engineering, Sydney, Australia.

Östlund L. (1971) *Momentutjämning i kontinuerliga balkar* (*Redistribution of bending moment in continuous timber beams*) (in Swedish). Nordisk Symposium om Traekonstruktioner, June 1970. Statens Byggeforskningsinstitut, Copenhagen, Denmark.

5

Mechanical Performance and Modelling of Glulam

Erik Serrano

5.1 BACKGROUND AND INTRODUCTION

5.1.1 History

Glued-laminated timber, or glulam, is basically what one obtains by stacking a number of boards or laminations on top of each other and gluing them together, so that they form a beam cross-section of the shape desired. Glued timber constructions have been used for centuries, but the breakthrough of glulam can be said to occur when Otto Hetzer (1846–1911) obtained his patent on glued timber constructions in 1906, the so-called Hetzer-binder. This means that for approximately a century, glulam has been used as a material with enhanced performance as compared with solid wood. Among the most often cited advantages of using glulam are:

- Improved strength and stiffness properties.
- Freedom in the choice of geometrical shapes.
- Possibility to match the lamination qualities in relation to expected stress levels.
- Improved accuracy of dimensions and stability of shape during exposure to moisture.

Early examples of glulam structures still in use are found at the railway stations in (Malmö, 1922) and (Stockholm, 1925), Sweden.

5.1.2 Glulam Production

In principle, any wood species can be used for glulam production, as long as a suitable adhesive is used. In practice, however, the use of softwoods predominates since the use of hardwoods is often associated with difficulties in gluing. Planks, or laminations, of approximately 40–50 mm in thickness and 1.5–5.0 m in length are commonly used. Thinner laminations (20–30 mm) may be required for curved beams. The laminations are dried to uniform moisture content (typically 12–15%) and strength graded. The strength grading makes it possible to match the lamination quality to the expected stress levels in bending members, such that high quality laminations are placed in the outer parts of the cross-section and lower quality laminations in the inner zones. To form a continuous lamination with high strength, finger-joints are used for the lengthwise splicing. The continuous lamination is then cut into the lengths required. The friction between the fingers in the joint makes it possible to perform the cutting directly after the jointing process. After the laminations have cured, the curing time depending on the adhesive type used, the laminations are planed to remove the unevenness at the finger-joints. Following this, the laminations are glued together to form the desired cross-section. The glulam is kept under pressure during curing. After curing the complete beam is usually planed to achieve a smooth and clean surface. For glulam manufacturing, the most frequently used adhesives are Phenol-Resorcinol-Formaldehyde (PRF) and Melamine Urea Formaldehyde (MUF) adhesives. High frequency techniques (microwaves) can be used to speed up the process.

Theoretically, glulam can be produced in almost any size. For practical reasons related to transportation and factory layouts, the maximum length is commonly approximately 16–20 m. Another limiting factor is the open time, the so-called pot life, of the adhesive. Glulam comes in various shapes, straight prismatic beams and columns being the most common, but curved or tapered beams are also in general use.

Glulam is a highly engineered product, which, due to the industrialised production method, allows for quality controls to be performed in the production process. The quality controls include bending or tensile tests of finger-joints, and delamination tests or shear tests of the glue line. Such quality control methods are important parts of the glulam production involving both internal control performed by the producer and external control performed by an independent third party.

5.1.3 Aim

Going beyond the simplest possible beam theory approach to predict the strength properties of glulam, two additional methods are possible:

1. Testing of a number of glulam members, using a sample size sufficiently large to determine the properties on a statistical basis.
2. Using a model, verified by testing, linking the properties of the glulam to the properties of the individual laminations.

Taking into account the many parameters influencing the properties of glulam (e.g. timber species and grade, lamination thickness, the properties of finger-joints, member depth) and the number of specimens needed, the former method is clearly too expensive as a general method. A model which makes it possible to estimate the properties of glulam from those of the laminations is therefore desirable. In addition, a rational model adds to the understanding of glulam behaviour. This chapter represents an effort to introduce the reader to the field of such mechanical modelling of glulam.

5.2 MECHANICAL PERFORMANCE OF GLULAM

5.2.1 General Remarks

In this section some general remarks on the performance of glulam are discussed, the performance mainly relating to the bending strength. Many experimental studies have been carried out to investigate the mechanical performance of glulam. In an early German study from 1924, quoted by (Colling, 1990), it was concluded that the Hetzer-binder beam was not stronger than the solid beams tested, but that the variability in strength was smaller. This conclusion has remained. The lesser variability in strength is often explained in terms of smearing-out-effects, i.e. the cutting of solid wood into smaller pieces leads to low-strength defects being more uniformly distributed, and each defect being of less importance than compared with the case of solid wood.

Studying the performance of glulam often involves the comparison of glulam performance with that of solid wood, and here two approaches are possible:

1. To compare the bending strength of glulam beams with the bending strength of solid timber beams.
2. To compare the bending strength of glulam beams with the bending or tensile strength of laminations.

The first approach is not relevant since glulam is used, not instead of large solid beams, but because large solid beams are not available on the market. The second approach is the one most frequently used, and the bending strength of glulam beams is often compared with the tensile strength of laminations, which are matched to those used in the beams. If the bending strength of a glulam beam is mainly influenced by the strength of the laminations on the tensile side, such an approach is indeed appropriate. The bending strength of a glulam beam is correlated with the tensile strength of its laminations, but of course not identical.

5.2.2 Failure Modes

A key issue to understand and model the performance of glulam is to understand the different failure modes that occur. Apart from instability failures, normally the failure modes that can occur are: bending, tension failure parallel to grain, shear failure and failure due to high tensile stresses perpendicular to grain. Since wood under compression has a large yield capacity, the bending strength of a glulam beam is often determined by the tensile strength of the outer lamination. The failure often originates at a knot or a finger-joint. In this chapter, emphasis is placed on the description of failure in bending tension parallel to the laminations, which is the most commonly investigated in the literature. For certain types of notched beams, beams with a hole or curved beams, failure may very well be caused by tension perpendicular to the grain. However, this failure mode is not dealt with here.

5.2.3 Laminating Effect

A typical feature of glulam is that the bending strength of a beam is higher than the tensile strength of its laminations. The ratio of the lamination tensile strength to the beam bending strength is termed the laminating effect, often expressed by a laminating factor, k_{lam}, defined according to:

$$k_{lam} = \frac{f_m}{f_t} \tag{5.1}$$

Here, f_m is the bending strength of the beam evaluated using ordinary beam theory, and f_t is the tensile strength of the outer tension lamination.

The laminating factor as expressed in Equation (5.1) is a formal definition, linked to the assumption that the load-bearing capacity of a glulam beam is essentially governed by the tensile strength of its outer laminations. Furthermore, it is assumed that the stiffness and the strength of the laminations are positively correlated, i.e. a weak zone is also a zone with low stiffness. The laminating effect has been explained by Foschi and Barrett (1980), Larsen (1982) and Colling and Falk (1995):

1. A single lamination tested in pure tension by applying a centric tensile force, may bend laterally due to knots and unsymmetrically placed anomalies. This is due to the stiffness not being constant across the cross-section of the lamination. If the same lamination was contained in a glulam beam exposed to bending, the rest of the beam would prevent such lateral bending. This is referred to as an 'effect of test procedure'.
2. In a glulam beam, the defects are smeared out resulting in a more homogeneous material than solid wood. The probability of a defect having a serious influence on the strength of the beam is less than it is in a single lamination. This is referred to as a 'dispersion effect'.

3. A lamination that contains knots or other zones of low stiffness will be reinforced by adjacent laminations when it is contained in a glulam beam: the stiff and strong laminations take up a larger part of the tensile stresses. This is sometimes referred to as a 'reinforcing effect', although the term 'stress redistribution effect' is perhaps more adequate.

The dispersion effect may be questioned, however. Typical lamination thicknesses are in the range of 20–50 mm and strength-reducing defects, such as knots, are of the same order of magnitude in size, or smaller, at least not larger. The dispersion effect can, however, be expected to contribute to the lesser variability in products like Laminated Veneer Lumber (LVL) or OSB, where the wood is cut into thinner pieces.

The above three explanations are related to the beam being built up of laminations. In addition to this, $k_{lam} \neq 1.0$ can be caused by a nonlinear stress-strain performance of the material, or due to different strength in compression and in tension.

An effort to quantify the different contributions to the laminating effect has been presented by Falk and Colling (1995). Experimental data showing laminating effects are found in the works of Larsen (1982) and Falk *et al.* (1992). Larsen found that for different beam compositions the laminating factor varied from 1.06–1.68. The investigation of Falk *et al.* yielded laminating factors in the range of 1.35–1.65.

If the laminating effect is explained in terms of differences in stiffness for different laminations (cf. item (3) above), it is of importance to investigate the stiffness of finger-jointed laminations. The stiffness of finger-joints is high, and often comparable to the clear-wood stiffness of the lamination (Colling, 1990). This would indicate that a laminating effect according to item (3) above is not present in the finger-joint zone of a glulam beam lamination.

To take advantage of the laminating effect, several studies on the use of laminations that in turn are laminated have been performed. The use of such sub-lamination technique, however, implies that a larger number of finger-joints are introduced. If, however, the sub-laminations are

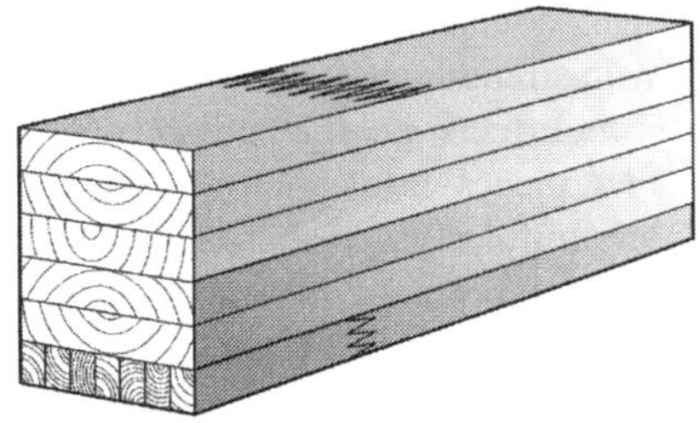

Figure 5.1 Using a glulam beam with laminated, outermost lamination, takes advantage of the laminating effect

oriented as depicted in Figure 5.1, each finger-joint has a lesser influence on the strength, since each sub-lamination only covers a part of the beam width.

In the case of Laminated Veneer Lumber (LVL), further advantage of the laminating effect has been taken by using extremely thin laminations (3–5 mm).

5.2.4 Size Effect

It is a well-established fact that strength can depend upon the absolute size of a structure, at least for structures showing brittle failures. This can be explained by a weakest link concept such as the Weibull theory (Weibull, 1939), which states that the strength of a structure depends solely upon the strength of its weakest link. Since the probability of encountering a severe defect is greater in a larger volume than in a smaller one, the strength decreases with increasing size. The size effect can be expressed in terms either of mean values or of characteristic values of strength, and the results are different.

The Weibull theory is appropriate to use for materials exhibiting brittle failures such as wood under tension. The failure in bending is different than failure in pure tension, since bending involves both tension and compression, and the compressive failure is extremely ductile. The size effect or size factor, k_{vol}, is often expressed as:

$$k_{vol} = \left(\frac{V}{V_0}\right)^{\eta}, \quad \eta < 0 \tag{5.2}$$

where V is the current volume and V_0 is the reference volume for which the material strength was determined. Several investigations of the size effect have been carried out over the years for determining the dependence of the strength of glulam on volume. Typical values of η are in the range of −0.1 to −0.4.

The size effect is often expressed in terms of a depth or a length factor, since beams that are tested according to some standard procedure tend to have a predefined depth to length ratio, typically L/h = 18. The width of the beam is usually not expected to have any influence. The volume effect as expressed in Equation (5.2) can then be expressed in terms either of beam length or of beam depth.

5.2.5 Stress Distribution Effect

Another reason why the beam bending strength may be larger than the tensile strength of the laminations is the effect of stress distribution. It is a well-established fact that for materials exhibiting brittle failure, according to Weibull theory, the strength in bending is higher than in pure tension. A simple way to explain this is by comparing the case of pure, uniform tension of a lamination with the case of pure bending of a beam. We assume that the strength of the material is stochastic, i.e. it varies spatially, and that the lamination and the beam fails when a single point is stressed to a level corresponding to the strength at that point. For the uniform tensile stress case, we obtain an average tensile strength, which we denote f_t. For the case of bending, the stress distribution is non-uniform, and it is not likely that the beam fails when the edge stress equals f_t, since all interior points are less stressed. Since the stress distribution influences the probability of failure, not only the geometry (e.g. beam depth) and loading conditions (tension vs. bending) but also the load configuration (e.g. three-point bending vs. four-point bending) has an impact.

5.2.6 Importance of Finger-joints

A summary of a large number (1767) of bending tests of glulam beams can be found in (Colling, 1990). Of the 1767 beams examined, there were 277 for which the region of maximum bending moment had finger-joints in the outer tension lamination. Of the 277 beams, there were 220 (79%) that failed in the finger-joints. According to tests performed by (Johansson, 1990), the finger-joints were of lesser influence: of the 32 beams tested with a finger-joint in the outer tension lamination in the region of maximum bending moment, only 10 (31%) failed due to the finger-joint. A reasonable estimation of the influence of finger-joints on the bending strength of glulam is that in about 40–60% of the cases where finger-joints are placed in the region of maximum bending moment, the finger-joint is also the main cause of beam failure.

5.2.7 Load Sharing Between Laminations

An important issue in mechanical performance of glulam is the load sharing between laminations. Load sharing is related to the reinforcing effect mentioned above, and is of utmost importance if a realistic glulam model is to be derived. The term 'load sharing' can also refer to the situation when a lamination, finger-joint or other weak section has failed. The release of strain energy can either cause a catastrophic failure of the entire beam, or only a minor drop in load if neighbouring laminations are able to take up the released stresses. For the case of an outer lamination of a glulam beam failing, the load-bearing capacity of the entire beam is normally completely lost. To delaminate in a stable way, the lamination thickness of the outer lamination should be small, in the range of 10 mm or less (Serrano and Larsen, 1999).

5.3 MODELLING OF GLULAM

5.3.1 General Remarks

Having pointed out the most important features in the mechanical performance of glulam, the following sections deal with aspects of mechanical modelling of glulam behaviour. In such modelling the more important features to take into account include: the variability of material properties such as strength and stiffness, influence of finger-joints and the interlaminar bond lines. Furthermore, depending on the modelling techniques used, special criteria for load sharing between laminations and global failure criteria must also be included.

5.3.2 Special Modelling Techniques

The vast majority of theoretical glulam models for the determination of mechanical performance are based on stochastic analysis using Monte Carlo simulations, i.e. a series of simulations is performed, and in each run the material parameters are taken from statistical distributions. This technique simulates the variability experienced when performing experiments.

Many of the glulam models are based on the use of finite element programmes, and often plane stress analysis is performed. Some of the proposed models use so-called 'transformed section methods', based on beam theory. For the transformed section methods, typical beam theory assumptions (linear axial strain distribution) are made, and the response of the beam is then determined by investigating a number of sections along the beam.

The constitutive models used in most of the glulam beam models can be said to be fairly simple. Most of the effort has been put on the description of the variability of the basic material parameters such as strength and stiffness. However, often these parameters have been determined in one-dimensional tests, and typically only the strength and stiffness parallel to grain is known. It is a challenging task to perform all the necessary tests to obtain all the material data needed even for the simplest possible constitutive model of, for example, the behaviour of clear wood. In some of the models described here, analyses that go beyond the simplest possible linear elastic material behaviour have been included.

5.3.3 Glulam Models for Hand Calculation

The simplest possible glulam beam model is the one used most often in engineering applications. It involves the use of deterministic, conventional beam theory (Bernoulli–Euler), and is often combined with a linear elastic material behaviour and a single point maximum stress failure criterion. This means that for pure bending the outermost lamination is the one that fails first if no account is taken of the stiffness differences between the laminations. Transformed section methods, which can be used to account for stiffness differences, allow for failures in other than the outermost lamination.

The main drawback of a deterministic model is that no information regarding the distribution of, for instance, beam bending strength can be obtained. A way to obtain such information is to use the classic theory of Weibull (1939) for the rupture of brittle materials. At least for the case of failure due to axial tensile stresses in a beam, the behaviour is indeed brittle. By using Weibull theory, it is possible to explain such phenomena as the laminating effect, the size (volume) effect and the influence of the stress distribution, for example the increase of bending strength relative to the tensile strength of a brittle material.

An early empirical model developed and used for glulam, is the so-called I_k/I_g-method, which uses a reduction in the moment of inertia due to knots as a way of taking the strength-reducing effects of knots into account. I_k and I_g are the moment of inertia of the knots at a cross-section, and the moment of inertia of the beam cross-section, respectively. This model was developed in the late 1940s, and has been used mainly in the US and Canada. The method assigns the same strength reducing effect for knots, whether they are placed in the tensile zone or in the compressive zone, which is unrealistic. The I_k/I_g-method does not allow the distribution of beam strength to be calculated.

5.3.4 Glulam Models for Computer Analysis

Several theoretical and more or less complex models for predicting the load-bearing capacity of glulam beams in bending have been proposed. Here a short summary and overview of some of the most well known models is given. The models described represent slightly different approaches in terms of mechanical modelling, but are all based on the same principle: divide the beam into zones which each is assigned material properties, including their variability. These models all rely on computer simulations.

Model of Foschi and Barrett

In 1980 Foschi and Barrett presented a model for the simulation of the strength of a glulam beam (Foschi and Barrett, 1980). Their model uses Monte-Carlo simulations to predict the strength and the distribution of the strength. A glulam beam is subdivided into cells, each 150 mm long and having the depth and width of a lamination. Each cell is assigned properties in terms of density, Knot-Area-Ratio (KAR-value), modulus of elasticity and strength, according to probability density functions. Using a finite element program the beam response is calculated, and the strength of the beam is determined. Among the most important assumptions and simplifications made in the model are:

1. The material is linear elastic in both tension and compression.
2. No account of the periodicity of knots and density is taken.
3. A brittle weakest link failure criterion is used.
4. The strength and stiffness properties are based on test results from tests on specimens larger than the cell size.
5. No account is taken of the influence of finger-joints.

The strength and stiffness parameters used for each cell in the glulam beam were calculated from test results from tests on long laminations which were free to bend laterally. Since this lateral movement is restrained when a lamination is contained in a beam, the strength assigned to each cell was corrected. The correction factor was calculated using finite element simulations of a lamination with an edge knot. The simulation results were then calibrated to test results in order to obtain correction factors for other knot positions. One type of glulam beam lay-up was used to calibrate the glulam model, and seven other beam lay-ups were then used to evaluate the model prediction capabilities. Figure 5.2 shows the results from the calibration and evaluation of the model in terms of bending strength.

The main drawbacks of the model of Foschi and Barrett are the assumption of linear elastic behaviour of the wood in both tension and compression, and the fact that the influence of finger-joints is neglected. Another serious simplification made is the assumption of a weakest link failure, so that no progressive failures can be evaluated.

The Karlsruhe Model

In the mid-1980s, the so-called Karlsruhe model (Ehlbeck *et al.*, 1985a–c, Colling, 1990) was developed. The Karlsruhe model uses a subdivision of a glulam beam into cells, similar to that used by Foschi and Barrett, each cell being assigned random material properties. The Karlsruhe calculation model is based on two computer programs, one that simulates glulam beam lay-up, and one that performs finite element calculations.

The beam lay-up simulation program assigns values to each cell along a continuous lamination. The lamination is assumed to consist of two 'materials': wood and finger-joints. First, the position of each finger-joint is determined by sampling from statistical distributions determined by measuring the distance between finger-joints in glulam beams. The density is assumed to be constant within a lamination, but is allowed to vary between laminations. The KAR-values and the density value are then used to calculate the stiffness (modulus of elasticity) using regression equations. For each lamination, i.e. for all the cells between two finger-joints, a single basic KAR-value is assigned. This KAR-value is then used within the lamination to assign each cell a specific KAR-value by multiplying the characteristic lamination-KAR-value by a factor, taken from statistical distributions, its aim being to simulate the influence of multiple knots within a limited zone (= the length of the cell) of the lamination. Following the assigning of KAR-values, each lamination in the beam is assigned a modulus of elasticity in the cells of that lamination. The procedure is shown in Figure 5.3. For a given lamination, i,

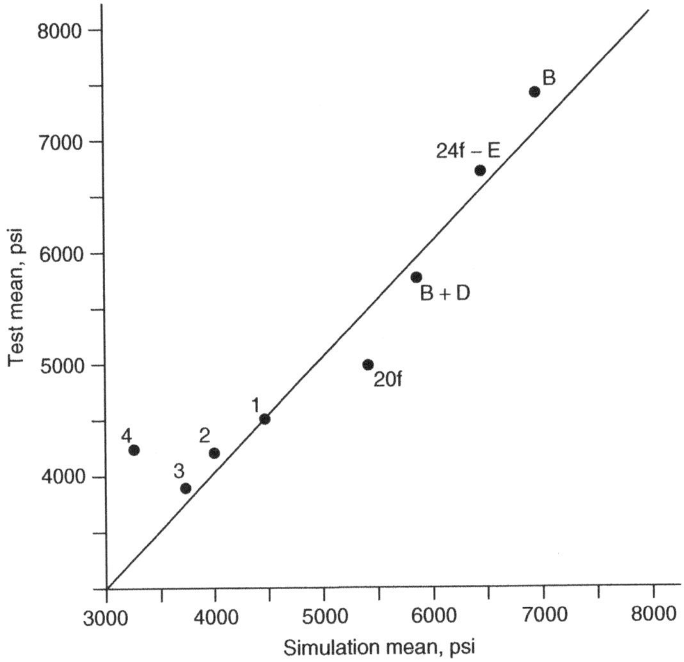

Figure 5.2 Tested versus simulated bending strength of glulam (Foschi and Barrett, 1980)

and a given KAR-value, the modulus of elasticity is calculated from:

$$\ln(E_i) = \ln(E_{reg}) + \Delta_{B,i} + X_i \qquad (5.3)$$

where E_i is the modulus according to the regression equation for lamination i. Each such equation gives an expression for the modulus of elasticity as a function of density and KAR-value. Δ_B is defined in Figure 5.3, and is a measure of the difference from the average regression equation. Δ_B is assumed to be normally distributed. X_i is a measure of distance from the average regression line to the regression line of the current lamination, and is also assumed to be normally distributed. As an example, the variability for a variable X_1 is shown in Figure 5.3 as $S_{R,B,1}$.

This procedure is capable of simulating the correlation between KAR-values found within a single lamination. Also, the procedure used means that even if two cells are assigned the same KAR-value, their respective modulus of elasticity need not be the same.

The continuous lamination also contains finger-joints, which are modelled in the same way as described above, but by assigning 'finger-joint properties' instead of 'wood properties' to the cell that contains a finger-joint. The Karlsruhe model includes some obvious improvements as compared with the original model of Foschi and Barrett:

1. The influence of finger-joints. Although the finger-joints were not modelled in detail, at least experimental surveys to obtain material data for the finger-joints were performed.
2. Nonlinear behaviour in compression. The wood was assumed to behave nonlinear in compression, using a simple plasticity model.
3. Progressive failures. The Karlsruhe model allows for progressive failure to be taken into account. The beam response is calculated in an iterative manner: elements that have failed in tension are removed and the beam response is re-calculated.

One of the drawbacks of the Karlsruhe model is that the numerical modelling is relatively

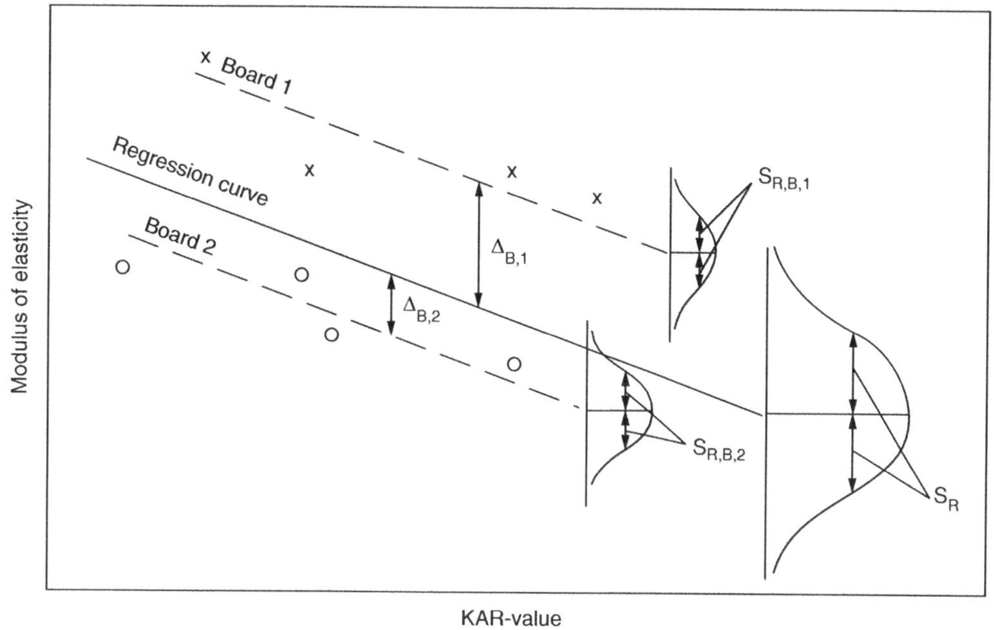

Figure 5.3 Derivation of modulus of elasticity for various laminations (Colling, 1990)

simple in terms of element mesh. Each finite element has the same size as the region of for which the basic material properties have been determined (150 mm length and a depth equal to the lamination thickness). Although material data is available only with this resolution, the finite element discretisation could be made much finer.

Another drawback is related to the failure criterion of the beam. The progressive failure of a beam is not monitored until a maximum load is reached. Instead, four different failure criteria are used. These criteria are based on 'experience gained during the calibration of the model to beam bending tests':

1. If the stress redistribution due to the removal of a failed element leads to the failure of another element, the beam is assumed to fail. This simulates a brittle failure in tension.
2. If an element fails in tension within a predefined region (see Figure 5.4) of a previously failed element, the beam is assumed to collapse. This simulates a failure due to the high shear stresses in that region, although the model does not include the shear strength of the wood as a parameter.
3. If a finger-joint fails in tension the beam fails. This is motivated by the fact that finger-joints induce a failure across the complete lamination width, and not only a part of it as is the case for knots.
4. If none of the above criteria have been fulfilled, the beam is assumed to fail when the fifth element fails in tension.

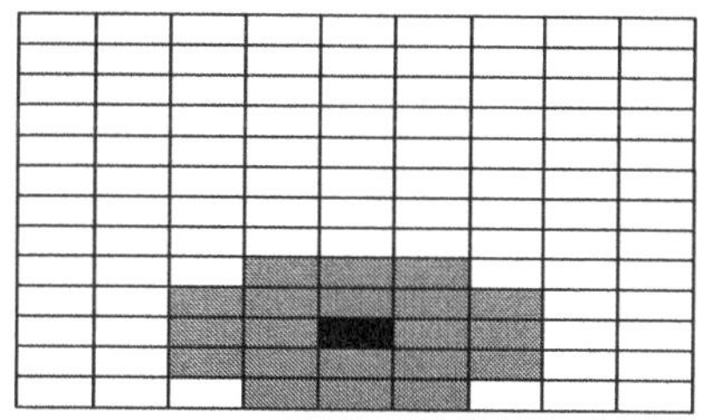

Figure 5.4 If an element fails close to (grey region) a previously failed element (black) the beam fails (Colling, 1990)

Note that all of the above criteria relate to the failure of an element in 'tension', and are therefore one-dimensional criteria. Furthermore, since the stresses are evaluated at the centroid of the elements, the criteria use the average (one-dimensional) tensile stress in the laminations.

PROLAM Model

The calculation model, PROLAM, was developed in the mid-1980s and early 1990s (Bender *et al.*, 1985, Hernandez *et al.*, 1992). This model uses a transformed section method in combination with beam theory to predict the load-bearing capacity of a number of sections along the beam. The beam capacity is then estimated as the capacity of the weakest cross-section. Progressive failure of finger-joints and laminations are allowed, so that a single failed finger-joint or lamination not necessarily leads to global failure of the beam.

Model of Renaudin

The glulam beam model described in Renaudin (1997) is in some respects similar to the Karlsruhe calculation model. Instead of using standard finite elements, beam theory assumptions of plane cross-sections remaining plane are used together with two types of one-dimensional elements: 'wood' elements assumed to account for the influence of knots and finger-joints; and interlaminar elements assumed to account for the mechanical behaviour of the interlaminar bond line. Failure of a wood element leads to the redistribution of stresses according to explicit rules, which were set up using finite element simulations of the stress distribution close to a crack. After failure of an interlaminar element, the beam theory assumption is relaxed so that a displacement discontinuity is allowed across the interface (see Figure 5.5). The main improvements in the model of Renaudin are the failure criterion and the modelling of the interlaminar bond line. The work of Renaudin also includes a comprehensive experimental study on timber, with special emphasis on the stochastic nature of the basic material properties such as density, stiffness and strength.

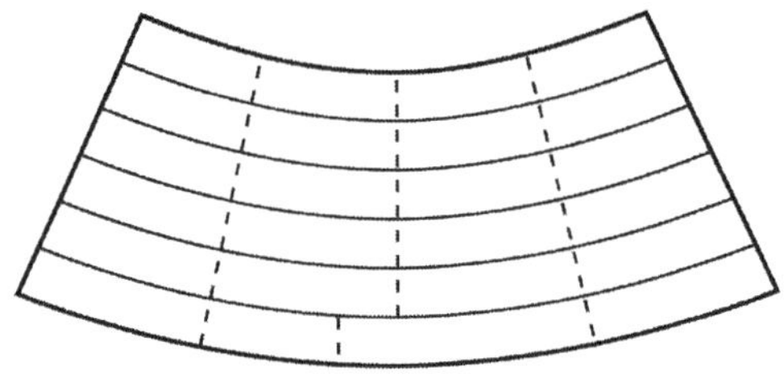

Figure 5.5 Displacement discontinuity after interlaminar failure

Serrano *et al.*

A refined way to model the important finger-joint failure in glulam beams is given in Serrano *et al.* (2000). This model is more detailed than those used in the above-described models, in the sense that it makes it possible to take into account the variability across the lamination depth, and also has the possibility of simulating a progressive failure and crack propagation within the lamination. In principle, it is possible to use the same type of detailed modelling approach for the failure of the wood and for the bond line between the laminations. The aim of the study was to establish a better fundamental understanding of the failure mechanisms of finger-joints in glulam by:

- setting up a detailed model of a finger-joint, using an approach based on Non-Linear Fracture Mechanics (NLFM), assuming that the material parameters are stochastic; and
- to use the finger-joint model in finite element simulations, and investigate the behaviour of the finger-joint and laminated beams in terms of (a) size effect, (b) depth effect, (c) laminating effect, and (d) effect of finger-joint ductility.

To isolate the effect of the modelling approach used for the finger-joint, the wood outside the finger-joint region was regarded as a homogeneous material with deterministic properties. Consequently, it was not possible, nor the aim of the study, to validate the glulam beam analyses with independent test data. However, the modelling approach can be included into existing glulam beam models such as those discussed above, which indeed use sophisticated modelling of the wood variability.

A finger-joint was modelled with a number of parallel nonlinear spring elements of initial zero length. Due to the zero initial length, the elongation of such a spring is associated with crack formation, and represents the local additional deformation due to the presence of the finger-joint. Each spring is assumed to represent the response of a finger-joint area defined by the lamination width times the distance between neighbouring nodes in the height direction of the lamination. The modelling is thus closely related to a so-called fictitious crack model, a fracture mechanics model for describing the gradual localised damage of a material. The finger-joint spring characteristics are chosen in line with results from tests and from detailed FE-modelling of finger-jointed laminations. The detailed FE-modelling included the complex geometry of a complete finger-jointed lamination, with the fingers visible on the wide face.

One of the results obtained concerns the laminating effect. As expected, the beam bending strength was predicted to be higher than the tensile strength of the lamination. In this case, the strength increase is mainly due to the fact that the outer lamination is exposed to a non-uniform stress distribution, and that when evaluating the strength of the beam, the nominal stress at the outer edge is used as a reference. However, due to the stochastic nature of the strength distribution, the actual failure position is not necessarily the outer tensile edge. So, instead of interpreting the results as a 'strength increase' of the lamination, it is perhaps more natural to assume that the position of failure in the finger-joint changes if the beam depth is changed. The strength increase can then be expressed as an equivalent position of failure in the outer lamination. As mentioned previously, a common assumption made in previous models is that the mid-depth is the characteristic point of failure initiation, independent of beam depth and lamination thickness. Another important result from the study (Serrano *et al.*, 2000) is that the finger-joint ductility is of great importance for the beam strength. To evaluate the influence of the ductility on the beam strength, clearly a perfectly brittle model cannot be used.

5.4 SUMMARY AND CONCLUDING REMARK

In the following a summary of the key issues in the modelling of glulam performance, and thus also key issues for the mechanical performance of glulam, are given. These important issues relate to the modelling of localised wood properties, to the choice of failure criteria and to the modelling of finger-joints.

5.4.1 Localised Material Properties of Wood

To predict the strength distribution of glulam, the distribution of the input parameters such as stiffness and strength are needed. Here 'distribution' refers to the statistical distribution (mean value and variability), as well as to the spatial variation and correlation. The key issues regarding localised material properties are the lengthwise, within-member, variability and the correlation of strength and stiffness as well as the possible correlation between members. The variability in material properties mainly relate to the presence of knots.

5.4.2 Finger-Joints

In experimental studies, it has been shown that the finger-joints are important for the strength of glulam. Numerical studies also have shown that there is a large potential for increasing the strength of finger-joints by increasing their ductility. To fully understand the behaviour of a glulam beam, one must also understand the behaviour of its finger-joints, and it is therefore crucial that a model, which can account for the flexibility and ductility of a finger-joint, is used.

5.4.3 Failure Criteria

Although a locally brittle response, or failure criterion, of the material in tension is assumed in some of the above models, the assumption of redistribution of stresses after failure introduces a considerable amount of ductility in the model at a global level. The properties of the outer laminations, especially the strength of the finger joints, are very important for the modelling of glulam (Colling, 1990). Using as a failure criterion the fact that the mid-depth stress in the outer lamination corresponds to the tensile strength of the lamination, defined as the average stress at failure in a uniaxial tension test, is far too primitive. Nevertheless this is the failure criterion used in previously proposed models. Instead a failure criterion, which includes the response of the complete lamination depth, and permits a progressive failure through the lamination, is thought to be more appropriate. Using such a failure criterion, the study by Serrano *et al.* (2000) shows that the mid-depth stress criterion is obtained only as a special case for large beam depths. The use of the mid-depth stress in a failure criterion is one of the reasons why the theoretical models are not used in practice to determine the properties of glulam. Instead, semi-empirical models (e.g. that proposed in Schickhofer (1996)) are used. They are, however, only valid within a small parameter range. The use of a more sophisticated constitutive model for the finger-joints as the one employed in Serrano *et al.* (2000) means that no special failure criterion for the lamination or beam is needed.

5.4.4 Concluding Remark on Adhesive Bond Lines

A timber engineering structural element, like a glulam beam, has a typical dimension in the range of 0.1–10 m. At this scale it may seem irrelevant to account for effects that relate to the adhesive bond lines present, since these represent a very small geometrical part of the structure (typically 10^{-4}–10^{-3} m in thickness). With the use of extremely brittle adhesives, like those commonly used today, it is indeed possible that the mechanical performance of the bond line is not very different from that of the wood. However, if adhesive bonds with properties markedly different from those of the wood are to be used, a detailed modelling of the adhesive bond lines is of great importance.

REFERENCES

Bender D.A., Woeste F.E., Schaffer E.L. and Marx C.M. (1985) Reliability Formulation for the Strength and Fire Endurance of Glued-laminated Beams. Research paper FPL460, Forest Products Laboratory, Madison, WI, USA.

Colling F. (1990) Tragfähigkeit von Biegeträgern aus Brettschichtholz in Abhängigkeit von den festigkeitsrelevanten Einflußgrößen. Doctoral thesis, University of Karlsruhe, Germany.

Ehlbeck J., Colling F. and Görlacher R. (1985a) Einfluß keilgezinkter Lamellen auf die Biegefestigkeit von Brettschichtholzträgern. Teil 1. *Holz als Roh- und Werkstoff*, **43**, 333–337.

Ehlbeck J., Colling F. and Görlacher R. (1985b) Einfluß keilgezinkter Lamellen auf die Biegefestigkeit von Brettschichtholzträgern. Teil 2. *Holz als Roh- und Werkstoff*, **43**, 369–373.

Ehlbeck J., Colling F. and Görlacher R. (1985c) Einfluß keilgezinkter Lamellen auf die Biegefestigkeit von Brettschichtholzträgern. Teil 3. *Holz als Roh- und Werkstoff*, **43**, 439–442.

Falk R.H. and Colling F. (1995) Laminating effects in glued-laminated timber beams. *J. Structural Engineering*, **121**(12), 1857–1863.

Falk R.H., Solli K.H. and Aasheim E. (1992) The performance of glued laminated beams manufactured from machine stress graded Norwegian spruce. Rep. no. 77, Norwegian Institute of Wood Technology, Oslo, Norway.

Foschi R.O. and Barrett J.D. (1980) Glued-laminated beam strength: a model. *J. Structural Division*, **106**(ST8), 1735–1754.

Hernandez R., Bender D.A., Richburg B.A. and Kline K.S. (1992) Probabilistic modeling of glued-laminated timber beams. *Wood and Fiber Science*, **24**(3), 294–306.

Johansson C.J. (1990) Strength and stiffness of glulam with laminations of machine stress graded timber. SP RAPPORT 1990:22. Swedish National Testing and Research Institute, Borås, Sweden (in Swedish).

Larsen H.J. (1982) Strength of glued laminated beams. Part 5. Report no. 8201, Institute of Building Technology and Structural Engineering, Aalborg University, Aalborg, Denmark.

Renaudin P. (1997) Approche probabiliste du comportement mecanique du bois de structure, prise en compte de la variabilite biologique. Doctoral thesis, LMT, ENS Cachan, Paris, France.

Schickhofer G. (1996) Development of efficient glued laminated timber. *Proc. CIB-W18, Meeting 29*, Bordeaux, France International Council for Building Research Studies and Documentation, University of Karlsruhe, Germany, pp. 12.1.1–12.1.17.

Serrano E., Gustafsson P.J. and Larsen H.J. (2000) Modelling of finger-joint failure in glued-laminated timber beams. In: Adhesive joints in timber engineering. Modelling and testing of fracture properties. Report TVSM-1012, Division of Structural Mechanics, Lund University, Sweden. (Also submitted for publication, *ASCE J. Structural Engineering*.)

Serrano E. and Larsen H.J. (1999) Numerical investigations of the laminating effect in laminated beams. *J. Structural Engineering*, **125**(7), 740–745.

Weibull W. (1939) A statistical theory of the strength of materials. Nr. 151, Proc. Royal Swedish Institute for Engineering Research, Stockholm, Sweden.

6

Engineered Wood Products for Structural Purposes

Frank Lam

Helmut G.L. Prion

6.1 INTRODUCTION

6.1.1 Background

In the recent past, forest product industries from many parts of the world relied on the abundant supply of high quality fibres from old growth timber to successfully and profitably produce large volumes of high quality solid sawn timber products. The production of these renewable timber products served the society by the provision of building and structural materials for shelters and employment opportunities for strong workforces.

Sustenance of forest resources is critical to maintaining and improving the economic development for many forest product producing countries. With the increase in world population and the improvement in general living standards, there is an associated increase in demand for structural wood products. For example, in 1998 the consumed wood and fibre products in the United States reached an all-time high level, requiring approximately 0.505 billion m^3 of roundwood production. The current demand for wood worldwide (approximately 3.5 billion m^3) has doubled in the last 30 years. Despite efforts to improve sustainable forest management worldwide, the gap between the demand and supply continues to increase drastically so that the resource and its available use are both stressed. For example, by 2050 the projected future consumption of wood will increase to 5.2 billion m^3. This imbalanced supply and demand situation has particularly stressed the forestry supply in many developing countries with emerging economies, where many of the traditional wood species are not managed in a sustainable manner.

As the demand for commercial use of the forests increases, there is a strong need to balance other demands on the forest. The recognition of the forests as ecosystems and their impact on the environmental quality of the world is particularly important. The forest is part of an ecosystem that protects the watershed, allows nutrients to be recycled from the tree to the soil, provides a storage for carbon, regenerates oxygen, influences global climate and provides shelters and food sources for animals. The forests also serve as parks and recreational areas, and provide spiritual inspiration for many people. The competing use of the forest, as a source of commercial fibre versus the societal needs for recreation and a healthy environment, has led to continued debates. In some parts of North America, these concerns on environmental and sustainability issues have led to restrictions of the availability of old growth fibre supply.

One solution to the changing timber resource is by reforestation and forest plantation. Currently, we are growing more wood than we are cutting. However, the supply of high quality wood fibre suitable for engineered structural uses are limited because of reduced access to old growth timber supply. Another solution to the changing timber supply is by improving the efficiency of wood fibre utilisation from the forest to the final product. This includes the development of new technologies to improve the harvesting and production process to reduce material losses. Also, from the utilisation perspective, the development of new technologies and new products that can lead to the commercial use of smaller diameter, lower quality logs and previously underutilised species and residues is a positive approach to address the changing timber supply and demand situation.

The combined effects of changing timber resource and increased demand on high quality fibre for structural wood products have led to the development of new products that can better use the available timber resource. In fact, it can be argued that the continual competitiveness of the forest products industry will depend upon the development of new value added products using non-traditional resources and underutilised species.

6.1.2 Scope

Engineered wood products (EWP), by definition, consist of a broad class of structural wood products that are commonly used as structural materials in construction. These products differ from dimension lumber or solid sawn timber that is obtained by sawing the logs or cants into individual single solid members. Instead, engineered wood products are made from veneers, strands or flakes that have either been peeled, chipped or sliced. These flakes, strands or sheets of veneer are arranged or formed for structural purposes and then bonded together with adhesives under heat and pressure to make panels or timber-like or shaped structural products.

Under this definition, examples of engineered wood products that utilise flakes or veneers to form panels include plywood and Oriented Strand Board (OSB). Other composite panel products that may have limited structural applications include particleboard, hardboard and medium density fibreboard. In terms of timber- or lumber-like composite wood products, structural composite lumber is a generic term used to describe a family of Engineered Wood Products that combine wood veneer or strands with exterior structural adhesives to form timber-like structural members. In structural composite lumber, the wood veneer or strands are typically aligned. In most cases, the grain angle of the strands or veneers are principally oriented along the length of the member. In some cases, members can also be made with orthogonally arranged layers of flakes. By the nature of their manufacturing process, large defects such as knots and other strength reducing characteristics are either eliminated or dispersed throughout the cross-section to produce a more homogeneous product. Structural composite lumber products that utilise wood veneer sheets are referred to as laminated veneer lumber while those utilising wood veneer strands are referred to as parallel 'strand lumber'. Finally, structural composite lumber products that utilise flakes are referred to as 'laminated strand lumber' and 'Oriented Strand Lumber' (OSL). The production of OSL is an extension of the OSB manufacturing technology for thick oriented strand board.

Engineered wood products in the broad sense also include products that are made by bonding individual solid sawn pieces into larger structural members. Some examples include finger-jointed lumber, glued-laminated timber, wood I-joists, laminated solid panels and structural insulated panels.

Engineered wood products can be used in a wide range of structural applications, including residential, agricultural and commercial structures, to form essential components in floors, walls and roofs. These products can also increase the structural efficiency of wood frame construction when used alone or in combination with solid sawn timber by improving building performance and reducing cost. Figure 6.1 shows the progression of development and commercial introduction of some engineered wood products for building applications in the last century.

Traditionally, wood-based composite panels such as plywood and OSB are used as sheathing for walls, floors and formwork, where composite action and load sharing behaviour are developed within the structural system to allow the panel and framing members to effectively carry the applied loads.

Since the late 1970s, research and development on structural composite lumber products such as parallel strand lumber, laminated strand lumber, laminated veneer lumber and thick oriented strand board have led to their successful introduction into the construction industry. Together with glued-laminated beams and I-joists, EWPs have been substituted for traditional solid sawn timber components as beams, headers, columns and chord members. Moreover, structural composite lumber products are starting to be used in applications typically dominated by steel or concrete (i.e. long span commercial roof truss and shell structures). In the future, with the reduced availability of large size solid sawn timber due to environmental concerns, EWPs will play an even more important role as engineered structural materials.

One of the key features of EWPs is the reconstitution manufacturing process of forming a larger member from smaller pieces. This process tends to disperse natural macro defects in the wood, resulting in more consistent and uniform mechanical and physical properties compared to those of solid sawn timber. Increase in uniformity can lead to more efficient utilisation of the fibre resource, as illustrated by the following simple example.

Suppose a normally distributed axial load with a mean of 13 kN and a standard deviation of 2.6 kN is imposed onto 38 × 89 mm members that have a normally distributed resistance. From the theory of structural reliability, when both the load effects (U) and the resistance (R) are normally distributed and uncorrelated, the probability of failure, P_f, can be estimated through the calculation of a so-called reliability or 'Hasofer–Lind' safety index, β, as:

$$\beta = \frac{\overline{R} - \overline{U}}{\sqrt{R_{\text{s.d.}}^2 + U_{\text{s.d.}}^2}}$$

$$P_f = \Phi(-\beta)$$

$\overline{U}$ and $U_{\text{s.d.}}$ are the mean and standard deviation of load effect, $\overline{R}$ and $R_{\text{s.d.}}$ are the mean and standard deviation of resistance, and $\Phi(\cdot)$ denotes standardised normal probability distribution function (Thoft-Christensen and Baker, 1982). More detailed treatment on the subject of reliability of

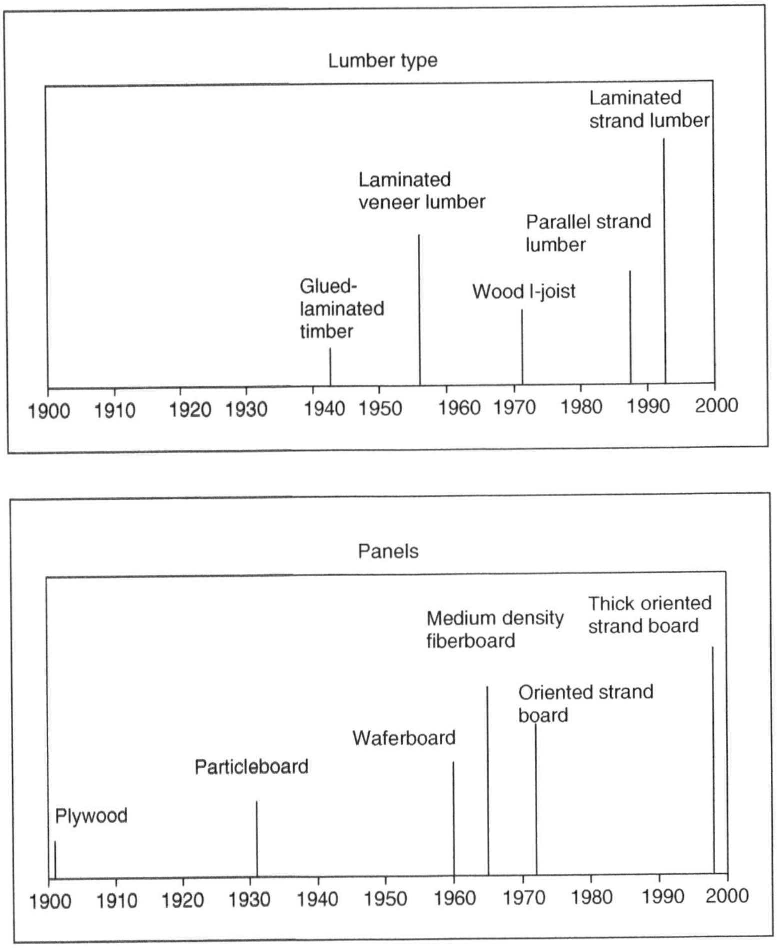

Figure 6.1 The progression of development and commercial introduction of engineered wood products for building applications in the last century

Table 6.1 Comparisons of performance of members with different variability

Case	Mean resistance (MPa)	Coefficient of variation of resistance (%)	Standard deviation of resistance (MPa)	β	P_f
1	35	35	12.25	2.540	≈0.0055
2	35	20	7.00	4.484	≈0.000005
3	35	10	3.50	8.711	≈0

wood structures can be found in Foschi *et al.* (1989) and in Chapter 11.

Now, consider three cases of member resistance with identical mean resistance $\overline{R}$ of 35 MPa, but different coefficients of variation, as shown in Table 6.1. In these cases, the strength variability is in the expected range for visually graded timber (35%), machine stress rated lumber (20%) and structural composite lumber (10%), respectively. The performance of the members can be compared in terms of β and P_f based on reliability assessments.

Improvements in performance via the reduction of variability of the resistance can clearly be observed in the example. In reality, producers of EWPs would take advantage of the reduction in variability of the mechanical properties of EWPs by the assignment of higher characteristic strengths for these products, while aiming to maintain the reliability in relation with other products. Table 6.2 shows the comparison of the bending strengths of 286 mm deep solid sawn visually stress rated and machine stress rated dimension lumber, parallel strand lumber and laminated veneer lumber. Although the mean strengths of these products are only slightly different, the more uniform products such as parallel strand lumber and laminated veneer lumber have significantly higher fifth percentile strength properties compared to machine stress rated lumber, which in turn has higher fifth percentile strength properties compared to visually stress rated lumber.

As manufactured products, tight quality control is needed to ensure consistency of the product's physical and mechanical quality. For example, Bledsoe (1990) presented a paper on testing and evaluation of structural composite timber products, and Bledsoe and Craig (1990) summarised their work on quality assurance of parallel strand lumber. A standard for the evaluation of structural composite lumber is available in American Standards of Testing and Material ASTM D5456. This standard outlines procedures for establishing design properties of structural composite lumber. It also provides guidelines for checking the effectiveness of properties assignment and quality assurance programs of structural composite lumber. These include monitoring and re-evaluating the structural capacities of structural composite lumber, and minimum requirements for establishment of quality control, assurance and audit. In North America, the acceptance of the evaluation of proprietary products is available from the building officials such as Canadian Construction Materials Centre (CCMC), the International Conference of Building Officials (ICBO), and the National Evaluation Service Inc. (NES). In Europe and Japan similar approvals can be obtained from the respective building officials.

In general, the structural performance of EWPs is governed by the properties of the wood species, the manufacturing strategy and methods, the quality control processes, and the final application (Forest Products Laboratory, 1999, Canadian Wood Council, 1991, Bledsoe, 1990, Bledsoe and Craig, 1990). This chapter will discuss the manufacturing processes, structural properties, physical attributes and common applications of some of the major modern EWPs, including glued-laminated timber, parallel strand lumber, laminated strand lumber, laminated veneer lumber and thick oriented strand board.

6.2 GLUED-LAMINATED TIMBER

Glued-laminated timber, commonly referred to as glulam, is a structural timber product made of elements glued together from smaller pieces of wood. In Europe, North America and Japan, glued-laminated timber is used in a wide variety

Table 6.2 Comparisons of the bending strengths of 286 mm deep members

Material	Grade	Species	Fifth-percentile bending strength (MPa)
Visual stress rated	Select Structural	Douglas fir	20
	No. 2	Douglas fir	13
Machine stress rated	2400f-2.0E	Douglas fir	35
	1650f-1.5E	Douglas fir	24
Parallel strand lumber	2900f, 2.0E	Douglas fir	42
Laminated veneer lumber	2600f, 1.9E	Douglas fir	38

of applications, ranging from headers or supporting beams in residential framing to major structural elements in non-residential buildings as girders, columns and truss members. Many studies on the performance of glulam members are available. Foschi and Barrett (1980), Falk and Colling (1995), Serrano and Larsen (1999) as well as Chapter 5 provide useful reviews on the subject.

Figure 6.2 shows some examples of structural applications of glued-laminated timber. As glued-laminated timber can be produced to almost any desired shape and size, ranging from long straight beams to complex curved–arch configurations, it offers the builder/architect many opportunities to express their artistic concepts in a building while satisfying the strength requirements. Finally, with the warm and pleasing appearance of glued-laminated timber, it is often purposely left exposed in the building as an architectural feature.

In glued laminated timber manufacturing, typically a special grade of timber, the laminating stock (lamstock), is end jointed using finger jointing or scarf jointing techniques to form the laminates. These are then arranged in layers or laminations, and glued together to form the structural member. Glued-laminated timber can be produced in either straight or curved form, with the grain of all the laminations essentially parallel to the axis of the member. This laminating technique allows high strength lumber of limited size to be manufactured into larger structural members. Furthermore, higher grade lamstock can be used in high stress zones, thereby maximising the strength-to-cost ratio.

As a non-proprietary product, glued-laminated timber is manufactured according to production standards at certified plants. These standards govern the process of qualification/certification of the plants, as well as the day-to-day production procedures of grading the lumber, end joining the lamstock, gluing procedures, finishing techniques and quality control process.

In Canada, for example, glued-laminated timber manufacturers are required to be qualified under Canadian Standards Association Standard 0177, which provides guidelines for equipment, manufacturing and record keeping procedures. In addition, the manufacturing of Canadian glued-laminated timber is in accordance to the requirements of Canadian Standards Association Standard 0122, *Structural Glued-Laminated Timber*. Under these manufacturing procedures, representative glue bond and end joint samples are tested for strength adequacy.

Figure 6.2 Structural applications of glued-laminated timber

The fabricated glued-laminated timber members are also required to have a comprehensive quality assurance record on the lumber grading, glue bond and end-joint test results. Also, information on the laminating conditions including glue spread rate, assembly time, curing conditions and curing time are recorded. Independent certification agencies perform scheduled mandatory audits to ensure that in-plant procedures meet the requirements of the manufacturing standard.

In the US, glued-laminated timber is manufactured in accordance to the requirements of the American Institute of Timber Construction AITC 103 *Standard for Structural Glued-Laminated Timber* (ANSI/AITC A190.1). In Japan and Europe, glued-laminated timber is manufactured in accordance to the requirements of JAS 111 and EN 386 and 1194, respectively. Typically, the glued-laminated timber producers can make available a certificate of product conformance upon request by their customers.

Glued-laminated timber can be manufactured with many available species. The most common species for glued-laminated timber production include Douglas fir, larch, spruce, pine, southern pine, yellow poplar, radiata pine and Norway spruce. Figure 6.3 shows a schematic view of the manufacturing sequence of glued-laminated timber. During fabrication, the lamstock is dried to a moisture content range of 7–15%. The surfaces of the lamstock are then machine finished (not sanded) to a uniformly smooth surface with a thickness tolerance of ±0.4 mm within a piece or within a lamination composed of more than one piece. These steps are required to allow the quality of the glue-line to be met. The lamstock is then visually graded and subjected to a non-destructive bending stiffness test to estimate its modulus of elasticity. Based on the visual grade and stiffness assessments, the placement of the piece within a glued-laminated timber will be determined. Typically, higher stiffness and better

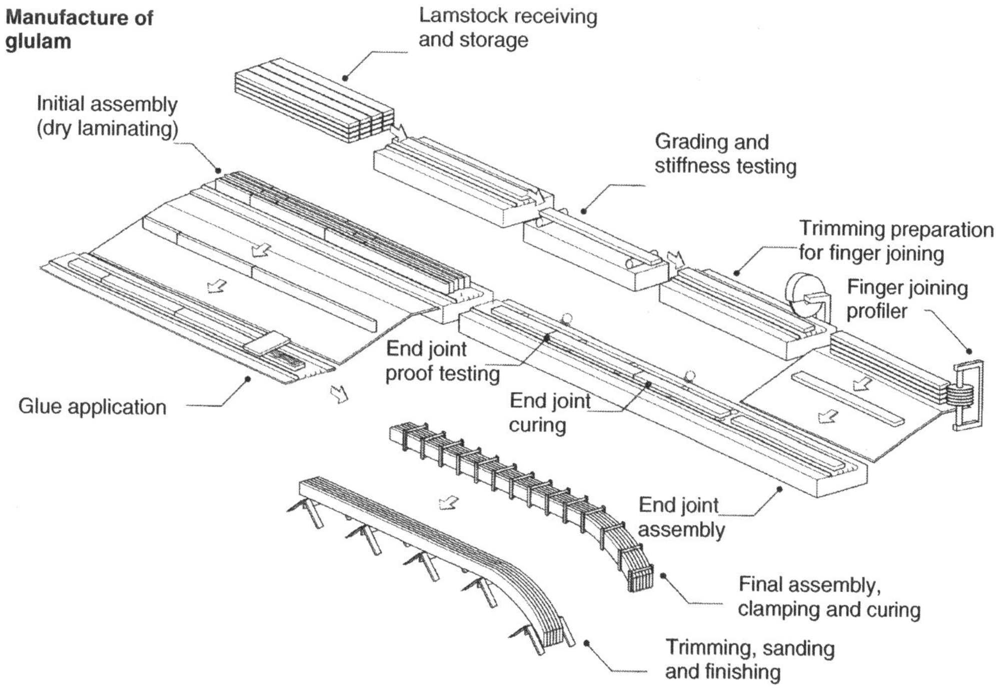

Figure 6.3 Manufacture of glued-laminated timber. (*Source: Picture obtained from Wood Reference Handbook, Canadian Wood Council, 1991.*)

quality pieces will be used as the outer laminates in the construction of beam members to maximise its efficiency.

The assessed materials are then end-trimmed and profiled for end-jointing into full length laminates required for a beam lay-up. Typically, finger-jointing or scarf-jointing techniques are used. A plane sloping scarf where the tapered surfaces of the laminations are glued together can develop 85–90% of the tensile strength of the controlled specimens. Finger jointing is less wasteful, but requires detailed control of quality of the profiling of the fingers and gluing process. In a finger joint, it is desirable to have a combination of thin tip, flat slope, and narrow pitch for high strength. The longer fingers will yield higher joint strength, although shorter fingers offer more savings. The endjoint profile can be achieved with the scarf or fingers formed by cuts perpendicular to either the wide or narrow faces of the board. Phenol-Resorcinol and melamine-based resins are examples of adhesives used in glued-laminated timber production. Typically, the glued endjoints are cured with heat to produce full-length laminates. Radio-frequency based techniques are commonly used to heat up the end-joints and to speed up the curing process.

The laminates are then arranged in layers as a preliminary lay-up in terms of the grade and stiffness of the laminates and in accordance with the product requirements. When members wider than the available lamstock are required, lamination may consist of multiple boards side by side with the edge joints staggered in the member. Adhesives are then applied to the individual laminates and reassembled under pressure using clamps (>0.7 MPa) to create either straight beams or members with complicated curvatures. The assembly under pressure is then placed in position for curing. The condition of the curing including the curing time, temperature and relative humidity in the curing location, temperature of the glue-line and other pertinent controls that adhere to the glue manufacturer's specifications. Sometimes heaters are also installed in the curing location to keep the temperature at the desired level.

After completion of the curing process, the product is finished with surface planing, patching, and end-trimming. Some facilities also provide special drilling and notching for connections, and special sanding and staining for finish. The final glued-laminated timber product receives a coat of protective sealer before shipping, and is wrapped for protection during shipping and erection.

The manufacturing process of glued-laminated timber allows members of a wide variety of shapes and sizes to be produced. Standard sizes have also been developed for glued-laminated timber to allow optimum utilisation of the timber that are multiples of the dimensions of the lamstock used for the production of glued-laminated timber. The standard sizes offer designers economy and fast delivery that is suitable for a wide range of typical applications. In many European countries, glulam manufacture is highly automated for standard sizes and lengths, resulting in a high strength wood products at competitive prices. Non-standard dimensions may also be specially ordered and produced by the manufacturer.

Based on the above attributes, glued-laminated timber products can be specified according to the stress grades available from individual suppliers. Usually, an appearance grade should also be specified to match the intended use of the product.

6.3 PARALLEL STRAND LUMBER

Parallel Strand Lumber (PSL) was invented and developed in Canada by MacMillan Bloedel Limited. Since its introduction and launch as a commercial proprietary product in 1986 under the trade name Parallam® PSL, the product has steadily gained an important market share in timber construction in North America. As a result of changes in company ownership, Parallam® PSL is now produced by Trus Joist™ A Weyerhaeuser Business.

In the original product concept in the 1970s, parallel strand lumber was intended to utilise forest wastes such as branches and trimmings to form a structural composite lumber product to eventually replace larger dimension solid sawn timbers. After many iterations and more than a decade of research and development, the final product concept utilised veneer strands by bonding them together with a phenolic based adhesive under heat and pressure to form parallel strand lumber as a high strength structural composite lumber product. Currently, parallel strand lumber products are manufactured in facilities in both Canada and the United States (Barnes, 1988a, Churchland, 1988). Relatively few journal publications are available on the mechanical properties of parallel strand lumber, whereas computer modelling of this type of strand-based product has been studied by several research groups (Clouston and Lam, 2001, Triche and Hunt, 1993, Wang and Lam, 1998a).

A general schematic of the manufacturing process of parallel strand lumber is shown in Figure 6.4. In Canada Douglas fir veneers and in the United States southern pine and yellow poplar veneers are primarily utilised in the production of parallel strand lumber. Typically sheets of 3.2 mm × 1.2 m × 2.4 m veneer are directly peeled from logs or obtained from veneer mills. The veneers are dried to target moisture content and then clipped into strands of approximately 19 mm in width. The maximum length of these strands is obviously 2.4 m. However, at this stage it is still possible to have short flakes or strands in the line resulting from the clipping process. Since clipped strands from veneer sheets are used, it is not necessary to use only sheets that are precisely 1.2 × 2.4 m in size. Imperfect veneer sheets obtained from the start of peeling of a log that are generally not suitable for plywood manufacturing or laminated veneer lumber production can also be used in parallel strand lumber. This use of undesirable veneer allows parallel strand lumber to increase the conversion efficiency of a log to a structural member.

A defect removal process is introduced next to remove the short and/or weak fibres. The strands are transported lengthwise on a conveyer belt

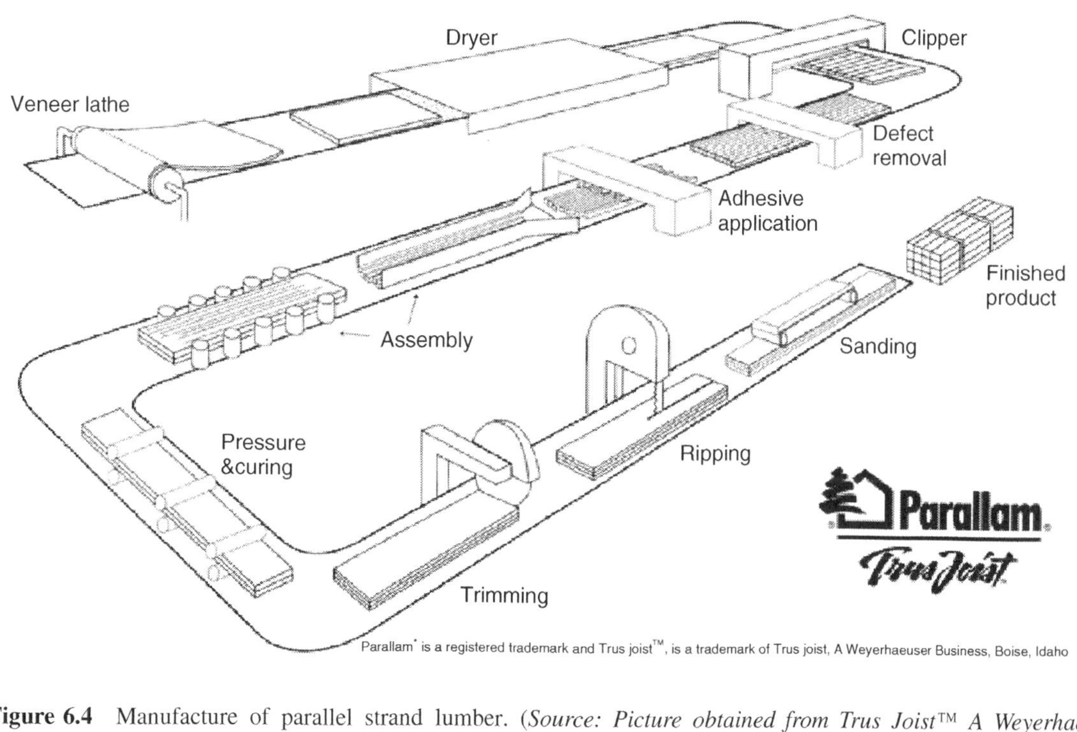

Figure 6.4 Manufacture of parallel strand lumber. (*Source: Picture obtained from Trus Joist™ A Weyerhaeuser Business Media Kit.*)

system towards the next production step. Within the strand transportation line, a gap is introduced in the conveyer belt system. Individual strands have to be able to bridge this gap to be moved to the next production phase, and short strands will drop into a waste collection bin below the conveyers. Furthermore, strands that are too weak to bridge the gap may also break off and drop to the waste collection line. This simple defect removal system effectively eliminates short fibres and/or weak fibres that are undesirable in terms of some of the product strength attributes. The target is to include strands longer than 305 mm in the product.

The next step is coating the strands with phenol-formaldehyde resins with either a roll spreader or dipping the strands in an adhesive bin. Dipping the strand will completely coat it with adhesive, but the excess adhesive needs to be eliminated via an air blowing system. The strands are then further transported to an assembly station where a continuous loose billet of resinated strands is formed. From the transportation and strand deposition system, the strands tend to align themselves along the length of the billet. Throughout the strand transportation process, mass flow systems are used to allow the operators to monitor and carefully control the amount of fibres to be included in the formation of the loose billet.

The loose billet is then fed continuously into a belt press. Under confinement, the billet is heated to the curing temperature of the adhesive using microwave energy. Finally, a finished continuous billet of 280 mm × up to 482 mm in cross-section is produced. Because of the continuous press process, the billet length is limited only by transportation constraints, typically 20 m. The finished billet is then cross cut to desired lengths, rip sawn to produce rough dimensions or custom sizes, and sanded down to finished dimensions. It is also possible to produce larger dimension members by edge gluing techniques commonly used in the production of glued-laminated beams. Individual beam weighing is also performed as a final step to control the shipped density.

The available stock member sizes for parallel strand lumber are comparable with established wood framing material sizes. Flexibility of custom sizing is also possible through the manufacturing process.

Parallel strand lumber is produced at a moisture content range of 8–12%. This manufacturing

Figure 6.5 Applications of parallel strand lumber. (*Sources: Pictures obtained from Parallam® PSL Headers, Beams and Columns, Trus Joist MacMillan, 1998. Abbotsford Recreation Centre–Engineered wood brings a new standard to ice arenas, Canadian Wood Council.*)

process for parallel strand lumber results in both a strong and consistent material that is less prone to shrinking, warping, cupping, bowing or splitting resulting from seasoning stresses. In residential construction, parallel strand lumber is well suited for use as beams or columns in post and beam construction, and as beams, headers and lintels in wood frame construction. In heavy timber commercial building, parallel strand lumber is also well suited for use as intermediate and large members. Examples of applications of parallel strand lumber are shown in Figure 6.5. Finally, the voids in the cross-section of parallel strand lumber allow the member to accept preservative treatment with a very high degree of penetration depth. Therefore, when chemical protection against decay is needed, treated parallel strand lumber can be considered. The allowable mechanical properties of 2.0E Parallam PSL are shown in Table 6.3.

Nail withdrawal and lateral resistance connection properties of Parallam® PSL are deemed to be equivalent to that of Douglas fir sawn timber with specific gravity of 0.5. Standard bolted connection design values are provided in the adopted code for Douglas fir with specific gravity of 0.5. Fire tests on parallel strand lumber demonstrated that it is appropriate for use in all applications for which solid sawn timbers are suited.

Table 6.3 Allowable mechanical properties of 2.0E Parallam® PSL (standard term loading and for dry service conditions) (Adapted from Trus Joist™ A Weyerhaeuser Business Commercial Products Design Manual)

Shear modulus of elasticity	$G = 860$ MPa
Modulus of elasticity	$E = 13790$ MPa
Bending Strength for 305 mm depth	$F_b = 20.0$ MPa[a]
Compression perpendicular to grain parallel to wide face of strands	$F_{C\perp} = 5.17$ MPa[b]
Compression parallel to grain	$F_{C\parallel} = 20.0$ MPa
Horizontal shear perpendicular to wide face of strands	$F_v = 2.00$ MPa

[a]For depth d other than 305 mm apply a size factor of $(305/d)^{0.111}$.
[b]Do not adjust for load duration effect.
Note: The allowable design values are for use with the US National Design Standards. Approved specified strengths for use in Limit State Design Codes for a number of other countries are also available from the product manufacturer. Allowable properties and grades of the products are subject to change; therefore, the readers should refer to the manufacturer's literature for the full range of grades and latest values.

6.4 LAMINATED STRAND LUMBER

Laminated Strand Lumber (LSL) is a proprietary product invented and developed in Canada by MacMillan Bloedel Limited. It was introduced and commercially launched in the early 1990s under the trade name TimberStrand® LSL. As a result of changes in company ownership, TimberStrand® LSL is now produced by Trus Joist™ A Weyerhaeuser Business.

Laminated strand lumber aims to replace some larger dimension solid sawn timber products using underutilised commercial species (Barnes, 1988b). The final product concept utilises strands from fast growing hardwood species such as aspen-poplar by bonding them together with Isocyanate as diphenylmethane di-isocyanate (MDI) adhesive under heat and pressure to form the laminated strand lumber. Currently, laminated strand lumber products are manufactured in facilities in the United States. A new facility in Kenora, Canada was planned for start up by the end of 2002.

A general schematic of the manufacturing process of laminated strand lumber is shown in Figure 6.6. Medium length strands made with aspen-poplar strands are utilised in the production of laminated strand lumber. These wood elements have a thickness range of 0.9–1.3 mm. The strand width ranges from 13–25 mm, and strand length is up to 300 mm.

Fresh cut logs are soaked in hot water ponds to soften the wood prior to debarking and stranding to reduce the amount of generated fines and slivers. After conditioning, the logs are debarked and fed into a stranding machine to obtain the desired strands. The strands are screened to remove fines and unwanted particles.

A cylindrical dryer is then used to dry the strands to a moisture content range of 3–7%. It is desired to minimise breakage and maintain

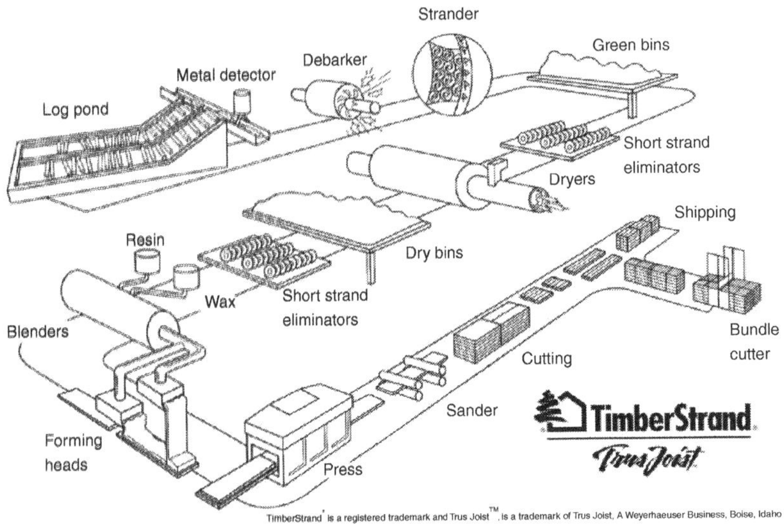

Figure 6.6 Manufacture of laminated strand lumber. (*Sources: Picture obtained from Trus Joist™ A Weyerhaeuser Business Media Kit.*)

uniform moisture distribution of the strands during drying. After drying the strands are resinated in a blender with MDI resins and wax. Resin content is proprietary information for the product. A careful monitoring process is in place to regulate the relative weights of the strands and the resin. A former/orienter machine is used to arrange and deposit the strands into a loose mat with the strands primarily oriented parallel to the long axis of the mat. The footprint of the loose mat is approximately 2.4 m wide × 10.7 m or 14.6 m long. The weight of the loose mat is also monitored to check for uniformity of distribution of the strands.

The loose mats are then pressed under heat and pressure for the curing of the adhesives to form solid panels. A steam injection pressing process is utilised to enhance the heating of the core material such that relatively thick members can be formed with a uniform density profile through the thickness of the member. The steam injection process also shortens the required press time by enhancing the heating of the core, which leads to cost savings from reduced press time. Non-destructive testing devices are used to scan the panels for defects such as blows. The finished panels are cross-cut and rip-sawn to produce rough dimensions or custom sizes, and sanded down to finished dimensions.

Member sizes for laminated strand lumber are comparable with established wood framing material sizes. Flexibility of custom sizing is also possible through the manufacturing process. TimberStrand® members are available up to 140 mm in thickness; 1.2 m in width, and 14.6 m in length.

Laminated strand lumber is produced at a moisture content range of 6–8%. The manufacturing process for laminated strand lumber results in both a strong and consistent material that is less prone to shrinking, warping, cupping, bowing or splitting resulting from seasoning stresses. In residential wood frame construction, laminated strand lumber is well suited for use as headers, lintels, studs, and rimboard. Examples of applications of laminated strand lumber are shown in Figure 6.7. The allowable mechanical properties of 1.3E and 1.5E Timber Strand® LSL are shown in Table 6.4.

Nail withdrawal and lateral resistance connection properties of TimberStrand® LSL are deemed to be equivalent to that of Douglas fir sawn timber with specific gravity of 0.5. Standard bolted connection design values are provided in the

Figure 6.7 Applications of laminated strand lumber. (*Sources: Pictures obtained from TimberStrand® LSL Headers and Columns, Trus Joist MacMillan (now Trus joist, A Weyerhaeuser Business), 1998.*)

Table 6.4 Allowable mechanical properties of 1.3E and 1.5E TimberStrand® LSL (standard term loading and for dry service conditions) (adapted from Trus Joist™ A Weyerhaeuser Business Commercial Products Design Manual)

	1.3E TimberStrand®	1.5E TimberStrand®
Shear modulus of elasticity	$G = 560$ MPa	$G = 646$ MPa
Modulus of elasticity	$E = 8960$ MPa	$E = 10345$ MPa
Bending Strength for 305 mm depth	$F_b = 11.7$ MPa[a]	$F_b = 15.5$ MPa[a]
Compression perpendicular to grain parallel to wide face of strands	$F_{C\perp} = 4.69$ MPa[b]	$F_{C\perp} = 5.34$ MPa[b]
Compression parallel to grain	$F_{C\parallel} = 9.6$ MPa	$F_{C\parallel} = 13.4$ MPa
Horizontal shear perpendicular to wide face of strands	$Fv = 2.76$ MPa	$Fv = 2.76$ MPa

[a]For depth d other than 305 mm apply a size factor of $(305/d)^{0.092}$.
[b]Do not adjust for load duration effect.
Note: the allowable design values are for use with the US National Design Standards. Approved specified strengths for use in Limit State Design Codes for a number of other countries are also available from the product manufacturer. Allowable properties and grades of the products are subject to change; therefore, the readers should refer to the manufacturer's literature for the full range of grades and latest values

adopted code for Douglas fir with specific gravity of 0.5. Fire test information on laminated strand lumber demonstrates that TimberStrand® LSL has a comparable char rate and flame spread class as solid sawn timber.

6.5 LAMINATED VENEER LUMBER

Laminated Veneer Lumber (LVL) is made by gluing layers of wood veneer sheets together using an exterior type adhesive to form a structural composite lumber product. It was first used to make airplane propellers during World War II. Since the mid-1970s, laminated veneer lumber has been used as a structural composite lumber product as beams, headers, flange stock for wood I-joists, and scaffolding planks to take advantage of its attributes including high strength, dimension stability and uniformity.

In most laminated veneer lumber products, the grain orientation of each veneer layer is aligned parallel to the long direction of the member. This provides the member with strong directional strength properties along the member length. In some special laminated veneer products, a few sheets of veneer can also be introduced into the lay-up in the direction perpendicular to the long direction of the member to enhance the strength properties orthogonal to the member length. This special arrangement is intended to particularly improve connection properties of the members (Cha and Pearson, 1994). Similar to other engineered wood products, the reconstitution process of laminated veneer lumber creates a structural element of higher reliability and lower variability through defect removal and dispersal.

A general schematic of the manufacturing process of laminated veneer lumber is shown in Figure 6.8. Typically, in North America, the production of LVL uses species or combinations of species such as Douglas fir, larch, southern pine, yellow poplar, western hemlock, lodgepole pine and spruce. In some products, mix species can be introduced in the member lay-up. The initial production step of laminated veneer lumber is similar to that of plywood, where logs are rotary peeled on a lathe to yield 2.6 m wide veneer sheets. The veneer, which is between 2.5 mm and 4.8 mm thick, is cut into sheets 686 mm or 1372 mm in width and 2565 mm in length.

After the veneers are dried to target moisture content, major strength reduction defects can be removed by a clipping device. Visual grading of the veneers is then performed. Some manufacturers also use non-destructive testing devices, such as ultrasound systems and density verification equipment with a visual override to grade the veneers.

The graded veneer sheets are then cut to the desired width as required for the billet. Typically, phenol-formaldehyde resins are applied on the veneer. The resinated veneer sheets are then built up either by hand or by an automatic lay-up system in the assembly line to the desired number of layers and pattern in form of a long loose billet. The veneer sheets are typically end-jointed with lap or butt joints. High quality/grade of veneers are usually placed at the outer faces of a member during manufacturing. This is especially important if the member is used on the flat as is the case of scaffolding planks where higher quality material on the member faces improves the strength efficiency of the member, and provides a better wearing surface.

After forming, the billet is fed continuously through a series of roller presses that provide heat and pressure to cure the adhesive and consolidate the billet. Billets of up to 25 m in length (limited by handling and transportation issues) can be made, which are then cross cut and rip-sawn

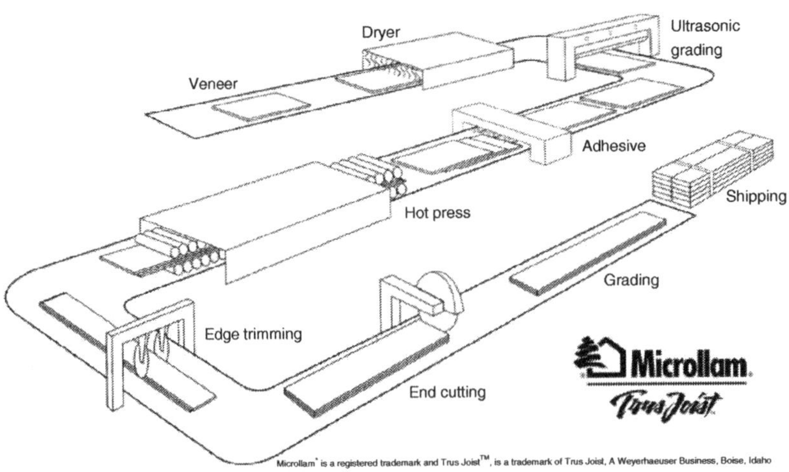

Figure 6.8 Manufacture of laminated veneer lumber. (*Source: Picture obtained from Trus Joist™ A Weyerhaeuser Business Media Kit.*)

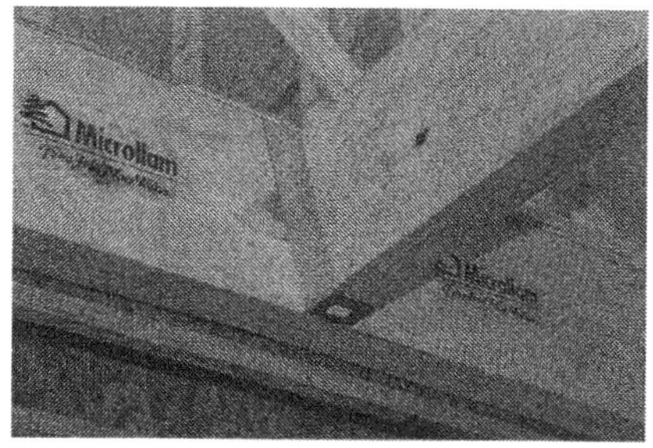

Figure 6.9 Applications of laminated veneer lumber. (*Sources: Pictures obtained from 1.9E Microllam™ LVL Headers and Beams, Trus Joist MacMillan (now Trus joist, A Weyerhaeuser Business), 1998.*)

Table 6.5 Allowable mechanical properties of typical LVL products (Standard term loading and for dry service conditions) (adapted from Trus Joist™ A Weyerhaeuser Business Commercial Products Design Manual)

Shear modulus of elasticity	$G = 820$ MPa
Modulus of elasticity	$E = 13100$ MPa
Bending Strength for 305 mm depth	$F_b = 17.9$ MPa[a]
Compression perpendicular to grain parallel to wide face of strands	$F_{C\perp} = 5.17$ MPa[b]
Compression parallel to grain	$F_{C\parallel} = 12.3$ MPa
Horizontal shear perpendicular to wide face of strands	$F_v = 1.97$ MPa

[a]For depth d other than 305 mm apply a size factor of $(305/d)^{0.136}$.
[b]Do not adjust for load duration effect.
Note: The allowable design values are for use with the US National Design Standards. Approved specified strengths for use in Limit State Design Codes for a number of other countries are also available from product manufacturers. Allowable properties and grades of the products are subject to change; therefore, the readers should refer to the manufacturer's literature for the full range of grades and latest values

into the desired length and width, as required by the customer.

Laminated veneer lumber is available in thicknesses of 19.1–89 mm, depths of 63.5–1219 mm and lengths of up to 25 m. The product is used primarily as beams and headers in structural framing for residential and commercial construction. Other uses include scaffold planks and flanges of wood I-joists. Figure 6.9 shows some examples of structural applications of LVL.

A list of allowable mechanical properties of typical laminated veneer lumber products is given in Table 6.5. Other grades of laminated veneer lumber are available via veneer selection process during manufacturing. The properties of other laminated veneer lumber products can be obtained from individual product manufacturers.

Nail withdrawal and lateral resistance connection properties of Microllam® LVL are deemed to be equivalent to that of Douglas fir sawn timber with specific gravity of 0.5. Standard bolted connection design values are provided in the adopted code for Douglas fir with specific gravity of 0.5. Fire test information on LVL demonstrates that it has a comparable char rate and flame spread class as solid sawn timber.

6.6 WOOD I-JOISTS

The wood I-joist is primarily a bending member with an I cross-section. It is made by gluing flanges to a web to form an engineered wood product. Typically, the flanges are made with either solid sawn timber (either visually stress graded or machine stress rated lumber), laminated veneer lumber or laminated strand lumber. The web of a wood I-joist is typically made with either oriented strand board or plywood panels. The wood I-joist was successfully introduced in the construction market in the 1970s as an alternative to larger dimension solid sawn timber, to be used as joists and beams in both residential and commercial buildings. Compared to solid sawn timber, wood I-joists have the advantages of more efficient structural shape, higher strength to weight ratio, better dimensional stability, and lower variability of mechanical properties. Being a factory made prefabricated wood product, wood I-joists can also be supplied with predrilled holes for electrical services, mechanical ductwork and plumbing as required in a building. A review of the research related to the strength properties of Wood I-joists is available from Forest Products Society (1990) and Foschi and Yao (1993).

One of the critical elements in wood I-joists is the flange to web joint and the joints between adjacent panels in the web. There are different variations of the flange to web connection. Often individual manufacturers have proprietary flange to web joints protected by patents. In the web, adjacent panels are typically joined together using scarf jointing, or tongue and groove jointing techniques. Phenol-resorcinol or phenol-formaldehyde resins are usually used for the web-to-web and web-to-flange joints. Long length wood I-joists can also be manufactured by finger-jointing shorter members together to form the flanges. Example applications of wood I-joists are shown in Figure 6.10.

Manufacturing of wood I-joists may vary between different producers. Figure 6.11 shows a schematic of the generic production process of wood I-joists. Flange stock with a moisture content range of 8–18% is obtained from mills. Shorter length flange stock, either solid sawn timber or laminated veneer lumber, will be end jointed using finger-joining techniques to form a longer member needed for the flange. A central groove will be routed into the flange member to house the web material. The shape and size of this groove may differ between products. The key is to make the groove such that sufficient contact area between the web and the flange is available to transfer the longitudinal shear stresses via the adhesive when the member is under load without unduly weakening the flange itself.

The web stock with approximate moisture content of 5–10% is obtained from OSB or plywood mills. The web is cut into the size required for the depth of the wood I-joist. Sometimes the ends of the web members are profiled for the web-to-web connection. Often, butt connection between adjacent web members is used. The edge of the web member will be profiled according to the details of the flange-to-web connection so that it can effectively mate with the flange. Again, these profiles are designed to increase the contact area required for the transfer of the longitudinal shear stresses via the adhesive between the flange and the web. Phenolic resin is typically used for the web-to-web and web-to-flange joints. Other resins such as polymeric MDI can also be used. Presses are used to create good contact between the members. The members are then cut to specific lengths and cured either in an oven or curing space at temperatures of 21–65 °C. Final product inspection is performed,

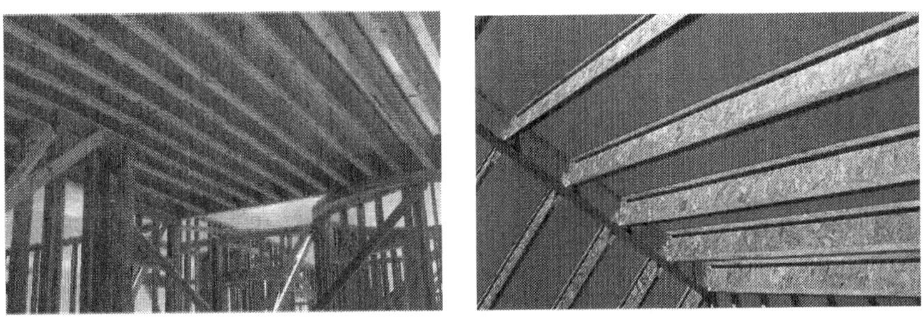

Figure 6.10 Applications of wood I-joists. (*Sources: Pictures obtained from TJI® /Pro™ 150TS Joists Featuring the Silent Floor® System for Residential Applications, Trus Joist MacMillan (now Trus joist, A Weyerhaeuser Business), 1998.*)

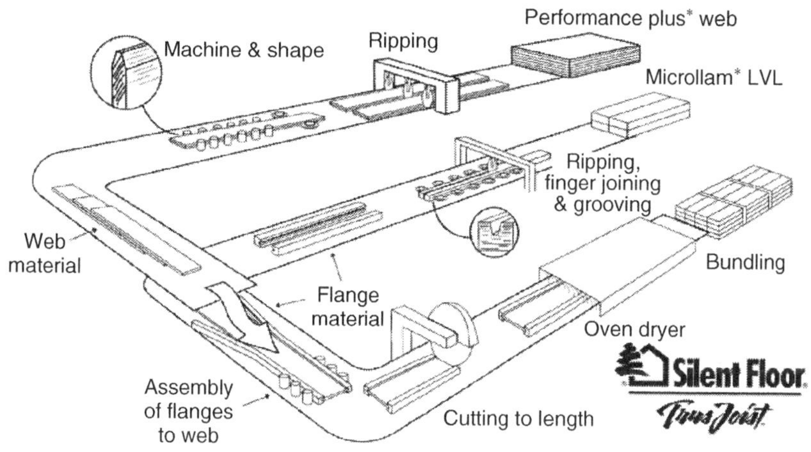

Figure 6.11 Manufacture of wood I-Joist. (*Source: Picture obtained from Trus Joist™ A Weyerhaeuser Business Media Kit.*)

and the product is wrapped for weather protection during shipment.

Procedures for establishing strength values and monitoring the structural capacities of prefabricated wood I-joists are outlined in ASTM Standard D5055. Following these procedures and meeting the acceptance criteria of individual agencies, manufacturers of wood I-joists can obtain product evaluations on their wood I-joist products. On-going in-house quality control programs and independent third party quality audits conducted by an accredited certifying agency on a regular basis are also necessary to ensure product quality.

Fire performance information of assemblies with prefabricated wood I-joists is available from the manufacturers. Such information is also available in listings books of the accredited certifying agencies such as the Underwriters' Laboratories of Canada (ULC), Intertek Testing Services in Canada or Underwriters' Laboratories Inc. (ULI) in the US.

6.7 THICK ORIENTED STRAND BOARD

Oriented Strand Board (OSB) is a product commercially introduced in North America in the early 1970s. It aims to replace plywood as a panel product by using underutilised commercial species. The product uses flakes from fast growing hardwood species such as aspen-poplar by bonding them together with phenol-formaldehyde and/or Isocyanate as diphenylmethane di-isocyanate (MDI) adhesives under heat and pressure to form the structural panel members.

A substantial amount of literature based on empirical studies exists on wood composite panel products (Kelly, 1977). Many researchers studied the influence of the pressing conditions in wood-based panel production on panel properties (Suchsland, 1967, Geimer *et al.*, 1985, Wolcott *et al.*, 1990, Barnes 2000a, b). Other work also focused on the flakeboard mat consolidation occurring during the pressing process (Lang *et al.*, 1996, Lenth *et al.*, 1996, Oudjehane *et al.*, 1998a, b), as well as panel physical and mechanical properties as influenced by its structure (Dai and Steiner, 1994, Garcia *et al.*, 2001, Lu and Lam, 1999, 2001a, b, Lu *et al.*, 1998, Oudjehane and Lam, 1998, Wang and Lam, 1998b, 1999, Xu, 2000, Xu and Suchsland, 1997).

Typically, Oriented Strand Boards are made in three orthogonal layers. The flakes in the face layers have a larger percentage of flakes that are aligned parallel to the long axis of the panel, whereas, the core layer is either aligned perpendicular to the long axis of the panel or completely randomly deposited. The core to face ratio is typically in the range of 60:40 to 40:60. Short length flakes, made with aspen, yellow poplar, or southern pine, are utilised in the production of OSB. The wood elements have a thickness, width and length of approximately 0.8 mm, 13 mm and 100 mm, respectively. In some facilities, depending on the availability of the log supply, sometimes a mixture of aspen-poplar logs with small fir and pine logs is also used in the production of OSB.

Fresh cut logs are soaked in hot water ponds to soften the wood prior to debarking and flaking/stranding to reduce the amount of generated fines and slivers. After conditioning, the logs are debarked and fed into a stranding machine to obtain the desired flakes, after which the flakes are screened to remove fines and unwanted particles. In some facilities, the fines can be recovered, resinated, and reintroduced into the core layer for the panel product. A certain amount of fines has the benefit of filling the voids between larger flakes, and thus reduce the density variation in a panel.

A cylindrical dryer is used to dry the flakes to a moisture content range of 3–7%. This is desired to minimise breakage and maintain uniform moisture distribution of the flakes during drying. After drying, the flakes are resinated in blenders. Typically, modern facilities will have a blender with phenolic resins to resinate the flakes for the face layers of the product and a blender with MDI resins to resinate the flakes used in the core of the product. The advantage of MDI resins is the reduced press time for curing and thus significant cost savings. The disadvantage of MDI resins is that it can adhere to steel, and therefore requires careful application and chemical release agents. In some mills it is also possible to produce all-MDI panel products. In these cases, the MDI bonded panels are made in mills with continuous roller presses, whereas mills with the multi-daylight opening presses typically use the combined phenolic/MDI system. Resin formulation and content are proprietary information for individual producers. A careful monitoring process is in place to regulate the relative weights of the flakes and the resin. Typically, a 2–2.75% ratio of resin by weight is used in the production of OSB. Hot wax of approximately 1.5% by weight is also added to improve water resistance and reduce thickness swelling.

So-called former or orienter machines are used to arrange and deposit the flakes into a loose mat. For a three-layer system, both a core former and a face former are needed. In the process of mat forming, the flakes are deposited through alignment devices and then fall freely from a set height onto the mat. The lower the drop, the higher the percentage of orientation of the flakes, although a certain amount of randomness still prevails in the placement of the flakes. For the multiple opening press facilities, the foot print of the loose mat can be up to 3.6 m × 7.4 m.

The loose mats are pressed under heat and pressure for the curing of the adhesives to produce a solid panel. The press platens are heated and the face layers of the loose mat heats up via the conduction of heat. As the temperature and pressure increases, the moisture in the flakes turns into steam, which migrates towards the core of the panel, and therefore increases the temperature of the core. The heat and pressure polymerize the resin and binds the flakes together. Since the process of heat and mass (steam) transfer is non-uniform through the thickness of the panel, the consolidation and densification of the oriented strand board is typically also non-uniform through its thickness. The face and core layers may have densities of 130% and 70% of the average density of the product, respectively. After the product comes out of the press, it is heat stacked for 24 hours to allow full curing of the adhesives. The finished panels are then cross cut and rip sawn to produce rough dimensions or custom sizes, and sanded down to finished dimensions. Figure 6.12 shows a schematic of the generic production process of OSB.

The quality of the product is influenced by the process control of the production. Careful and continuous monitoring of the processing variables is needed to maintain the product quality and properties. The most important processing factors are: log sorting by species, size, and moisture content; flake

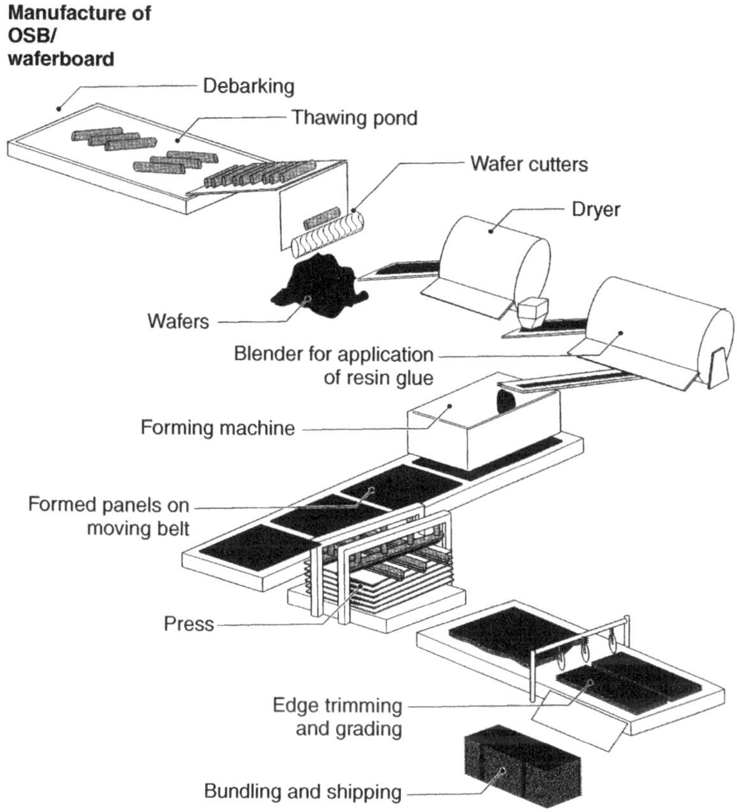

Figure 6.12 Manufacture of oriented strand board. (*Source: Picture obtained from Wood Reference Handbook, Canadian Wood Council, 1991.*)

size, quality, thickness, and post drying moisture content; the consistency of blending flakes, resin and wax; the flake forming and orientation processes including the uniformity of flake deposition in the loose mat; the press temperature, pressures, closing speed, thickness control and pressure release control; quality of panel faces and edges; panel dimensions and the appearance of the finished panel.

Based on these manufacturing technologies, thick (31.75 mm) OSBs can also be made economically. These members can be further manufactured into a deep lumber type products that is suitable for use in some structural applications. Examples of application of thick OSB include rimboard that forms the perimeter of a floor and short span headers. In the rimboard applications, long and deep members can be made where the capacity of the member to transfer vertical loads without crushing or buckling is important. Rimboards are also required to transfer lateral loads imposed by earthquakes or high winds. Standard tests are available for these evaluations. In the short span header applications, the shear strength of the member will be important. In general, the shear strength to bending strength ratio of a product such as thick Oriented Strand Board/rimboard tends to be higher than conventional solid sawn products; therefore, the product may be suitable for use in short span headers applications. The OSBs are produced at moisture content of approximately 5%. It's manufacturing process results in a product with relatively consistent properties. Figure 6.13 shows an example of application of rimboard.

As a manufactured product, tight quality control is needed to ensure consistency of the product's

Figure 6.13 Applications of rimboard. (*Source: Pictures obtained from Ainsworth Lumber Ltd.*)

physical and mechanical quality. The individual mills must establish in-house quality control programs for the requirements specified in the applicable standard. Physical testing of the products is needed to verify production conformance. Independent quality assurance organisations are also involved to certify and monitor manufacturing performance and mill quality control programs.

6.8 SUMMARY

In North America, engineered wood products are relative new structural members that have been successfully introduced in the construction industry. With the anticipated increase in the future demand of structural wood products and changes in accessibility and quality of the timber resource, the development and application of engineered wood products will continue to increase in the future. These products will need to utilize non-traditional resources, and at the same time possess improved physical and mechanical properties, compared to traditional structural products.

Increased market opportunities have already led to the research and development of even newer EWPs using plant-based fibre resources such as bamboo and straw (Rowell and Young, 1997, Youngquist *et al.*, 1994). Hybrid structures that are made by combining wood with plastic or by reinforcing wood members with high performance fibres such as glass, carbon and Kevlar fibres are also being developed and introduced into the market (Youngquist, 1995). Details relating to the processing/cost/application issues have yet to be fully worked out for these new products to gain full acceptance and a firm market place in the construction industry.

Research and development efforts are needed to fully understand the interacting relationship between the raw material, the manufacturing and processing variables, the engineering and physical properties of the product, and the end-use applications. Research on the application of EWPs with respect to their connection details in heavy timber construction is another important topic. The performance and behaviour of EWP systems under both ultimate and serviceability conditions must be studied to expand their applications.

REFERENCES

Barnes D. (1988a) Parallam–a new wood product; invention and development to the pilot plant scale stage. Proceedings Marcus Wallenberg Foundation Symposium, Söderhamn, Sweden, pp. 5–24.

Barnes D. (1988b) Waferboard lumber. US Patent No. 4,751,131.

Barnes D. (2000a) An integrated model of the effect of processing parameters on the strength properties of oriented strand wood products. *Forest Products Journal*, **50**(11/12), 33–42.

Barnes D. (2000b) A model of the effect of strand length and strand thickness on the strength properties of oriented wood composites. *Forest Products Journal*, **51**(2), 36–46.

Bledsoe J.M. (1990) Testing and evaluation of a structural composite timber product. Proceedings International Timber Engineering Conference, Tokyo, Japan, **1**, 203–206.

Bledsoe J.M. and Craig B.A. (1990) Quality assurance for Parallam PSL. Proceedings International Timber Engineering Conference, Tokyo, Japan, **3**, 895–897.

Canadian Wood Council (1991) *Wood Reference Handbook: A guide to the architectural use of wood in building construction.* Ottawa, Canada, Canadian Wood Council.

Cha J.K. and Pearson R.G. (1994) Stress analysis and prediction in 3-Layer laminated veneer lumber:

response to crack and grain angle. *Wood and Fiber Science*, **26**(1), 97–106.

Churchland M.T. (1988) Parallam–Commercial process development. Proceedings of the Marcus Wallenberg Foundation Symposium, Söderhamn, Sweden, pp. 25–42.

Clouston P.L. and Lam F. (2001) Computational modeling of strand-based wood composites. *Journal of Structural Mechanics*, **127**(8), 844–851.

Dai C. and Steiner P.R. (1994) Spatial structure of wood composites in relation to processing and performance characteristics. *Wood Science and Technology*, **28**, 135–146.

Falk R.H. and Colling F. (1995) Laminating effects in glued-laminated timber beams. *Journal of Structural Engineering*, **121**(12), 1857–1863.

Forest Products Laboratory (1999) Wood handbook – Wood as an engineering material. Gen. Technical Report FPL-GTR-113, Madison, WI: U.S. Department of Agriculture, Forest Service, Forest Products Laboratory.

Forest Products Society (1990) *Wood I-Joists. Wood Design Focus*. Forest Product Society Madison, WI, USA, 1990: 1.

Foschi R.O. and Barrett J.D. (1980) Glued-laminated beam strength: A model. *Journal of Structural Division*, **106**(8), 1735–1754.

Foschi R.O., Folz B.R. and Yao F.Z. (1989) Reliability-based design of wood structures. Structural Research Series Report No. 34, Department of Civil Engineering, University of British Columbia, Vancouver, Canada.

Foschi R.O. (2002) Reliability of structures with timber and wood-based products. In: *Advanced Topics in Timber Engineering*, Wiley, London, UK.

Garcia P. Avramidis S. and Lam F. (2001) Internal temperature and pressure responses to flake alignment during hot-pressing. *Holz als Roh und Werkstoff*, **59**(4), 272–275.

Geimer R.L., Mahoney R.J. and Meyer R.W. (1985) Influence of processing-induced damage on strength of flakes and flakeboards. Research Paper FPL-463, U.S. FPL, Madison, WI.

Kelly M. (1977) Critical review of the relationship between processing parameters and physical properties of particleboard. General Technical Report FPL-10, U.S. FPL, Madison, WI.

Lang E.M. and Wolcott M.P. (1996) A model for viscoelastic consolidation of wood-strand mats. Part II. *Wood and Fiber Science*, **28**(3), 369–379.

Lenth C.A. and Kamke F. (1996) Investigations of flakeboard mat consolidation. Part II. *Wood and Fiber Science*, **28**(3), 309–319.

Lu C., Steiner P.R. and Lam F. (1998) Simulation study of wood-flake composite mat structures. *Forest Products Journal*, **48**(5), 89–93.

Lu C. and Lam F. (1999) Study on the X-ray calibration and overlap measurements in robot-formed flakeboard mats. *Journal of Wood Science and Technology*, **33**(2), 85–95.

Lu C. and Lam F. (2001a) Random field representation of horizontal density distribution in partially oriented flakeboard mat. *Journal of Wood and Fiber Science*, **33**(7), 437–449.

Lu C. and Lam F. (2001b) Relationship between thickness swelling and mat structure in robot-formed flakeboard mats. *Holz als Roh und Werkstoff*, **59**(3), 201–210.

Oudjehane A., and Lam F. (1998) On the density profile within random and oriented wood based composite panels: Horizontal Distribution. Composites Part B. *Engineering Journal*, **29**(6), 687–694.

Oudjehane A., Lam F. and Avramidis S. (1998a) A continuum model of the interaction between manufacturing variables and consolidation of wood composite mats. *Wood Science and Technology*, **32**, 381–391.

Oudjehane A., Lam F. and Avramidis S. (1998b) Forming and pressing processes of random and oriented wood composite mats. Composites Part B *Engineering Journal*, **29**(3), 211–215.

Rowell R.M. and Young R.A. (1997) *Paper and Composites from Agro-based Resources*. CRC Lewis Publishers, Boca Raton, FL.

Serrano E. and Larsen H.J. (1999) Numerical investigations of the laminating effect in laminated beams. *Journal of Structural Engineering*, **125**(7), 740–745.

Suchsland O. (1967) Behavior of a particleboard mat during the press cycle. *Forest Products Journal*, **17**(2), 51–57.

Thoft-Christensen P. and Baker M.J. (1982) *Structural Reliability Theory and its Applications*. Springer-Verlag, New York, NY.

Triche M.H. and Hunt M.O. (1993) Modeling of parallel-aligned wood strand composites. *Forest Products Journal*, **43**(11/12), 33–44.

Wang Y.T. and Lam F. (1998a) Computational modeling of material failure for parallel-aligned strand based wood composites. *Computational Material Science*, **11**, 157–165.

Wang K. and Lam F. (1998b) Robot based research on a three-layer oriented flakeboard. *Journal of Wood and Fiber Science*, **30**(4), 339–347.

Wang K. and Lam F. (1999) Quadratic RSM models of processing parameters for a three-layer oriented flakeboard. *Journal of Wood and Fiber Science*, **31**(2), 173–186.

Wolcott M.P. Kamke F.A. and Dillard D.A. (1990) Fundamentals of flakeboard manufacture: viscoelastic behavior of the wood component. *Wood and Fiber Science*, **22**(4), 345–361.

Xu W. (2000) Influence of percent alignment and shelling ratio on modulus of elasticity of oriented strandboard: a model investigation. *Forest Products Journal*, **50**(10), 43–47.

Xu W. and Suchsland O. (1997) Linear expansion of wood composites: A model. *Wood and Fiber Science*, **29**(3), 272–281.

Youngquist J.A. English B.E. and Scharmer R.C. (1994) Literature review on use of non-wood plants fibers for building materials and panels. General Technical Report FPL-GTR-80. Madison, WI: U.S. Department of Agriculture, Forest Service, Forest Products Laboratory.

Youngquist J.A. (1995) Unlikely partners? The marriage of wood and nonwood materials. *Forest Products Journal*, **45**(10), 25–30.

7

Fracture Perpendicular to Grain – Structural Applications

Per Johan Gustafsson

7.1 INTRODUCTION

Wood has drastically different strength properties across the grain than it has parallel to the grain. Across the grain wood is weak, both with respect to a homogenous tensile state of stress and with respect to crack propagation. The tensile strength is only a few % of that parallel to grain, and there is a difference in resistance to crack propagation of a similar magnitude.

In structural design efforts are therefore made to avoid tension perpendicular to grain and shear. It is, however, not possible to completely avoid such stresses and several cases of timber structural damage have occurred due to fracture perpendicular to grain. It is probable that more cases of structural damage have occurred due to tension perpendicular to grain than due to tension parallel to grain, although the prime modes of structural loading are bending, tension and compression along the structural member.

A reason for this can be lack of knowledge. Comparatively little material testing and research has been carried out and available methods for engineering strength analysis are comparatively crude. A complication is that several non-trivial factors are significantly affecting both the perpendicular to grain strength and the perpendicular to grain stress, making load-carrying predictions difficult and uncertain. In addition, in practical design it is not possible to distinguish the fibre orientation from the orientation of the structural member. Commonly, it is tacitly assumed that these orientations coincide, which in some cases is far from true and adds to uncertainties and scatter.

7.2 SCOPE AND LIMITS

This chapter deals with material strength properties and calculation methods for analysis of fracture perpendicular to grain. Causes and some examples of fracture are also surveyed.

In codes of timber design, consideration to risk of perpendicular to grain tensile fracture is commonly made by empirically found figures for the load-bearing capacity of the structural element or joint, or by empirical rules of thumb, such as rules for minimum edge distance in joints. In contemporary timber engineering research, there is, however, a strive for development of more general and rational methods of strength analysis, i.e. methods following foundations of sciences such as strength of materials and continuum, structural and fracture mechanics. Although the purely empirically found rules are of great importance in timber design, here the discussion will focus on the more rational methods.

The discussion will, moreover, focus on strength parameters and strength models that are used together with linear elastic orthotropic or transversely isotropic models for the stress versus strain performance of the material outside the fracture process region. Characterisation of the fracture region can be made according to some conventional stress-based failure criterion, or according to classic or generalised linear elastic fracture mechanics, or according to nonlinear fracture zone modelling. Time-dependent effects such as creep fracture and the corresponding strength reduction at long duration of load, as well as the eigenstresses and possible fracture due to transient moisture conditions, are important issues in relation to fracture perpendicular to grain, but only briefly dealt with in this chapter.

7.3 EXAMPLES OF PERPENDICULAR TO GRAIN FRACTURE

Some cases and causes of perpendicular to grain fracture are illustrated in Figure 7.1.

- *The geometrical shaping of the structural member* can imply tension perpendicular to the grain when the structure is loaded. This tensile stress may be fairly *uniformly distributed* and comprise a large volume of material (Figure 7.1a), more *local* (Figure 7.1b) or *concentrated* to a small region where the stress theoretically is singular, approaching an infinite value even at low external loading (Figure 7.1c).
- *Mechanical and adhesive joints* commonly give rise to tension across grain in the local vicinity of the joint. One example (Figure 7.1d) is the risk of cleavage of the wood in the vicinity of a steel rod glued in parallel to grain. In other examples (Figures 7.1e, 7.1f and 7.1g), there is direct loading across grain from the connector to the wood. Also, loading along grain (Figure 7.1h) is known to produce tensile stress and risk for tensile fracture across the grain. In nail joints (Figure 7.1i), fracture may develop even at zero external load due to the *wedge action* of the nails if the edge distance or the distance between the nails is too small. In adhesive joints (Figure 7.1j), tension perpendicular to grain may develop as a result of the geometry of the joint.
- *Eigenstresses* of various kinds may result in fracture perpendicular to grain. One family of eigenstresses is due to *structural redundancy*, which in combination with imposed deformations, e.g. due to displacement of supports or moisture variation induced stresses or strains in the material, may give large structural stressing at zero external load. According to one philosophy of design, timber structures should therefore be designed to be statically determinate. *Growth stress* is a kind of eigenstresses in the living stem. Along grain these stresses may be of significant magnitude and for some hardwoods they may result in fracture as suggested

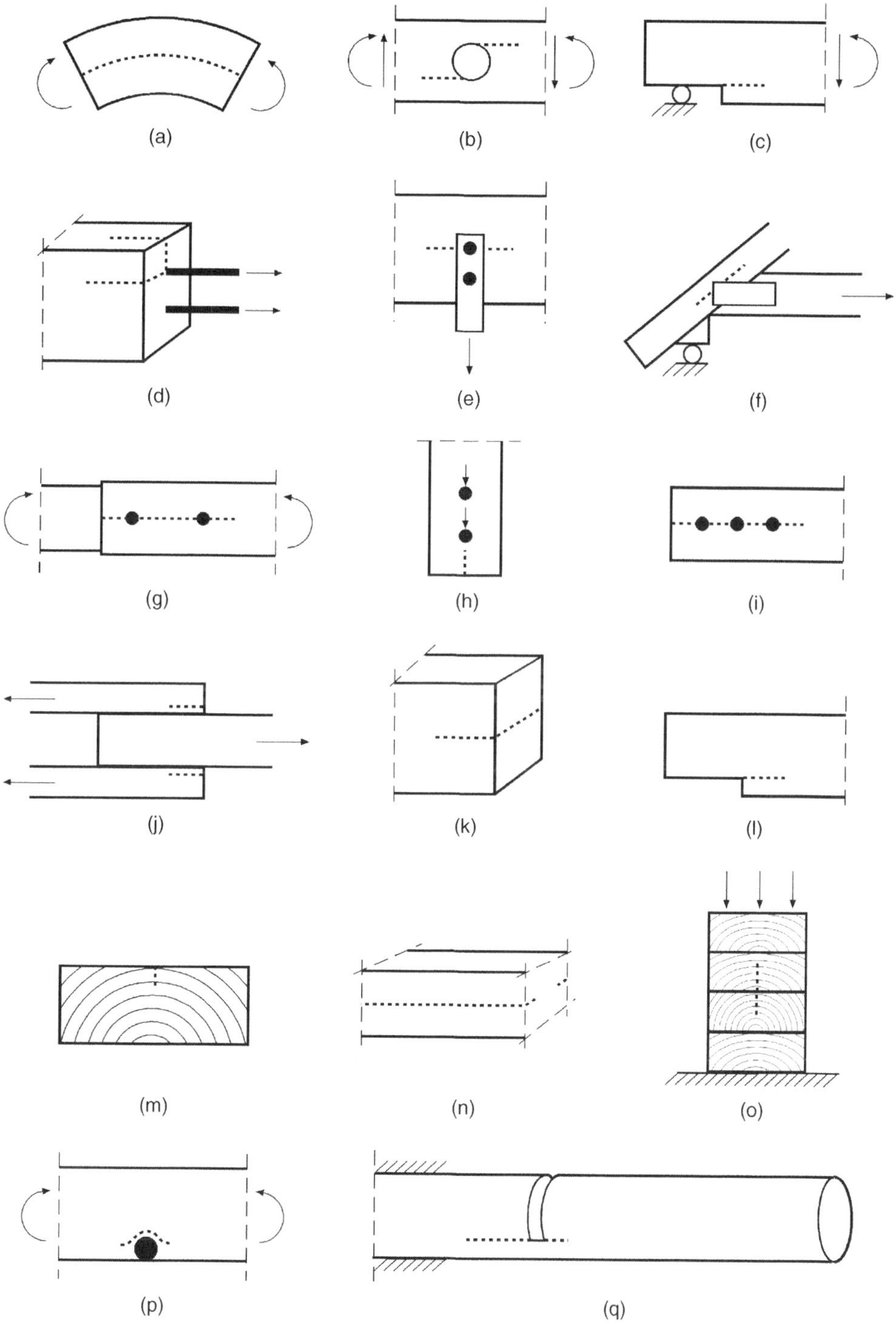

Figure 7.1 Various types and causes of fracture perpendicular to grain

in Figure 7.1k, when the log is sawn and the stresses released. *Change of the moisture* content to a lower level may give eigenstress and fracture as indicated in Figures 7.1m and 7.1n due to the different shrinkage in the tangential and the radial directions. The *non-uniform moisture content* in a structural member during change of the moisture level can give eigenstresses in terms of high perpendicular to grain compressive and tensile stresses (Figures 7.1k, 7.1l), and thereby very much effect the load carrying capacity of the member or produce cracking already at zero external load.

- *The non-homogenous character of wood and defects* such as knots and abnormal anatomy may give tension perpendicular to grain, stress concentrations and strength reduction. As an example, Figure 7.1o) shows how the growth ring induced variation of the orthotropic material orientation may result in tensile fracture when a piece of glulam is loaded in compression perpendicular to grain. Figure 7.1p illustrates how grain deviation around a knot in the lover edge of a beam may yield perpendicular to grain tensile fracture, and as a result, reduction of the bending strength of the beam.
- *As a special and troublesome example* of perpendicular to grain fracture, Figure 7.1q shows how a cleavage crack develops during cutting of the stem into pieces when the tree is being harvested by use of a forest machine.
- In traditional *carpentry joints*, not illustrated in Figure 7.1, there are commonly high local shear stresses that may yield perpendicular to grain fracture and govern the strength of the joint (Ehlbeck and Kromer, 1995).

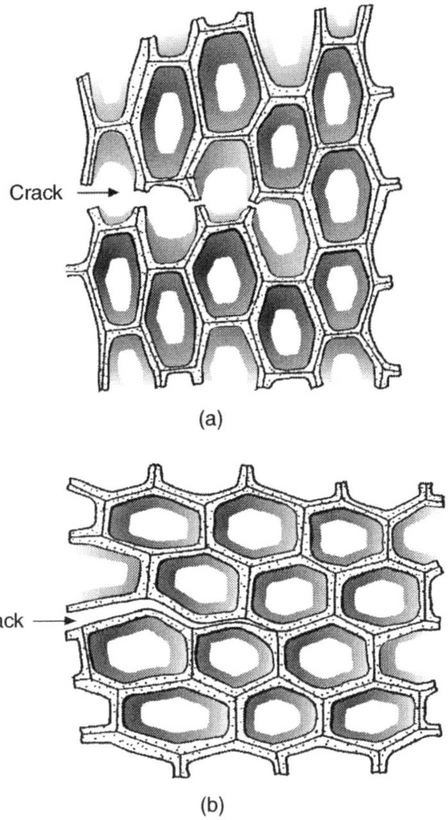

Figure 7.2 Perpendicular to grain fracture: (a) through cell walls, (b) along cell walls

7.4 MATERIAL PERFORMANCE AND PROPERTIES

7.4.1 Micro-Scale Fracture Performance in Clear Wood

At the fibre scale the perpendicular to grain tensile fracture may develop through the cell walls or along cell walls, as shown in Figure 7.2 (Gibson and Ashby, 1987). Crack propagation along the cells walls (Figure 7.2b) is, according to Uhmeier (1995), favoured by:

- Crack propagation in the radial direction (Figure 7.2b); see also Figure 7.5.
- Thick cell walls (see Figure 7.5).
- High temperature and chemical treatment.

Fracture along the cell walls is at room temperature mainly in the primary cell wall layer or in the S_1 cell wall layer (Bodig and Jane, 1982). These two cell wall layers are the two thin outer layers that cover the thick S_2 layer. When the temperature is increased to 150 °C, the fracture path is gradually shifted towards the middle lamella, which predominantly is made up of lignin.

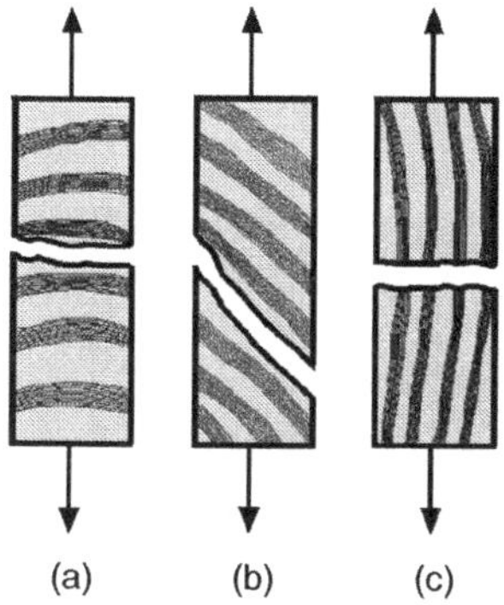

Figure 7.3 Fracture localisation at growth ring scale for a small clear specimen

At the growth ring scale (Figure 7.3), the most commonly observed failure modes are for small clear specimens affected by the orientation of the growth rings with respect to the direction of the tensile load: Tension failure in the early wood (Figure 7.3a), shear failure in the early wood (Figure 7.3b), and tangential tensile failure (Figure 7.3c), can be identified (Bodig and Jane, 1982).

In larger pieces of wood with varying growth ring angle or when the crack plane orientation is governed by an initial crack, the fracture plane may develop at an angle to the orientation of the growth rings. In Figure 7.4 is shown how the fracture plane may develop in a glulam lamellae when loaded perpendicular to grain. When the fracture plane versus growth ring is inclined there is a wavy pattern, as shown in Figure 7.4b. This typical perpendicular-to-late-wood fracture surface pattern may develop even at radial tensile load (see Figure 7.5) (Stefansson, 1995).

7.4.2 Timber versus Clear Wood

The natural structural variations and possible defects in the stem of a tree can make a big difference between the perpendicular to grain properties of a small clear wood specimen and the properties perpendicular to the lengthwise direction of boards and other timber structural elements. Natural variations include heterogeneities due to *varying properties from pith to bark* (varying density, stiffness, strength and shrinkage), the *circular shape of the growth rings* making the direction of the material axes vary, *spiral grain* and the *two kinds of knots* from dead and living branches, respectively. In timber the *eigenstresses* due to drying or growth can be large, while in small clear wood specimens they can be expected to be smaller because of the stress release during cutting. *Compression wood* is a common and natural structural abnormality, and *shakes* and *pitch pockets* are common defects. Shakes are separations of wood tissue and pitch pockets are separations filled with resin.

The heterogeneities give nonuniform stress distribution under load and when the timber is dried the heterogeneity give eigenstresses and perhaps cracking. Knots and grain orientation deviations have in general a positive influence on the perpendicular to timber element strength, since the perpendicular to grain direction in general is the weakest direction. Shakes and pitch pockets may be

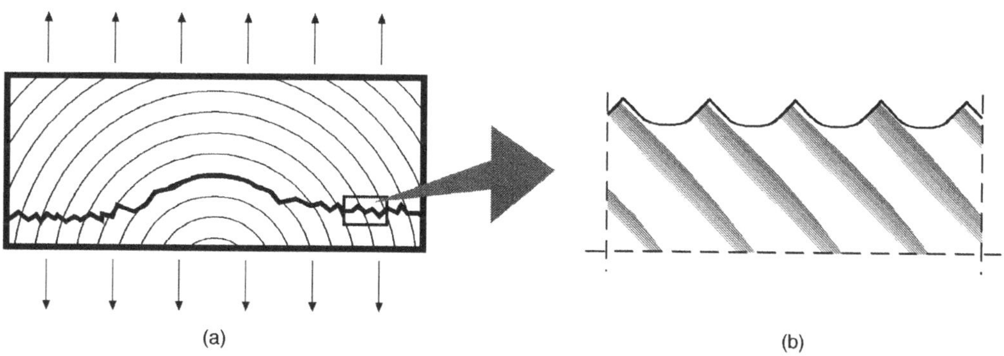

Figure 7.4 (a) Possible tensile fracture in a glulam board, (b) fracture surface pattern, grey areas indicating late wood

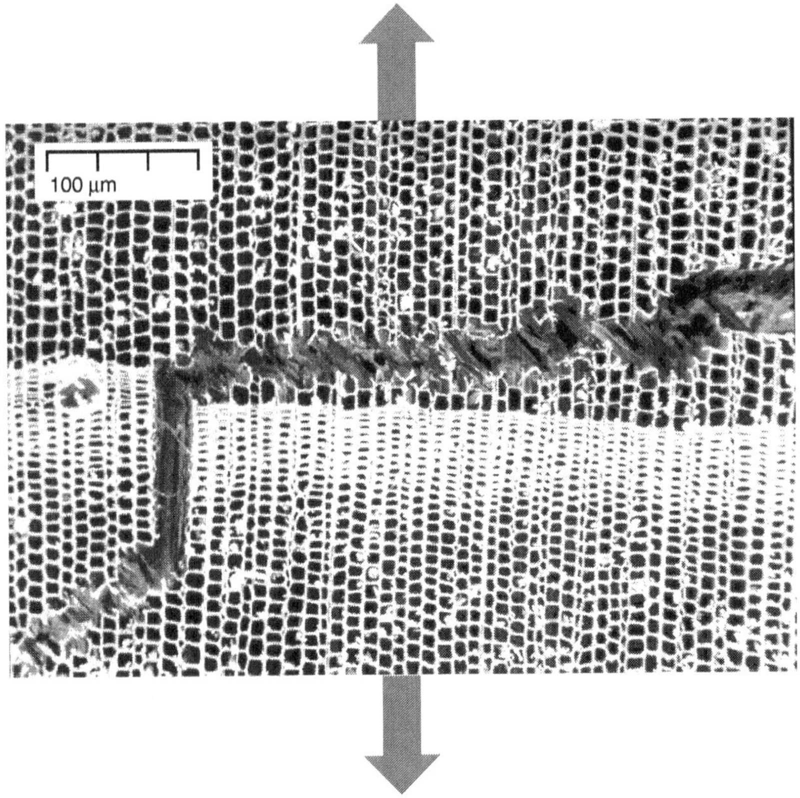

Figure 7.5 Tensile fracture in Spruce (*Picea abies*) loaded in the radial direction

hidden, giving a surprising and perhaps dangerous reduction of the perpendicular to timber strength.

The failure initiation in clear wood specimens when subjected to bending is the formation of wrinkles in the compression zone. Commercial timber most often has a bending failure initiation created by tension perpendicular to grain stresses (Madsen, 1992). Such stresses develop where the fibres have been disturbed in the vicinity of a knot and localised slope of grain has been created, for instance as shown in Figure 7.1p.

In timber engineering strength design, timber is commonly regarded as a homogeneous or statistically homogeneous material. This basis will also be adopted below.

7.4.3 Material Parameters

The strength parameters used for material characterisation are in general different for different models. Models for wood and timber at the engineering structure level of size assumes that clear wood, as well as timber are continuous and statistically homogeneous materials, and the strength parameters involved with analysis fracture perpendicular to grain relate to two kinds of properties: failure stress and fracture toughness. The former property is decisive in cases of simultaneous high stressing of a large volume of material, in which case start of development of a fracture zone immediately leads to global failure. The fracture toughness is decisive in the case of a local stress concentration, which, for instance, can be found in front of a sharp notch or crack. In general, both properties affect fracture and load-bearing capacity.

In transversely isotropic models, the properties in the tangential and radial directions are assumed to be equal. This assumption is used in timber engineering. When testing small clear wood specimens, the tangential and radial directions are, on the contrary, commonly distinguished. The stiffness properties of a linear elastic orthotropic material is fully defined by nine independent parameters, and the transversely isotropic material by six parameters, which in the case of 2D analysis is reduced to four or two in-plane parameters, depending upon the orientation of the plane.

A state of stress is in the general case defined by six stress components, and a general stress failure criterion for orthotropic materials may include several independent material parameters. In the present context, the most important stress is the perpendicular to grain normal stress and the corresponding tensile strength parameter, here denoted $f_{t,90}$. Other stress components, in particular shear, may in some situations be large, and then affect the perpendicular to grain failure.

The fracture toughness characterisation requires, to start with, knowledge or assumptions about the stiffness properties of wood. Knowing the stiffness, the fracture toughness for tensile fracture of any given orientation is within linear elastic fracture mechanics defined either by a critical value of the stress intensity factor, K_c, or by the corresponding critical value of the energy release rate, G_c, or by the corresponding energy dissipation during the development of the fracture surface, G_f. Within linear elastic fracture mechanics $G_c = G_f$, and with the use of the elasticity parameters, K_c can also be related to G_c.

For orthotropic materials there are, for all possible orientations of a crack plane, three possible states of loading of the crack: tensile loading, often called mode 1 or the opening mode, in-plane shear and transverse shear. Here the discussion focuses on crack planes with the normal vector perpendicular to grain and to the tensile loading mode. With these limitations, only one parameter for G_c or G_f is needed in the case of transversely isotropic material properties. If consideration of different properties for the tangential, radial and intermediate fracture plane orientations must be made, more parameters are needed. While G_c and G_f have a single value for a given mode of loading and orientation of the fracture plane, K_c in general has different, but not independent, values for different directions of the crack propagation.

The parameter G_f can be experimentally determined from the energy dissipation during a stable bending or tensile fracture test of a specimen with or without an initial crack. The parameters G_c and K_c can be determined from the load-bearing capacity of some specimen with a deep crack or sharp notch.

The above-mentioned material properties (failure stress and fracture toughness) relate to the resistance to fracture. Other material parameters may affect the magnitude and distribution of the stress at given external load. This has reference in particular to the shrinkage and swelling properties of the material, (Jönsson and Svensson, 2000) giving stress and perhaps cracking even at zero external load.

7.4.4 Tensile Strength

Table 7.1 shows a compilation of values from literature of *tensile, compressive and shear strengths of various species*. The shear strength at rolling shear is of the same order as the perpendicular to grain tensile strength. The compilation is from Götz *et al.* (1978) and Larsen and Riberholt (1983). The most commonly found literature values are given together with the range of values. The relative difference between the species and modes of loading are, due to the influence of specimen volume on strength, of greater interest than the absolute strength values.

Table 7.1 Tensile, compressive and shear strengths of various species

Species	$f_{t,90}$ MPa	$f_{t,0}$ MPa	f_v MPa	$f_{c,90}$ MPa	$f_{c,0}$ MPa
Spruce	3 (1–4)	90 (20–250)	7 (4–12)	6 (2–10)	30 (45–80)
Scots pine	4 (2–4)	100 (35–200)	10 (6–15)	4 (8–14)	30 (50–90)
Larch	2	100	9 (4–10)	8	35 (55–80)
Beech	7	140 (60–180)	11 (6–19)	9	40 (60–100)
Oak	4 (2–10)	90 (50–180)	11 (6–13)	8 (11–19)	40 (55–90)

Table 7.2 Strength $f_{t,90}$, MPa, of Scots pine at various load directions and MC

	Radial	45°	Tangential	Mean
8% MC	5.3	5.3	4.1	4.9
10% MC	5.0	5.3	4.2	4.8
14% MC	4.5	4.5	4.0	4.3
Mean	4.9	5.0	4.1	4.7

Test results showing the influence on tensile strength of the *orientation of the growth rings* are presented in Boström (1992) (Table 7.2). Tests were made at three moisture content levels corresponding to 33% RH, 59% RH and 75% RH. The material tested was Scots pine (*Pinus silvestris*), with a mean dry density of 463 kg/m^3. The test results were obtained for very small prismatic specimens with b*h*l = 20*12*3 mm^3 glued in between two steel plates and loaded in the l-direction so that the cross-section area was b*h = 20*12 mm^2. The total number of tests was 83, and the time to peak stress 1–3 minutes. The mean value of the coefficient of variation for the test series was 20.1%. The strength in the radial direction was found to be about 20% higher than in the tangential direction. This difference is much less than the difference in elastic stiffness, where the radial direction roughly is about twice as stiff as the tangential direction, which in turn roughly is about twice as stiff as the intermediate 45°-direction. Testing at 8% and 10% MC gave about 10% higher strength than at 14% MC. The higher strength of Scots pine indicated in Table 7.2 than in Table 7.1 can be explained by the very small size of the specimens represented in Table 7.2.

For specimens of almost exactly the same small size as tested by Boström (1992), test results on the tangential tensile strength of Spruce (*Picea abies*) at various *strain rates* are presented in Holmberg (1998). The stroke rates was varied from 0.001 mm/s to 10 mm/s, corresponding strain rates from 0.03–330%/s, and time to peak stress from about 50 s to about 0.005 s. The results suggest about a 12% increase of the tensile strength for each 10-fold increase of the strain rate. Further test results on the rate effect are presented in Olejniczak and Gustafsson (1994).

The *duration of load* effect on tensile strength has been tested for large specimens made up of several boards glued together, forming a specimen similar to a short part of a glulam beam (Madsen, 1992). Tests were made for both defect-free and commercial material. The time to failures investigated being from 1 minute to 70 days, the failure stress was reduced by about 15% for the defect-free material, and about 7% for commercial material for each 10-fold increase of time to failure. Within each group there was, however, no difference in the duration of load effect between the weak and strong specimens. More information about the duration of load effect can be found in other chapters of this book, and in Morlier and Ranta-Maunus (1998), Aicher, Dill-Langer and Ranta-Maunus (1998), Gustafsson, Hoffmeyer and Valentin (1998) and Svensson, Thelandersson and Larsen (1999).

A very comprehensive experimental work on the tangential tensile strength and stiffness of Scots pine (*Pinus silvestris*), Spruce (*Picea excelsa*) and Birch (*Betula verrucosa*) at various *densities*, *MCs* and *temperatures* is presented in Siimes (1967). Several thousand tests were made with specimens of size b*h*l = 10*40*50 mm^3, where l = 50 mm is the effective length of the specimen. Tests results in terms of average values from some of the test series are compiled in Figure 7.6. It is evident that the density affects $f_{t,90}$, but not as much as it affects the elastic modulus, $E_{t,90}$. For the same density, birch has a much higher strength than the softwoods, but the stiffness is somewhat less. In relation to the effect of MC, green wood was also tested. The test results for green wood are in the figure represented as wood with a MC of 28%, intended to reflect the fibre moisture saturation point beyond which mechanical properties commonly are not much affected by further increased moisture content. For birch the strength is about 60% less at 28% MC than at 4% MC, while for the softwoods the corresponding reduction is about 40%. Increase of the temperature above 40 °C has a significant influence on the strength. The stiffness is significantly reduced, by about 20%, already by an increase in temperature from 20–40 °C. The effect of increased temperature

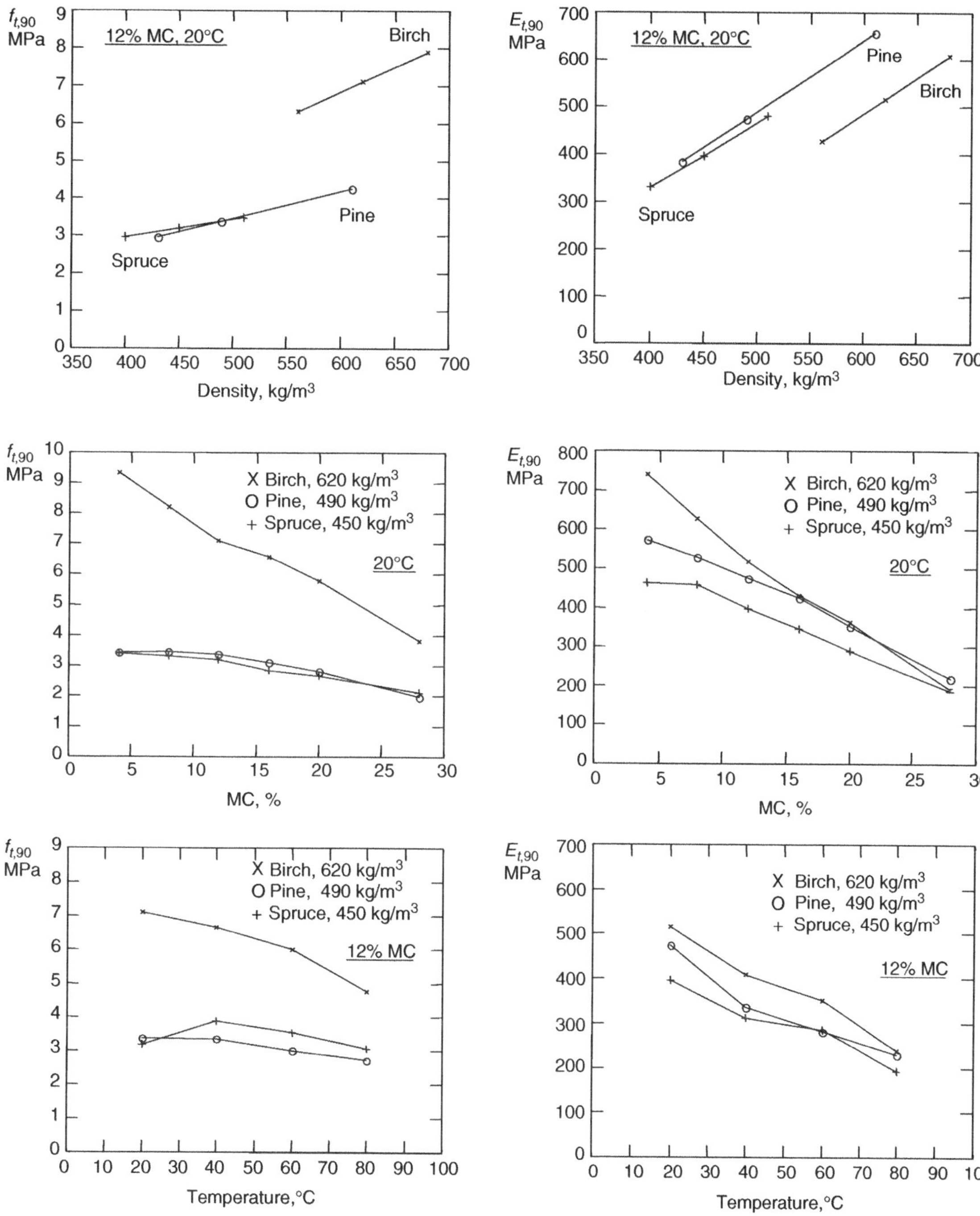

Figure 7.6 Tensile strength and elastic tensile stiffness in the tangential direction of birch, pine and spruce of various densities and at various moisture content and temperature. Stated density is valid at 12% MC. Green wood is indicated by 28% MC. (From test results of Siimes, 1967)

may for real structures be reduced by a simultaneous decrease of the MC.

The perpendicular to grain tensile strength is strongly *size*-dependent. This is shown in Table 7.3 and Figure 7.7, where test results obtained for spruce are compiled. The results for glulam were obtained by tensile loading of specimens with the kind of geometry shown in Figure 7.1o and by four-point bending of curved beams (Figure 7.1a). The length measure, l, given for the curved beams is the length between the two loads, i.e. the length of the part of the beam exposed to constant bending moment. The density is given in kg/m^3 and the MC was about 12%. The tests were made as short-term ramp load tests after conditioning in a constant climate. Reference G-88 is Gustafsson (1988), LR-83 is Larsen and Riberholt (1983) and ADR-98 is Aicher, Dill-Langer and Ranta-Maunus (1998). The results for the small solid specimens were obtained for growth ring orientations fairly close to tension in the tangential direction. The higher tensile strengths recorded for curved beams than for prismatic specimens of the same volume is because only a small part of the volume in the curved beams is exposed to high stress.

Table 7.3 Perpendicular to grain tensile strength of spruce for specimens of various volume

Specimen: geom., mat., dens.	Specimen: b*h*l, mm^3	Volume, dm^3	Strength, MPa Mean (5%-fract.)	Reference
Prismatic, solid, 467	15*16*20	0.005	4.0 (2.7)	G-88
Prismatic, glulam, –	–	0.027	2.4	LR-83
Prismatic, glulam, –	–	0.28	1.8	LR-83
Prismatic, glulam, –	–	2.5	1.0	LR-83
Prismatic, glulam, –	–	26	0.63	LR-83
Prismatic, glulam, 530	90*275*400	10	0.89 (0.74)	ADR-98
Prismatic, glulam, 493	140*405*528	30	0.67 (0.55)	ADR-98
Curv. beam, glul., 470	90*400*1000	36	1.21 (0.95)	ADR-98
Curv. beam, glul., 496	90*600*2000	108	0.85 (0.72)	ADR-98
Curv. beam, glul., 503	90*600*4000	216	0.71 (0.59)	ADR-98
Curv. beam, glul., 493	140*600*4000	336	0.61 (0.46)	ADR-98

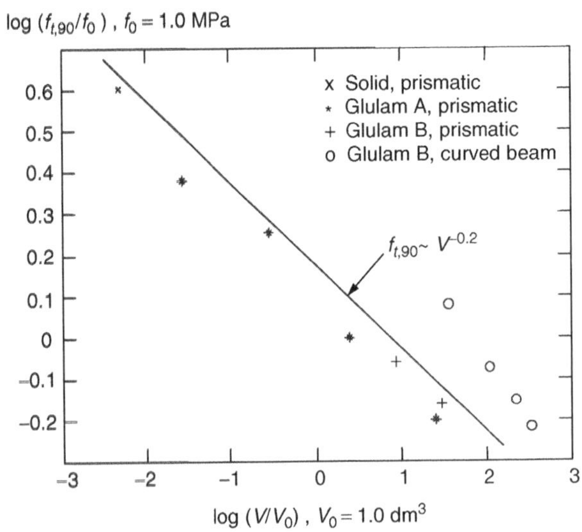

Figure 7.7 Tensile strength versus volume of specimen

The logarithmic linear relation shown in Figure 7.7 is

$$(f_{t,90}/f_0) = 1.5(V/V_0)^{-0.2} \tag{7.1}$$

where $f_0 = 1.0$ MPa and $V_0 = 1.0\ \mathrm{dm}^3$. Since V is proportional to the length scale raised to 3, Equation (7.1) suggests that $f_{t,90}$ is proportional to a length measure of the structural element raised to -0.6.

The influence of volume is commonly explained by the *weakest link theory* of Weibull. In a large volume of a material there are more weak points, and the weakest point, in Weibull theory assumed to be decisive, is probably weaker the more weak points there are. Another explanation to the volume effect is the *heterogeneous stiffness*. The approximately 2–4 times higher stiffness in the radial direction in the middle of a board than in the sloping growth ring direction at the edge of a board implies a region in middle where the stress may be much higher than the nominal mean stress adopted in the evaluation of the tensile strength test results. This effect is small for narrow boards and strong for wide boards.

A third explanation to the volume effect is that defects such as drying cracks tend to be larger in large specimens. According to linear *fracture mechanics*, if the depth of a crack is proportional to the size of the specimen, then a doubling of the length scale i.e. an 8-folded increase of the volume of the specimen produces a tensile strength reduction by the factor $(1/2)^{1/2}$, i.e. a 30% reduction.

A typical 5%-fractile characteristic value, $f_{t,90,k}$, for wood and glulam given in *codes of timber design* practice is 0.50 MPa (from the Swedish code BKR-94). The characteristic value is in codes reduced to a design strength value by a factor γ_n for structural safety level, a factor γ_m for material uncertainties, and a factor κ_r that depends upon climatic conditions and the duration of the loads. In the special case of tension perpendicular to grain, there may be additional reduction factors. As an example, for curved glulam beams there may be an additional reduction factor κ_{rad} that reflects the influence of the built-in eigenstresses due to the curving of the beam during the manufacturing, and a reduction factor κ_{vol} that reflects the influence of volume on strength. Typically, from a Swedish glulam design manual,

$$\kappa_{\mathrm{vol}} = k_{\mathrm{shape}}(V/V_{\mathrm{ref}})^{0.2} \tag{7.2}$$

where V is the volume of the structural element, $V_{\mathrm{ref}} = 10\ \mathrm{dm}^3$ and $k_{\mathrm{shape}} = 1.4$ for curved beams of constant height.

It is interesting that code values for the perpendicular to grain tensile strength, unlike the bending strength and other mechanical properties, are often assigned a constant or only slightly varying value with respect to the grading or strength class of the timber or glulam. In the Swedish BKR-94, $f_{t,90,k}$ is constant (0.5 MPa), although the corresponding values for the bending strength varies from 12–38 MPa, and the tensile strength in the parallel to grain direction varies from 8–27 MPa for the various strength classes defined in the code. In the European EN 338 $f_{t,90,k}$ is from 0.3–0.4 MPa for conifers and poplar, with characteristic density values from 290–420 kg/m^3, and from 0.6–0.9 MPa for deciduous species with characteristic density values from 530–900 kg/m^3. The corresponding relative variation of the characteristic bending strength is for the conifers and poplar much greater, from 14–40 MPa, and for the deciduous species from 30–70 MPa.

A high strength class corresponds to high density, few and small knots, and/or high bending stiffness and high bending strength. There is, apart from known positive influence of density and knots, no evidence available for correlation between high strength class and high perpendicular to timber tensile strength. In tests of notched beams (Figure 7.1c), it was on the contrary found that the load-bearing capacity was significantly higher for the beams having a knot in the vicinity of the notch, both with respect the mean value and the 5%-fractile value (Larsen and Riberholt, 1972). Positive influence of density on tangential clear wood strength is indicated in Figure 7.6. Defects like pitch pockets that may reduce the perpendicular to grain tensile strength may not be visible from the outside, nor be revealed by influence on bending stiffness or bending strength.

7.4.5 Fracture Energy from Bending Tests

The fracture toughness of a material is governed by its stiffness and fracture energy. The fracture energy is the energy dissipated in the fracture process region during the development of one unit area of a fractured cross-section or crack. The fracture energy can be determined directly from recorded work of fracture, giving the parameter that is denoted here as G_f. Within the framework of linear elastic fracture mechanics, fracture property characterisation can alternatively be made from the failure load of some test specimen with a pre-fabricated sharp crack. From the failure load, it is possible to evaluate the critical energy release rate, G_c, and as an alternative, the critical value of the crack tip stress intensity factor, K_c. Linear elastic fracture mechanics is valid for large specimens where the depth of the prefabricated crack is large as compared to the size of the fracture process region.

There is a Nordic standard for the determination of G_f of wood by testing notched beam specimens in three point bending, as shown in Figure 7.8a. The specimen is made up of three pieces of wood that are glued together. At testing according to the standard, the length a is $a = d = 60$ mm, the depth of the notch is $0.6d$ and the width of the specimen can be chosen according to the purpose of the test, e.g. 45 mm. The notch may be sawn, and to reduce scatter in the test results, should be finished with a sharp razorblade cut. Unless the testing equipment is very weak, it is possible to record the entire load, P, versus stroke, δ, curve as shown in Figure 7.8b, without the occurrence of any instability. G_f is obtained as

$$G_f = W/A, \quad W = \mathrm{mg}\delta_0 + \int_0^{\delta_0} P\,d\delta \qquad (7.3)$$

where A is the cross-section area of the ligament, mg the weight of the specimen and δ_0 the deflection when the specimen finally falls down by its own weight. The integral is the work carried out by the machine and obtained as the area under the curve. $\mathrm{mg}\delta_0$ is the work of the dead weight, half of it made before $\delta = \delta_0$. A basic assumption in

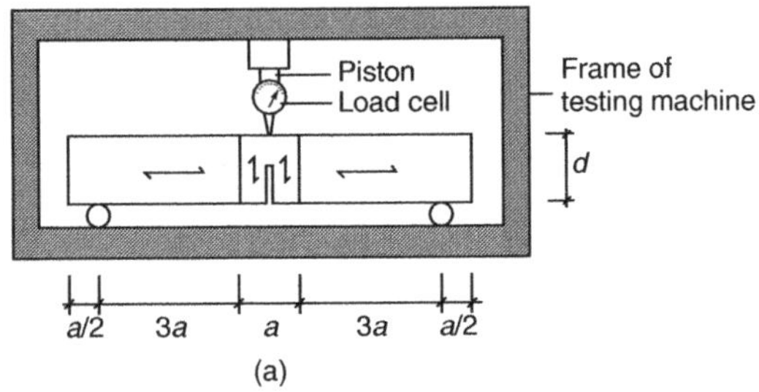

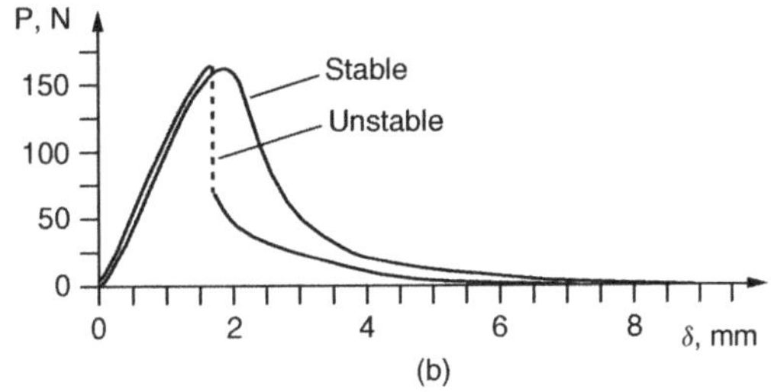

Figure 7.8 Specimen for fracture energy test (a) and examples of test recordings (b)

this evaluation is that energy is dissipated only in the fracture zone. This is a reasonable assumption for small specimens. For large specimens, having a large volume of material to ligament area ratio, G_f can be overestimated due to plastic straining of material outside the fracture section.

During development of the test method, various woods were tested for their fracture energy at a dozen laboratories throughout the world. Wood density was found to significantly effect G_f (Figure 7.9a showing the results from test series with Nordic redwood (Scots pine) and Figure 7.9b showing the results for all species tested (Larsen and Gustafsson, 1990, 1991)). Other parameters varied were crack propagation direction, fracture surface orientation and MC.

For a given orientation of the fracture surface, there should be no influence of the crack propagation direction, and neither was any significant influence on G_f found in the tests. The load capacity, on the other hand, is strongly affected as a result of different stiffnesses in different directions. Taking all species tested as one group, no decisive fracture surface orientation dependence could be concluded. According to a small set of tests

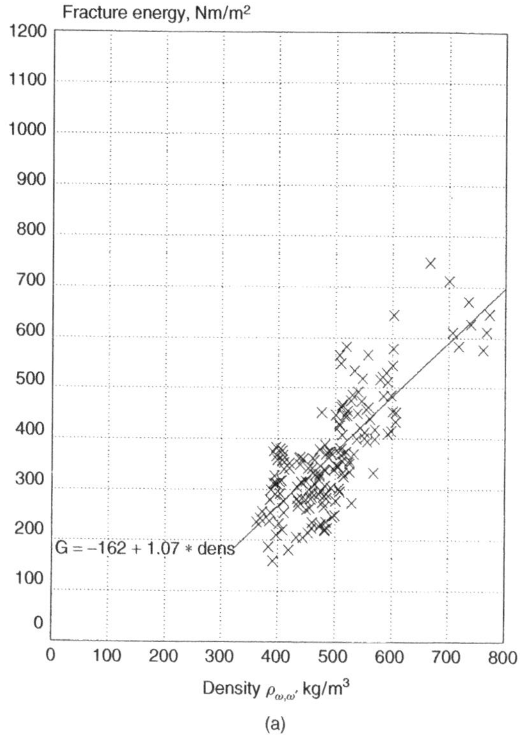

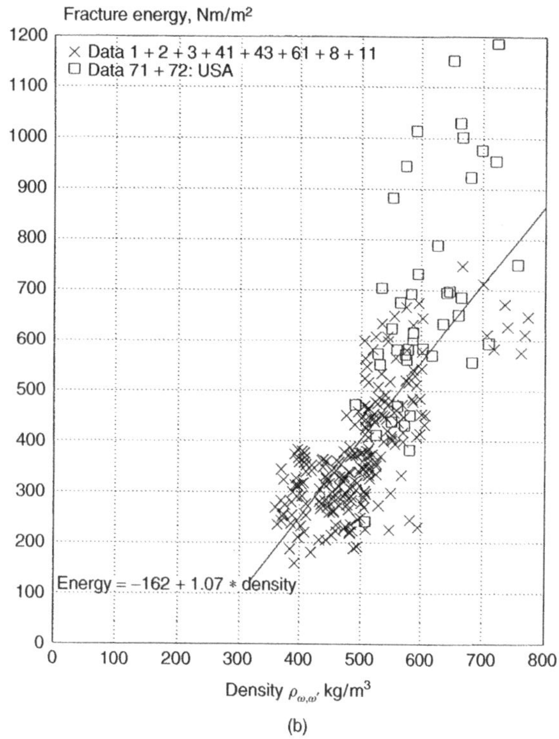

Figure 7.9 G_f vs. density at about 12% MC for (a) Nordic redwood, and (b) all test results

specifically devoted to an orientation investigation, the fracture energy is higher for a surface with its normal vector in the tangential direction than for a surface with radial normal direction. The fracture toughness (i.e. the crack growth or cleavage resistance), which is also affected by the orthotropic stiffness of the material, is for cleavage along grain in general greater for tension in the radial direction than for tension in the tangential direction (Valentin *et al.*, 1991). The force needed for cleavage creating a fracture surface in the tangential plane is, for softwoods, 10–30% higher than for the radial plane; see the material property survey in Dinwoodie (1981).

Tests of G_f at various MC between 10% and 20% did not indicate any evident influence of the MC. For wet wood, the G_f is known to be less than for dry wood: Helmersson (1978) found for Scots pine, $G_f = 330$ J/m^2 at both 9% and 14% MC, but was 180 J/m^2 at 90% MC for wood that originally was dry and then kept in water for a week.

7.4.6 Stress versus Deformation Performance of Wood During Fracture

By testing a small specimen, it is possible to record the gradual damage and fracture of wood in terms of stress versus deformation across a fracture zone, including the descending branch of the curve. Figure 7.10 shows such specimens used for testing the fracture performance in tension and shear (Gustafsson, 1992). The specimens were glued to steel parts and loaded in a testing machine with a stiff load cell. If using too large a specimen or a flexible machine, the amount of elastic energy becomes large, giving instability with a rapid and brittle failure just after peak load. The test recordings shown in Figure 7.10 were obtained for Scots pine.

After peak load there is unloading of the material outside the fracture zone, and the deformation of this material decreases as the stress decreases,

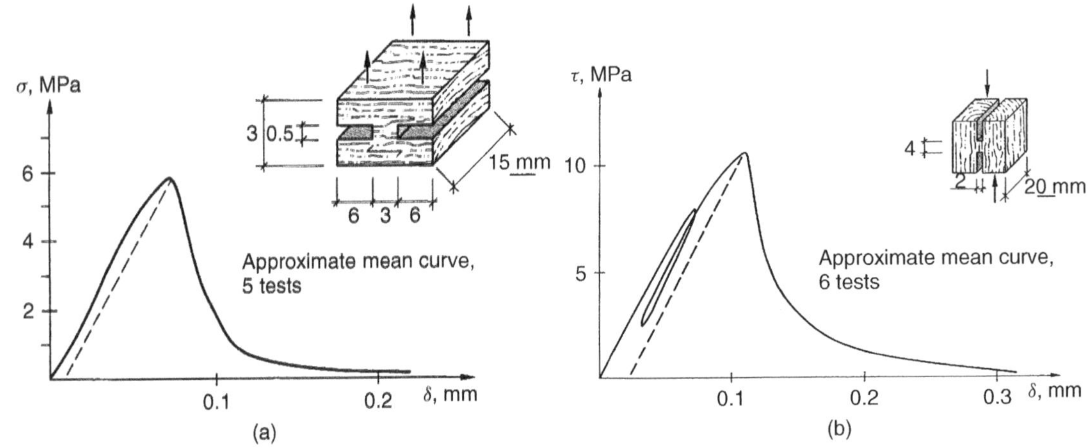

Figure 7.10 Specimen and test recordings for (a) tension and (b) shear

as indicated by the broken lines in Figure 7.10. The total deformation, or elongation, is made up of deformation of the material outside the fracture region and the deformation within the fracture region. This means that the fracture deformation at any given stress can be obtained as the difference between the total deformation and the deformation indicated by the broken line. By this it is possible to characterise the fracture properties of the material by a curve that shows stress versus fracture zone deformation as the material is being damaged and the stress decreasing from peak to zero. This curve can be used as a basis for both linear and nonlinear fracture mechanics analysis. The described material test and evaluation method is fundamental in the 'fictitious crack model' of Hillerborg *et al.* (1976).

Figure 7.11 shows stress versus deformation curves for perpendicular to grain tensile fracture in Scots pine (Boström, 1992). RL-orientation denotes tension in the radial direction. From the curves the fracture energy, G_f can be obtained from the area under the curves, as indicated in the figure. Using a biaxial testing machine, tests have been made for pure and simultaneous tension and shear. Figure 7.12, taken from Wernersson (1994), shows the tensile and shear stresses versus time, recorded during testing of a Spruce specimen for an equal rate of deformation in the normal and shear directions. The normal deformation and normal to fracture surface directions coincide with the tangential direction of the wood. It can be noted that peak tensile and shear stresses do not occur at the same time. If the test had been made at

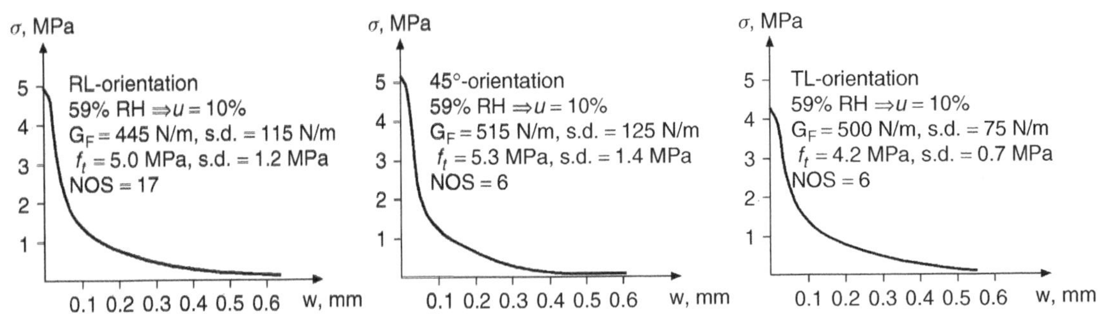

Figure 7.11 Fracture performance of wood in tension perpendicular to grain for various orientations

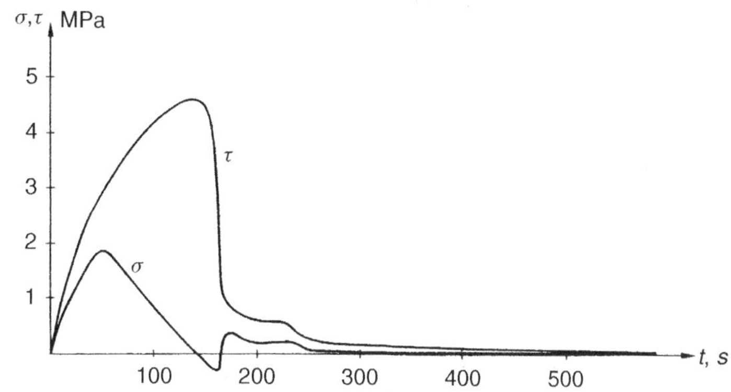

Figure 7.12 Tensile and shear stress vs. time recorded at a mixed mode test

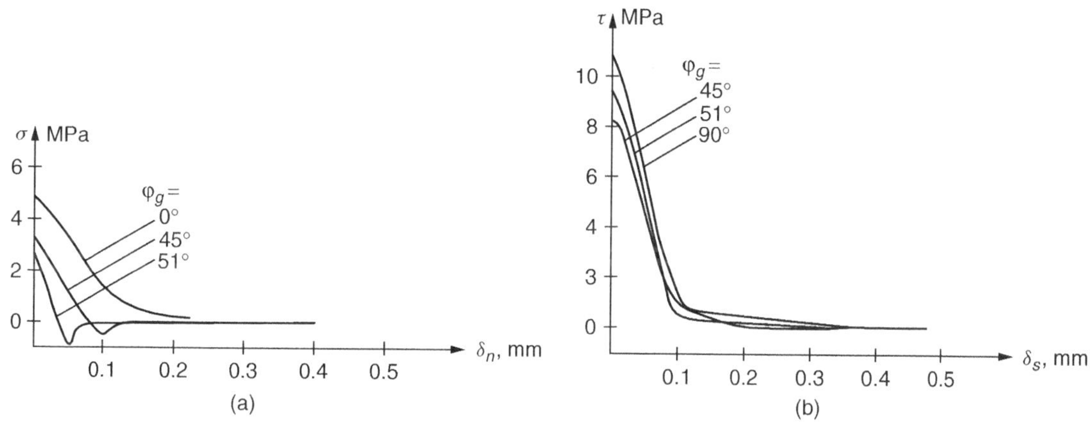

Figure 7.13 Mean curves for (a) tensile stress vs. normal deformation, and (b) shear stress vs. shear deformation obtained for four different mixed mode displacement ratios

force control, as opposite to stroke or deformation control, brittle failure of the specimen would have occurred at the instant of peak tensile stress.

Figure 7.13 shows the results of similar mixed mode tests in terms of tensile stress versus normal deformation, and shear stress versus shear deformation for four values of the mixed mode ratio, indicated by the inclination of the stroke displacement vector. The mixed mode ratio within the fracture zone may be somewhat different, and vary during the course of fracture. The peak tensile and shear stresses, and the normal, shear and total fracture energies for the various mixed modes, are shown in Table 7.4. Even for pure

Table 7.4 Mixed mode fracture properties (Wernersson, 1994)

Mixed mode angle	σ MPa	τ MPa	$G_{f,\sigma}$ J/m^2	$G_{f,\tau}$ J/m^2	$G_{f,\mathrm{tot}}$ J/m^2
0°	4.8	0.9	475	0	475
45°	3.3	8.1	146	481	627
51°	2.7	9.4	75	552	627
90°	0.2	10.9	0	1120	1120

normal stroke displacement, some non-zero shear stress was recorded at some instance during the course of fracture. The standard deviation was about 0.5 MPa in recorded peak stresses, and about 70 J/m^2 in recorded energies. The recorded peak tensile and shear stresses agree reasonably well with the simple criterion:

$$(\sigma/f_{t,90})^2 + (\tau/f_v)^2 = 1.0 \qquad (7.4)$$

Additional test results on the stress versus deformation properties of a fracture zone can be found in Boström (1992), Wernersson (1994) (adhesive joints), Olejniczak and Gustafsson (1994) (specimen geometry, growth ring orientation, rate), Holmberg (1998) (rate), Serrano (2000) (adhesive joints) and Stefansson (2001) (mixed mode).

7.5 METHODS FOR STRENGTH ANALYSIS

7.5.1 An Overview of Models

The conventional model in timber engineering for rational strength analysis is linear elastic stress analysis together with a stress-based failure criterion, assuming timber to be a continuous, homogeneous, transversely isotropic and brittle material with deterministic properties. Looking at the examples of perpendicular to grain failure shown in Figure 7.1, it is obvious that rational analysis by the conventional model cannot be used or expected to give poor results in most of the examples. The only example where the conventional analysis – although modified to include the volume effect – is known to be used for perpendicular to grain strength design is the curved glulam beam (Figure 7.1a). According to Madsen (1992), there is no tension perpendicular to grain strength property stated for timber, only for glulam, in the American codes. This reflects the lack of good rational methods in design practice with respect to risk for perpendicular to grain fracture.

Steps towards more accurate and rational strength design modelling in relation to fracture perpendicular to grain may include:

- Consideration of the volume effect and the stochastic nature of the material properties. Such consideration is already, by use of Weibull theory, partly made in some contemporary codes of practice, at least tacitly.
- Consideration of the non-zero fracture energy of wood, making it possible to analyse the strength of structural elements and details of stress concentrations at joints and due to cracks and notches, etc.
- Consideration of moisture-induced stresses and other eigenstresses as loads rather than by reduction of design strength values of the material.

In the ideal case, the assumptions of statistically homogenous and transversely isotropic properties should also be abandoned. This may, however, be almost impossible to realise generally in practice.

In Table 7.5 models are organised with respect to assumptions on the material properties: deterministic (homogeneous) versus stochastic (heterogeneous) strength, and zero versus non-zero fracture toughness. The models mentioned will be discussed below, throughout with the assumption of a linear elastic and time-independent stress-strain performance of the material. The assumption of linear elasticity can be reasonable for the pre-peak stress performance of wood when subjected to short-term tensile stress.

Table 7.5 Models for timber engineering strength analysis

	Deterministic (homogeneous)	Stochastic (heterogeneous)
Brittle ($G_f = 0$)	• Conventional stress analysis	• Weibull weakest link theory
With fracture ductility ($G_f \neq 0$)	• Linear fracture mechanics • Generalised lin. fracture mechanics • Nonlinear fracture mechanics	• Probabilistic fracture mechanics

7.5.2 Example of Contemporary Design According to a Code of Practice

An example of strength design according to a contemporary code, Eurocode 5, is given as an introduction before discussing the calculation methods indicated in Table 7.5. The example is taken from Ehlbeck and Görlacher (1995), and relates to the kind of joints shown in Figure 7.14, 7.1e and 7.1f. For these joints, the design condition with respect to risk for perpendicular grain to tensile fracture is

$$V_d \leq \frac{2f_{v,d}b_et}{3} \tag{7.5}$$

where $f_{v,d}$ is the design shear strength of the wood, b_e and t are as indicated in Figure 7.14, and V_d is the design shear force representing the maximum of $V_{1,d}$ and $V_{2,d}$. It is required that $b_e > 0.5$ h. This design procedure substitutes the design for the concentrated tensile stress perpendicular to grain at the base of the joint with a more or less fictitious shear strength design of the cross-section tb_e with respect to the shear stress at the distance $b_e/2$ from the edge. Important factors such as the perpendicular to grain tensile strength and fracture toughness of the wood are not taken into account. The basis for the design procedure can be difficult to understand, but it has the advantage of being simple. If $b_e < 0.5$ h, then some more detailed calculation is required, but the method of calculation is not provided by the code.

7.5.3 Conventional Stress Analysis Failure Criteria

The failure criterion for a general state of stress, $\boldsymbol{\sigma}$, may be written as

$$F(\boldsymbol{\sigma}, \Psi) = 0 \tag{7.6}$$

where Ψ is a set of material parameters. Assuming an ideal brittle performance, as is commonly made in conventional timber engineering strength analysis, the structural element or detail under consideration is assumed to break as soon as the state of stress in some point fulfils the criterion. Making no distinction between the radial and tangential directions and considering a state of plane stress, a commonly adopted criterion appears to be one of Norris criteria (Norris, 1962):

$$(\sigma_{90}/f_{90})^2 + (\sigma_0/f_0)^2 + (\tau/f_v)^2 = 1.0 \tag{7.7}$$

in which five material parameters are involved: the tensile or compressive strengths f_0 and f_{90} and the shear strength f_v. Often, the first term is small and therefore ignored, for tensile stress, giving Equation (7.7).

A state of stress of particular interest is uniaxial tension, σ_α, at an angle α to the grain, giving the stress components

$$\begin{cases} \sigma_{90} = \sigma_\alpha(1 - \cos 2\alpha)/2 \\ \sigma_0 = \sigma_\alpha(1 + \cos 2\alpha)/2 \\ \tau = -(\sigma_\alpha/2)\sin 2\alpha \end{cases} \tag{7.8}$$

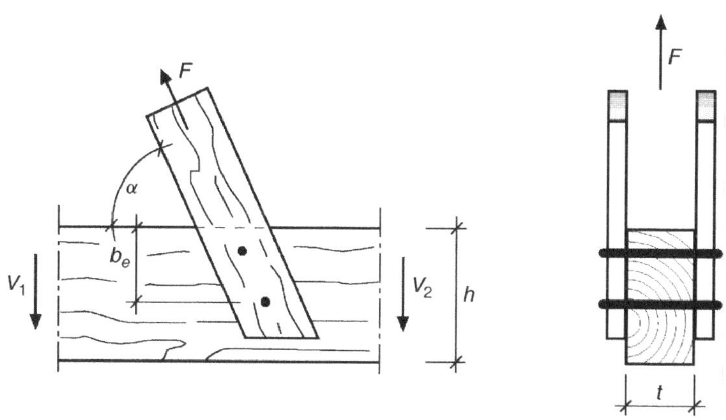

Figure 7.14 A joint

A failure criterion for this uniaxial state of stress is a formula commonly called Hankinson's equation:

$$\sigma_\alpha = \frac{f_{t,90} f_{t,0}}{f_{t,90} \cos^n \alpha + f_{t,0} \sin^n \alpha} \qquad (7.9)$$

where the exponent n according to Kollmann and Cote (1968) gives satisfactory agreement with test results for values from 1.5–2.0. Even at combined states of stress, it can in some cases be sufficient to consider the stress components separately, for plane stress giving the failure criteria

$$\begin{cases} \sigma_{90} = f_{90} \\ \sigma_0 = f_0 \\ \tau = f_v \end{cases} \qquad (7.10)$$

so that the different failure modes of tensile fracture perpendicular to grain, shear fracture and tensile fracture in the grain direction are considered separately. The material parameters f_{90} and f_0 may have different values for tension and compression.

Criteria (7.7), (7.9) with $n = 2$, and (7.10) are for the uniaxial tensile state of stress of Equation (7.8) compared in Figure 7.15. In this comparison, $f_{t,90} = 4.5$ MPa, $f_{t,0} = 89$ MPa and $f_v = 12$ MPa. The test results are from Kollmann and Cote (1968), and they were obtained for Basswood (*Tilia americana*). Similar influence of grain angle is found for other species and also for loading in bending and compression. For Scots pine, Table 7.1 and Ullman (1997) suggest parallel/perpendicular strength ratios in the order of 20 for tension, 15 for bending and 8 for compression.

For the general 3D state of stress, Norris (1950) proposed a criterion with separate considerations to possible failure for the three principal planes of the orthotropic material:

$$\begin{cases} (\sigma_1/f_1)^2 + (\sigma_2/f_2)^2 + (\tau_{12}/f_{v12})^2 \\ \quad - (\sigma_1/f_1)(\sigma_2/f_2) = 1.0 \\ (\sigma_1/f_1)^2 + (\sigma_3/f_3)^2 + (\tau_{13}/f_{v13})^2 \\ \quad - (\sigma_1/f_1)(\sigma_3/f_3) = 1.0 \\ (\sigma_3/f_3)^2 + (\sigma_2/f_2)^2 + (\tau_{32}/f_{v32})^2 \\ \quad - (\sigma_3/f_3)(\sigma_2/f_2) = 1.0 \end{cases} \qquad (7.11)$$

where 1, 2 and 3 represent the stresses and corresponding strength parameters for the tangential, radial and longitudinal directions. Equation (7.11) represents for plane stress, an alternative or extension to the Norris criterion of Equation (7.7).

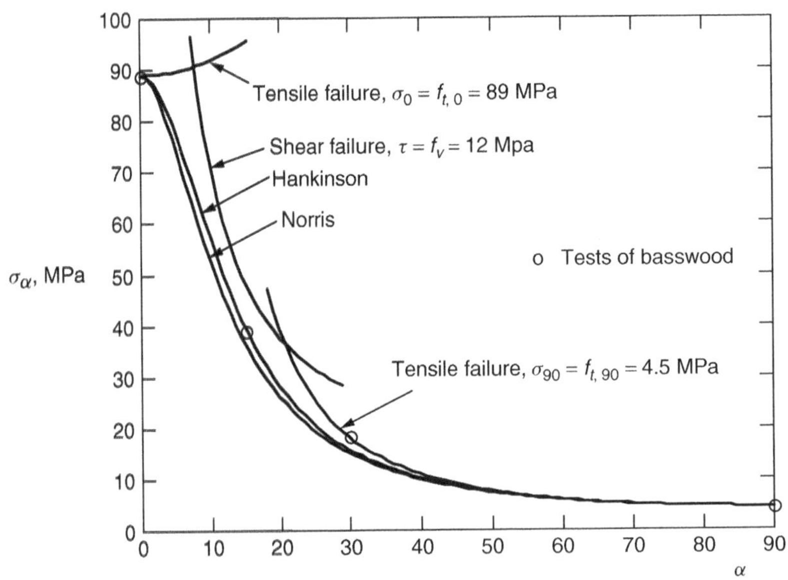

Figure 7.15 Uniaxial tensile failure stress versus load to grain angle

7.5.4 Weibull Weakest Link Theory

If the probability of failure of a link number i in a chain with n links is s_i, then the probability that no link in the chain will break is $P = (1 - s_1)(1 - s_2)(1 - s_3)\ldots(1 - s_n)$. Accordingly, the probability of failure of the chain is $1 - P$. Using the Weibull distribution for the characterisation of the strength of a link (or the strength of a unit volume of a material),

$$S = 1 - e^{-(\sigma/f)^m} \tag{7.12}$$

where S is the probability of failure, σ is the stress in the link and f and m are material parameters, the probability of failure of a chain made up of V/dV links (or the probability of failure of a material volume made of V/dV unit volumes) is, by multiplication, found to be

$$S = 1 - e^{-I}, \quad \text{where } I = \int_V (\sigma/f)^m dV \tag{7.13}$$

From Equation (7.13), it is possible to calculate the median and mean strengths of the volume of material under consideration. This volume can, for instance, be a test specimen in homogeneous tension or a structural element in which the stress distribution is heterogeneous. Indicating the mean strength of the tensile test specimen of volume V_0 by $f_{t,90}$, and representing the magnitude of the load acting on the structural element by the magnitude of the stress σ_{90} in the most stressed point, the mean failure value of σ_{90} becomes

$$\sigma_{90,\text{failure}} = f_{t,90}(V/V_0)^{-1/m}\gamma^{-1/m},$$
$$\text{where } \gamma = \int_V (\sigma/\sigma_{90})^m dV/V \tag{7.14}$$

The parameter m is governed by the magnitude of the scatter in the strength of the material. m has a one-to-one relation to the coefficient of variation, as well as to the volume effect in strength. $m = 5$ corresponds to 22.9% COV in strength and to a 13% reduction in $\sigma_{90,\text{failure}}$ for a doubling of the volume, and $m = 10$ corresponds to 12.0% COV and to a 7% reduction in $\sigma_{90,\text{failure}}$ for a doubling of the volume. For other values of m and for more details in the derivation of Equation (7.14) (see Gustafsson, 1985). The coefficient $\gamma = 1.0$ if the state of stress is homogeneous throughout the element, and $\gamma < 1.0$ in all other cases. A heterogeneous stress distribution with high stress only in a small local region gives a small value of γ and a high maximum stress at failure, i.e. a high $\sigma_{90,\text{failure}}$.

Equation (7.14) can be compared with Equation (7.1) and Figure 7.7. The apparently higher tensile strength $f_{t,90}$ of the material in the curved beams (see Figure 7.7), than in the prismatic specimens corresponds within Weibull theory to a smaller γ for the geometry and stress distribution of a curved beam. The influence of volume and stress distribution on the probability of fracture has lead to definition in analogy with Equation (7.14) of an 'effective Weibull stress', so that the loading level of different structural elements can be compared (Ranta-Maunus, 1998).

In the present context, $f_{t,90}$, σ and σ_{90} are tacitly assumed to represent the perpendicular to grain tensile stress, but may equally well be interpreted as representing some other stress component or an effective stress as defined by some failure criterion, e.g. some of those in Section 7.5.3.

The above two-parameter Weibull distribution has some nice features. It produces an extreme value distribution (Equation (7.13)) of the same kind as the distribution itself (Equation (7.12)), and the COV for the strength of the link (the unit material volume) is the same as the COV of the element, independently of its size and shape. The distribution yields no strength values below zero, and the shape is for large m fairly similar to that of the Gauss distribution. There is also a three-parameter Weibull distribution, the third parameter shifting the location of the curve so that there will be no strength value below some arbitrary value greater than zero.

The Weibull weakest link modelling was proposed by Weibull (1939a, b). For application of the theory, the stresses σ in actual volume has to be calculated and then integrated according to Equation (7.14). The calculation can in some cases be made by beam theory, while numerical calculations by the finite element method and numerical integration may be needed in other cases. The volume and shape factors proposed in Eurocode 5 for use in strength design of curved, tapered and pitched cambered glulam beams are based on Weibull theory considerations (Ehlbeck and Kurth, 1995).

An objection to general application of the weakest link theory is that the performance of wood and wooden structural elements is not ideally brittle, i.e. a structural element is not in general acting as if all points in the volume of the element were links in a chain. Wood, just as all other solid materials, has on the contrary fracture toughness greater than zero.

The conventional stress criteria discussed in Section 7.5.3 are known to break down when there is a crack or sharp notch producing a point where the stress approaches an infinitely large value already at very small external load. Weibull theory is found to break down in a similar way in the presence of a stress singularity (Gustafsson and Enquist, 1988). For $m < 4$, Weibull theory predicts that the tip of a crack will not fracture, no matter the magnitude of the load; for $m = 4$ no prediction is obtained, and for $m > 4$ the crack is found to have zero strength.

7.5.5 Linear Elastic Fracture Mechanics

Linear Elastic Fracture Mechanics (LEFM) can be used for calculating the load that gives propagation of a pre-existing crack or sharp notch, having 'infinite' stress at the tip of the crack or notch. Basic assumptions are that the material is linear elastic and of infinite strength in terms of stress. LEFM gives results that are in good agreement with reality if the length of the initial crack or notch is large as compared to the size of fracture process region. In wood the size of the fracture process region is roughly from a few millimeters for crack propagation in the radial-tangential plane to several centimeters for shear fracture along grain.

LEFM can be formulated as an analysis of the magnitude of the stress intensity at the tip of the crack or as analysis of the energy balance during crack propagation. Here only the latter formulation will be dealt with. The energy balance during an extension of the cracked area by dA is

$$G_f = W - dT/dA - d\phi/dA \qquad (7.15)$$

where G_f is the dissipation of energy during the crack propagation, W is the work by external loads during the crack propagation dA, T is the kinetic energy and ϕ is the elastic strain energy (Hellan, 1985). For quasi-static conditions and loading only by a single load, Equation (7.15) gives, for a structural element with the compliance $C = 1/K = \delta/P$,

$$G_f = (1/2)(P_f^2/b)(dC(a)/da) \qquad (7.16)$$

where da is the extension of the length, a, of the crack, and where $b = dA/da$ is the width of the fracture area. P_f is the magnitude of load that is just about sufficient to start a propagation of the crack. The crack propagation becomes unstable and rapid at prescribed load, P, if P_f decreases as a is increased, and vice versa.

To illustrate the energy balance analysis further, the double cantilever specimen shown in Figure 7.16 can be analysed from the energy potential point of view. The energy potential of the system, U, is the sum of the strain energy

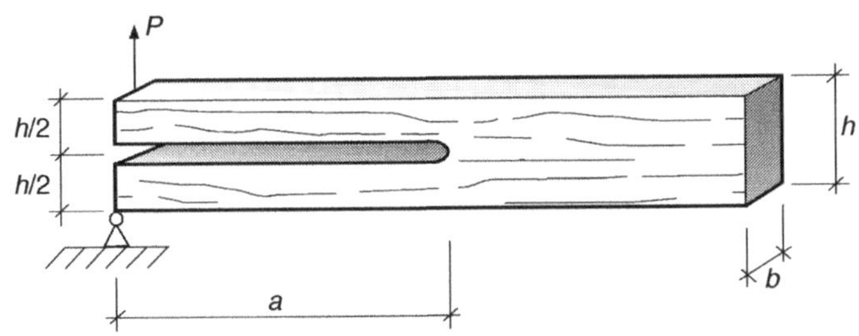

Figure 7.16 A double cantilever specimen

in the body, $P\delta/2$, and the potential of the load, $-P\delta$, giving

$$U = -P\delta/2 \qquad (7.17)$$

Making the decrease in U equal to the fracture energy G_f times the fracture area bda

$$-dU = G_f bda \qquad (7.18)$$

it is by use of Equation (7.17) found that

$$P_f = \sqrt{2bG_f/(dC/da)} \qquad (7.19)$$

which is the same result as in Equation (7.16). For application to the particular example of Figure 7.16, the compliance C must be calculated. Elementary beam theory gives an approximation of C:

$$C = \delta/P = 2a^3/(3E_0 I) \qquad (7.20)$$

where E_0 is the longitudinal modulus of elasticity and the moment of inertia, I, is $b(h/2)^3/12$. Thus, for the double cantilever beam, by Equation (7.19):

$$P_f = \sqrt{bG_f E_0 I/a^2} \qquad (7.21)$$

This result is general with respect to the influence of stiffness, E, and fracture energy, G_f.

By normalising P_f with respect to the size of the element, i.e. by expressing the failure load by stress at failure:

$$\sigma_{\text{failure}} = P_f/(bh) = \sqrt{E_0 G_f/h}(h/a)/\sqrt{96} \qquad (7.22)$$

a general and important result is also found with respect to the influence of the absolute size: the stress at failure is not constant but inversely proportional to the square root of the absolute length scale of the structural element. It can also be noted that σ_{failure} is not affected at all by the perpendicular to grain tensile strength of the wood, $f_{t,90}$.

The key information needed for application to other structural elements is the compliance, C, as a function of the crack length, a. For some geometries it is fairly simple to calculate C, for instance by use of beam theory, while numerical computation by use of the finite element method may be used in other cases.

For an end-crack, as shown on the right-hand side support of the beam in Figure 7.17, it is by LEFM and calculation of C by beam theory found that the beam shear force at the root of notch at failure is

$$V_f = b\alpha h\sqrt{E_0 G_f/h} \Big/ \left[\sqrt{0.6(E_0/G)(\alpha - \alpha^2)} + \beta\sqrt{6(1/\alpha - \alpha^2)}\right] \qquad (7.23)$$

where G is the shear stiffness of the wood (Gustafsson, 1988). The same result is obtained for all the different kinds of notches shown in Figure 7.17. The beam depth measures h and αh then indicate the measures at the notch considered, and β represents the bending moment at the notch, i.e. β is in the general case replaced by $M/(Vh)$, where M is the bending moment.

Equation (7.23) has, for the kind of end-notch shown on the left-hand support, been compared with all the test results found in the literature, for beams of structural sizes showing a good agreement between LEFM theory and the test results. For small beams or specimens, it is found that the condition of a small fracture process zone as compared to the depth of the notch is not fulfilled, giving failure load predictions somewhat higher than found experimentally.

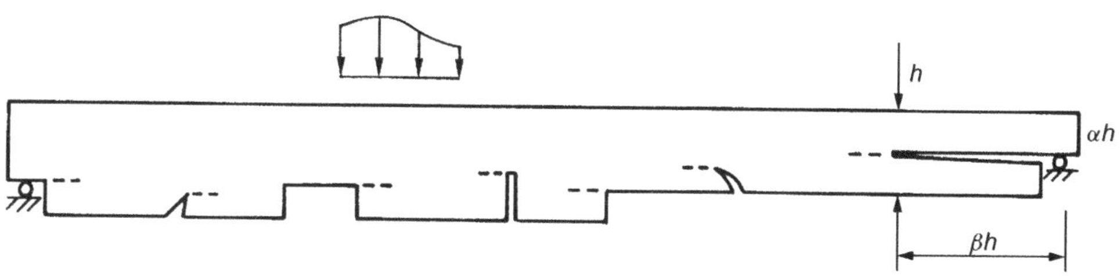

Figure 7.17 Notches of various kind

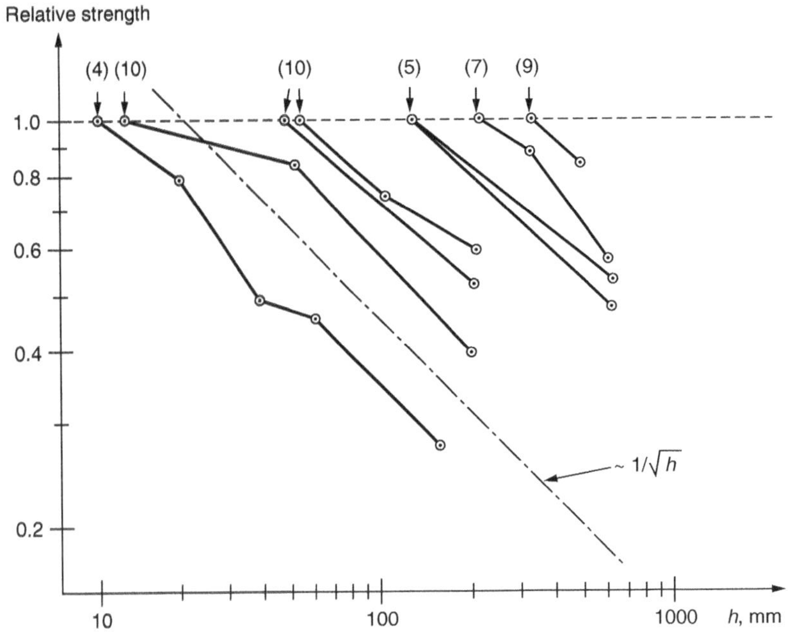

Figure 7.18 Influence of beam size on the strength of end-notched beams

In Figure 7.18 is a comparison made between the LEFM theory and tests with respect to the influence of absolute size on strength. The eight test series relate to end-notched beams of various shape and they are from seven independent experimental investigations (Gustafsson, 1991). In earlier timber engineering design codes, based on conventional failure stress calculation, there was no consideration of any size effect caused by the presence of a crack or notch. This size effect is in more general terms caused by the fracture process region having an absolute size that is governed by the properties of the material, not by the size of the structure. The effect of the size of the fracture process region becomes evident when there is a stress gradient in the high stress region, this gradient being especially large in the vicinity of the tip of a notch or crack.

Examples on research on applications of LEFM to the analysis of mechanical joints can be found in Jorrisen (1998), Yasumura and Daudeville (1999), Larsen and Gustafsson (2000) and van der Put and Leijten (2000). For wood adhesive joints, LEFM applications can be found in state-of-the-art-reports by Valentin *et al.* (1991), Wernersson (1992, 1994), Gustafsson and Serrano (2000) and Serrano (2000), and in references given in these reports.

7.5.6 Generalised Linear Elastic Fracture Mechanics

Basic limitations for application of the LEFM are that

- The size of the fracture process region at the tip of the crack must be small as compared to the length of the crack, or else the strength predictions become poor.
- There must be a pre-existing sharp crack or notch in the specimen or structural element, giving a square root singularity in the stress, meaning that the linear elastic stress is proportional to $r^{-1/2}$ in the vicinity of the crack tip, r being the distance to the crack tip.

To overcome the first limitation, various modifications or additions to the basic LEFM analysis have been proposed. A simple and widespread modification is to assign the crack a somewhat *increased length*, so that the crack tip computationally becomes located close to the centre of gravity of the stresses acting across the fracture region. The length of the fracture zone at essentially tensile stress fracture crack propagation along the grain is, according to a set of nonlinear fracture calculations by the FEM, roughly about 0.25 $G_f(E_0G)^{1/2}/(f_{t,90})^2$, where the length for softwoods is commonly in the order of 10–15 mm, and as a rough estimation it has been proposed that the addition to the length of the crack or notch may been taken as 0.2 $G_f(E_0G)^{1/2}/(f_{t,90})^2$ (Gustafsson, 1985, 1988). This addition to the crack length becomes significant when making strength predictions for small specimens with a short crack.

A further complication when the initial crack is short is that it may become unstable before the fracture zone is completely developed to its full size, i.e. the specimen or structural element may break before the fracture energy of the material is fully activated. This can be considered in an approximate way by so-called *R-curve* characterisation of the material properties, the R-curve showing how the crack propagation resistance increases as the length of the fracture zone increases. R-curve analysis has been applied to steel and concrete.

The second of the above two limitations of LEFM application is of a more fundamental character. However, generalisations, still within linear elastic theory, have been proposed. One proposal is a method sometimes called the *mean stress method*, and the other proposal that will be discussed here is sometimes called the *initial crack method*.

In the mean stress method, stress and strength analysis is made in an almost conventional way. But in the stress criterion used, for instance Equations (7.3) or (7.7), the stresses σ_{90} and τ (and σ_0) are not taken as the stresses in a point, but as the mean stresses for an area of a certain size. In this way, there will be no 'infinite' stress. The size of the area can be derived from the condition that the strength prediction obtained for specimens with a very long sharp crack coincides with the prediction of LEFM. For mode I fracture (i.e. for pure tensile stress across a potential fracture surface), the length of the area is found to be

$$x_I = \frac{2}{\pi}\frac{E_I G_f}{f_{t,90}^2} \tag{7.24}$$

where stiffness parameter E_I is the short for

$$E_I = \left\{\frac{1}{E_{\parallel}}\sqrt{\frac{E_{\parallel}}{2E_{\perp}}}\sqrt{\sqrt{\frac{E_{\parallel}}{E_{\perp}}}+\frac{E_{\parallel}}{2G}-\nu_{\perp\parallel}\frac{E_{\parallel}}{E_{\perp}}}\right\}^{-1} \tag{7.25}$$

$E_{\perp}$ is the stiffness of the wood perpendicular to the fracture plane, and $E_{\parallel}$ is the stiffness of the wood in the direction of the fracture propagation. G and $\nu_{\perp\parallel}$ indicate shear stiffness and Poisons ratio, respectively. The specimen is assumed to be in a state of plane stress, and the width of the area is accordingly equal to the width of the specimen. A similar, but lengthier expression can be derived for mixed mode loading (Gustafsson, Petersson and Stefansson, 1996). There is, however, a significant influence on the length x of the shear stress component only when $\arctan(\sigma/\tau)$ is less than about 20°. The length x_I is for softwood typically in the order of 20 mm for fracture propagation along grain, and 4 mm for fracture propagation in the radial direction.

In the other proposal, the initial crack method, the strength analysis is made by LEFM, assuming the existence of an initial and additional crack of a certain length where fracture may develop. If there is already a crack in the element, the additional crack adds to the length of the initial real crack. The length of the extra crack is derived from the condition that the predicted strength of an element without any stress singularity coincides with the prediction obtained by a conventional stress-based failure criterion, which in the case of pure tension perpendicular to grain is $\sigma_{90} = f_{t,90}$. By this condition, the length of the additional crack is, for pure mode I loading, found to be

$$a_I = x_I/2 \tag{7.26}$$

where x_I is given according to Equation (7.24). With parameters typical for softwood, a_I is typically about 10 mm for fracture propagation along the grain and 2 mm for fracture propagation in the radial direction. The length for fracture along the grain can be compared with the above rough estimate, 0.2 $G_f(E_0G)^{1/2}/(f_{t,90})^2$, of the crack length increase to improve the results of LEFM analysis when the crack is short as compared to the length of the fracture zone. Equation (7.26) is also valid at mixed mode loading, i.e. $a = x/2$.

Further information on the above two methods and examples of applied strength analysis can be found in Landelius (1989), where the mean stress method is dealt with for mode I loading, in Morris, Gustafsson and Serrano (1995), where the methods are applied to the strength analysis of I-shaped lightweight beams with a circular hole, and in Gustafsson, Petersson and Stefansson (1996), where applications are made to timber and glulam beams of two sizes and with different kinds of holes and notches. It seems that the methods give reasonable results. The prime limitation is that the relevant dimensions of the structural element analysed should be large as compared with the lengths a and x. As compared with nonlinear fracture mechanics analysis, one may in general expect the mean stress method to overestimate the load capacity, and the initial crack method to underestimate the load capacity if the lengths a and x are not sufficiently small so that the results obtained become more or less approximate. A similar but different method, called the finite small area theory, is presented in Masuda (1988). Using this theory, the means stress is calculated for a volume of the material, the thickness and length of the volume being estimated from the dimensions of wood fibres.

7.5.7 An Example of Nonlinear and Quasi-nonlinear Fracture Analysis

Glued joints, as shown in Figure 7.19, were tested by Glos and Horstman (1989) to find design criteria for glued truss heal joints, shown as a detail in Figure 7.21b. One of the glues tested was a resorcinol-phenol adhesive. This glue is strong and brittle, so that strength properties of the wood (*Picea abies*) most probably were decisive. In a theoretical study of the joint (Gustafsson and Serrano, 1998), nonlinear fracture mechanics plane stress analysis by the FEM and a quasi-nonlinear fracture mechanics method were conducted. In the nonlinear analysis, fracture zone properties according to Figure 7.20 were assumed (the same for all α), together with linear elastic transversely isotropic stiffness properties typical for Spruce: $E_0 = 11\,000$ MPa, $E_{90} = 300$ MPa, $G_0 =$

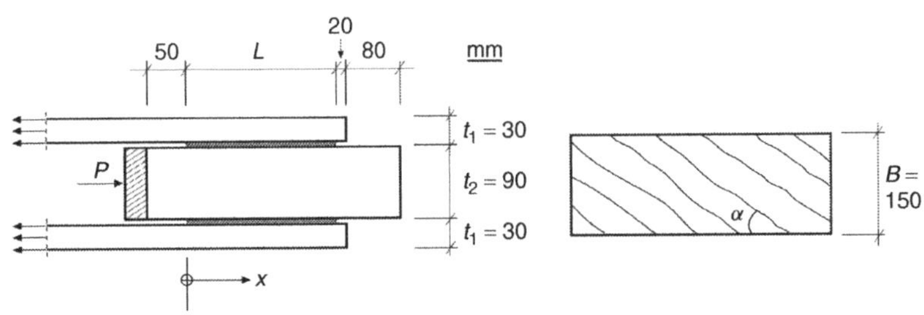

Figure 7.19 A glued joint (Presses polytechniques et universitaires romandes. Reproduced with publisher's authorization.)

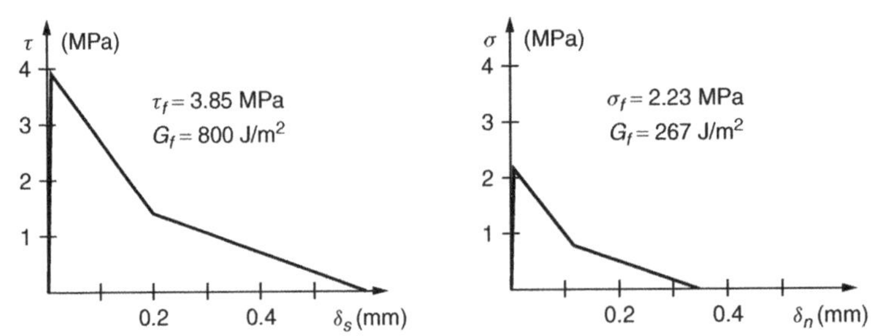

Figure 7.20 Fracture zone properties at pure shear and at pure tension (Presses polytechniques et universitaires romandes. Reproduced with publisher's authorization.)

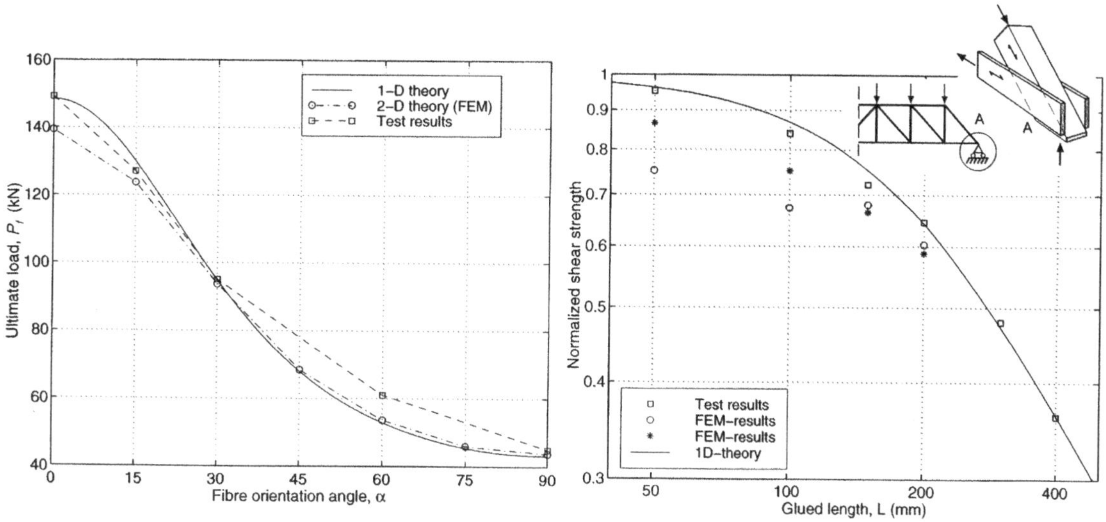

Figure 7.21 (a) Joint strength versus α. The joint length is 200 mm; (b) normalised joint strength versus joint length. Angle $\alpha = 0^0$ (Presses polytechniques et universitaires romandes. Reproduced with publisher's authorization. All rights reserved)

600 MPa, $G_{90} = 60$ MPa, $\nu_{0,90} = 0.3$ and $\nu_{90,0} = 0.0082$. Figure 7.20 shows the fracture zone properties for pure tension and pure shear. For mixed mode conditions, the model used is of the type

$$\begin{cases} \sigma = \sigma(\delta_n, \delta_s) \\ \tau = \tau(\delta_n, \delta_s) \end{cases} \tag{7.26}$$

with a mixed mode interaction corresponding to Equation (7.3) (Wernersson, 1994). The quasi-nonlinear analytical analysis was carried out analytically as a shear lag analysis, where the three timber parts are modelled as elastic bars and the fracture zone as a shear lag layer defined by a linear relation between shear stress and shear deformation. The strength of this linear elastic layer was 3.85 MPa, and its stiffness was assigned a value so that the work for bringing the layer to peak stress and fracture became 800 J/m^2 (cf. Figure 7.20). The effect of the grain angle α enters the analyses, both the analytical analysis and the FEM analysis, only through its influence on the stiffness of the middle timber part.

Figures 7.21a and b show the theoretical results compared with the test results for different lengths of the joint and for different grain angles, respectively. The FE-analyses were made for two different boundary conditions. A reduced joint strength was predicted when the loaded ends of the outer timber parts were free to move in the direction perpendicular to the load. In that case, failure was for the short joints initiated by perpendicular to grain tensile fracture at the right-hand end of the joint.

Figure 7.21b illustrates a general feature of the size effect in strength. For large structures with a notch there is a strong size effect equal to that predicted by LEFM, and corresponding to a constant slope equal to -0.5 in a log-log diagram, but as the size is made smaller the size effect gradually decreases, approaching constant strength for very small structural elements. For the joint studied it appears that LEFM would give reasonable strength predictions down to joint lengths of about 200 mm. The normalised shear strength shown in Figure 7.21b is the mean shear stress at failure divided by the shear strength of the material, 3.85 MPa.

For joint lengths greater than about 200 mm, i.e. when LEFM is valid, there is very little influence of the strength of the material on the failure load of the joint, P_f. Instead, P_f is governed by the toughness of the material, i.e. the stiffness and the fracture energy. For very short joints it is the opposite situation: P_f is governed by the tensile

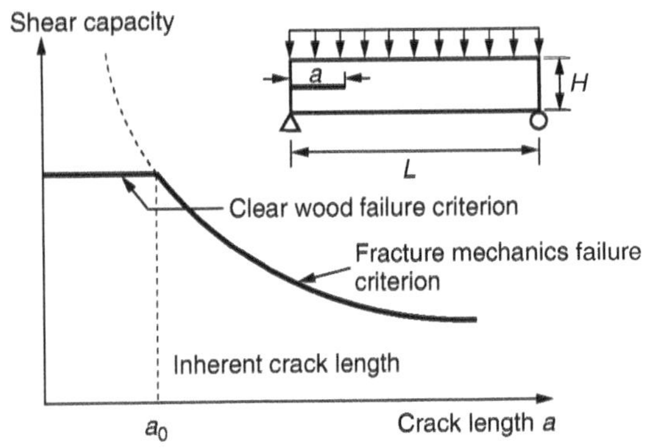

Figure 7.22 Shear strength of an end-cracked beam (Reproduced by permission of RILEM)

and shear strength of the material, and almost not affected at all by its fracture toughness.

7.5.8 Probabilistic Fracture Mechanics

Probabilistic fracture mechanics refers in this context to rational calculation models, where the stochastic nature of the material in a timber structural element, as well as the non-zero fracture toughness of the material is taken into account (see Table 7.5). In Foschi, Folz and Yao (1989), a combination of LEFM and stochastics is applied to the shear force strength design of timber beams. End-cracks (see Figure 7.22) may develop during drying, and these cracks are assumed to govern the beam shear strength. The length of the end cracks is defined as a stochastic variable according to empirical observations. For cracks deeper than a certain length, a_0, LEFM is assumed to be valid, and gives the beam shear strength in terms of the length of the crack and the fracture toughness of the material. In this way, a reliability strength design method is achieved, in which both a fracture mechanics size effect and a scatter size effect can be included.

In Gustafsson and Serrano (1999), another combination of fracture mechanics and probabilistic theory was very briefly outlined. Generalised LEFM based on the mean stress concept was combined with Weibull analysis. For the special case of plane stress and consideration in the fracture criterion only to the perpendicular to grain tensile stress component, the sketched proposal involves the calculation of

$$\varphi = \left\{ \frac{1}{V_{\text{ref}}} \int_V \left(\frac{\overline{\sigma_{90}}}{f} \right)^m dV \right\}^{1/m} \tag{7.27}$$

where $\overline{\sigma_{90}}$ is the mean perpendicular to grain tensile stress for each parallel to grain area of size bx_I, where the length x_I is given as according to Equation (7.24). In numerical calculation and discretisation by the FEM, $\overline{\sigma_{90}}$ may as an approximation be taken as the stress in the centre of a finite element of length x_I along the grain. The integration is then a summation over the elements in the volume. The coefficient φ is an effective stress. Equal effective stress φ in the structural element as in a tensile test specimen indicates equal failure probability.

7.6 MEASURES FOR STRENGTH IMPROVEMENT

The perpendicular to grain strength can be improved by reinforcement. Curved glulam beams and glulam beams with holes or end-notches can be reinforced by gluing in rods or threaded bolts perpendicular to the beam (Johansson, 1995). Use of a screw placed without glue in a hole drilled through the beam and tightened by a nut should, in general, be avoided. Creep and shrinkage entail loss of action of the screw.

Good reinforcement can also be achieved by gluing, or by gluing and nailing, plywood to the sides of the beam. Effective reinforcement can, moreover, be achieved by gluing glass fibre to the wood by means of a polyester matrix (Dahlblom *et al.*, 1993, Larsen, Gustafsson and Traberg, 1994). The glass fibres can be in the form of a mat with randomly orientated short fibres or as woven material, where most of the fibres may be oriented in a single direction. The glass fibre-polyester reinforcement has the advantage of being almost invisible because of transparency and small thickness. Reinforcement by glued-in rods and by plywood is used in engineering practice, while glass fibre reinforcement still is in a stage of development. Reinforcement can be used also locally in the vicinity of a bolt (Figure 7.1e), and if using glass fibres, at a nailed joint (Figure 7.1i), to reduce edge and nail distances, and also increase the load capacity of the individual nail. Reinforcements glued to the sides of the glulam or timber have the advantage of preventing or slowing down moisture changes in the wood, and thus reduce the development of moisture-induced stresses.

REFERENCES

Aicher S., Dill-Langer G. and Ranta-Maunus A. (1998) Duration of load effect in tension perpendicular to grain of glulam in different climates. *Holz als Roh- und Werkstoff*, **56**, 295–305.

Bodig J. and Jayne B. (1982) *Mechanics of Wood and Wood Composites*. Van Nostrand Reinhold, New York.

Boström L. (1992) Method for Determination of the Softening Behaviour of Wood and the Applicability of a Nonlinear Fracture Mechanics Model. Doctoral thesis, Report TVBM-1012, Div of Building Materials, Lund University, Sweden.

Dahlblom O., Enquist E., Gustafsson P.J., Knudson R., Larsen H.J., Ormarsson S. and Traberg S. (1993) Fibre reinforcement of glulam – Summary report and report 1–7. Division of Structural Mechanics, Lund University, Sweden.

Dinwoodie J.M. (1981) *Timber, Its Nature and Behaviour*. Van Nostrand Reinhold, London.

Ehlbeck J. and Görlacher R. (1995) Tension perpendicular to grain in joints. Lecture C2, *Timber Engineering STEP 1*, Centrum Hout, The Netherlands.

Ehlbeck J. and Kurth J. (1995) Tapered, curved and pitched cambered beams. Lecture B8, *Timber Engineering STEP 1*, Centrum Hout, The Netherlands.

Ehlbeck J. and Kromer M. (1995) Carpentry joints. Lecture C12, *Timber Engineering STEP 1*, Centrum Hout, The Netherlands.

Foschi R.O., Folz B.R. and Yao F.Z. (1989) Reliability-Based Design of Wood Structures. Structural Research Series, Report No. 34, Department of Civil Engineering, University of British Columbia, Canada.

Gibson L. and Ashby M. (1987) *Cellular Solids*. Pergamon Press, Oxford.

Glos P. and Horstman H. (1989) Strength of glued lap timber joints. Paper CIB-W18A/22–7-8, *Proc. of CIB-W18A Meeting 22*, East Berlin, pp. 1–17.

Gustafsson P.J. (1985) Fracture mechanics studies of non-yielding materials like concrete. Doctoral thesis, Report TVBM-1007, Division of Building Materials, Lund University, Sweden.

Gustafsson P.J. and Enquist B. (1988) Träbalks hållfasthet vid rätvinklig urtagning. Report TVSM-7042 (Bilaga C), Division of Structural Mechanics, Lund University.

Gustafsson P.J. (1988) A study of strength of notched beams. Paper CIB-W18A/21-10-1, *Proc. of CIB-W18A Meeting 21*, Vancouver Island, Canada, p. 25.

Gustafsson P.J. (1991) Eurocode 5 draft design criterion for notched beams. *Proc. of 1991 International Timber Engineering Conference*, London, UK, Vol. 1, 66–74.

Gustafsson P.J. (1992) Some test methods for fracture mechanics properties of wood and wood adhesive joints. *Proc. of RILEM TC Workshop on Determination of Fracture Mechanics Properties of Wood*, Bordeaux, France, p. 14.

Gustafsson P.J., Hoffmeyer P. and Valentin G. (1998) DOL behaviour of end-notched beams. *Holz als Roh- und Werkstoff*, **56**, 307–317.

Gustafsson P.J. and Serrano E. (1998) Glued truss joints analysed by fracture mechanics. *Proc. 5th World Conf. on Timber Engineering*, **1**, Lausanne, Switzerland, pp. 257–264.

Gustafsson P.J. and Serrano E. (1999) Fracture mechanics in timber engineering – some methods and applications. *Proc. 1st RILEM Symposium on Timber Engineering*, Stockholm, Sweden, pp. 141–150.

Gustafsson P.J. and Serrano E. (2000) Predicting the pull-out strength of glued-in rods. *Proc. 6th World Conference on Timber Engineering*, B.C., Canada, pp. 7.4.4.1–7.4.4.8.

Götz K.H., Horr D., Möhler K. and Natterer J. (1978) *Holzbau Atlas*. Inst fur Internationelle Architektur-Dokumentation,Munchen.

Hellan K. (1985) *Introduction to Fracture Mechanics*. McGraw-Hill, New York.

Helmersson H. (1978) Materialbrott för olika byggnadsmaterial. Examensarbete, Avd. för Byggnadsmaterial, Lunds Universitet, Sweden.

Hillerborg A., Modéer M. and Petersson P.-E. (1976) Analysis of crack formation and crack growth in concrete by means of fracture mechanics and finite elements. *Cement and Concrete Research*, **6**, 773–782.

Holmberg S. (1998) A numerical and experimental study of initial defibration of wood. Doctoral thesis, Report TVSM-1010, Division of Structural Mechanics, Lund University, Sweden.

Kollmann F. and Cote W. (1968) *Principles of Wood Science and Technology – 1 Solid Wood*. Springer-Verlag, Berlin.

Jönsson J. and Svensson S. (2000) Internal stresses in cross-grain direction of wood induced by climate variations. Paper CIB-W18A/33-12-1, *Proc. of CIB-W18A Meeting 33*, Delft, The Netherlands.

Johansson C.J. (1995) Glued-in bolts. Lecture C14, *Timber Engineering STEP 1*, Centrum Hout, The Netherlands.

Jorissen A. (1998) Double shear timber connections with dowel type fasteners. Doctoral thesis, Civil Engineering Department of Delft University of Technology, The Netherlands.

Larsen H.J. and Gustafsson P.J.H. (1990) The fracture energy of wood in tension perpendicular to the grain – results from a joint testing project. Paper CIB-W18A/23-19-2, *Proc. of CIB-W18A Meeting 23*, Lisbon, Portugal.

Larsen H.J. and Gustafsson P.J.H. (1991) The fracture energy of wood in tension perpendicular to the grain. Paper CIB-W18A/24-19-1, *Proc. of CIB-W18A Meeting 24*, Oxford, UK.

Larsen H.J., Gustafsson P.J. and Traberg S. (1994) Glass fibre reinforcement perpendicular to grain. *Proc. Pacific Timber Engineering Conf.*, Surfers Paradise, Australia.

Larsen H.J. and Gustafsson P.J. (2000) Timber joints loaded perpendicular to the grain – Long-term strength; theory and experiments. *Proc. of NATO Advanced Research Workshop: 'The Paramount Role of Joints into the Reliable Response of Structures. From the rigid and Pinned Joints to the Notion of Semi-rigidity'*, Ouranoupolis, Greece.

Larsen H.J. and Riberholt H. (1983) Traekonstruktioner – Beregning. SBI Anvisning 135, SBI, Denmark.

Larsen H.J. and Riberholt H. (1972) Forsög med Uklassificeret Konstruktionsttrae. Rapport nr R31, Avd. for Baerende Konstruktioner, Danmarks Tekniske Höjskole, Denmark.

van der Put T. and Leijten A. (2000) Fracture of beams loaded by joints perpendicular to the grain. Paper CIB-W18A/33-7-7, *Proc. of CIB-W18A Meeting 33*, Delft, The Netherlands.

Madsen B. (1992) *Structural Behaviour of Timber*. Timber Engineering Ltd, North Vancouver, Canada.

Masuda M. (1988) Theoretical consideration on fracture criteria of wood – proposal of finite small area theory. *Proc. Int. Conf. on Timber Engineering*, Seattle, WA, **2**, 584–595.

Morlier P. and Ranta-Maunus A. (1998) DOL effect of different sized timber beams. *Holz als Roh- und Werkstoff*, **56**, 279–284.

Morris V., Gustafsson P.J. and Serrano E. (1995) The shear strength of light-weight beams with and without a hole. *Proc. of COST 508 Wood Mechanics Workshop on Mechanical Properties of Panel Products*, Watford, UK, pp. 198–214.

Norris C.B. (1950) Strength of orthotropic materials subjected to combined stresses. Report No 1016, Forest Products Laboratory, Madison, WI.

Norris C.B. (1962) Strength of orthotropic materials subjected to combined stresses. Report No 1816, Forest Products Laboratory.

Olejniczak P. and Gustafsson P.J. (1994) Rate effect in tangential tension fracture softening performance. *Proc. of COST 508 Wood Mechanics Workshop on Service Life Assessment of Wooden Structures*, Finland, pp. 137–148.

Ranta-Maunus A. (1998) The effect of weather changes and surface coating on the load bearing capacity of curved glulam structures. *Proc. 5th World Conf on Timber Engineering*, Montreux, **1**, 541–548.

Serrano E. (2000) Adhesive Joints in Timber Engineering – Modelling and Testing of Fracture Properties. Doctoral thesis, Report-TVSM-1012, Division of Structural Mechanics, Lund University, Sweden.

Siimes F.E. (1967) The Effect of Specific Gravity, Moisture Content, Temperature and Heating Time on the Tension and Compression Strength and Elasticity Properties Perpendicular to the Grain of Finnish Pine, Spruce and Birch Wood and the Significance of these Factors on the Checking of Timber at Kiln Drying. Julkaisu 84 Publication, VTT – The Technical Research Centre of Finland.

Svensson S., Thelandersson S. and Larsen H.J. (1999) Reliability of timber structures under long term loads. *Materials and Structures*, **32**, 755–760.

Stefansson F. (1995) Mechanical Properties of Wood at Microstructural Level. Masters thesis, Report TVSM-5057, Division of Structural Mechanics, Lund University, Sweden.

Stefansson F. (2001) Fracture analysis of orthotropic beams – linear and nonlinear methods. Licentiate thesis, Report TVSM-3029, Division of Structural Mechanics, Lund University, Sweden.

Uhmeier A. (1995) Some Aspects on Solid and fluid Mechanics of Wood in Relation to Mechanical Pulping. Report 1995:12, Division of Paper Technology, Royal Institute of Technology, Stockholm.

Ullman E. *et al.* (1997) Materiallära. Karlebo, Liber AB, Sweden.

Valentin G., Boström L., Gustafsson P.J., Ranta-Maunus A. and Gowda S. (1991) Application of Fracture Mechanics to Timber Structures – RILEM State-of-the-art Report, Research Note 1262, VTT – The Technical Research Centre of Finland.

Weibull W. (1939) A statistical theory of the strength of materials. Proceedings no 151, Ingenjörsvetenskapsakademien, Sweden.

Weibull W. (1939) The phenomenon of rupture in solids. Proceedings no 153, Ingenjörsvetenskapsakademien, Sweden.

Wernersson H. (1992) Wood adhesive bonds – fracture softening properties in shear and in tension. Report TVSM-3012, Division of Structural Mechanics, Lund University, Sweden.

Wernersson H. (1994) Fracture Characterization of Wood Adhesive Joints. Doctoral thesis, Report TVSM-1006, Division of Structural Mechanics, Lund University, Sweden.

Yasumura M. and Daudeville L. (1999) Design and analysis of bolted timber joints under lateral force perpendicular to grain. Paper CIB-W18/32-7-3. *Proc. of Meeting 32 of CIB Working Commission W18 – Timber Structures.*

8

Strength Under Long-term Loading

Preben Hoffmeyer

8.1 INTRODUCTION

Wood and wood products experience a significant loss of strength and stiffness when loaded over a period of time. The effect of ten years load duration may amount to a strength loss of approximately 40% for solid wood and approximately 80% for some wood-based panel products. This phenomenon of creep-rupture – often called the Duration Of Load (DOL) effect – has therefore been the subject of particular interest for everyone in the timber engineering community concerned with safe and efficient engineering design.

The scientific treatment of DOL effects has been going on for more than 200 years. However, the start of the present era of duration of load research took place around 1970, when Professor Borg Madsen (Department of Civil Engineering, University of British Columbia, Canada) pioneered DOL research work on timber rather than on small clear specimens. Much effort has since been devoted to DOL research, and much new knowledge has been gained. Even so, the understanding and modelling of the duration of load behaviour of wood and wood products is still far from complete.

The study of DOL behaviour of wood and wood products introduces the scientist to a variety of challenging research problems: by the mere nature of the phenomenon, experiments involve long periods of time. The experimental data are confined to simple load histories – often a constant load or a ramp load – while the in-service conditions involve much more complex load histories. DOL experiments are performed by assessing the time to failure of a specimen loaded to a defined percentage of the short-term strength, and yet the assessment of the short-term strength is out of reach because a specimen cannot be broken twice. The anisotropy of wood introduces a variety of failure modes depending on the loading mode; each failure mode may have its own DOL behaviour. Moisture variations may significantly reduce time to failure. This so-called 'mechanosorptive' behaviour consequently adds a size effect to the complexity of DOL behaviour, because large dimension beams experience less moisture variations than small dimension beams. As a result of such complexities, duration of load behaviour is still a very active and relevant area of research. In addition, the practical implications of new research results may be of direct and significant importance to the competitiveness of timber structures.

This chapter contains a brief summary of the state-of-the-art of DOL research. Attention is focused on the most important parameters influencing DOL behaviour. The most frequently used DOL models are presented. The chapter will not deal with the probabilistic analysis involved in assessing the duration of load factors (k_{mod}) to be included in timber design codes. For the treatment of DOL effects in connections, reference is made to this chapter.

8.2 BACKGROUND

This chapter gives a historical perspective of DOL research. The presentation takes the reader from the early research mainly on the behaviour of small clear specimens, to today's full-scale tests, including large glulam beams. The chapter focuses attention on the DOL behaviour of beams subjected to bending. Other loading modes are treated in a subsequent chapter. We do not attempt to give an account of all previous studies, but rather to concentrate on some of the most important works. This background chapter is based largely on the state-of-the-art presentation by Barrett (1996).

The first recorded observation that wood under load experiences a loss of strength was made by the French naval architect Georges Louis Le Clerc, Compte de Buffon (1740). He observed the behaviour of structural oak beams under sustained bending load, and concluded:[1]

> ainſi dans des bâtiments qui doivent durer long-temps, il ne faut donner au bois tout au plus que la moitié de la charge qui peut le faire rompre, & il n'y a que dans des cas preſſants & dans des conſtructions qui ne doivent pas durer, comme lorſqu'il faut faire un Pont pour paſſer une Armée, ou un Echaffaud pour ſecourir ou aſſaillir une Ville, qu'on peut hazarder de donner au bois les deux tiers de ſa charge.
>
> *Mem. 1740.* . Nnn

Buffon's observation that the maximum long-term load for structural timber in bending should not exceed 50% of the short-term strength then fell into oblivion. It would take almost 250 years before his observation was confirmed.

8.2.1 Small Clear Specimens

After Buffon (1740) and until Madsen (1973), almost all load duration experiments were carried out using small, clear straight-grained wood specimens. From the late 1800s until the mid-1900s, studies carried out in the United States provided similar estimates of the long-term strength of wood. It was cautioned that long-term failure may occur even at the 60% stress level, and at the same time suggested that long-term loading does not affect short-term strength and elasticity provided that the load is kept below the proportional

[1] ... thus, in building structures, which are supposed to last a long time, the timber should not be subjected to more than half of the load which causes it to break, and only in the most urgent cases, and in constructions not supposed to last, like building a bridge to let an army cross, or a scaffolding structure to defend or attack a city, can one risk to subject the timber to two thirds its strength.

limit. The proportional limit in bending was taken to be 9/16ths of the short-term bending strength. This value of 9/16ths (56%) formed the basis for DOL adjustments in the US timber design code, and also served as a guideline for most European timber codes. The hypothesis that the proportional limit is a threshold level of stress below which no failure will ever occur was questioned by many researchers, but the existence of a threshold level is still today open to question. Friedley *et al.* (1995) provide an in-depth treatment of the history of DOL research.

Among the numerous early investigations, the work at FPL, Madison stands out. Wood (1947, 1951) published his results on 25 × 25 × 410 mm beams subjected to a constant three-point load in bending. He also included short-term ramp load test results from Markwardt and Liska (1948) and Liska (1950), as well as impact tests by Elmendorf (1916). Wood's results were to become not only the basis for the assignment of DOL factors in the US timber design code, but even the reference for all later DOL research results. The master curve relating all three types of experiments is now commonly known as 'The Madison Curve'. This empirical curve (Figure 8.1) describing the time to failure as a function of the stress ratio (stress in ratio to the ultimate short-term strength) was chosen to fit three pre-selected points, corresponding to:

- Impact loading: stress ratio is 150% at a time-to-failure of 0.015 seconds.
- Short-term ramp loading: stress ratio is 100% at a time-to-failure of 7.5 minutes.
- Long-term loading: stress ratio is 69% at a time to failure of 3750 hours.

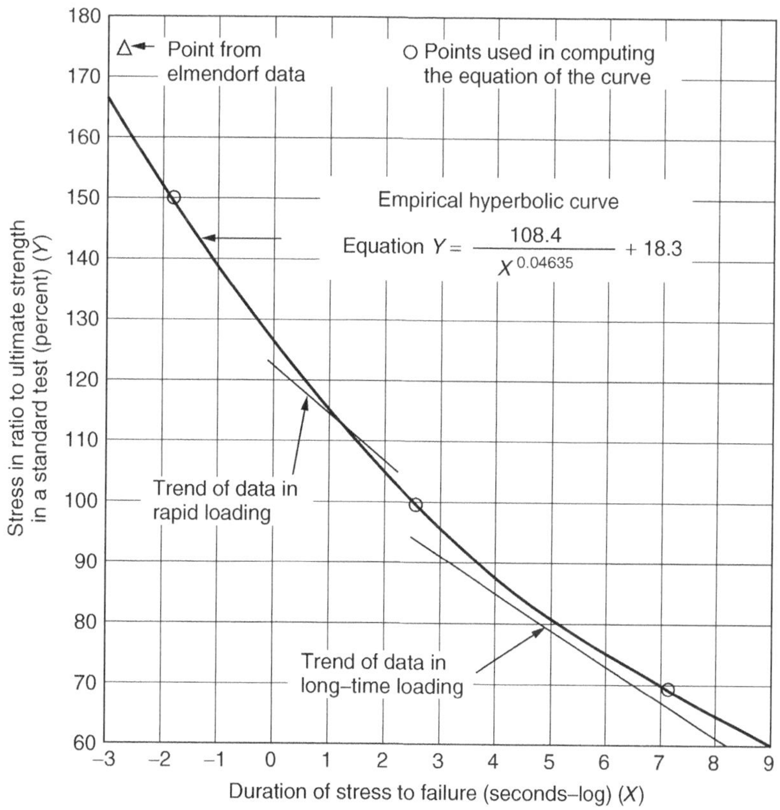

Figure 8.1 DOL behaviour of clear wood. Hyperbolic curve fitted to three different series of data (Wood, 1951), here after Madsen (1992) (Reproduced by permission of Timber Engineering Ltd.)

The hyperbolic Madison Curve has the equation:

$$SL = 18.3 + 108.4t_f^{-0.0464} \qquad (8.1)$$

where SL = actual stress level over predicted short-term strength, and t_f = time to failure in seconds.

The equation predicts failure after 10 years at a constant load, corresponding to 62% of the short-term strength.

Analysis of Wood's constant stress level data shows the following relationship to fit the results better than Equation (8.1):

$$SL = 90.4 - 6.3\log_{10} t_f \qquad (8.2)$$

where t_f = time to failure in hours.

The equation predicts failure after 10 years at a constant load, corresponding to 59% of the short-term strength (Figure 8.2).

Subsequent studies tended to reinforce the philosophy underpinning Wood's work. Pearson (1972) reviewed duration of load papers from eight investigations, and found good agreement in spite of differences of species, dimensions and moisture content (Figure 8.3). The best exponential relationship to fit the results below $SL = 100\%$ was shown to be

$$SL = 91.5 - 7\log_{10} t_f \qquad (8.3)$$

where t_f = time to failure in hours.

The equation predicts failure after 10 years at a constant load, corresponding to 58% of the short-term strength (Figure 8.3). Pearson's 'composite curve' is thus slightly steeper than Wood's straight line relationship (Equation (8.2)).

The use of the Madison Curve (Equation (8.1)) as a reference curve may be questioned. Wood's hyperbolic curve is not representative of his dead load data (Equation (8.2)). The upward bend in a log(time) depiction is the result of mixing two different test procedures, ramp load and dead load, and particularly of trying to fit the curve through data for impact loading. Thus, data for ultra fast ramp loading wags the tail of the curve for dead load behaviour at the longer times relevant for DOL design. The broader foundation of Pearson's composite curve (Equation (8.3)) should form the proper reference, rather than the hyperbolic Madison Curve.

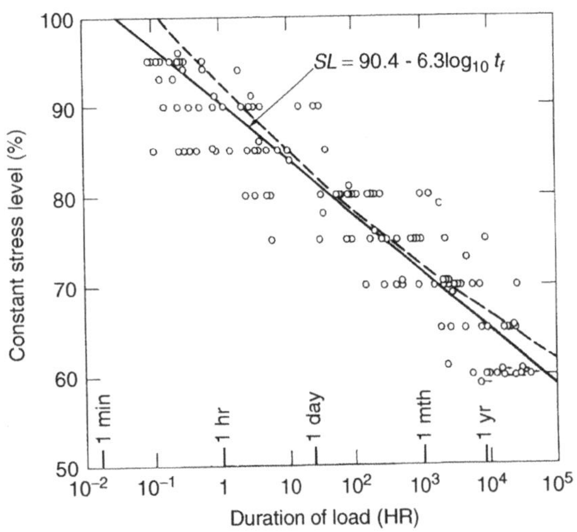

Figure 8.2 Constant stress level data with Equation (8.1) (dashed line) and Equation (8.2) (solid line) (Wood, 1951), here after Gerhards (1977)

8.2.2 Structural Lumber

The pioneering North American work on DOL of structural lumber is reviewed in Barrett (1996). This section is largely based on Barrett's paper.

The use of strength data from small clear specimen testing for the prediction of the behaviour of structural lumber had already been abandoned in some European countries by the early 1960s. In North America, the movement 'from clear wood to lumber' was pioneered by Madsen, who introduced the 'in-grade-testing' philosophy.

In 1970 Madsen initiated a series of studies to assess the long-term behaviour of structural lumber representative of commercial production. In his early studies, Madsen challenged the clear wood philosophy, introduced new sampling and matching techniques, new testing and loading procedures and new data analysis methods.

Madsen's new techniques were to have a profound impact in the design of all future DOL research programs. First, the test material was selected to be representative of a range of commercial production. Secondly, matched groups of

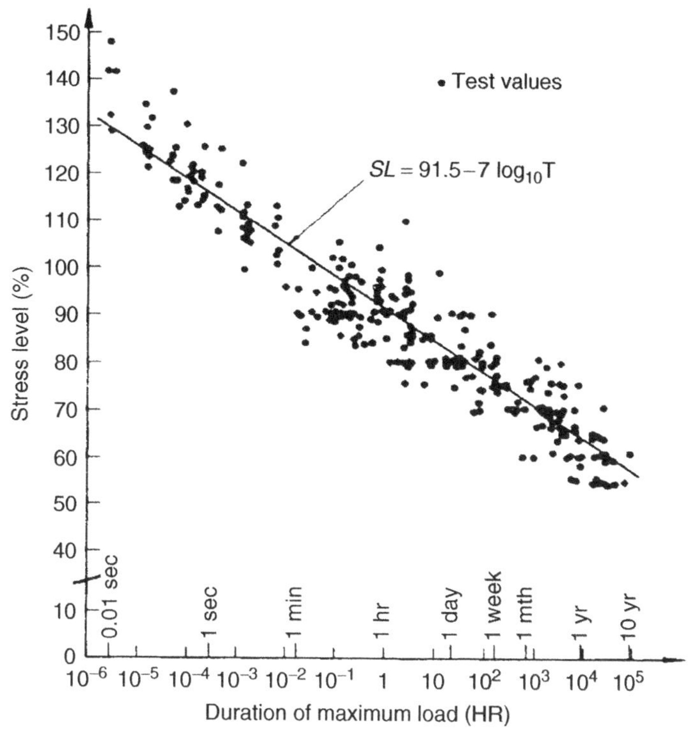

Figure 8.3 DOL behaviour of clear wood in bending. Eight different investigations including a variety of dimensions, species and moisture levels (Pearson, 1972) here after Madsen (1992) (Reproduced by permission of Timber Engineering Ltd.)

specimens replaced the traditional matched specimen technique adopted in earlier small clear test programs. Groups of specimens were matched on the basis of visual and/or non-destructively measured properties, and subsequently subjected to different loading regimes. The behaviour was evaluated at selected percentiles of the resulting time-to-failure (or stress) distributions. Thirdly, a step-wise ramp-load procedure was adopted, where loads were gradually increased until failure occurred (Figure 8.4). Finally, he reported his data on an absolute load level vs. rate of loading basis.

Based on this methodology, Madsen (1973) reported results from step-wise ramp loading of western hemlock lumber. These results suggested that the DOL effect varied with material quality. Low quality (i.e. low strength) material seemed to exhibit less DOL effect than high quality material of the same sample. For low quality material, the DOL effect was significantly smaller than predicted from the Madison Curve for both dry and wet material (Figure 8.5).

The step-wise ramp loading results raised obvious questions: Would constant load tests result in a different DOL response? Would species have an influence? Madsen and Barrett (1976) addressed some of these issues in a large study of Douglas-fir subjected to constant bending loads. Matched groups of specimens were subjected to constant loads at selected percentiles of the short-term strength. The results confirmed the findings from the step-wise ramp load tests, suggesting that DOL effects in lumber are less severe than predicted from the Madison Curve for load durations up to almost one year (Figure 8.6). The results were supported from results of another early Canadian study of DOL effects in western hemlock (Foschi and Barrett, 1982).

The results of the early DOL research on structural lumber initiated by Borg Madsen stimulated

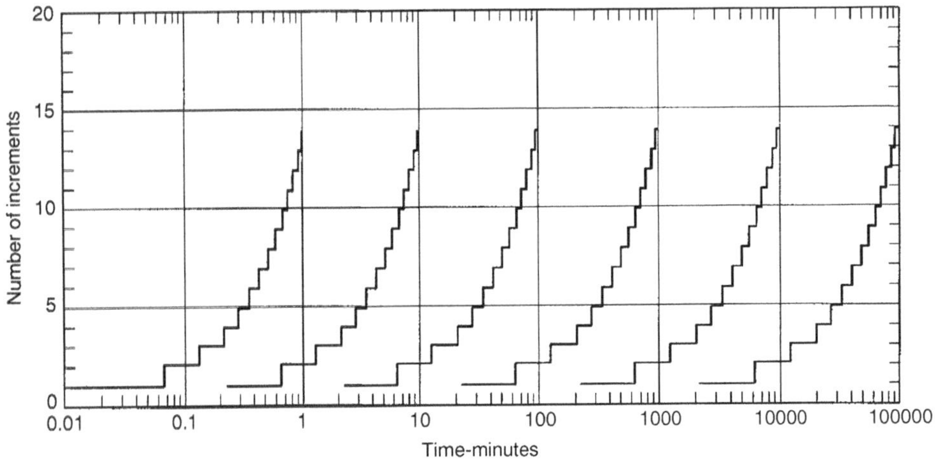

Figure 8.4 Stepwise ramp-loading tests. A constant load increment is applied at constant time intervals. In real time these curves would be stepped, equidistant straight lines. With this loading regime all the specimens will fail and contribute to the information of the DOL behaviour (Madsen, 1992) (Reproduced by permission of Timber Engineering Ltd.)

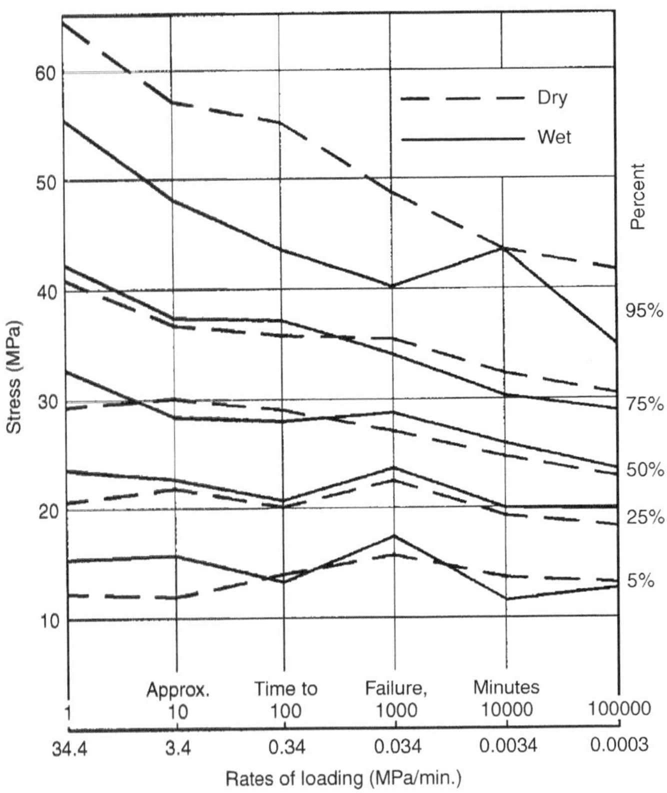

Figure 8.5 Wet and dry bending tests with #2 Grade material subjected to a stepwise ramp loading. The behaviour is very similar for the wet and dry tests, except for the very strong material where dry specimens are stronger than wet specimens (Madsen, 1992) (Reproduced by permission of Timber Engineering Ltd.)

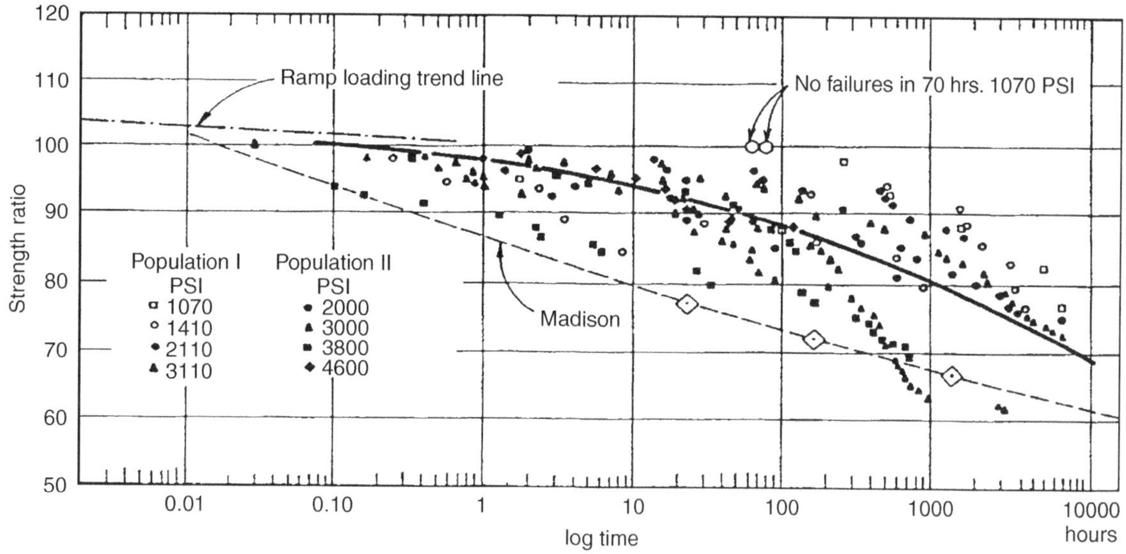

Figure 8.6 Relation between stress level and time-to-failure for Douglas-fir (Madsen and Barrett, 1976)

interest in research on DOL effects, and in the early 1980s DOL research programs were running in Canada, the US and Europe. The hope, of course, was to be able to confirm that the DOL effect for structural lumber really was less than that identified in the various timber codes that were based on the findings of Wood (1951). The 1980s saw large coordinated DOL programs in Canada and the US. This work had two centres: Gerhards and co-workers at Forest Products Laboratory in Madison, WI, USA; and Karacabeyli and co-workers at Forintek, BC, Canada. The results from this phase of DOL research tended to weaken the conclusions of results from the 1970s, both with regard to the Madison Curve being conservative and with respect to an effect of lumber quality on the DOL effect. A detailed account of North American DOL studies from this period can be found in Karacabeyli and Soltis (1991). DOL studies in Europe also seemed to be more in agreement with the Madison Curve, and with no clear effect of quality demonstrated (Fewell, 1986, Glos, 1986, Hoffmeyer, 1987, 1990).

Stress ratio plots for a majority of the North American DOL studies are shown in Figure 8.7 and Figure 8.8 for logarithmic time and real time, respectively (Sharp and Craig, 1996). The real-time depiction is included so as not to forget that DOL effects are particularly significant at short- and medium-terms. For long-term behaviour the time scale does not influence load capacity very much. An error of 2–3 times the correct value in the engineers' assessment of how long a time the design load will last will only cause an error of a few percentage points change in strength (Madsen, 1992).

Figure 8.9 shows the average, minimum and maximum stress ratios for the same data with the Madison Curve superimposed on it. The lumber average trend line is similar to the Madison Curve for the time period of one hour to approximately one year, after which the lumber curve drops below the Madison Curve. An extrapolation of results to ten years duration suggests a DOL factor in the range 0.4–0.6. Such a variation is to be expected, since the results represent a large range of species tested at different laboratories (different procedures) and under different climatic conditions; moisture variations alone can cause such a variation (Figure 8.13).

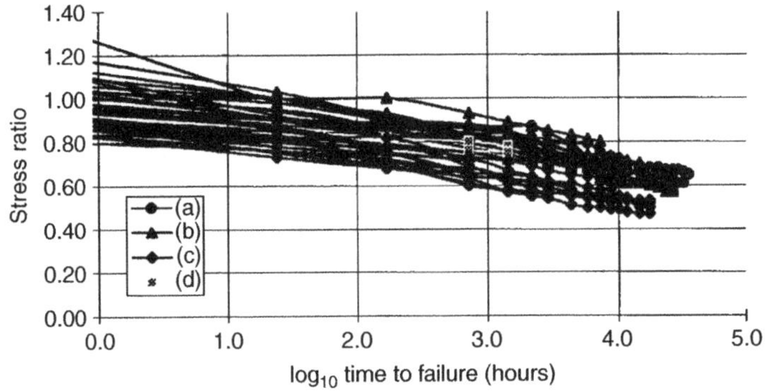

Figure 8.7 Bending DOL response taken from a large number of North American programmes. (a) Western Hemlock – No. 2&Btr, 38 × 140 mm; (b) White Spruce – Quality 1, 38 × 184 mm; Quality 2, 38 × 184 mm; Quality 3, 38 × 89 mm; (c) Douglas-Fir – Sel.Str., No. 2&Btr, 38 × 89 mm and 38 × 140 mm; (d) Southern pine – No. 2&Btr, 38 × 89 mm, high temp. dried, CCA treated (Sharp and Craig, 1996. Reproduced by permission of Vijaya K.A. Gopu)

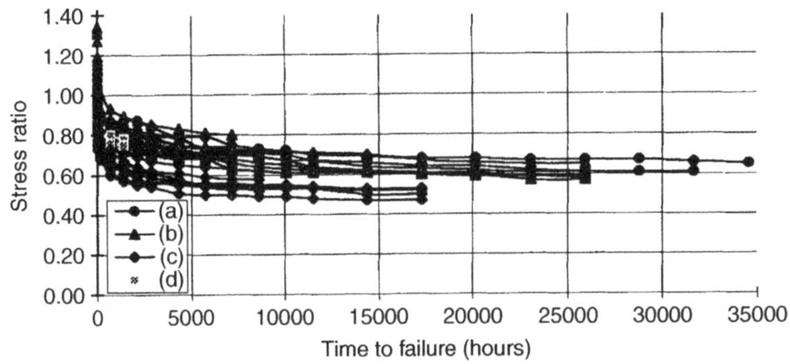

Figure 8.8 Bending DOL response taken from a large number of North American programmes. Species, grades and sizes same as Figure 8.7 (Sharp and Craig, 1996. Reproduced by permission of Vijaya K.A. Gopu)

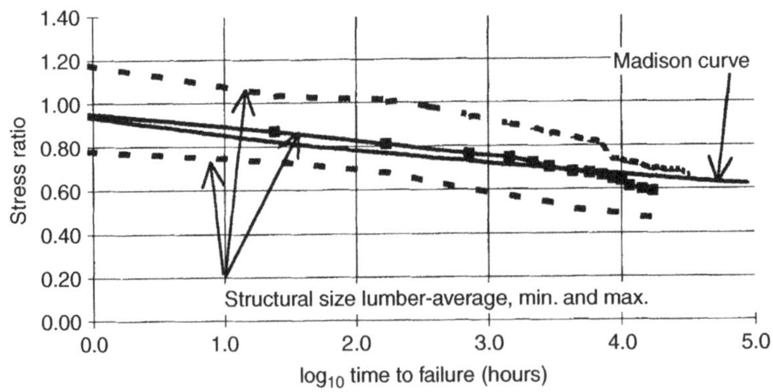

Figure 8.9 Madison Curve compared to lumber average, minimum and maximum stress. Species, grades and sizes same as Figure 8.7 (Sharp and Craig, 1996. Reproduced by permission of Vijaya K.A. Gopu)

8.3 ASSESSMENT OF THE SHORT-TERM STRENGTH OF A DOL SPECIMEN

For the interpretation of results from DOL experiments to be reliable, the assessment of the actual stress level of individual specimens must be reliable. A specimen cannot be broken twice, and since the long-term strength is the target, the short-term strength must be assessed without breaking the specimen. Several methods have been tried in the past. One method is pair matching, which operates by sampling specimens in pairs cut next to each other. One specimen is then tested in a short-term test while the other is long-term loaded. The method works for specimens free from knots and other anomalies, but still has to allow for substantial short-term strength differences between the two specimens of a pair.

To be able to use pair matching for commercial size timber would make DOL research much easier. This was attempted by Norén (1986) by very carefully placing the saw cut not only through the pith of a piece of lumber, but even having it divide the critical knot into two equally sized halves. Despite all his efforts, it was not possible to find a means of pair matching lumber with sufficient accuracy.

Pair matching of lumber has been used with some success for DOL studies of finger joints. This is because two almost identical cross-sections may be obtained from two different boards A and B by cross cutting into two A/A surfaces and two B/B surfaces, and subsequently assembling two identical A/B surfaces by finger jointing. Even such a technique results in substantial variation.

The predominant technique for estimating the short-term strength of structural lumber for DOL tests is based on the Equal Rank Assumption. A detailed description of the technique can be found in Madsen (1992). An adequate number of specimens are ranked according to strength by means of a non-destructive assessment, e.g. based on machine grading or knot size measures. These specimens are sub-divided into two groups of equal size and equal distribution of predicted strength. One group – the control group – is tested in a short-term strength test, while the other group is long-term loaded to a specific percentile of the short-term strength distribution. Depending on the load level, some of the specimens Break On Loading (BOL). The time-to-failure of the specimens surviving the uploading is monitored. Specimens that are still unbroken at the time of termination of the test are taken down to be tested in a short-term strength test. Such a procedure provides three sets of information:

- Short-term strength of specimens failing during uploading.
- Time-to-failure data for specimens failing under constant load.
- Short-term strength of specimens surviving the long-term test.

A typical data set would be represented as shown in Figure 8.10. Triangles represent the short-term strength distribution of 102 specimens. When the matched group is long-term loaded (e.g. to the 5-percentile of the short-term strength), approximately five specimens out of 102 should break on loading. Some of the remaining specimens will then fail under the constant long-term load, while others may survive the test. Let us assume a one-to-one correspondence between short-term and long-term strength. This means that the specimens will fail in order of increasing short-term strength. Once

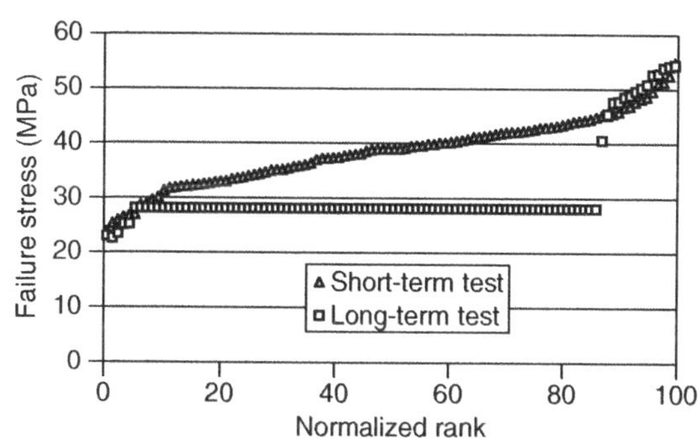

Figure 8.10 Predicting the short-term strength of constant-load specimens using the Equal Rank Assumption. 102 specimens are loaded to the 5-percentile of the short-term strength; five BOL specimens should be expected

the cumulative percentile (or rank) of the long-term data is known, the short-term strength may be established from the short-term strength distribution of a matched control group. The stress ratio applied to a constant load test specimen is established as the ratio of the applied stress to the estimated short-term strength.

The Equal Rank Assumption is an elegant way of 'breaking a specimen twice'. However, the resultant shape of the DOL curve does not necessarily reflect the DOL behaviour of the single specimens; it may only reflect the DOL behaviour of the particular distribution.

As an example, consider the DOL behaviour to be quality-dependent, and assume that the DOL behaviour of specimen no. i follows the relationship $SL = A_i + B_i \cdot \log_{10}(time)$. The quality dependency may take either of two forms: (1) low quality material may exhibit longer times to failure, e.g. due to relaxation of stress peaks around knots; or (2) low quality material may exhibit shorter times to failure, e.g. because failure around knots is influenced by the tension perpendicular to grain mode, where the DOL effect is particularly significant. The Equal Rank Assumption will produce different DOL curves in the two cases. This is illustrated in a simplified example of five specimens of different quality subjected to the same long-term load. Low quality material consequently experiences high stress levels, and vice versa. The DOL behaviour of each of the specimens follows its own straight-line relationship. However, the DOL behaviour of the group of five specimens based on the Equal Rank Assumption is described by a nonlinear curve: case (1) produces a curve with a downward tendency, while case (2) produces a curve with an upward tendency (Figure 8.11).

DOL curves based on the Equal Rank Assumption may thus be sensitive not only to the actual short-term strength distribution, but even to the way the long-term strength distribution may develop as a result of quality, environmental factors, etc. Because of this, DOL results are sometimes depicted solely on the basis of two points: the (SL, log(time to failure)) in the short-term test of the reference population, and the (SL, log(time to failure)) of the 'average' specimen in the long-term tested population (e.g. Hanhijärvi *et al.*, 1998).

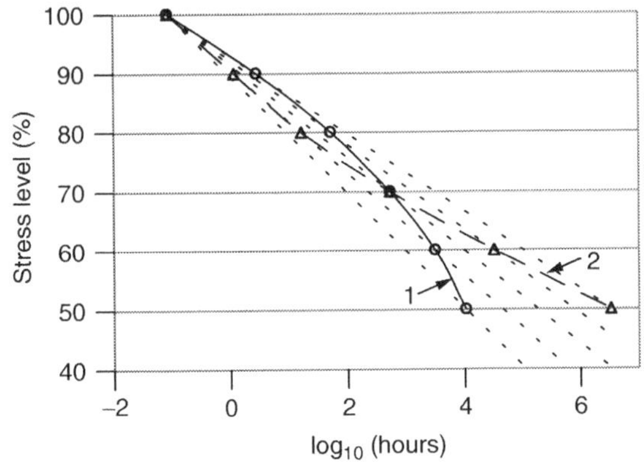

Figure 8.11 A quality dependency of DOL behaviour will influence the shape of DOL curves that are based on the Equal Rank Assumption: (1) lower quality ⇒ longer time-to-failure; (2) lower quality ⇒ shorter time-to-failure

Such use of short-term strength inflicts the problem of relating time under ramp load to time under dead-load. Failure in a short-term ramp load test is normally accomplished in a time, $t_{short,ramp}$ of about five minutes. However, the ultimate load, $\sigma_{ult,ramp}$, in a ramp procedure has only been experienced by the specimen for a very short time; the damage was also accumulated at the lower stress levels, leading to the ultimate load. To equal ramp load with dead load, it is necessary to consider dead load composed of a very short period ($t \approx 0$) of ramp loading up to a stress of $\sigma_{ult,ramp}$, followed by a dead load at stress $\sigma_{ult,ramp}$ acting over a period of time, $t_{short,dead}$ shorter than $t_{short,ramp}$. The ratio $t_{short,dead}/t_{short,ramp}$ may be derived from the models presented in the chapter on modelling.

The rate of loading in the short-term testing of the reference strength is important. The literature reveals times to short-term failure in the range from 30 seconds to 10 minutes. Such a time difference corresponds to about 10% difference in SL! The test standards normally define target time to short-term failure as (300 ± 120) seconds. Such a modest tolerance alone corresponds to about a 2.5% variation in SL. In practice, this means that weak specimens are allocated a relatively higher short-term strength (shorter time to failure) than stronger specimens (longer time to failure)

8.4 THE INFLUENCE OF LOADING MODE

Wood is an anisotropic cellular material which exhibits specific failure mechanisms for specific loading modes. Tension parallel to grain failure of clear wood is brittle, and on the cellular level caused by the rupture of the microfibrils of the cell wall, while compression parallel to grain failure is highly ductile, and on the cellular level caused by the on-set of stability failure in the microfibrils followed by the creation of macroscopically visible folding of the cells. Tension perpendicular to grain failure of clear wood is brittle, and on the cellular level caused by the tearing away of the compound middle lamella of cells or the mere rupture of earlywood cell walls. Compression perpendicular to grain failure, on the other hand, is highly ductile, and on the cellular level caused by the gradual collapse of the tubular structure, eventually making cell lumens vanish. Such different failure modes may well result in DOL effects that are also dependent on loading mode. The various loading modes are known to result in different creep functions. This is an additional reason to expect that creep rupture – or time to failure – is dependent on loading mode.

DOL studies including loading modes other than bending are scarce. However, those available surprisingly suggest that in general there is no significant influence of loading mode for stresses parallel to grain. Karacabeyli and Soltis (1989) report a slightly stronger DOL effect for the tension parallel to grain than for compression and bending parallel to grain. Glos *et al.* (1987) found the Madison Curve applicable also for DOL effects in glulam laminations subjected to tension and bending. For shear parallel to grain, the behaviour seems to follow the same patterns as for bending parallel to grain, while for rolling shear there seems to be a more serious DOL effect (Norlin *et al.*, 1999).

Tension perpendicular to grain strength seems to be more sensitive to the duration of load than bending strength. This is the case not only for small clear specimens (Bach, 1975), but even for specimens cut from glulam beams (McDowall, 1982, Madsen, 1972). Extrapolation of the results indicates a DOL effects in the neighbourhood of 30–50% residual strength after 10 years load duration. A significant DOL effect was also demonstrated by Aicher *et al.* (1998) (Figure 8.14).

8.5 THE INFLUENCE OF MOISTURE

The DOL effect is dependent on moisture content. The higher the moisture content, the shorter the time to failure. The DOL effect is also susceptible to variations in moisture. Such variations are known to significantly increase creep, and therefore should be expected to shorten the time to creep-rupture. The classical illustration of this so-called mechanosorptive effect is shown in Figure 8.12 for small clear, straight-grained specimens subjected to bending. Note how creep at constant high humidity settles at a relatively low level of deformation, while a variation of moisture between a low level and the high level makes the specimen creep much more, eventually leading to creep-rupture.

Moisture content has a similar effect on DOL for structural lumber (Hoffmeyer, 1987, 1990, Fridley *et al.*, 1991). Figure 8.13 shows the results of a European DOL programme including 50 × 100 mm Norway spruce subjected to four-point bending. The programme was terminated in 2000 at a time when the surviving specimens had been loaded for a period of more than 13 years. Three matched distributions (samples) of approximately 100 boards each were subjected to the fifth percentile of the short-term strength of similarly matched distributions (Hoffmeyer, 1990).

Sample I was kept at 11% moisture content, sample II at 20% moisture content, and the moisture content of sample III was made to vary between the low and high moisture content levels. In addition, a fourth sample (sample IV) of constant high moisture content was included and

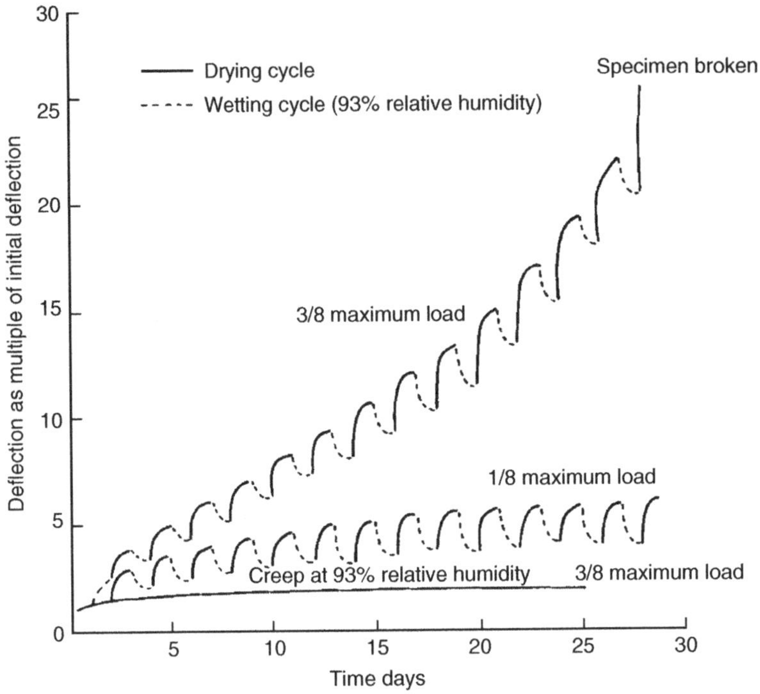

Figure 8.12 The effect of cyclic variations in moisture content on relative creep of samples of beech loaded to $^1/_8$ and $^3/_8$ of ultimate load (Hearmon *et al.*, 1964; here after Dinwoodie, 2000) (Reproduced by permission of BRE)

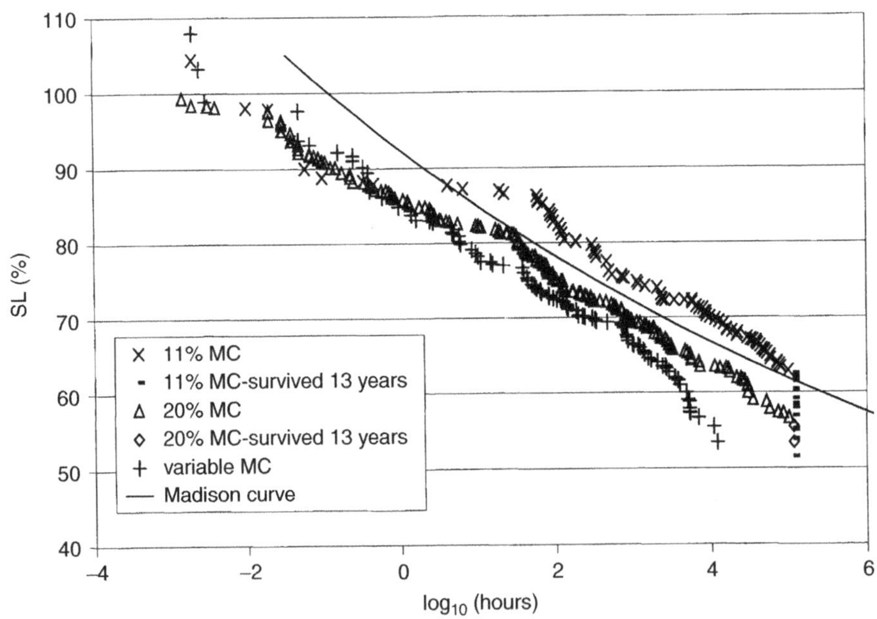

Figure 8.13 Stress level (%) against logarithmic time to failure (hours) for 50 × 100 mm beams of Norway spruce subjected to bending at MC = 10%, MC = 20% and MC varying between 10% and 20%

loaded to the fifteenth percentile of the short-term strength. The stress level was based on the short-term strength at a moisture content level corresponding to the long-term tests, i.e. 11% moisture content for sample I and 20% moisture content for samples II–IV. The results show that the Madison Curve is conservative for constant dry conditions (11%) and slightly non-conservative for constant moist conditions (20%). As soon as the moisture is made to vary, the mechanosorptive effect accelerates the creep, and thus shortens time to failure. A similar very significant mechanosorptive effect is reported for glulam specimens subjected to tension perpendicular to grain (Aicher *et al.*, 1998) (Figure 8.14).

The influence on strength of the combined action of time and moisture is masked when expressing the stress ratio as a function of the short-term strength at the actual moisture content. To express the combined action of time and moisture, it is necessary to base the short-term strength used for calculating the stress ratio on the reference condition (65% RH, 20 °C). This would make the DOL curve for 20% MC (Figure 8.13) steeper, because the short-term strength at the reference condition (app. 12% MC) is higher than that at 20% MC.

Although, as a rule, high moisture and moisture fluctuations shorten time to failure, there are exceptions. One example is the DOL behaviour of notched beams subjected to varying relative humidity. A notched beam subjected to bending fails as a result of a crack developing from the root of the notch and running parallel to grain. The predominant limiting factor in such a failure mode is the tensile strength perpendicular to grain. Because of the structural anisotropy, moisture travels much easier along the grain than across the grain. This means that moisture variations manifest themselves very rapidly through the end grain exposed at the root of the notch. An increasing moisture content therefore rapidly makes the wood volume swell in the vicinity of the notch. As a result, compression stresses perpendicular to grain develop. Such compression stresses are superimposed the tension stresses perpendicular to the grain resulting from the load, and the crack tends to close. The opposite mechanism results from a decreasing relative humidity (Jensen *et al.*, 1996, Gustafsson *et al.*, 1998). The result is a lower probability of failure during periods of increasing relative humidity than during periods of decreasing relative humidity.

For certain loading modes such as tension and bending along the grain, the influence of moisture is a function of timber quality (Hoffmeyer, 1978). Low grade material may therefore act differently from high grade material also with respect to the influence of moisture on DOL effects. In a recent study Gerhards (2000) reported the results of DOL experiments in an uncontrolled environment and of 12-plus years of duration. Load levels were low, and only the low quality specimens had failed. Gerhards results show no effect of moisture fluctuations on the duration of load. He also concluded that after 12-plus years of load duration, there were no signs of approaching a threshold level of stress below which no damage accumulates.

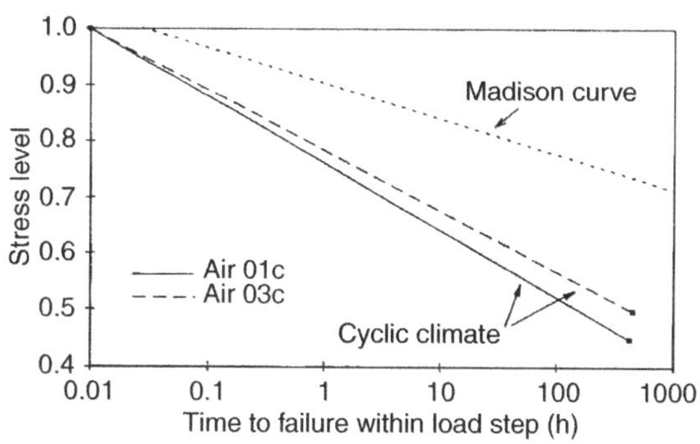

Figure 8.14 Stress level vs. time to failure (hours) for tension perpendicular to grain specimens tested in cyclically varying climate. Specimens from samples AIR 01c and AIR 03c have volumes of 0.01 m^3 and 0.03 m^3, respectively (Aicher *et al.*, 1998) (reproduced by permission of Springer-Verlag)

8.6 THE INFLUENCE OF SIZE

From the fact that the mechanosorptive effect is triggered by moisture variation, it follows that large cross-sectional members are less influenced by such effects than small cross-sectional members. This is because climate variations takes longer to manifest themselves as moisture changes in large dimension lumber. Such an influence of size was confirmed in a major EC research programme concluded in 1997 (see Figure 8.14). Reference is made to Morlier *et al.* (1998), Hanhijärvi *et al.* (1998), Aicher *et al.* (1998) and Gustafsson *et al.* (1998) for further details.

8.7 THE INFLUENCE OF PULSATING LOAD

Most materials exhibit fatigue behaviour, i.e. failure due to repeated load oscillations below the static strength of the material. Normally, fatigue is described by number of oscillations to failure at a given stress level. Test results are depicted in a Wöhler diagram showing Number Of Cycles (NOC) to failure as a function of stress level. Such curves are normally taken to be independent of frequency. This approach may not be representative of the behaviour of wood under pulsating load. The viscoelastic properties of wood make the difference. At high frequencies the time to failure may be so short that the viscoelastic properties do not have the necessary time to manifest themselves; in such cases, NOC is the decisive parameter. On the contrary, at sufficiently low frequencies, it is possible to have failure after only a few oscillations because the Time Under Load (TUL) is the decisive parameter.

As an instructive first approximation, it may be assumed that fatigue failure may be described by a failure surface in a coordinate system spanned by SL, NOC and TUL (Clorius, 2000). Such a failure surface is shown in Figure 8.15. The plane spanned by SL and NOC contains the Wöhler diagram for 'unit frequency'; the curve is assumed to have a fatigue threshold for stresses below a certain non-critical value. The plane spanned by SL and TUL contains a schematic of the Madison Curve, including a hypothetical threshold[2] for stresses below a certain non-critical value. The surface includes the following three regimes:

I. In this regime the failure is dependent on both TUL and NOC, i.e. the fatigue resistance cannot alone be quantified by NOC, because NOC will decrease as the load cycle frequency is lowered.

[2] The existence of such a threshold has not been experimentally verified.

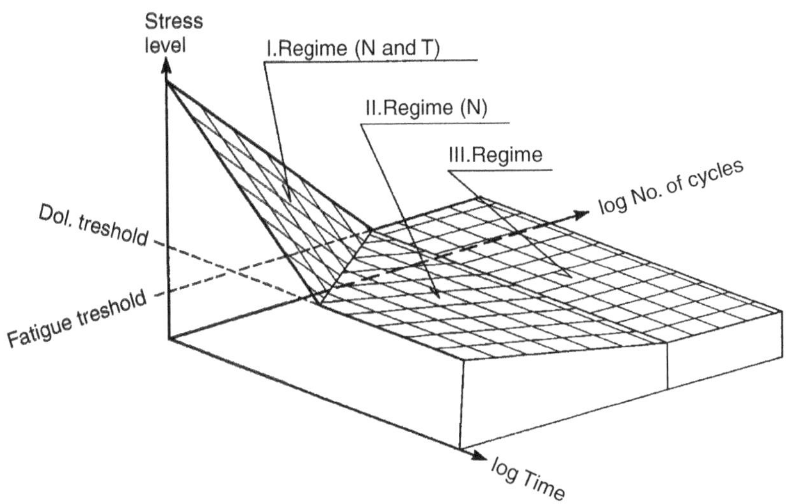

Figure 8.15 Combined fatigue and duration of load failure surface divided into three possible failure regimes (Clorius, 2001)

II. In this regime the surface lies below the TUL threshold, and the failure is solely dependent on NOC.

III. In this regime the surface lies below both the TUL threshold, and the NOC threshold, and no failures are encountered.

Bach (1975) investigated the influence of frequency. Compressive square wave shaped load cycles were imposed on small clear specimens in compression parallel to grain. Frequencies ranged from 10^{-1} Hz to 10^{-5} Hz. Bach suggested that at frequencies from 10^{-1} Hz and lower, fatigue reduces to a pure duration of load phenomenon. In a related experiment, Clorius *et al.* (2000) took small clear specimens to failure in compression parallel to grain at frequencies ranging from 10^{-2} Hz to 10 Hz. It was shown (Figure 8.16) that while NOC to failure increases for increased frequency, TUL decreases. This holds for both of the two moisture levels used in the experiment (RH = 65% and RH = 85%). Hence, the specimens fail as a result of a combined influence of NOC and TUL (regime I in Figure 8.15). In a recent investigation (Clorius, 2001) including tension perpendicular to grain tests at frequencies from 10^{-2} Hz to 1 Hz, a regime II behaviour was found at RH = 65%, while a regime I behaviour was found for RH = 85%.

A comprehensive review of experiments with wood subjected to pulsating loads is given in Clorius (2001) and Kreuzinger *et al.* (1994).

8.8 MODELLING TIME TO FAILURE

For the interpretation of experimental results, the development of a mathematical DOL model is required. Conclusions from constant load tests have to be linked to results from ramp load tests and stepwise ramp load tests. Results from high load experiments have to be extrapolated to the lower loads typical of design situations. The DOL model should be capable of describing the results of a stochastic loading. Fatigue developing from repetitive loadings at varying frequencies should be included in the model, and even the influence of varying moisture conditions should be accounted

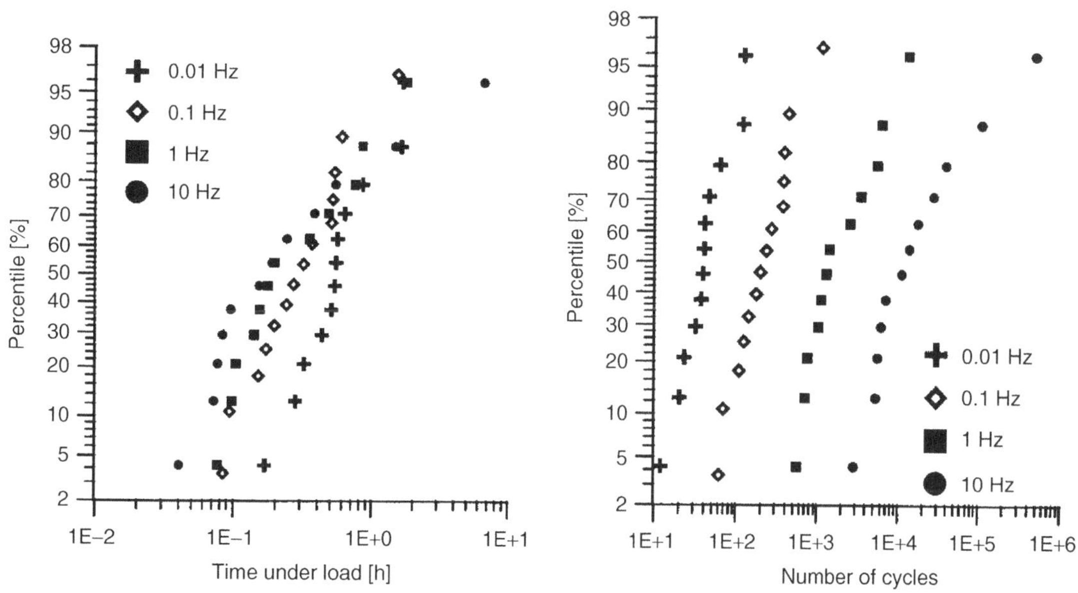

Figure 8.16 Log-normal probability ranking of failures at different frequencies for failure in compression parallel to grain at SL = 80% and RH = 85% (Clorius, 2001)

for. A summary of DOL models for wood can be found in Morlier *et al.* (1994).

Two approaches have proven particularly successful in the development of time-to-failure modelling: accumulation of damage; and fracture mechanics of viscoelastic materials.

DOL models based on chemical kinetics have also been proposed (Caulfield, 1985, van der Put, 1986), as have energy-based models (Bach, 1973, Philpot *et al.*, 1994). However, such models have still not been fully developed to account for the various aspects of DOL mechanisms.

8.8.1 Accumulation of Damage

In this approach, damage is seen as a state variable ranging from 0 at the beginning of load application to 1 at failure.

The works of Gerhards (1977, 1979) and Gerhards and Link (1987) use the concept that the rate of damage accumulation is exponentially dependent on stress level:

$$\frac{d\alpha}{dt} = \exp\left[-a + b \cdot \frac{\sigma(t)}{\sigma_s}\right] \quad (8.4)$$

where $d\alpha/dt$ is the rate of damage accumulation, $\sigma(t)$ represents applied load history, σ_s is short-term strength, and a and b are empirical constants for a given wood population.

Equation (8.4) can be integrated for any stress history $\sigma(t), t \geq 0$, to determine the accumulated damage α. The time at $\alpha = 1$ is identical to time to failure, t_f.

For constant load conditions, and times much larger than the initial ramp loading time, Equation (8.4) becomes

$$t_f = \exp[a - b \cdot SL] \quad (8.5)$$

where SL is the applied stress level.

Thus, Gerhards' cumulative damage equation is seen to be similar to the phenomenological two parameter Equation (8.3) used by Pearson (1972).

Barrett and Foschi (1978a, b) introduced a four parameter damage accumulation model which expresses the rate of damage growth as

$$\frac{d\alpha}{dt} = a\left[\frac{\sigma(t)}{\sigma_s} - \frac{\sigma_o}{\sigma_s}\right]^b + \lambda\alpha(t) \quad (8.6)$$

where a, b and λ are empirical constants for a given population of wood, σ_o is a threshold stress below which no damage can occur, and σ_s is short-term strength.

In this model, damage rate is dependent on the degree of damage as expressed by the term $\lambda\alpha(t)$. The model exhibits the peculiar nature of letting stresses infinitesimally higher than the threshold stress result in exponential growth to failure. This was rectified in a new model (Foschi and Yao, 1986) by letting λ be a function of stress level:

$$\frac{d\alpha}{dt} = a[\sigma(t) - \sigma_o]^b + c[\sigma(t) - \sigma_o]^n \cdot \alpha(t) \quad (8.7)$$

where c and n are empirical constants for a given population of wood.

The parameter a may be approximately obtained as a function of the other model constant as:

$$a = k(b+1)/[\sigma_s - \sigma_o]^{(b+1)} \quad (8.8)$$

where k is the ramp loading rate of the initial loading.

The damage α_o done during this ramp loading can be approximately expressed as

$$\alpha_o = [(\sigma_a - \sigma_o)/(\sigma_s - \sigma_o)]^{(b+1)} \quad (8.9)$$

The time-to-failure t_f for the stress history of Figure 8.17 is obtained from the integration of

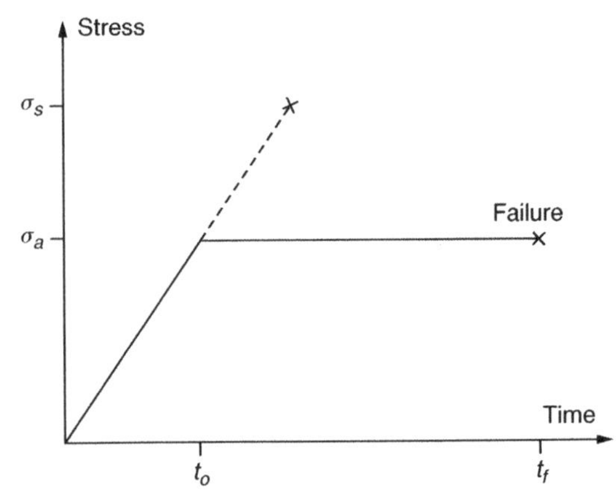

Figure 8.17 Experimental stress history

Equation (8.7) as

$$t_f = t_o + \frac{1}{c(\sigma_a - \sigma_o)^n} \times \ln\left[\frac{1 + (a/c)(\sigma_a - \sigma_o)^{(b-n)}}{\alpha_o + (a/c)(\sigma_a - \sigma_o)^{(b-n)}}\right] \quad (8.10)$$

This damage model has four independent parameters, (b, c, n, σ_o), to be estimated from experiments as compared to the two parameters of Gerhards' model. The two parameters a and α_o are assessed from Equations (8.8) and (8.9). This model is often termed the 'Canadian Model'. It has been used in the reliability-based Canadian timber design code. The general appearance of the model is shown in Figure 8.18.

8.8.2 Fracture Mechanics of Viscoelastic Materials

Damage accumulation models may be seen entirely as appropriate curve fitting tools. Such models do not necessarily reflect any physical/mechanical mechanisms leading to failure. A different approach is to try to model the mechanisms behind the fact that, for most loading modes, wood fails as a result of crack growth.

Viscoelastic fracture mechanics was introduced to the area of wood by Nielsen (1979). A condensed treatment of his Damaged cracked Viscoelastic Material (DVM) theory can be found in Madsen (1992). Nielsen originally developed his DVM model for a linear viscoelastic material by introducing an elastic – viscoelastic analogy, i.e. introducing a time-dependent modulus of elasticity in the Dugdale crack model. Over the years, the model has been further developed and refined by Nielsen to even include modelling of the frequency dependent DOL behaviour under pulsating load (Nielsen, 2000a, b).

The time-dependency of the modulus of elasticity is described by a power law expression, i.e. creep is modelled by the normalised creep function $C(t)$:

$$C(t) = 1 + (t/\tau)^b \quad (8.11)$$

where τ and b are material parameters depending, for example, on loading mode and moisture history.

Nielsen's model for static load considers crack development with time in three phases (Figure 8.19):

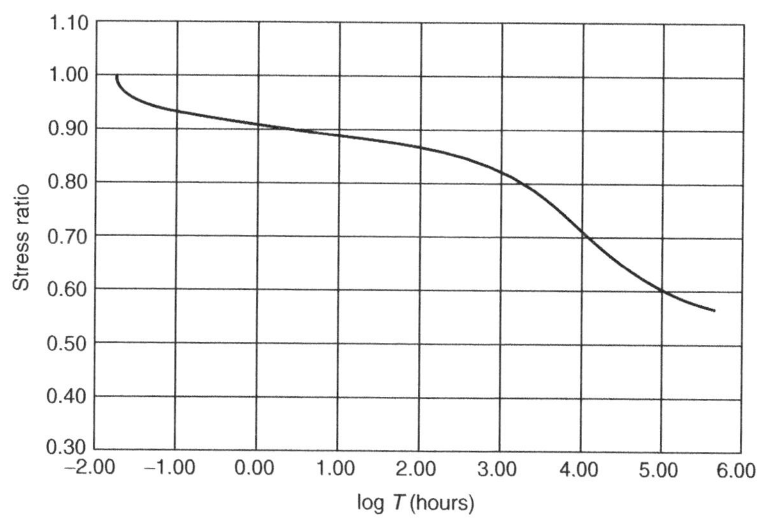

Figure 8.18 The Canadian DOL model (Foschi and Yao, 1986), here after (Madsen, 1992) (Reproduced by permission of Timber Engineering Ltd.)

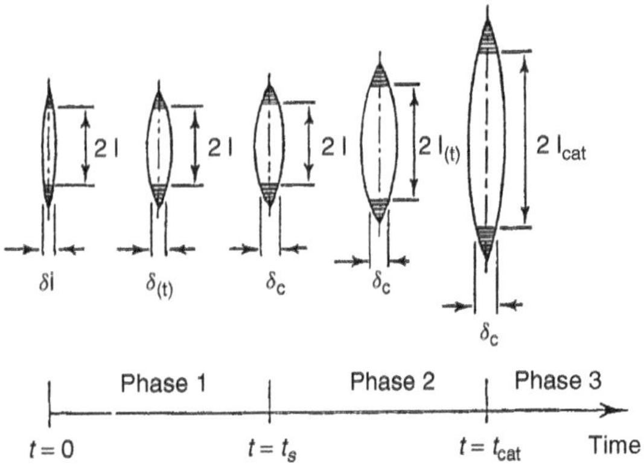

Figure 8.19 Illustration of how crack growth takes place by going through three phases (based on Johns and Madsen (1982) and Madsen (1992))

Phase 1: the crack length stays constant, but due to creep the crack gradually opens to its critical displacement.

Phase 2: from time t_s the crack opening displacement still stays at its critical value, while the crack now starts to grow longer. The short-term strength will diminish.

Phase 3: ultimate failure (*cat*astrophic failure) occurs at time t_{cat} when the crack has grown to the critical length corresponding to the applied long-term load. During Phase 3, tertiary creep may occur.

Time to start of crack growth, t_s, and time to catastrophic failure, t_{cat}, are expressed as

$$t_s = \tau \cdot [SL^{-2} - 1]^{1/b} \tag{8.12}$$

and

$$t_{cat} = \tau[SL^{-2} - 1]^{1/b} + \frac{8\tau q}{(\pi \cdot FL \cdot SL)^2} \times \int_1^{SL^{-2}} \frac{(\theta - 1)^{1/b}}{\theta} d\theta \tag{8.13}$$

where SL = the stress level defined as the ratio σ/σ_{cr} between applied stress and short-term strength measured in a very fast ramp test (actually under an infinite rate of loading). Nielsen's definition of stress level thus differs from the usual definition which relates long-term strength to short-term strength as obtained in a 1–5 minutes test. FL is the strength level defined as the ratio σ_{cr}/σ_L between short-term strength as defined above, and the intrinsic strength of the (hypothetical) non-cracked material. FL may be seen as a measure of the actual quality of the wood. θ is a time variable, and

$$q = [(b + 1)(b + 2)/2]^{1/b} \tag{8.14}$$

Equation (8.13) is seen to reflect an influence of timber quality on time to failure. Timber of low quality (low FL) will have a longer time to failure than high quality timber (high FL).

The parameters required to estimate time-to-failure from Equation (8.13) are the creep parameters b and τ and the intrinsic strength of a non-cracked material, σ_L. Although physical in nature, none of these parameters are readily accessible to measurements. The creep parameters are related to the viscoelastic behaviour around the crack tip, and thus depend upon very localised conditions of grain orientation and morphology. For these creep conditions on the microscopic level to be consistently reflected in macroscopic creep tests, the material has to exhibit balanced creep. Wood does not exhibit balanced creep. The strength of the (hypothetical) non-cracked material can only be assessed theoretically. This means that the materials parameters of Nielsen's model must be obtained by calibration, as is the case for the damage accumulation models. However, the advantage of Nielsen's model over other models is that it offers a consistent theory based on a physical reality (creep and crack growth), thereby giving more immediate clues with respect to the relative influence on DOL behaviour of loading mode, moisture history, density, species, etc.

Figure 8.20 illustrates the quality of fit for Foschi's (Equation 8.7) and Nielsen's (Equation 8.13) models when used on some of the experimental data (11% MC) shown in Figure 8.13. Both models are producing acceptable fits. For very short times, Foschi's model seems to better describe the first steep part of the DOL curve, while for medium times to failure Nielsen's model produces the best fit. For times longer than 20 months, both models produce accurate fits. A threshold level of 50%

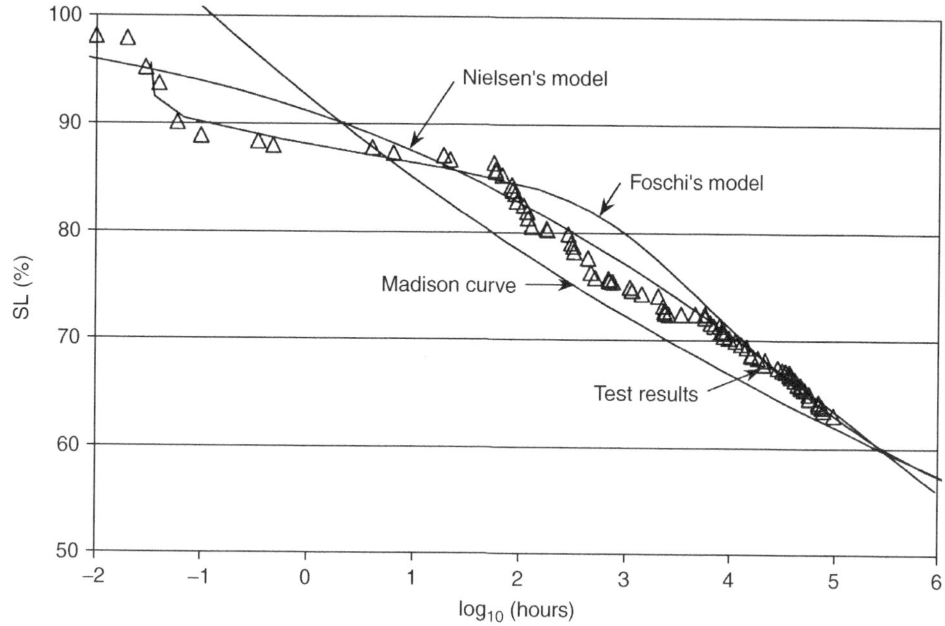

Figure 8.20 DOL results from Figure 8.13 (11% MC) modelled by Foschi's and Nielsen's models

was used for the Foschi model. The three material parameters of Nielsen's model to fit the experimental values were $FL = 0.3$, $b = 0.20$ and $\tau = 50$ days. The bending creep parameters obtained from the actual creep measurement were $b \approx 0.21$ and $\tau \approx 6.3 \cdot 10^5$ days. The fact that the theoretical and experimental values of τ (doubling time) are so different emphasises what was mentioned earlier regarding the experimental difficulties in assessing the parameters independently from the DOL curve itself.

All the models presented above work for variable load histories, but none of the models is able to include the frequency dependency of the load. However, the Nielsen model was recently developed to also accommodate the frequency dependency (Nielsen, 2000a, b). Although it is beyond the scope of this chapter to introduce the theory, an example of its potential is included. In a recent study of DOL in tension perpendicular to grain under pulsating loads (Clorius, 2001), the theory was put to use. The time to failure equation includes five material parameters that need calibration on the basis of experiments. By using material parameters already defined in Nielsen (2000b), a remarkable agreement with new test results was demonstrated (Figure 8.21).

8.9 DOL BEHAVIOUR OF ENGINEERED WOOD PRODUCTS

The duration of load behaviour of high quality plywood and engineered wood products such as glulam and Laminated Veneer Lumber (LVL), Parallam® Parallel Strand Lumber (PSL) is similar to that of solid wood (e.g. see Hanhijärvi *et al.*, 1998). However, for reconstituted wood products such as Oriented Strand Board (OSB), particle board and fibre board, the situation is different. Because such materials are produced by first breaking down solid wood into strands, particles or fibres and subsequently gluing these together to form wood-based panels, the DOL behaviour becomes a function of a variety of factors, such as:

- size, geometry and orientation of the wood elements,

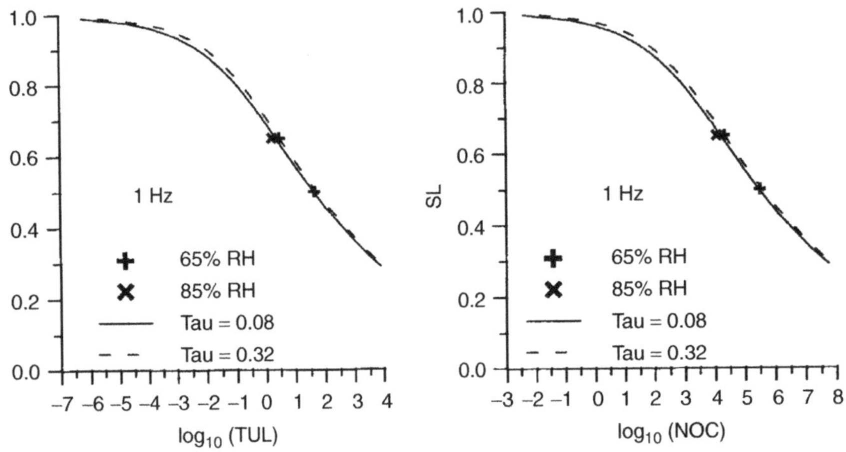

Figure 8.21 Stress level against Time Under Load (TUL) and stress level against Number Of Cycles (NOC) for wood subjected to tension perpendicular to grain failure at 1 Hz. Test results from Clorius (2001) and modelling by the DVM theory. The five material parameters necessary for the model are taken directly from standard values as laid down in Nielsen (2000b) (Clorius, 2001)

- amount of glue and glue type,
- hot press parameters.

Generally, panel products of reconstituted wood have a poorer DOL behaviour compared to solid wood. Panels made from large, oriented strands (e.g. Laminated Strand Lumber (LSL), OSB) exhibit better DOL performance than small, randomly oriented particles or fibres (e.g. particleboard, MDF). For reconstituted wood products with large, parallel oriented strands (e.g. Parallam® Parallel Strand Lumber) the DOL behaviour in the strand direction may approach that of solid wood (Sharp and Craig, 1996).

Research results are scarce and contradictory with respect to DOL behaviour of panel products. Often, between mill variation of a panel type is so large (e.g. see Laufenberg *et al.*, 1999) that it does not make sense to assign a global DOL factor to a particular type of panel product. Instead, it is necessary to refer to a certain standardised quality class (or strength class) when allocating DOL factors.

A quantification of the duration of load behaviour of panel products is presently being introduced as part of the requirements for acceptance into the quality system defined by the European harmonised standard for panel products. The EN-standard provides the quality class system that makes DOL factor assignments meaningful. However, for many of the products there are insufficient research results to allow an acceptance into the system. This scarcity of results therefore makes DOL research for panel products of prime importance.

Eurocode 5 (ENV 1995-1-1) contains information on DOL effects for panel products. Although some of the numbers are not firmly rooted in research results, they still represent the best possible 'educated guess'. The numbers form the basis for Table 8.1, which shows the strength reduction factors to be used for permanent load in dry conditions (service class 1).

Table 8.1 Strength reduction factors to be used for permanent load in dry conditions (ENV 1995-1-1)

Product	DOL factor
Solid wood, glulam, LVL, plywood	0.6
Particleboards (EN 312-6 and 7)	0.4
Particleboards (EN 312-4 and 5)	0.3
OSB (EN 300, grades 3 & 4)	0.4
OSB (EN 300, grade 2)	0.3
Fibreboards (Hardboards, EN 622-2)	0.3
Fibreboards (Medium boards, EN 622-3)	0.2
MDF (EN 622-5)	0.2

REFERENCES

Aicher S., Dill-Langer G. and Ranta-Maunus A. (1998) Duration of load effect in tension perpendicular to the grain of glulam in different climates. *Holz als Roh- und Werkstoff*, **56**, 295–305.

Bach L. (1973) Reiner–Weisenberg's theory applied to time-dependent fracture of wood subjected to various modes of mechanical loading. *Wood Science*, **5**(3), 161–171.

Bach L. (1975) Frequency-dependent fracture in wood under pulsating loading. *Proc. FPRS-Annual Meeting*, Portland, OR..

Bach L. (1975) Failure perpendicular to the grain in wood subjected to sustained bending loads + Letters to the editor. *Wood Science*, **7**(4), 323–327 (**8**(2), 85–87).

Barrett J.D. and Foschi R.O. (1978a) Duration of load and probability of failure in wood. Part I. Modelling creep rupture. *Canadian J. Civil Engineering*, **5**(4), 505–514.

Barrett J.D. and Foschi R.O. (1978b) Duration of load and probability of failure in wood. Part II. Constant, ramp and cyclic loadings. *Canadian J. Civil Engineering*, **5**(4), 515–532.

Barrett J.D. (1996) Duration of load. The past, present and future. *Proc. International COST 508 Wood Mechanics Conference*, Stuttgart, Germany.

Buffon G.L. Le Clerc, Compte de (1740) Experiences sur la Force du Bois. Paris L'Academie Royale des Sciences. *Histoire et Memoires*, pp. 453–467.

Caulfield D.F. (1985) A chemical kinetics approach to the duration-of-load problem in wood. *Wood and Fiber Science*, **17**(4), 504–521.

Clorius C.O., Pedersen M.U., Hoffmeyer P. and Damkilde L. (2000) Compressive fatigue in wood. *Wood Science and Technology*, **34**, 21–37.

Clorius C.O. (2001) Fatigue in Wood – An Investigation in Tension Perpendicular to Grain. PhD Thesis, Department of Structural Mechanics and Materials, Technical University of Denmark.

Dinwoodie J.M. (2000) *Timber, Its Nature and Behaviour*. E & FN Spon, London.

Elmendorf A. (1916) Stresses in Impact. *J. Franklin Institute*, **182**(6).

Fewell A.R. (1986) Testing and analysis carried out as part of the Princes Risborough Laboratory's programme to examine the duration of load effect on timber. *Proc. International Workshop on DOL in Lumber and Wood Products*, Richmond, BC, Canada.

Foschi R.O. and Barrett J.D. (1982) Load duration effects in Western hemlock lumber. *ASCE J. Structural Engineering*, **108**(7), 1494–1510.

Foschi R.O. and Yao Z.C. (1986) Another look at three duration of load models. *Proc. 19th CIB/W18 Meeting*, Florence, Italy.

Fridley K.J., Tang R.C. and Soltis L.A. (1991) Environmental effects on the load-duration behaviour of structural lumber. *Proc. International Timber Engineering Conference, London, UK*, **4**, 180–187.

Fridley K.J., Hunt M.O. and Senft J.F. (1995) Historical perspective of duration-of-load concepts. *For. Prod. J.* **45**(4), 72–74.

Gerhards C.C. (1977) Time-related effects of loads on strength of wood. *Proc. Conference on Environmental Degradation of Engineering Materials*, Blacksburg, VA..

Gerhards C.C. (1979) Time-related effects of loading on wood strength. A linear cumulative damage theory. *Wood Science*, **11**(3), 139–144.

Gerhards C.C. and Link C.L. (1987) A cumulative damage model to predict load duration characteristics of lumber. *Wood and Fiber Science*, **19**(2), 147–164.

Gerhards C.C. (2000) Bending Creep and Load Duration of Douglas-Fir 2 by 4s Under Constant Load for up to 12-plus Years. *Wood and Fiber Science*, **32**(4), 489–501.

Glos P. (1986) Creep and lifetime of timber loaded in tension and compression. *Proc. CIB W/18/IUFRO S5.02*, Florence, Italy.

Glos P., Heimeshoff B. and Kelletshofer W. (1987) Load duration effect in tension and compression and its implications for reliability of structures. *Proc. CEC Seminar on Wood Technology*, Munich, Germany.

Gustafsson P.J., Hoffmeyer P. and Valentin G. (1998) DOL behaviour of end-notched beams. *Holz als Roh- und Werkstoff*, **56**, 307–317.

Hanhijärvi A., Galimard P. and Hoffmeyer P. (1998). Duration of load behaviour of different sized straight timber beams subjected to bending in variable climate. *Holz als Roh- und Werkstoff*, **56**(5), 285–293.

Hearmon R.F.S. and Paton J.M. (1964) Moisture content changes and creep in wood. *For. Prod. J.* **14**, 357–359.

Hoffmeyer P. (1978) Moisture content-strength relationship for spruce lumber subjected to bending, compression and tension along the grain. *Proc. IUFRO/S5.02 Timber Engineering meeting*, Vancouver, BC.

Hoffmeyer P. (1987) Duration of load effects for spruce timber with special reference to moisture influence. *Proc. CEC Conference on Wood Technology*, Munich, Germany.

Hoffmeyer P. (1990) Failure of Wood as Influenced by Moisture and Duration of Load. Doctoral dissertation, State University of New York, College of Environmental Science and Forestry, Syracuse, NY.

Jensen S.K. and Hoffmeyer P. (1996) Mechanosorptive behaviour of notched beams in bending. *Proc. International COST 508 Wood Mechanics Conference*, Stuttgart, Germany.

Johns K. and Madsen B. (1982) Duration of load effects in lumber. Part I: Fracture mechanics approach. *Can. J. Civ. Eng.* **9**.

Karacabeyli E. and Soltis L.A. (1989) Meeting of Duration of Load Research (Lumber) Program Review Committee. Forintek Canada Corp., Vancouver, BC, Canada (here after Barrett, 1996).

Karacabeyli E. and Soltis L. (1991) State-of-the-art report on duration of load research for lumber in North America. *Proc. International Timber Engineering Conference*, London, UK, **4**, 141–155.

Kreuzinger H. and Mohr B. (1994) Holz und Holzverbindungen unter nicht vorwiegend ruhenden Einwirkungen. Institut für Tragwerksbau, Technische Universität München.

Laufenberg T.L., Palka L.C. and McNatt J.D. (1999) Creep and Creep-Rupture Behaviour of Wood-Based Structural Panels. Research Paper FPL-RP-574, USDA Forest Products Laboratory.

Liska J.A. (1950) Effect of Rapid Loading on the Compressive and Flexural Strength of Wood. Report No. 1767, U.S.Forest Products Laboratory.

Madsen B. (1972) Duration of Load Tests for Wood in Tension Perpendicular to Grain. Structural Research Series, Report No. 7, UBC, Vancouver, BC.

Madsen B. (1973) Duration of load tests for dry lumber in bending. *For. Prod. J.* **23** (2), 21–28.

Madsen B. and Barrett J.D. (1976) Time-Strength Relationship for Lumber. Structural Research Series, Report No. 13, UBC, Vancouver, BC.

Madsen B. (1992) *Structural Behaviour of Timber*. Timber Engineering Ltd., Vancouver, BC.

Markwardt L.J. and Liska J.A. (1948) Speed of testing of wood. Factors in its control and its effect on strength. *Proc. ASTM*, **48**, 1139–1159.

McDowall B.J. (1982) The Duration of Load Effect in Tension Perpendicular to the Grain for Douglas Fir. Doctoral Dissertation, The University of British Columbia, Vancouver, BC, Canada.

Morlier P., Valentin G. and Toratti T. (1994) Review of the theories on long-term strength and time to failure. *Proc. COST 508 Workshop on Service Life Assessment of Wooden Structures*, Espoo, Finland.

Morlier P. and Ranta-Maunus A. (1998) DOL effect of different sized timber beams. *Holz als Roh- und Werkstoff*, **56**(5), 279–284.

Nielsen L.F. (1979) Crack failure of dead-. ramp- and combined loaded viscoelastic materials. *Proc. 1st International Conference on Wood Fracture*, Banff, Alberta, Canada.

Nielsen L.F. (2000a) Lifetime and residual strength of wood subjected to static and variable load – Part I: Introduction and analysis. *Holz als Roh- und Werkstoff*, **58**, 81–90.

Nielsen L.F. (2000b) Lifetime and residual strength of wood subjected to static and variable load – Part II: Applications and design. *Holz als Roh- und Werkstoff*, **58**, 141–152.

Norén B. (1986) Results of pair-matching of wood for long-term loading. International Workshop on Duration of Load in Lumber and Wood Products, Forintek Canada Corporation, Special Publ. No. SP 27.

Norlin L.P., Norlin C.M. and Lam F. (1999) Shear behaviour of laminated Douglas fir veneer. *Wood Science and Technology*, **33**, 199–208.

Pearson R.G. (1972) The effect of duration of load on the bending strength of wood. *Holzforschung*, **26**(4), 153–158.

Philpot T.A., Fridley K.J. and Rosowsky D.V. (1994) Energy-based failure criterion for wood. *J. Materials in Civil Engineering*, **6**(4), 578–593.

van der Put T.A.C.M. (1989) Deformation and Damage Processes in Wood. Doctoral Dissertation, Delft University Press, Delft, The Netherlands.

Sharp D.J. and Craig B.A. (1996) Load duration effects in structural composite lumber. *Proc. International Wood Engineering Conference*, New Orleans, LA, **4**, 113–119.

Wood L.W. (1947) Behavior of wood under continued loading. *Eng. News-Record*, **139**(24), 108–111.

Wood L.W. (1951) Relation of Strength of Wood to Duration of Stress. U.S. Forest Products Laboratory, Report No. (1916).

9

Effects of Climate and Climate Variations on Strength

Alpo Ranta-Maunus

9.1 INTRODUCTION

Wood is a hygroscopic material which takes water from surrounding air, and the moisture content of wood tends to settle down to equilibrium with the air humidity. As a consequence, the moisture content of wood is different under different climatic conditions. This is of importance because moisture content has a direct effect on the strength of wood: normally the strength is low when the moisture content is high. However, not only is the level of moisture content important: fast changes cause moisture gradients in wood, which induce stresses in a perpendicular to grain direction, and may cause splitting of the wood. In addition, varying moisture has negative effects on the strength and stiffness of timber under long-term loading; however, this is not the topic of this chapter.

This chapter first briefly describes the range of variation of the moisture content of wood under different climates and seasons. The next section is about the influence of moisture content level on the strength of timber, which is an area where a large amount of research has been done, and it will be covered only partly, to an extent which is thought to be relevant to the design of timber structures. The rest of the chapter illustrates cases where weather change and related transient moisture content in the wood increase the risk of failure, and should be considered as an extra loading of the structure. This is a fairly new topic in research, and has not yet been included in design standards.

9.2 MOISTURE CONTENT IN WOOD CAUSED BY NATURALLY VARYING CLIMATE

Long-term mean values of Equilibrium Moisture Contents (EMC) of wood exposed to outdoor air in different climates have been published in wood handbooks for different locations in the USA. In a northern climate the maximum is during winter, and the minimum during summer. In a continental climate in the North (Missoula, Minnesota), the range is from 9.8–17.6%. In southern areas the maximum is reached during summer and the minimum during winter, as in Los Angeles (EMC = 12–15%). In addition, there are zones with fairly constant moisture contents from dry desserts (below 10%) to humid coastal areas (14–15% in New Orleans). In heated or air conditioned rooms, the Relative Humidity (RH) is normally lower than outdoors. RH may be increased by moisture-producing activities such as cooking, bathing or drying laundry. American recommendations on moisture contents to which wood should be dried when used indoors are shown in Figure 9.1. In Nordic climates the average moisture content of wood is often 12% during summer and 6% during winter. Accordingly, wood should be dried to this range, and often 8% is recommended.

Conditions in heated and in sheltered, unheated buildings in Nordic countries are characterised by Figures 9.2–9.4. Figure 9.2 shows the RH and temperature measurement in heated (above) and unheated (below) buildings in Finland. Both

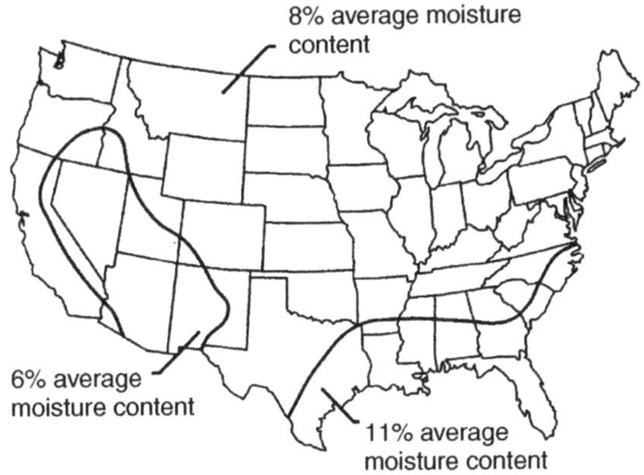

Figure 9.1 Recommended moisture content of wood for interior use in the USA (FPL Wood Handbook)

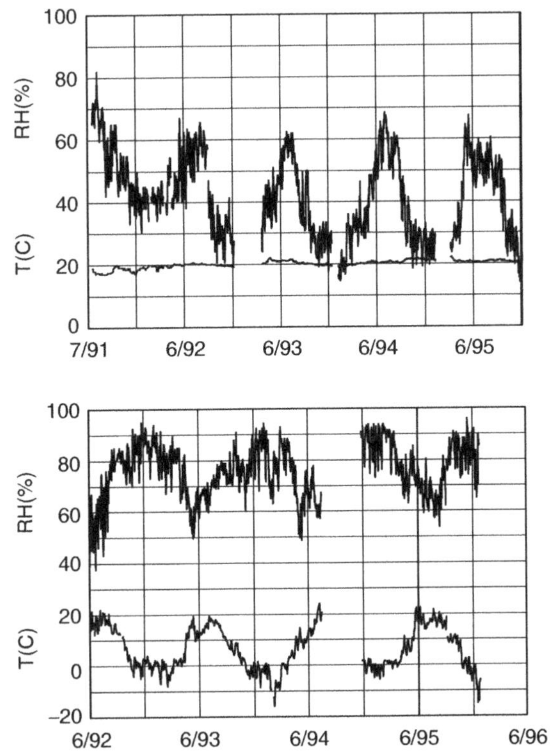

Figure 9.2 Changes in relative humidity and temperature (daily mid-day values) in a heated room in Espoo (above) and in a sheltered environment in Kirkkonummi (below)

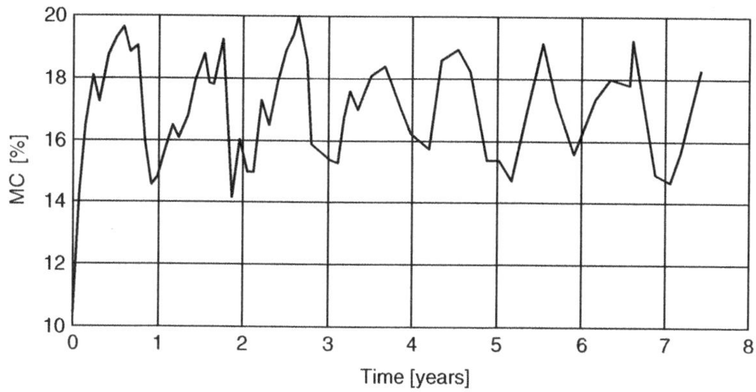

Figure 9.3 Mean moisture content of three end-sealed, untreated 16–22 mm thick boards in a sheltered environment in an unheated room in southern Finland (Kirkkonummi). Time scale begins in June 1992

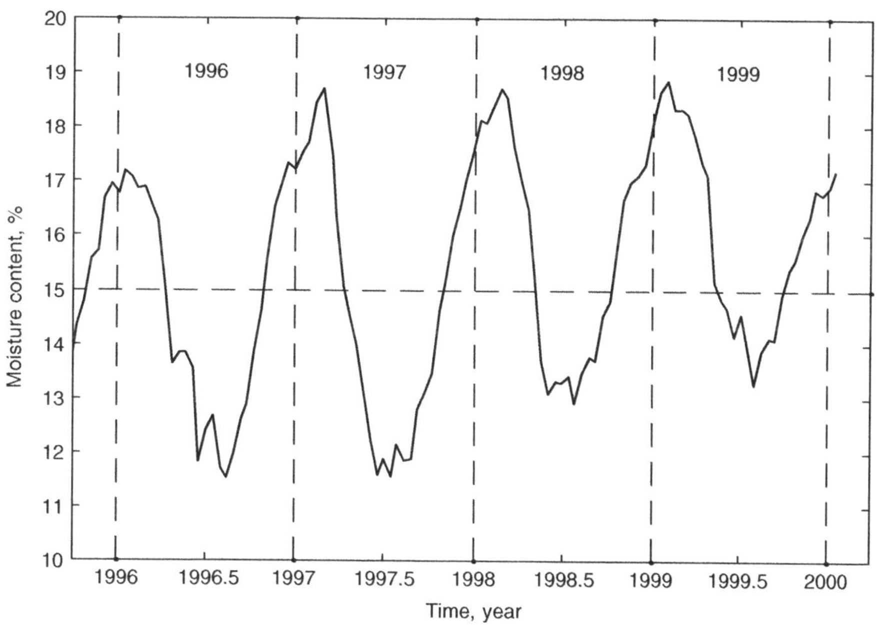

Figure 9.4 Mean moisture content in wood (glulam 90 × 100 × 600) versus time in a barn in Southern Sweden (Åsa) (Gustafsson *et al.*, 1998)

buildings are massive: the heated building is the VTT underground research hall in base rock, which is ventilated but the RH is not regulated, and the unheated building is an old granary made of stone. As part of long-term loading experiments, measurements of the average moisture content have been made by weighting test pieces.

Figure 9.3 shows the variation of the mean moisture content in thin timber in the Kirkkonummi building, and Figure 9.4 that of glulam in a barn in Åsa, Sweden. The latter measurements

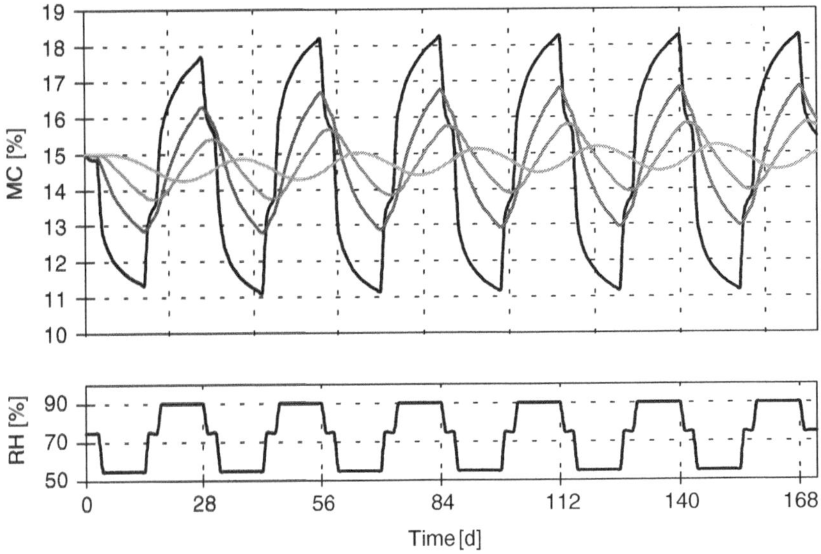

Figure 9.5 Calculated moisture history in cyclic tests in different depths from surface (0, 10, 20 and 45 mm) of 90 mm thick glulam (Figure 61, Gowda *et al.*, 1998)

are made more often and regularly, once in every two weeks, and give a more accurate picture of moisture variation, annual minimum and maximum values ranging from 12–19% MC.

Moisture content in glulam in a transient situation is illustrated in Figure 9.5, where calculated values of moisture contents in different depths are shown. The RH cycle has a length of four weeks with a maximum RH of 90% and a minimum of 55%.

9.3 STRENGTH AT DIFFERENT EQUILIBRIUM MOISTURE CONTENTS AND TEMPERATURES

Changes in the moisture content of wood products cause shrinkage (or swelling), as well as changes in strength and elastic properties. Shrinkage reduces member cross-sectional area, section modulus and moment of inertia.

The experimental results show that bending and compression strengths generally increase with decreasing moisture content below the fibre saturation point. For member capacities, a similar trend is observed in spite of the opposite effect of shrinking dimensions of the cross-section. Tension strength depends less upon moisture content, and this dependence is normally neglected.

Mechanical properties in high quality material are affected more by moisture than in low quality material. Numerical models for the strength and capacities of American and Canadian lumber are given in handbooks (Barrett *et al.*, 1994, *FPL Wood Handbook*, 1999). Moisture effects are often expressed in a simplified way: change in property (%) caused per 1% point change of moisture content. Table 9.1 shows such results for clear wood (Hoffmeyer, 1995), round timber (Ranta-Maunus, 1999) and values adopted in the CEN-standard EN 384 to adjust fifth percentile values of sawn timber. In the standard EN 384, no adjustment is used for bending and tension strengths. This reflects the fact that the lowest fifth percentile of the material is influenced little by moisture. More detailed illustration of the effect of moisture on different fractiles of strength of European and American timber is shown in Figures 9.6–9.8.

Even if dry wood has a higher strength than wet wood, and the trend is obvious, there is a limit in moisture content below which wood does

Table 9.1 Effect of moisture content change to mechanical properties of softwoods (%/%) between 8% and 20% MC. Values of clear wood and round timber are average effects, values of sawn timber are for characteristic values

Property	Clear wood	Round timber	Sawn timber (EN 384)
Compression strength (// and ⊥)	5	5	3
Bending strength	4	1	0
Tension strength (//)	2.5		0
Tension strength (⊥)	2		
Shear strength	3		
Impact bending strength (//)	0.5		
Modulus of elasticity (//)	1.5		2

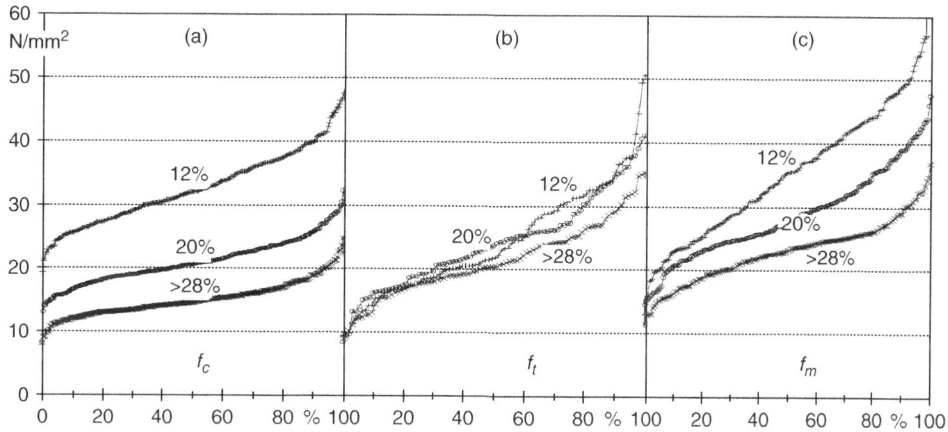

Figure 9.6 Compression (f_c), tension (f_t) and bending (f_m) strength vs. percentile of matched samples of spruce (*Picea abies*) at moisture contents 12, 20 and over 28, according to Hoffmeyer (1995)

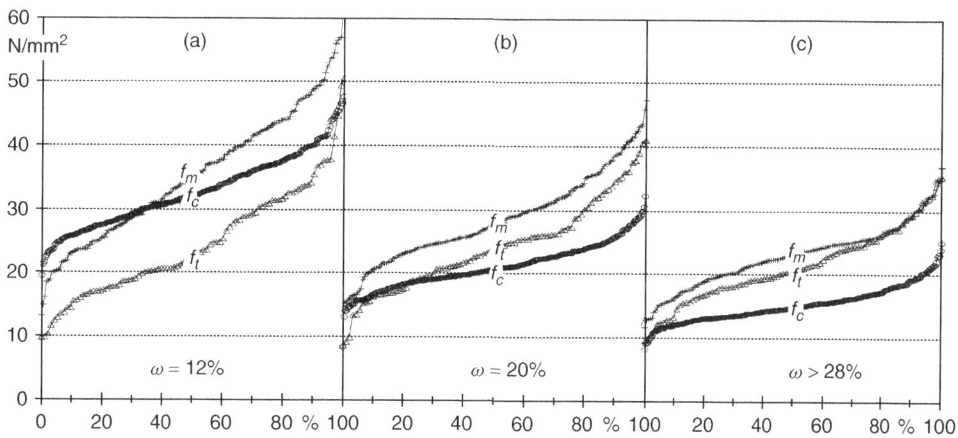

Figure 9.7 Bending (f_m), tension (f_t) and compression (f_c) strength vs. percentile of matched samples of spruce (*Picea abies*) at moisture contents 12, 20 and over 28, according to Hoffmeyer (1995). Reorganisation of results shown in Figure 9.6

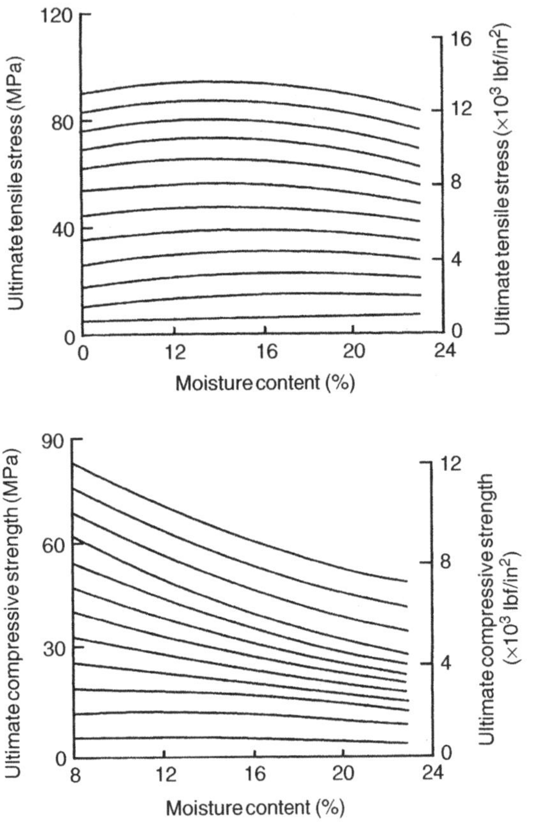

Figure 9.8 Effect of moisture content on tensile and compressive strength on different quality levels of American lumber (FPL Wood Handbook)

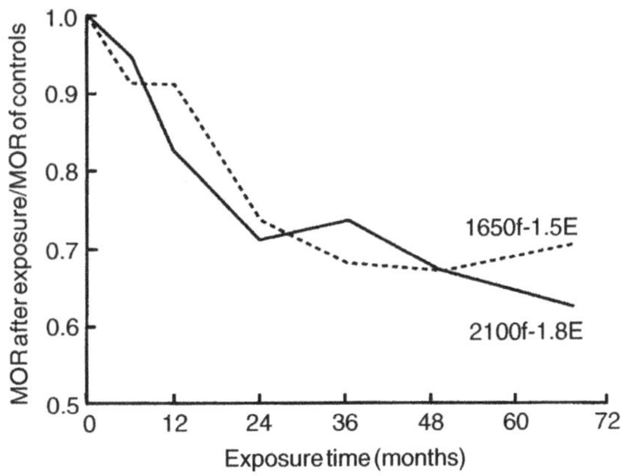

Figure 9.9 Permanent effect of exposure to 66 °C temperature to the bending strength of timber (FPL Wood Handbook)

not get stronger when drying; on the contrary, it may become weaker. This limit is not identical for all species and loading directions, but as a rule of thumb wood has its maximum strength around 10% moisture content. If the low moisture content is a result of long-term exposure to elevated temperatures, this may be an additional reason for reduced strength. When the temperature is above 60 °C, the exposure degrades wood, and the strength is permanently reduced (Figure 9.9). Also, the strength of normal timber tested at elevated temperatures is reduced, depending upon temperature and moisture content. Structures are, however, seldom at such high temperatures that the strength and stiffness reduction needs to be considered.

9.4 EFFECT OF MOISTURE GRADIENTS ON LOAD CARRYING CAPACITY

9.4.1 Experimental Results on Capacity of Curved Beams with Moisture Gradients

Two programs have been carried out to determine the duration of load effect on the tension strength perpendicular to grain in different sized curved beams exposed to cyclically varying humidity. The earlier project (VTT) was made between 1991–1993 (Ranta-Maunus *et al.*, 1994). The later, more comprehensive project (EU/AIR) was completed in 1997 (Aicher *et al.*, 1998, Gowda *et al.*, 1998). As part of the AIR-project, tensile tests with specimen volumes of 0.01 m^3 and 0.03 m^3 were made by FMPA in Germany.

In the experiments, with four weeks duration on one stress level, the ratio of failure load under changing humidity to the short-term strength, k_{DOL} ranges from 0.45 to 0.66 for uncoated specimens (Table 9.2), whereas the ratio in similar experiments at constant humidity is about 0.8, the reference DOL-curve at constant humidity being given in Figure 9.10 (Ranta-Maunus, 2001). The difference is caused by the moisture gradients. k_{DOL} is determined for the average beam in test series with variable humidity. The results show that

Table 9.2 Comparison of DOL factors obtained in cyclic humidity tests (Aicher *et al.*, 1998)

	VTT S2 painted curved beams	VTT S1&3 curved beams	AIR S2 curved beams	AIR S6 curved beams	FMPA small tensile	FMPA small tensile	FMPA large tensile	FMPA large tensile
Conditioning RH (%)	70	70	75	75	65	65	65	65
RH cycle (%)	40<->85	40<->85	55<->90	55<->90	55<->90	natural	55<->90	natural
Width (mm)	90	90	90	140	90	90	140	140
Time to failure (days)	13	20	28	17	18	2.6	19	25
k_{DOL}	0.76	0.55	0.60	0.66	0.45	0.60	0.50	0.64

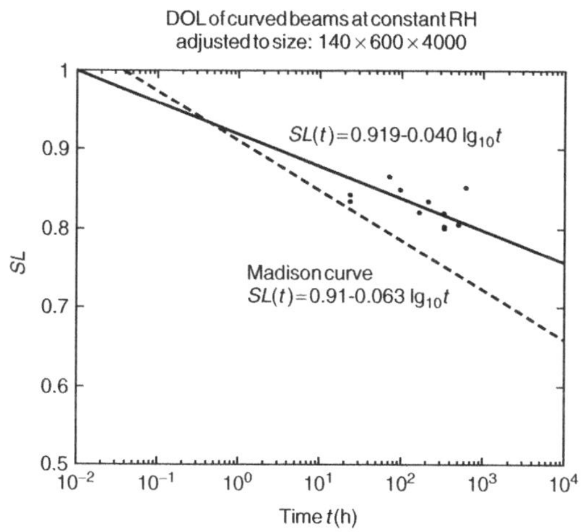

Figure 9.10 Relative stress vs. time-to-failure graph of curved beams at constant moisture content failing due to tension perpendicular to grain. The Madison curve is given here for comparison

the moisture cycles used will roughly double the effect of load duration. Wider cross-sections are less sensitive to moisture cycling than narrow ones. An effective protection against changing moisture appears to be normal surface treatment with alkyl paint. It prevents the major part of the effect of moisture cycling with a cycle length of four weeks.

Based on the first experiments, it was observed that when curved beams loaded under constant load experienced several similar moisture cycles, the beams surviving the first cycle did not fail during the following moisture cycles. The conclusion is that the magnitude of moisture-induced stresses is the primary reason for failure, not the number of cycles or duration of load. A combined moisture and structural analysis indicates that the stress distribution is essentially the same during successive moisture cycles, which supports the assumption that it is enough to analyse the most severe moisture change, not the whole history. Accordingly, the effect of cyclic moisture history affecting structures loaded by tension stresses perpendicular to grain may be seen rather as an extra load than part of the duration-of-load effect weakening the material under long-term loading. If this effect is considered as a DOL-effect, and results obtained during a few weeks loading are extrapolated on a logarithmic time scale to several years' duration, extremely low and unjustified strength values would be obtained.

Nearly all failures took place during the humid part of the moisture cycle, when the surface of the beam was under compression stress, and internal parts under increased tension stress.

9.4.2 Experiments with Notched Beams

Experiments similar to those of curved beams, described in Section 9.4.1, have been made with end-notched beams both in a cyclic laboratory and natural environments (Gustafsson *et al.*, 1998). The result is different from curved beams: the highest load capacities are obtained during a season when moisture content is increasing, and the lowest when wood is drying or at equilibrium moisture content.

Obviously, due to humidity variation, a moisture gradient is also created in the grain direction in the notched area, causing compressive or tensile stresses perpendicular to grain, depending upon the moisture transient. If the potential crack initiation area at the notch is compressed, a friction-like phenomenon is observed: load capacity is increased, but when the failure occurs, it is fast and brittle. However, if the load level at failure is lower and the fracture development is slower, the crack is initiated and kept open by a drying moisture transient. When the notch is coated, the effect of moisture transience is smaller, and the time to failure in long-term experiments is longer. The moisture-related effects to strength were more pronounced in large (300 mm high) than in small (100 mm) beams.

9.4.3 Connections

Some failures have taken place in timber structures because shrinkage of a glulam cross-section has been prevented structurally in a connection area. It is quite obvious that a beam may crack when it is fastened rigidly at two locations far from each other in a direction perpendicular to grain, and exposed to a moisture transient. Another reason for a failure of this type of structure is that after moisture movement, load distribution to connectors may change drastically. However, these important cases when the shrinkage of an entire cross-section creates stress resultants in a direction perpendicular to grain are not analysed in this paper.

Also, moisture gradients in wood may reduce the capacity of connections, even if the moisture movement of the cross-section is not restricted and moisture-induced stresses are self-balancing, as in case of curved beams. The author is, however, unaware of published results in which the effect of moisture changes could be separated from the pure mechanical load effect.

9.5 CALCULATION OF MOISTURE-INDUCED STRESSES

The state of stress in wood is affected not only by external loading, but also by moisture variation, because free moisture movement is restricted. Swelling and shrinkage are strongest in directions perpendicular to grain, and therefore moisture-induced stresses appear primarily in that direction. Most directly moisture-induced stresses are caused in the drying of wood: when green wood is dried, many cracks are often created even without any external load. Stress development and cracking during drying has been analysed by several researchers, but will not be elaborated on here. There are some basic differences to structural applications: the initial moisture content is above fibre saturation point, and the drying temperature is higher than in the normal environment of structures. This makes creep phenomena still more important in the analysis of industrial wood drying than in the analysis of structures. Also, the coupling of heat and mass transfer is more important in kiln drying: heat is transferred to wood in order to dry it via circulated air, which is also transporting the humidity. Accordingly, the needs in analysis of structures are different from those in wood drying, and an excellent engineering procedure in wood drying may be inadequate for structures, and vice versa.

9.5.1 Calculation Method

Calculation of moisture-induced stresses in directions perpendicular to grain includes the calculation of moisture distributions in wood at different times, and the calculation of stresses caused by restrained moisture deformations and external loads. The numerical calculation can use basic equations as follows.

Moisture transport inside wood in one dimension can be calculated using the diffusion equation with an effective diffusion coefficient D_{eff}:

$$\frac{\partial}{\partial t}\int_V u\,dV = \oint_{\partial V} D_{\text{eff}}\frac{\partial u}{\partial x}\,dS \qquad (9.1)$$

where u is moisture content. The equation states that the rate of moisture change in volume V is proportional to the gradient of moisture content, integrated over the boundary of the volume V. The effective diffusion coefficient can be determined as being dependent on moisture content u as

$$D_{\text{eff}} = \exp(a_0 + a_1 u) \qquad (9.2)$$

The mass flux density at the boundary is calculated using an analogy between heat and mass transfer, given by the boundary-layer theory:

$$F_u = k_{\text{paint}}\beta_l(p_v^* - p_v)\frac{\beta_w}{\beta_l} \quad (9.3)$$

where k_{paint} is the resistance caused by surface coating (where applicable), and β_l is the mass transfer coefficient from liquid water. p_v is the vapour pressure, and p_v^* the vapour pressure at the wood surface, which is characterised by the sorption curve. The last correction term is the difference between vapour emission from wood and liquid water surfaces, and is a function of the moisture content of the wood surface. More details on moisture transport can be found in the literature (e.g. Hukka, 1999).

The boundary condition can also be formulated in terms of moisture difference instead of vapour pressure difference. The equations can be solved by numerical methods like the finite element, finite difference or control volume methods.

The calculation of stress is based on a constitutive model including shrinkage, elastic, viscoelastic and mechano-sorptive strain component:

$$\varepsilon_{\text{tot}} = J_0 \cdot \sigma + \varepsilon_{ve}(\sigma) + \varepsilon_{ms}(\sigma) + \varepsilon_s \quad (9.4)$$

It is essential to include mechano-sorptive effect in the equation. Otherwise, we obtain stresses that are far too high. We have modelled the viscoelastic behaviour using the generalised Kelvin material model with seven Kelvin-units in series, and the elastic response is included within the viscoelastic model. Time-moisture content equivalence and a shift factor is used in modelling the dependency of creep rate on moisture content. Mechano-sorptive creep is also modelled using four Kelvin units in series, whose deformation depends upon the absolute value of the moisture content change, but not on time. As background reading on constitutive modelling of wood, a paper by Hanhijärvi (2000) is recommended.

In the one-dimensional case when only the moisture gradient through thickness is considered, and if additionally tension stress perpendicular to grain is caused by a bending moment loading a curved beam, boundary conditions have been applied as follows:

1. Condition of equilibrium of forces:

$$\int_A \sigma\, dx = \frac{3M}{2Rh} \quad (9.5)$$

   where A is the area through the width of a cross-section with unit length, M is the bending moment acting on the curved beam, R is the mean radius of curvature, and h is the height of the beam.
2. As a condition of compatibility, total strain is assumed to be constant throughout the cross-section.

In the cross-section of glued laminated timber, the different elastic properties in the radial and tangential directions have to be taken into account. In 2D analysis this is done by considering the pith location of each lamella and local material orientation accordingly. In 1D analysis this is done by using local effective E-values depending on pith location (Gowda *et al.*, 1998).

Numerical methods to solve the equations are not discussed here. Instead, some published results will be reported. Stresses perpendicular to grain in the curved glulam beam shown in Figure 9.11 caused by external load (bending moment) and varying humidity of air have been calculated, and an example of stress distribution at different times is shown in Figures 9.12 and 9.13. The great variability of stress in the width direction raises a question about the cracking criterion: does the failure take place when the critical value of stress is exceeded locally, or should a more developed criterion be adopted. At least in principle, the maximum stress can be anywhere in the cross-section, most likely in the middle of the beam or at the surface. Another complication is that strength is different in different orientations in the RT-plane. Accordingly, we should use a criterion which takes all these aspects into consideration, and is preferably based on fracture mechanics. We have instead applied Weibull theory, which is normally used for the analysis of different size effects and load configuration factors. Here it has been applied to analyse the severity of stress distribution within

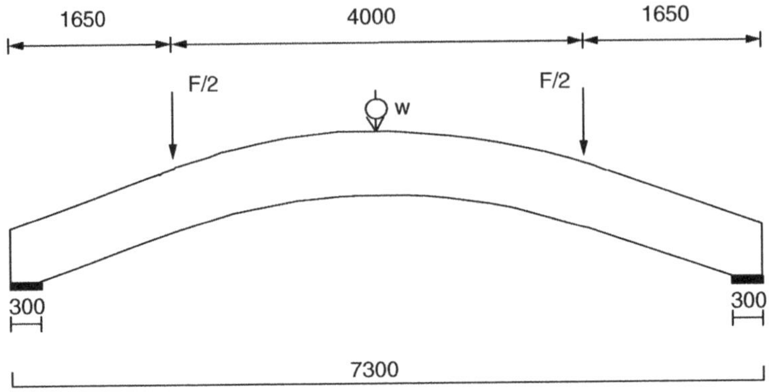

Figure 9.11 Example of the analysed and tested curved glulam beam

a cross-section. The effective Weibull stress caused by external mechanical loads and moisture effects is calculated as

$$\sigma_W = \left(\frac{1}{V_{\text{ref}}} \int_V \sigma_{t,90}^k \, dV \right)^{1/k} \tag{9.6}$$

where $\sigma_{t,90}$ means the tension stress perpendicular to grain. Equivalent stress calculated by Equation (9.6) gives the value of constant stress in the reference volume V_{ref}, causing the same probability of failure as the actual stress distribution in the actual volume V. More details on the calculation are reported in other papers (Gowda *et al.*, 1998, Aicher *et al.*, 1998, Ranta-Maunus, 1998).

9.5.2 Calculated Examples

The method described above has been used to analyse the experiments made with curved beams and tension specimens. Figure 9.12 shows that the stress in the middle of the beam is higher than at the surface, and the maximum is reached during the wetting part of the moisture cycle. This is caused by the cylindrical orthotropy of wood: the central part of the material carries most of the load because wood is much stiffer in the radial direction than in directions between R and T. In this case, the effective Weibull stress is nearly the same as the value of stress in the middle of the beam. The stress distribution at 126 days and 140 days is shown in Figure 9.14. Stresses depend upon the width of the beam: results for a 140 mm wide beam are shown in Figure 9.13 for comparison with the 90 mm wide beam in Figure 9.12. Obviously, the time needed for development of the maximum moisture-induced stresses depends upon the dimensions of the timber member.

The effect of moisture cycles is compared to the effect of mechanical loading at equilibrium moisture content by computing the value of mechanical load (average stress) which causes the same Weibull stress as a combination of mechanical and moisture load. The results are given in Table 9.3. Test cycles and single humidity changes have been analysed. The moisture load corresponds to an extra load of 0.15–0.35 MPa when acting simultaneously with a mechanical load of 0.2 MPa, when the beam is not surface coated. A good surface coating (vapour barrier) will decrease the moisture load from 0.15 MPa to 0.05 MPa. A single fast change from 65% RH to 90% RH seems to be more severe than the test cycles analysed. The conclusion is that fast changes of climate from dry weather to wet season with a duration of a few weeks is most harmful for structures loaded by tension stress perpendicular to grain.

A comparison of calculated moisture loads and observed failure loads is made in a simple way. The results in Table 9.3 (external load 0.5 MPa) are compared to test results: the difference of failure load at constant and cyclic humidity test. Results plotted in Figure 9.15 indicate that calculated stresses are normally higher than

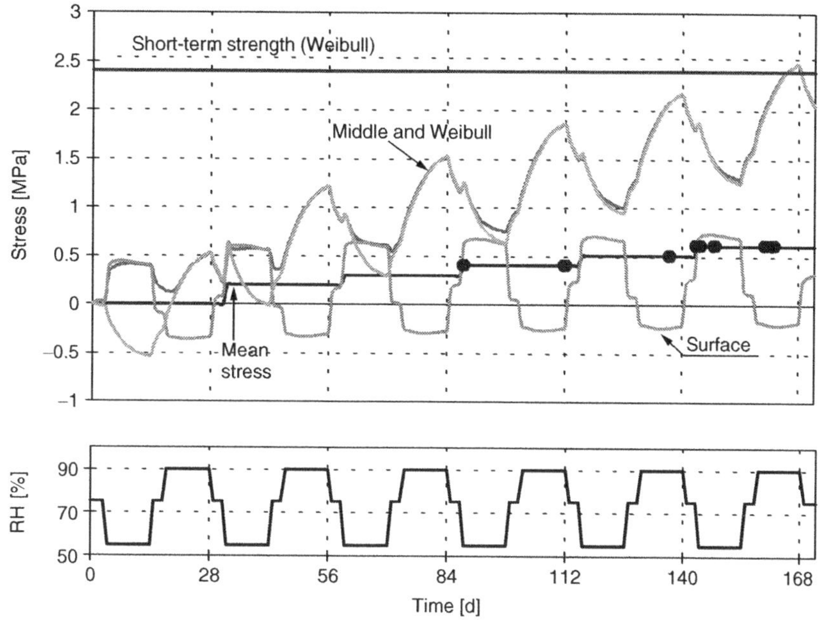

Figure 9.12 Calculated vertical stresses in the middle and at the surface of a 90 mm wide beam simulating a test at relative humidity cycling between 55% and 90%. Dots on the mean stress curve denote the times of failures of beams in the test (Figure 58, Gowda *et al.*, 1998)

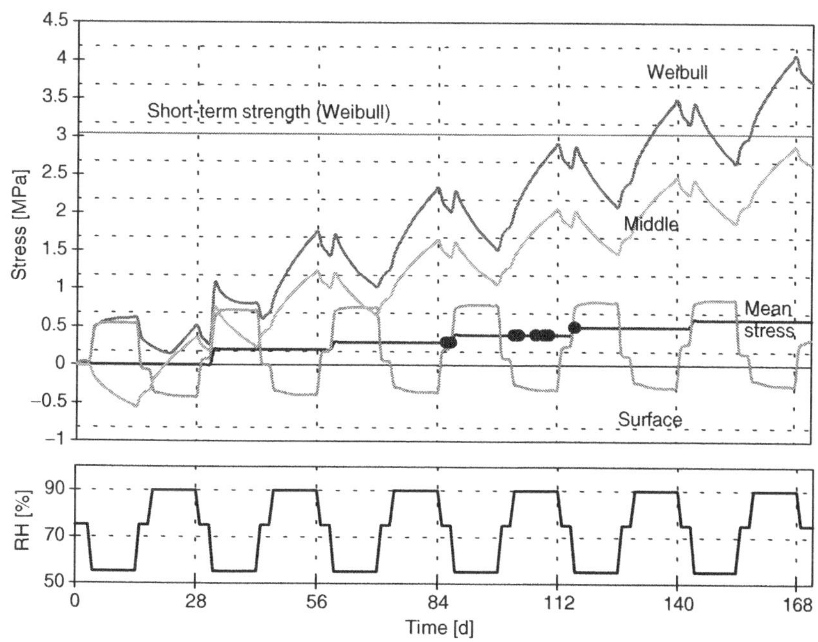

Figure 9.13 Calculated vertical stresses in the middle and at the surface of a 140 mm wide beam simulating a test at relative humidity cycling between 55% and 90%. Dots on the mean stress curve denote the times of failures of beams in the test (Figure 62, Gowda *et al.*, 1998)

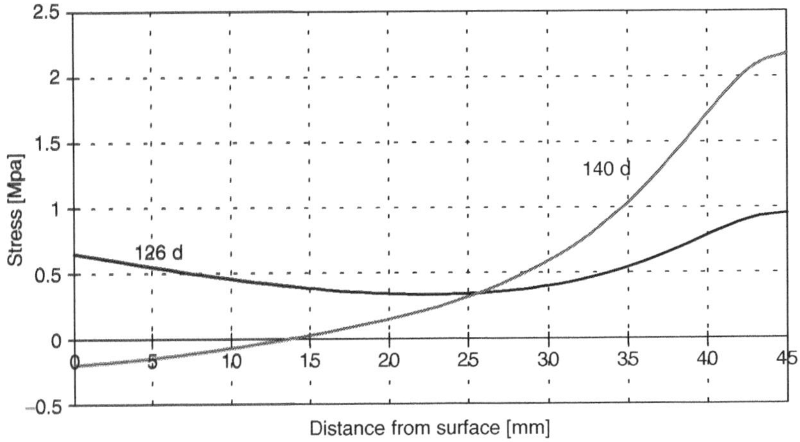

Figure 9.14 Calculated stress distribution for a half thickness of a 90 mm wide beam in test after a dry period (126 days) and a wet period (140 days). The calculated Weibull stresses at the same times are 1.01 MPa and 2.17 MPa, respectively (Gowda *et al.*, 1998)

Table 9.3 Calculated equivalent (mean) stresses for combinations of moisture cycling and load (Ranta-Maunus, 1998)

Thickness (mm)	RH cycle	Equivalent load for external load 0.2 MPa	Combined effect for external load 0.5 MPa
90	55%<->90%[a]	0.45	0.81
140	55%<->90%[a]	0.36	0.73
90	40%<->85%[b]	0.35	0.71
90	40%<->85%[b] painted	0.25	0.57
90	76%->90%[c]	0.40	0.73
90	65%->90%[c]	0.52	0.87
140	76%->90%[c]	0.41	0.75
140	65%->90%[c]	0.55	0.90

[a]Test cycle in AIR experiments at FMPA and VTT.
[b]Test cycle in earlier VTT study.
[c]Single fast change from equilibrium, lasting for four weeks.

observed ones. This discrepancy may be partly caused by differences between the real test conditions and the simplified ones used in the analysis: the cycle used at FMPA was calculated as a single change from 65% to 90% RH.

9.6 EFFECT OF MOISTURE GRADIENTS ON TEST RESULTS

Stresses caused by moisture gradients are also very important in test situations. When tested under tension in a direction perpendicular to grain, test specimens should be conditioned to test moisture content with special care. The moisture gradients have a considerable effect on the test results, which may be too low or too high depending on the gradient. This is important both in short-term and long-term testing. If a long-term test is made at constant humidity, but the conditioning humidity is not exactly the same as during loading, a moisture gradient will be created. If the difference is 3% in EMC, so that the test specimen is wetting during the experiment, as much as 35% lower strength values may be obtained. If the change is in the drying direction, the result is too high a

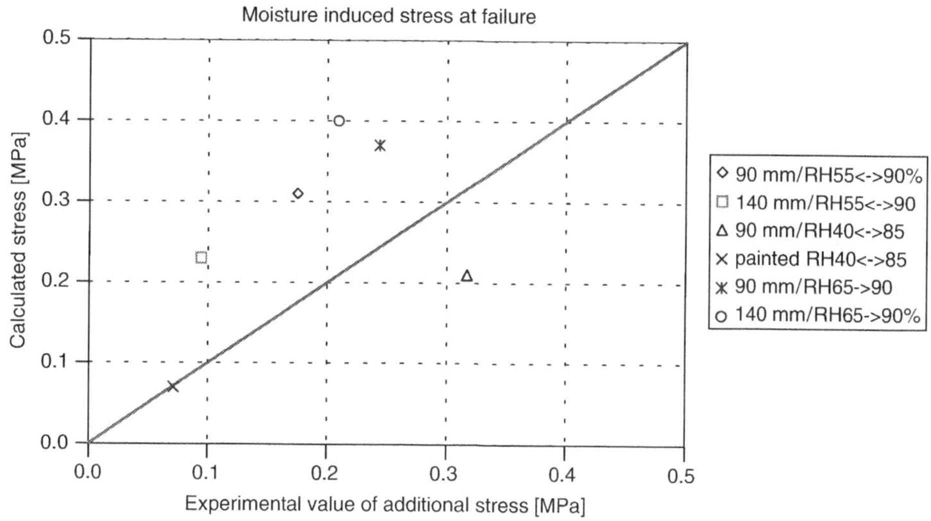

Figure 9.15 Comparison of calculated moisture stresses and experiments

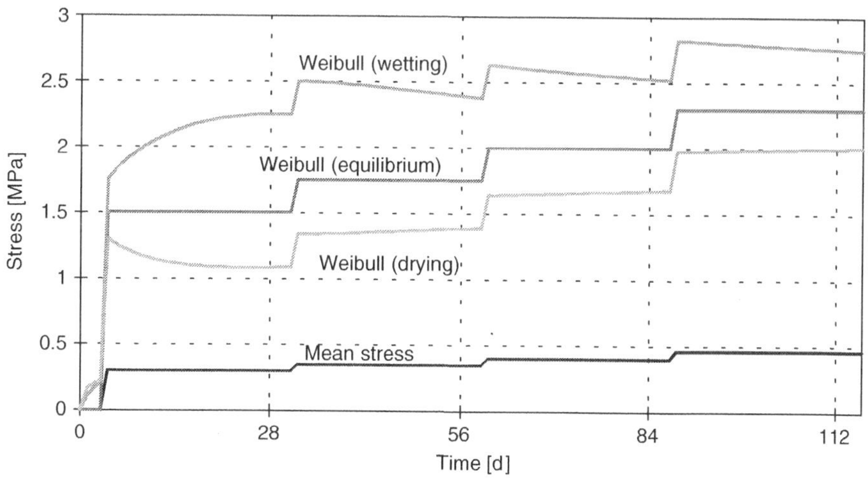

Figure 9.16 Calculated Weibull stresses in 140 mm wide test beams when pre-test conditioning moisture content is the same as during the test or 3% EMC lower or higher (Figure 57, Gowda *et al.*, 1998)

strength value. This is illustrated in Figure 9.16, where a long-term test under stepwise increasing load is analysed at constant humidity when the conditioning moisture content is correct, and when it is 3% higher or lower. The effective Weibull stress is shown for these three cases, indicating that the highest effective stress is nearly double the lowest one. This is expected to increase the variability of test results obtained.

9.7 CONSIDERATION OF MOISTURE GRADIENTS IN STRUCTURAL DESIGN

It is suggested that transient moisture conditions should be considered as a load case instead of as a strength reducing factor in the design calculation of timber structures when wood splitting is the failure mode.

The design equation for multiple loads is expressed in design codes in principle as follows:

$$\gamma_G \sigma_G + \gamma_Q(\sigma_{Q1} + \psi \sigma_{Q2}) \leq \frac{k_{\text{mod}} f}{\gamma_M} \qquad (9.7)$$

where σ_G is stress caused by permanent load, σ_{Qi} is stress caused by variable load Q_i, γ-factors are the partial safety factors for loads and material, ψ is the combination factor (less than 1) indicating that two different variable loads seldom have maximum values at the same time, and f is strength, modified to the appropriate service condition (load duration, moisture) by factor k_{mod}. Loads Q_1 and Q_2, for instance, can be snow and wind. When wood is loaded in tension perpendicular to the grain, moisture gradients should also be considered as another natural load. When doing so, two interesting questions arise:

1. Are the stresses additive, as assumed when writing Equation (9.7).
2. What should be the combination factor ψ when combining moisture loads with snow and wind loads.

The first question can be discussed by comparing the effective stress values obtained for different mechanical loads in the case of a curved beam (Table 9.3). The answer was 'no but yes', meaning that the theoretically correct method is to analyse all effects simultaneously, but due to severe problems in doing so in structural design, the effects have to be analysed separately. The error made when stress components are added can be tolerated, and overcome in the development of the design method.

The combination factor question has not yet been analysed. It seems likely that the humidity changes are not most severe when snow load has its maximum. It is, however, difficult to see any meteorological reason why maximum wind and humidity could not take place simultaneously. Accordingly, we can suggest that a combination factor of wind and moisture loads is about the same as the combination factor of wind and snow, whereas a combination factor for snow and moisture might be lower.

A simple approach to considering moisture-induced loads would, based on the experimental and calculated results presented here, be to give values for moisture stresses in design code unless a more precise analysis is made. Rough estimates for moisture-induced stresses perpendicular to grain could be 0.25 MPa for uncoated and 0.1 MPa for well coated beams. These values could be reduced by a factor ψ when combined with other stresses. In a similar way, moisture-induced stresses should be added to mechanical stresses in the design of end-notched beams and large mechanical connections.

9.8 SUMMARY

The strength of wood, especially compression strength, depends upon moisture content. Under normal climatic conditions, moisture gradients in wood enhance the risk of cracking, and can have an effect on load carrying capacity. This is true when the failure mode is splitting of wood caused by tension stress perpendicular to grain, or shear stress, or a combination of them.

The effect of moisture gradients is not taken into account in design calculations. Sometimes the effect of moisture gradients is considered as part of the duration of load effect, decreasing the strength of wood. The problem with this approach is that when test results obtained during a certain period of time are extrapolated to a long load duration, extremely low strength values are obtained. A more correct approach would be to consider stresses caused by transient moisture conditions as loads, and to combine these moisture-induced stresses with mechanical stresses.

REFERENCES

Aicher S., Dill-Langer G. and Ranta-Maunus A. (1998) Duration of load effect in tension perpendicular to the grain of glulam in different climates. *Holz als Roh- und Werkstoff*, **56**, 295–305.

Barrett J.D. and Lau W. (1994) *Canadian Lumber Properties*. Canadian Wood Council, Ottawa.

Forest Products Laboratory (1999) *Wood Handbook – Wood as an Engineering Material*. Gen. Tech. Rep. FPL-GTR-113, Madison, WI: US Department of Agriculture, Forest service, Forest products laboratory.

Gowda S., Kortesmaa M. and Ranta-Maunus A. (1998) *Duration of load effect on curved glulam beams. Part 2. Long term load tests and analysis*. VTT Publications 334, Espoo.

Gustafsson P.J., Hoffmeyer P. and Valentin G. (1998) DOL behaviour of end-notched beams. *Holz als Roh- und Werkstoff*, **56**, 307–317.

Hanhijärvi A. (2000) Advances in the knowledge of the influence of moisture changes on the long-term mechanical performance of timber structures. *Materials and Structures/Materiaux et Constructions*, **33**, 43–49.

Hoffmeyer P. (1995) Wood as building material. STEP Lecture A4, Centrum Hout, Holland.

Hukka A. (1999) The effective diffusion coefficient and mass transfer coefficient of Nordic softwoods as calculated from direct drying experiments. *Holzforschung*, **53**, 534–540.

Ranta-Maunus A. and Gowda S. (1994) *Curved and cambered glulam beams. Part 2. Long term load tests under cyclically varying humidity*. VTT Publications 171, Espoo.

Ranta-Maunus A. (1998) Duration of load effect in tension perpendicular to grain in curved glulam. CIB W18 Meeting in Savonlinna, Paper 31-9-1.

Ranta-Maunus A. (1999) Round small diameter timber for construction. Final report of project FAIR CT 95-0091, VTT Publications 383, Espoo.

Ranta-Maunus A. (2001) Moisture gradient as loading of curved timber beams failing in tension perpendicular to grain. *IABSE Conference Innovative Wooden Structures and Bridges*, Lahti, Finland, pp. 29–31.

Thelandersson S. (1994) Variation of moisture content in timber structures. In: *Creep in Timber Structures*, Ed. P. Morlier. RILEM Report 8, E&FN SPON, London.

PART TWO

Design Aspects of Timber Structures

10

Introduction: Safety and Serviceability in Timber Engineering

Sven Thelandersson

10.1 CHARACTERISTIC FEATURES IN THE DESIGN OF TIMBER STRUCTURES

Timber is in many ways very different from other structural materials, such as steel and concrete. The most important material characteristics governing the structural behaviour of timber have been described in Part One of this book. The following aspects are of special importance in design of timber structures:

- Low weight in relation to strength.
- Significant variability and inhomogeneity.
- Anisotrophy, i.e. properties depend upon direction.
- Interaction with moisture.
- Significant duration of load effects on strength.
- Significant time-dependent deformations.

For steel and reinforced concrete, design models based on plasticity can be applied for many of the most frequent failure modes. In contrast, the most common failure modes in timber elements are brittle or semi-brittle, which means that the ability to redistribute load effects in structural timber systems is limited. It should be noted, however, that mechanical timber connections with slender dowel type fasteners normally exhibits plastic behaviour.

Table 10.1 Strength and stiffness parameters for structural timber

Type	Mode	Orientation	Notation
Strength	Bending	Parallel	f_m
	Tension	Parallel	f_t
	Compression	Parallel	f_c
	Longitudinal shear	Parallel	f_v
	Tension	Perpendicular	$f_{t,90}$
	Compression	Perpendicular	$f_{c,90}$
	Rolling shear	Perpendicular	$f_{v,90}$
Stiffness	Tension/compression/bending	Parallel	E
	Longitudinal shear	Parallel	G
	Tension/compression	Perpendicular	E_{90}
	Rolling shear	Perpendicular	G_{90}

The anistropic behaviour of wood means that design strength values have to be defined for different material directions, and distinctions have to be made between tension and compression. Moreover, as described in Part One, the fact that timber is inhomogeneous with knots and other growth characteristics implies that strength in many cases has to be defined at an element level or cross-section level, rather than at the material level. The mechanical material response of structural steel can be characterised for engineering purposes by a few parameters–uniaxial yield strength f_y, elastic modulus E and Poissons ratio ν – whereas for structural timber, the corresponding characterisation requires at least seven strength parameters and four stiffness parameters (see Table 10.1). The situation is similar for many other wood-based structural products.

The above circumstances requires special attention by engineers designing timber structures. This will be further elaborated below, where design for ultimate limit states as well as for serviceability limit states will be discussed.

10.2 RELIABILITY AND LIMIT STATE DESIGN

One of the objectives for structural design is to fulfil certain performance criteria related to safety and serviceability. Such a performance criterion is usually formulated as a limit state, i.e. a mathematical description of the limit between performance and non-performance. Parameters used to describe limit states are, for instance loads, strength and stiffness parameters, dimensions and geometrical imperfections. Since these parameters are random variables, the outcome of a design in relation to a limit state is associated with uncertainty. The main issue is to find design methods ensuring that the relevant performance criteria are met with a certain desired level of confidence or reliability. Another way to describe this is to say that the risk of non-performance should be sufficiently low. A more detailed description of reliability concepts and analysis methods is given by Foschi in Chapter 11.

The question of reliability is especially complicated for timber because of the large natural variability of the material. A significant element of uncertainty is also introduced through lack of information about the actual physical variability.

The variability of strength *between* timber elements is significantly larger than for steel or reinforced concrete members. The coefficient of variation is of the order of 20–40%, with higher values for brittle type failure modes.

The variability of strength *within* timber members is also significant, but there are only limited statistical data to properly quantify this variability for various wood species, as described by Isaksson in Chapter 4. Since timber is treated as a homogeneous material (assuming constant strength along the length) in everyday engineering design, the within-member variability implies that the strength (when evaluated on the basis of standard engineering methods) depends upon the dimension of

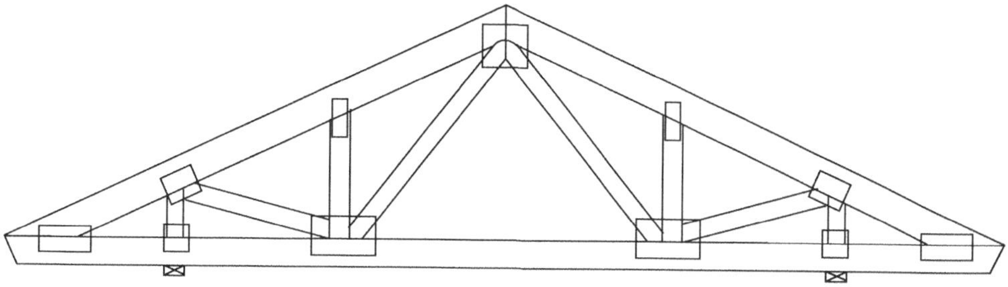

Figure 10.1 A timber roof truss is a system consisting of timber elements and joint elements. Locations with high stresses generally do not coincide with weak sections in the timber

the member and the spatial distribution of internal forces. These influences are called size and load configuration effects, respectively. The order of magnitude for these effects is such that the reliability of timber structures in the ultimate limit state is significantly affected.

In engineering design of multiple member wood structural systems, simple mechanical models are often used to determine the load effects in each member. Quite often, the actual behaviour of the system implies mechanical interaction between elements in the system, which is neglected in a simple engineering analysis, but may be described by more advanced models, or verified by testing. This may be called a structural system effect, which is often relevant for structural timber systems. Since the theory of plasticity cannot be applied in general for timber structures, structural system effects for timber cannot be quantified by simple engineering calculation methods.

The safety coefficients used in engineering design are based on calibration for single elements. This means that one single element in the system, which is identified to be the most stressed one by simple engineering analysis, determines the reliability of the system as a whole. Considering the real behaviour of the system and the fact that the properties of the elements in the system are random, it is clear that this procedure will not give a correct estimate of the reliability. A more detailed reliability analysis at the system level will often reveal that system performance improves safety. This difference, which is most often, but not always, on the conservative side, may be called the probabilistic system effect. It reflects the reduced probability that an element or a cross-section with highest stress will occur in a very weak element or a cross-section with very low strength. Figures 10.1 and 10.2 show two examples where probabilistic system effects are relevant. System reliability is further discussed by Foschi in Chapter 11.

A special feature for timber is that strength is affected in a significant way by load duration, as described by Hoffmeyer in Chapter 8. This means that a statistical correlation is introduced between strength and load effects. This will complicate the evaluation of safety for timber structures. Studies of reliability related to duration of load have been performed by Foschi *et al.* (1989) and Svensson *et al.* (1999). Duration of load factors k_{mod} were calibrated to obtain a consistent level of safety in terms of reliability. The performance criteria for structures can be classified into two categories, safety and serviceability, corresponding to the ultimate and serviceability limit states, respectively. An overview of key issues regarding

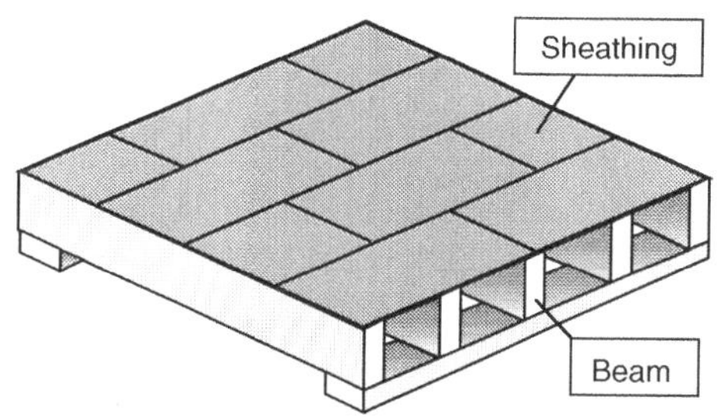

Figure 10.2 Timber floor with parallel joists. Loads may be transferred from weaker joists to strong and stiff companions

design for safety and serviceability of timber structures will be given below.

10.3 SAFETY-RELATED DESIGN OF TIMBER STRUCTURES

A major complication in the design of timber structures is that a number of different failure modes may occur. The low strength in directions perpendicular to grain requires special attention, and in many cases, specific design methods are needed. For large-scale structures based on glulam and other engineered wood products, special structural forms are used, such as curved beams, frames and arches, as well as tapered, pitched and cambered beams. Examples are given in Figures 1.14–1.16 in Chapter 1. Secondary stresses in directions perpendicular to grain will occur in this type of structure, and the risk of failure in tension perpendicular to grain should always be checked. Design methods for this are described by Larsen in Chapter 12.

As mentioned above, timber elements in themselves have limited ductility, which can make timber structures less forgiving than, for instance, steel and reinforced concrete (see Figure 10.3).

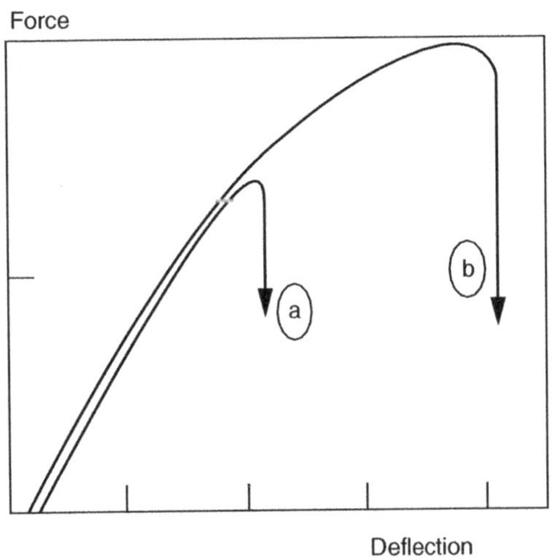

Figure 10.3 Typical load deformation relations for structural timber in bending, (a) weak board, (b) strong board within a sample, (after Isaksson, 1999)

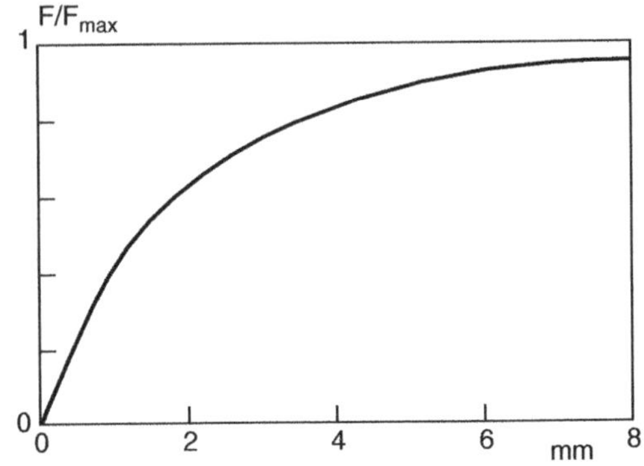

Figure 10.4 Typical load-deformation relation for joint with dowel type fasteners

The structural engineer has to be aware of this, and the design of more advanced types of structural systems in timber can be quite demanding. However, by careful design and detailing, it is possible to achieve robust structural systems based on timber. Ductility in structural systems can be introduced by proper use of mechanical connections, which if correctly designed have excellent ductility (see Figure 10.4).

Ductility is especially important in the case of earthquake loading. Timber structures actually have a record of being highly efficient to resist earthquake loading. One great advantage in this context is the low weight of timber, and properly designed timber frame systems are very attractive alternatives for buildings in earthquake regions. A very comprehensive description of the state-of-the-art for design of timber structures under earthquake loading is given by Karacabeyli and Popovski in Chapter 15.

10.4 DESIGN OF TIMBER STRUCTURES FOR SERVICEABILITY

Design in serviceability limit states is very often decisive for timber structures. Important factors here are low weight in relation to strength, moisture-induced deformations in all types of wood products, and significant creep deformations under load.

Floor and roof structures are commonly designed with beams as primary structural elements. For moderate spans, structural timber or lightweight I-beams are used, whereas for larger spans, glulam or other solid engineered wood products are relevant. Limitation of beam deflections often determines the dimensions of such structures. Deflections of beams made of wood or wood-based materials depend upon short-term bending stiffness and creep under long-term loads. Ambient climate conditions are important for deflections, since both stiffness and creep depends upon moisture content level. Moreover, the creep will increase significantly when the material is exposed to moisture variations. Short-term stiffness properties are relatively well-known for most wood-based materials, whereas information about the creep response and the influence of moisture is incomplete, in particular for wood-based panel products. Prediction of deflections in timber structures and design principles for deformation control are described by Mårtensson in Chapter 13.

One important aspect of design for serviceability is the choice of performance criteria. Deformation limits for structures may be motivated in terms of human perception, risk for damage or loss of functionality. For all these cases, the definition of specific deformation limits is associated with considerable uncertainties.

The loads acting on timber structures during its lifetime are, to a large degree, variable in time, which means that deflections will also be variable in time. This raises the question of defining the load combinations to be used for calculation of deflections in different situations. Furthermore, prediction of creep is usually based on test data under time-constant load, and the validity of creep predictions for time-varying loads may be questioned. The most common approach to this problem is to calculate creep deflections for a time averaged value for the variable loads. Various aspects on design for deformation control are discussed further in Chapter 13.

For lightweight timber floors the most critical issues are acoustics and limitation of vibrations from footfall and other dynamic loads. Regarding acoustics, the question of impact sound insulation is crucial, but this is outside the scope of this book. Floor vibration is, however, an important aspect of timber engineering. For lightweight floors, inertia forces to resist dynamic loads are small. Timber floors usually consist of joists spanning in

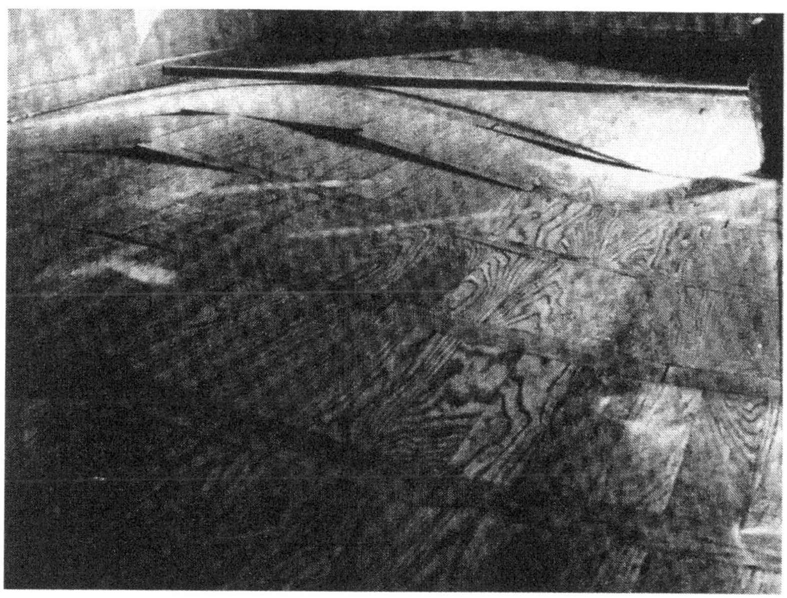

Figure 10.5 Effect of swelling deformations in a wooden floor due to lack of movement gaps at the edges

one direction sheathed with panel products (see Figure 10.2). The bending stiffness is therefore, as a rule, much higher in the main direction than in the transverse direction, whereas the vibration response depends upon the stiffnesses in both directions. A common way to improve vibration performance is to increase the transversal bending stiffness by blocking or cross-bracing between the joists. Apart from the properties of the floor itself, the vibration response depends upon the dynamic loading, which may be induced by humans or by mechanical equipment. The various aspects of vibration of timber floors are described by Smith in Chapter 14.

It is necessary to design floors so that vibrations are kept within a certain limit, based on human perception of motion, which in itself has a high variability within the human population (see Chapter 14).

Moisture-induced deformations in wood and wood-based products must be controlled to avoid serviceability problems. When wood-based products are used in wall linings and floor surfaces, appropriate measures have to be taken to allow moisture movements to occur without causing damage and cracking. A drastic example is shown in Figure 10.5. In large structural timber systems, moisture-induced movements may cause serviceability problems in the form of excessive settlements and horizontal movements. Vertical settlements in timber frame systems are further discussed in Chapter 13.

REFERENCES

Foschi R.O., Folz B.R. and Yao F.Z. (1989) Reliability-based design of wood structures. Structural Research Series, Report No. 34, Department of Civil Engineering, University of British Columbia, Vancouver, Canada.

Isaksson T. (1999) Modelling the variability of bending strength of timber. Length and load configuration effects. PhD dissertation. Report TVBK-1015, Division of Structural Engineering, Lund University, Lund, Sweden.

Svensson S., Thelandersson S. and Larsen H.J. (1999) Reliability of timber structures under long term loads. *Materials & Structures*, **32**, 755–760.

11

Reliability of Structures with Timber and Wood-Based Products

Ricardo O. Foschi

11.1 INTRODUCTION: RELIABILITY AND PERFORMANCE-BASED DESIGN

The objective in the design of any structure is to meet specified *performance criteria*, and to do so with a *desired confidence* or *reliability* for the intended service life. Performance criteria are associated with the different design situations, or *limit states*, which are usually considered by the engineer in the design process: for example, serviceability limits of tolerable deformations or vibrations, or bending or shear failure of a member, or sufficient residual strength after some exposure to fire, or connection failure due to crack propagation. The problem, thus formulated, is known as *performance-based design*.

The solution depends upon the quantification of performance for each limit state, and this, in turn, requires the availability of *calculation models*. For example, the consideration of floor vibrations requires a realistic model for estimation of the dynamic response of the floor to a specified excitation. In general, the structural behaviour represented by a calculation model will involve several variables, each with a different degree of uncertainty. Some of these variables may be known quite accurately (for example, the span of a beam), while others may be quite uncertain, or *random*, such as the loads due to wind or snow or earthquake shaking. Given the uncertainty in the input, the calculated performance output will also have a degree of uncertainty and, therefore, there will always be a chance that the actual performance will not meet the acceptance criterion. The structural design must then be adjusted to satisfy the criteria with *tolerable probabilities of non-performance*. These probabilities may also be called 'target probabilities of failure', although in general they may refer to limit states not involving actual failure or collapse. The complement of the probability of non-performance is the *reliability* of the structural design. It should be emphasised that the estimated reliability corresponds to the calculation model being used, and that it is important, therefore, that the models represent reality as close as possible. In any case, an additional random variable associated with *modelling error* should always be added.

In general, the performance of an engineering system can be described by a *performance or limit state function G(x)*, the form of which is

$$G(x) = C(x_c, d_c) - D(x_d, d_d) \qquad (11.1)$$

$G(x)$ is a function of the intervening random variables, and can always be written as the difference between two functions, a *capacity* C and a *demand* D. The variables may be divided into two groups: one, (x_c, d_c), related to the capacity C, and a second, (x_d, d_d), associated with the demand D. The vectors x_c, x_d include all those variables which are random or uncertain, while the vectors d_c and d_d include all those quantities which are *deterministic* or known with sufficient certainty. For example, the depth of a beam will affect its bending capacity and it will form part of d_c, while the span will affect the bending moment demand, and will form part of d_d. On the other hand, the bending strength will form part of x_c, while the actual applied load will be included in x_d.

When the performance function is written in the form shown in Equation (11.1), non-performance will correspond to situations where the variables combine to make $D > C$, or $G < 0$. Thus, estimating the probability of non-performance is equivalent to estimating the probability of the event $G < 0$. This will be the *probability of failure* P_f, and the reliability will be the complement $1.0 - P_f$.

The calculation of reliability requires statistical information on the variability of the random variables. For example, the bending strength of a laminated beam will vary from specimen to specimen, and this variability may be obtained through testing and represented by corresponding statistics. In a more sophisticated capacity calculation model, the statistics for the beam may be derived from those for individual laminations, and testing of laminations will then provide the required information. Some variables may have more influence than others and, accordingly, they will require a more precise statistical description. Some demand variables will depend upon available data, for example, for wind speeds, depth of snow packs, quantity of rain, historical earthquake intensities, or occupancy loadings. The statistics for such demands may be common to all materials. On the other hand, all those variables associated with the capacity are dependent on the material. In this regard, since wood tends to be generally more variable than steel, reliability theory should play a relatively more important role in wood engineering. Of course, there may be some variables for which no data or very little information are available. In this case, subjective estimates of the variability may be introduced to study the importance of such assumptions in the overall reliability of the design. Should a particular variable be found to be quite important but lacking detailed information, more effort should be spent in obtaining more data to avoid a corresponding penalty in the reliability estimate.

Reliability, in the context of performance-based design, offers many advantages not adequately addressed by more traditional, deterministic calculation methods. The calculated reliability in a given limit state will decrease if the information is sketchy or the variability is high. Conversely, a premium is received through better information on the material, better modelling, or better procedures for quality control. In this sense, these probabilistic methods facilitate innovation in applications and manufacturing.

How do we implement, in practice, reliability calculations and performance-based design? One approach will be to study each specific design situation, and to obtain values for the deterministic parameters d_c or d_d in such a way that a target reliability be achieved for the design life for the structure. This would be a *customised design approach*, perhaps required for important structures or for the development of prefabricated assemblies with wood-based products. Another approach will be the development of a *codified procedure*, by means of which a prescriptive design equation will be specified for each limit state. The codified approach could cover a large number of simpler applications for which the customised approach may not be necessary.

The code design equation normally contains a sufficient number of 'design factors' to be applied to nominal load demands or to nominal capacities. These factors would have been chosen or calibrated in a manner such that, when using the code, the resulting design would have achieved the desired target reliability. Of course, with just a few factors, the target cannot be satisfied uniformly for a large number of situations. As a result, the factors are calibrated by optimisation, minimising the difference between the target reliabilities and those achieved and over a large number of design conditions to which the code will apply.

One disadvantage of this optimised approach is that the reliability achieved varies across situations. For example, code prescriptions for design of houses in North America have been calibrated to achieve a fairly uniform reliability but, in reality, the level varies with geographical location in accordance to changes in the snow load statistics. The process of code calibration would only produce a code with an applicability range depending upon the choice of situations considered in the calibration exercise. A more detailed description of code calibration is shown in Foschi *et al.* (1990).

Performance-based design is then intimately linked to the calculation of structural reliability, and to the recognition of uncertainties in the design process. The required mathematical tool is provided by probability theory. For completeness, this chapter includes a brief review of basic probability concepts and methods for reliability estimation. The chapter then focuses on application examples to problems in timber engineering.

11.2 STATISTICAL REPRESENTATION OF VARIABLES AND BASIC PROBABILITY CONCEPTS

Data for each of the intervening variables must be provided in the form of statistical information. Basic background on statistics and probability concepts are very well discussed elsewhere (Benjamin and Cornell, 1970, Ang and Tang, 1984a). Here only a brief review is included.

To guide the discussion, let us consider the specific case of bending strength of a timber beam, of a given size and span. Suppose that N specimens are tested, maintaining constant certain characteristics deemed to influence the bending strength: for example, load distribution, rate of loading, temperature and moisture content. In all probability, the strengths f obtained from the N tests will all be different. In addition, suppose that each test is also used to obtain the modulus of elasticity E. Basic statistics to be obtained include the mean values and the standard deviations of both f and E, plus the covariance between them:

Mean values :

$$\overline{f} = \frac{1}{N}\sum_{i=1}^{N} f_i \quad \overline{E} = \frac{1}{N}\sum_{i=1}^{N} E_i \qquad (11.2)$$

Standard Deviations :

$$\sigma_f = \sqrt{\frac{\sum_{i=1}^{N}(f_i - \overline{f})^2}{N}} \qquad \sigma_E = \sqrt{\frac{\sum_{i=1}^{N}(E_i - \overline{E})^2}{N}} \tag{11.3}$$

Covariance $f - E$:

$$\sigma_{f-E} = \frac{\sum_{i=1}^{N}(f_i - \overline{f})(E_i - \overline{E})}{N} \tag{11.4}$$

with the associated parameters:

Coefficients of variations :

$$\mathrm{cov}_f = \frac{\sigma_f}{\overline{f}} \qquad \mathrm{cov}_E = \frac{\sigma_E}{\overline{E}} \tag{11.5}$$

Linear correlation coefficient $f - E$:

$$\rho_{f-E} = \frac{\sigma_{f-E}}{\sigma_f \sigma_E} \tag{11.6}$$

The standard deviation provides information on the dispersion of the variable about its mean, without reference to the sign of the deviation, while the coefficient of variation expresses that dispersion as a fraction of the corresponding mean. Two random variables are said to be *independent* if one can take any value within its range regardless of the value taken by the other. Two independent variables have zero covariance and, correspondingly, zero correlation coefficient. Variables which have a deterministic linear relationship between them are perfectly linearly correlated, and the correlation coefficient becomes +1 or −1, depending upon the slope of the linear relationship. In the bending test it would probably be found that the variables f and E are correlated, but not perfectly, and the correlation coefficient would be of the order of 0.60–0.70. This means that a stronger beam tends to be stiffer and, conversely, a weaker beam tends to be more flexible. Because of a positive correlation, it is unlikely to find combinations of a strong beam (high f) with flexible properties (low E).

For reliability calculations, it is not sufficient to provide only basic statistics. The data must also be ranked in order to construct a function, called a *cumulative probability distribution*, $F(x)$, defined as follows:

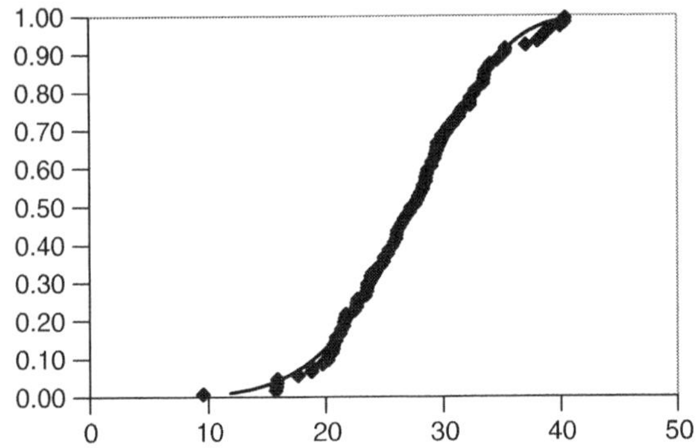

Figure 11.1 Example of discrete data points fitted with a Cumulative Distribution Function (2-parameter Weibull)

$$F(x_0) = Prob(x \leq x_0) \tag{11.7}$$

that is, $F(x_0)$ represents the probability of encountering values of x less than or equal to x_0. This function ranges between 0 and 1, as shown in Figure 11.1, for data corresponding to bending strength of a timber beam (in Mpa).

The discrete points obtained by constructing $F(x)$ from the ranked data can be fitted with a mathematical function, which is then used in the reliability estimations. Another useful function is the derivative $f(x)$ of the cumulative $F(x)$. The function $f(x)$ is called the *probability density*. Due to the relationship between $f(x)$ and $F(x)$, the probability $F(x_0)$ can also be written as

$$F(x_0) = \int_{-\infty}^{x_0} f(x)\,dx \tag{11.8}$$

The cumulative distribution $F(x)$ can be represented with a variety of mathematical functions. A detailed discussion of different probability distributions is presented by Benjamin and Cornell (1970), Ang and Tang (1984a), or Thoft-Christensen (1982). Here we will consider only some of the most useful distributions in wood engineering.

(a) *Normal* distribution (or Gaussian): this most famous distribution is best described by its density function $f(x)$:

$$f(x) = \frac{1}{\sqrt{2\pi}\sigma} \exp(-(x-\overline{x})^2/2\sigma^2) \quad (11.9)$$

It is symmetrical about its mean and it extends from $-\infty$ to $+\infty$. For the special case of zero mean and standard deviation $\sigma = 1$, the distribution becomes a *Standard Normal*, and the cumulative distribution $F(x)$ is noted with $\Phi(x)$, with the corresponding $f(x)$ becoming $\varphi(x)$. Although a Normal distribution can always be fitted to the data, it presents a fundamental problem when the variable x cannot be negative (as in the case of bending strength or modulus of elasticity). However, a Normal can be used if the mean value is positive and the coefficient of variation is small, resulting in negligible chances of x taking negative values. The Normal distribution has some very interesting properties: any linear combination of Normals is itself a Normal, a property which is not true for other distributions. Thus, a Normal of mean m and standard deviation σ can always be written as

$$x = m + \sigma R \quad (11.10)$$

where R is a Standard Normal variable. Since there are algorithms which generate random numbers with a Standard Normal distribution, Equation (11.10) can be used to generate random numbers with a general Normal distribution.

(b) *Lognormal* distribution: in this case, the logarithm of the variable x is Normal. Therefore, the variable x cannot be negative, and its range is from 0 to $+\infty$. This characteristic makes the Lognormal distribution a good one to use to represent positive data like strengths or moduli of elasticity. Another example of Lognormal application is in the representation of the variability in peak ground accelerations during an earthquake.

(c) *Extreme Type III or Weibull* distribution: the cumulative function $F(x)$ is, in this case,

$$F(x) = 1.0 - \exp\{-[(x-x_0)/m]^k\} \quad (11.11)$$

where three parameters are identified as follows: x_0, the location (there are no x smaller than x_0); m, the scale and k, the shape. A particular case of the Weibull distribution involves only two parameters for the special case of $x_0 = 0$. These parameters can be related to the mean and the standard deviation of the distribution: for example, the greater the variability the smaller the shape parameter k. A Weibull distribution cannot be negative, thus having the good characteristics of a Lognormal. Furthermore, it can be proven that a Weibull distribution is an asymptotic representation for the distribution corresponding to the *minimum* value among a large number samples of the variable x. This makes the Weibull a good choice to represent strengths, particularly for brittle materials where the capacity of an assembly of many components is determined by the capacity of the weakest component in the assembly. Thus, the behaviour of wood in shear or in tension perpendicular to the grain can be represented well by Weibull distributions. In general, it is found that a Weibull distribution is a good choice to fit data for wood strength, even when the assumption of a fully brittle material is not entirely justified (as in the case of bending strength). Equation (11.11) can also be written in a different form,

$$x = x_0 + m[-\log(1 - F(x))]^{1/k} \quad (11.12)$$

Since $F(x)$ ranges from 0 to 1, Equation (11.12) can be used to generate values of x, obeying a Weibull distribution, by utilising a random number generator which will generate unbiased numbers between 0 and 1.

(d) *Extreme Type I or Gumbel* distribution: the cumulative function $F(x)$ is, in this case,

$$F(x) = \exp(-\exp(-a[x-b])) \quad (11.13)$$

where a and b are parameters of the distribution, also related to the mean and the standard deviation of x. Since it can be shown that this distribution is an asymptotic representation of the *maximum* value among a large number of samples of x, it is a good distribution to use to represent demand data for which one is interested in the largest value within a certain time window. For example, statistics for maximum annual snow data in different Canadian locations are represented by Gumbel distributions. Equation (11.13) can also be re-written as

$$x = b + \frac{1}{a}\{-\log[-\log(F(x))]\} \quad (11.14)$$

permitting the generation of values of x, obeying a Gumbel distribution, when values of $F(x)$ are generated randomly between 0 and 1.

e) *Uniform* distribution: in this distribution the variable x can take any value between two limits, a and b, with equal probability. The density function is then

$$f(x) = \frac{1}{b-a} \quad (11.15)$$

This distribution is useful when no other information is available for a given variable except a subjective assumption about the variable range. In some cases, the Uniform distribution is used when one does not want to introduce biases: for example, the location of an earthquake epicenter is assumed to be uniformly distributed within a certain seismic zone when a detailed map of faults is not available.

The cumulative distributions $F(x)$ may need to be modified before they are used in reliability estimations. These modifications may include:

(a) *Modifications for Maximum*: in some cases, we may need to modify the original $F(x)$ to consider the distribution for the maximum value among N samples of x. For example, suppose that x is the annual maximum snow load in a location. We may be interested in the distribution $F_{100}(x)$, where x is now the maximum load that can be expected in a 100-year time window. It can be shown that

$$F_{100}(x) = F(x)^{100} \quad \text{or, in general,}$$
$$F_N(x) = F(x)^N \quad (11.16)$$

where N is the number of elemental windows (for which $F(x)$ is known) which are included in the new, desired interval. A basic assumption in the relationship is that the value of x in one elemental window is independent of the values in the others. Thus, if the elemental window is one year, the maximum snow load in one year is assumed to be independent of the maximum load in the following year. The result of the transformation in Equation (11.16) is a shift of the original distribution, increasing the mean value and decreasing the variability. If the original distribution is Gumbel, from Equation (11.13), raising it to the Nth-power will modify Equation (11.14) as follows:

$$x = b + \frac{\log N}{a} + \frac{1}{a}\{-\log[-\log(F_N(x))]\} \quad (11.17)$$

(b) *Modifications for Minimum*: in some cases, we may want to modify the original $F(x)$ to consider the distribution of the minimum value among N samples of x. This transformation is useful when analysing a brittle system, the strength of which is controlled by that of the weakest component. For example, a statically determined wood truss with toothed connection plates may be controlled by the strength of the weakest connection. It can be shown that, if N components are present, all with the same distribution $F(x)$, the distribution of minimum value among the N is given by

$$F_N(x) = 1.0 - [1.0 - F(x)]^N \quad (11.18)$$

The result of this transformation is also a shift of the original distribution, with a reduced mean and an increased variability. As a result, the strength of the assembly of connectors will decrease with the number of connectors (size effect).

(c) *Modifications for Lower Bound*: in some cases, it may be necessary to modify the original $F(x)$ to consider that x cannot take values lower than x_0, a lower bound. This transformation may be required when, for example, we want to introduce some measures of quality control: of all laminations with modulus of elasticity E available for manufacturing a glued-laminated beam, we may want to use only those satisfying $E > E_0$, a selection policy enforced by some quality control equipment. It can be shown that the modified cumulative distribution $F^*(x)$ satisfies

$$F^*(x) = \frac{F(x) - F(x_0)}{1.0 - F(x_0)} \tag{11.19}$$

(d) *Modifications for Upper Bound*: similarly, in some cases it may be necessary to modify the original $F(x)$ to consider that x cannot take values higher than x_0, an upper bound. This situation may occur when, for example, an applied pressure to a structure may be released through a valve should it exceed a certain threshold. It can be shown that the modified cumulative distribution $F^*(x)$ satisfies

$$F^*(x) = \frac{F(x)}{F(x_0)} \tag{11.20}$$

Finally, let us consider the probability of two events occurring *simultaneously*. This is called their *joint probability*, and can be expressed as

$$P(1,2) = P(1|2)P(2) \quad \text{or}$$
$$P(1,2) = P(2|1)P(1) \tag{11.21}$$

where 1 and 2 identify the two events, $P(1,2)$ is the joint probability and $P(1|2)$ is the *conditional probability* of 1 occurring when 2 has occurred. Conversely, $P(2|1)$ is the conditional probability of 2 occurring when 1 has already occurred. If the events 1 and 2 are *independent* of each other, then the conditional probability $P(1|2) = P(1)$, since the occurrence of 1 has nothing to do with the prior occurrence of 2. For independent events, therefore,

$$P(1,2) = P(1)P(2) \tag{11.22}$$

An interesting consequence of the definition of conditional probabilities is *Bayes' Theorem*. Let E be an event that can occur simultaneously with each of N mutually exclusive events called A_i. A practical example would be when E = failure of a beam, A_1 = failure in bending, A_2 = failure in shear, A_3 = failure due to crack extension, etc. Since we can write, from Equation (11.21),

$$P(E) = \sum_{i=1}^{N} P(E|A_i)P(A_i) \tag{11.23}$$

and

$$P(E|A_i)P(A_i) = P(A_i|E)P(E) \tag{11.24}$$

we conclude that

$$P(A_i/E) = \frac{P(E/A_i)P(A_i)}{\sum_{j=1}^{N} P(E/A_j)P(A_j)} \tag{11.25}$$

which is Bayes' Theorem. This equation can be used to update a-priori probabilities for each of the events A_i *given* the information that the event E has occurred. For the example of beam failure because of probable different causes, Bayes' Theorem can be used to estimate the probability of each cause given the information that the beam has indeed failed. *A priori* probabilities $P(A_i)$ can be subjective, for example, starting with all being equal to $1/N$, where N is the number of mutually exclusive events A_i. Later in this chapter, we shall discuss an application of Bayes' Theorem in the assessment of an existing structure.

11.3 METHODS FOR CALCULATING THE PROBABILITY OF NON-PERFORMANCE

There are several methods to calculate the probability of failure or non-performance. A very good discussion of different approaches is presented by Ditlevsen *et al.* (1996), Melchers (1987), Madsen (1986), Augusti *et al.* (1984) and Ang and Tang (1984b). The most basic and straightforward method is that of computer (Monte Carlo) simulation, while perhaps the most efficient are approximate methods based on the calculation of a reliability index β (FORM and SORM procedures). There are also methods for more efficient simulation, given the names of Importance Sampling or Adaptive Sampling Simulation. In addition, all methods can be implemented working either with the real performance function $G(x)$ or with a fitted surface approximating $G(x)$. This fitted surface is called the *response surface* for the problem. All of these methods have been implemented in software now available in the marketplace. For completeness, we include here only a brief description of the methods.

11.3.1 Monte Carlo Simulation

Suppose that, using appropriate random number generators, values of x are created according to, for example, Equations (11.10), (11.12), (11.14) or (11.17). Knowing x permits the calculation of the corresponding $G(x)$. If $G(x) > 0$, then the performance criterion is met. On the other hand, if $G(x) < 0$, the combination x leads to non-performance. If this calculation is repeated N times, and N_f is the number of non-performance events, the probability of failure can be approximated by the ratio

$$P_f \cong \frac{N_f}{N} \tag{11.26}$$

The result will be different if the process is repeated for another sample of N selections. The variability in the results depends upon N, diminishing as N increases, and the estimate will converge to the exact probability as N becomes large. This is an advantage of the simulation procedure. However, if the system being analyzed has a low probability of failure, for example, 10^{-6}, it would be necessary to perform, on average, $N = 10^6$ repetitions to expect to find one case of non-performance. Since this is the number of times that the performance function $G(x)$ would have to be evaluated, the procedure could be quite time consuming, particularly when this evaluation requires the running of a separate computer program to calculate either the capacity or the demand.

In a standard Monte Carlo simulation, the vectors x are chosen at random from their entire domain. To improve the efficiency of the simulation, techniques have been developed which permit a probability estimation with a reduced number of repetitions. These techniques, under the names of *Importance Sampling* or *Adaptive Sampling*, rely on selecting vectors x only from regions of importance, within which the combinations more likely to produce failure are located. These procedures are very effective and more detailed descriptions can be found in the literature (Ditlevsen, 1996, Schueller *et al.*, 1989).

11.3.2 First and Second Order Reliability Methods (FORM and SORM)

To provide an alternative to Monte Carlo simulation, very efficient approximate methods have been developed under the names of FORM or SORM, for (respectively) First Order or Second Order Reliability Methods. To understand the foundation for FORM, let us consider a simple problem where

$$G(x) = X_1 - X_2 \tag{11.27}$$

with the variables X being *Normal and uncorrelated.* Note also that the performance function is linear. If we express the variables X in terms of their means, standard deviations and *Standard Normals* x, following Equation (11.10), we can write

$$G(x) = \overline{X_1} - \overline{X_2} + \sigma_1 x_1 - \sigma_2 x_2 \tag{11.28}$$

Therefore, failure or non-performance will occur when $G(x) < 0$, or when

$$x_2 \geq (\overline{X_1} - \overline{X_2})/\sigma_2 + \sigma_1 x_1/\sigma_2 \tag{11.29}$$

Accordingly, Figure 11.2 shows the plane (x_1, x_2) with the corresponding region of failure

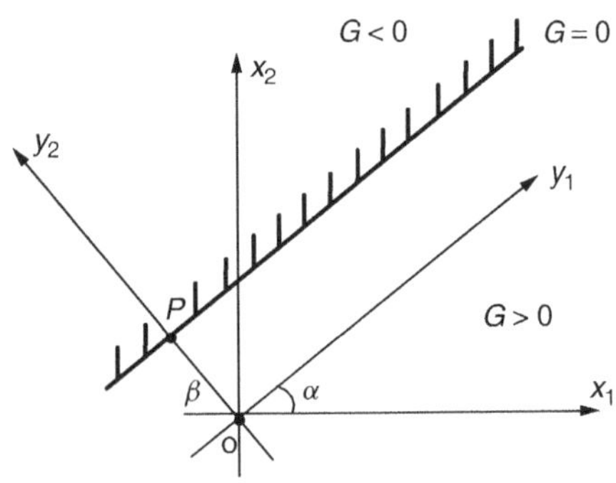

Figure 11.2 Failure surface and reliability index definition, $G = X_1 - X_2$

$G < 0$, defined by Equation (11.29), and the region of survival, $G > 0$. It is seen that the failure surface $G = 0$ is a straight line. A simpler characterisation of the failure region can be obtained by a rotation of the coordinate axes, from (x_1, x_2) to (y_1, y_2), since then $G < 0$ corresponds to

$$\textit{Failure} \text{ if} \triangleright y_2 \geq \beta \tag{11.30}$$

It can be shown that, given that the variables x were assumed to be *uncorrelated Standard Normals*, the variables y also result to be uncorrelated, Standard Normals. Therefore, the probability of non-performance, according to Equation (11.30), is the probability of the event $y_2 \geq \beta$. This can be obtained directly from the cumulative distribution function of a Standard Normal,

$$P_f = \Phi(-\beta) \tag{11.31}$$

Equation (11.31) implies that all that is required to compute the probability of failure is to find the distance β between the origin O and the point P. This is the minimum distance between the origin (corresponding to the mean values of X_1, X_2) and the failure surface $G = 0$. Finding this minimum distance is a geometric problem, easily solved by optimisation algorithms. The distance β is called the *reliability index*, and the point P is called the *design point*. Notice that the greater the reliability index, the farther the failure zone would be from the origin or mean point, implying lower probabilities of failure. Also, it is easily seen that, given the probability density of the variables y, the combinations of the variables more likely to lead to failure are grouped around point P. Thus, another name for P is the *most likely failure point*.

These conclusions can be generalised to any other function $G(x)$. However, the result of calculating β and then using Equation (11.31) will be exact only if the following three conditions are satisfied:

- All intervening variables are Normal.
- All variables are uncorrelated.
- The $G(x)$ function is linear (otherwise Equation (11.30) would not hold).

Through proper variable transformations (Ditlevsen, 1981, 1996, Rackwitz, 1978, Der Kiureghian, 1986, Melchers, 1987), the first two conditions can always be met. That is, non-Normal variables can be Normalised, and variables having a correlation structure can be uncorrelated. However, the linearity of the function G is related to the essence of the problem and cannot be eliminated through transformations. Thus, in general, using Equation (11.31) will provide only an approximate answer, and the error would be dependent on how nonlinear the performance function is in the neighbourhood of P. Fortunately, for many problems the error is quite acceptable for engineering applications. FORM has been modified, and with it, Equation (11.31), by assuming that the failure surface is a quadratic function with curvatures equal to those of the real surface at point P. The resulting method is called SORM, for Second Order Reliability. In many instances, the results show a slight improvement over FORM, but it is easy to construct examples where the answers would, in fact, be more erroneous.

Equation (11.31) can also be used in general to express the results either in terms of the reliability index β or the associated probability of failure P_f if the latter has been obtained by simulation.

It is also interesting to note that, for the simple case studied here, there is an equation for the reliability index. In fact, from Figure 11.2 and considering the slope and intercept of the straight failure surface $G = 0$, it can be concluded that

$$\beta = \frac{\overline{X_1} - \overline{X_2}}{\sqrt{\sigma_1^2 + \sigma_2^2}} \tag{11.32}$$

which is the correct β only for a two-variable problem, when the variables are Normal and uncorrelated. There is also an expression for the reliability index if the two variables were lognormal with small variability,

$$\beta \cong \frac{\log \overline{X_1} - \log \overline{X_2}}{\sqrt{V_1^2 + V_2^2}} \tag{11.33}$$

where V_1 and V_2 are the corresponding coefficients of variation.

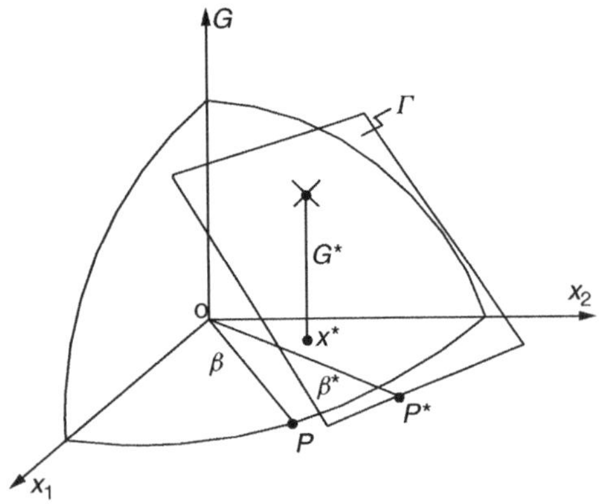

Figure 11.3 Iterative algorithm to find the reliability index β (Hasofer and Lind, 1974)

For the general case of several, non-Normal or correlated variables, there are no explicit expressions for the reliability index. Instead, it has to be calculated by numerical means. The important point here is that the FORM procedure is quite fast and efficient, since it only requires the geometric calculation of β.

Hasofer and Lind (1974) proposed a quasi-Newtonian iterative scheme which is commonly used in computer packages. Consider Figure 11.3, where the G function is plotted on the vertical axis as a function of the components of the vector x. The failure surface $G = 0$ is then the intersection of G with the horizontal plane, and the reliability index will be the length between point O and P, located at a minimum distance from O. Starting from an initial vector x^*, the algorithm replaces the real surface G with the tangent plane Γ at x^*. This plane intercepts $G = 0$ with a straight line, and the algorithm finds point P^*, at a minimum distance between that intercept and the origin O. Point P^* is then used as the new x^*, repeating the procedure until eventual convergence at P.

Convergence is dependent upon the adequacy of the initial choice x^*. However, when convergence is achieved, it is reached in a few iterations, accounting for the efficiency of the method. Computer packages implement special algorithms to find an adequate initial vector.

One of the very useful results from FORM is a unit vector in the direction OP, the components of which provide sensitivity coefficients for β in terms of the variables.

The accuracy of FORM is influenced by the nonlinearity of the performance function $G(x)$. The prediction from FORM can be improved by using Importance Sampling simulation around the design point P, the most likely failure point.

11.3.3 Response Surfaces

The methods previously discussed depend on being able to calculate the value of $G(x)$ for a vector x. Sometimes these values need to be the output of another computer program, for example, a dynamic analysis or a finite element calculation. In these cases, it may be inefficient to link directly the reliability index iteration with, for example, a nonlinear dynamic analysis. An alternative is to construct a *response surface* for the capacity or the demand, essentially evaluating $G(x)$ ahead of time, at a sufficiently large number of combinations x (Faravelli, 1989, Liu and Moses, 1994). The discrete points thus obtained can be used to fit a mathematical representation for the response, and this representation is then used as a substitute for the actual $G(x)$. The fitted response $G_F(x)$ is used for the reliability estimation to obtain a design point P and a reliability index β. The procedure can be stopped at that point or a new response surface may be constructed by sampling new vectors x around the point P just obtained. The new response surface is used to obtain an updated P and β, and the procedure can be continued until convergence in β has been achieved to a desired tolerance.

Different forms may be used for the fitted function $G_F(x)$. Bucher *et al.* (1997), for example, have considered polynomial expansions of the type

$$G_F(x) = a_0 + \sum_{i=1}^{N} a_i X_i + \sum_{i=1}^{N} b_i X_i^2 \qquad (11.34)$$

with the $2N + 1$ coefficients found by regression of the discrete data. Other forms may be chosen to represent better the influence of a variable. For example, suppose that one of the variables is the load Q and that the response variable is the deflection of a laminated plate. Other variables may be the moduli of elasticity of each lamination and their shear moduli. Obviously, if there is no load Q there must not be any deflection. In this case, it is perhaps better to propose

$$G_F(x) = Q\left[a_0 + \sum_{i=1}^{N} a_i X_i + \sum_{i=1}^{N} b_i X_i^2\right] + Q^2\left[c_0 + \sum_{i=1}^{N} c_i X_i + \sum_{i=1}^{N} d_i X_i^2\right] \quad (11.35)$$

which satisfies the condition of zero response when $Q = 0$.

11.3.4 Multiple Modes of Failure and Load Combination Problems

A system may fail in a number of modes and its reliability could be substantially different from that of the individual components (Thoft-Christensen *et al.*, 1986). For example, a beam may fail by bending or shear, or may not perform under a serviceability requirement. Each mode can be analysed separately, obtaining the individual reliability indices and corresponding probabilities of failure. If the system is considered to be a *series system*, when failure is deemed to have occurred when at least one mode has failed, the total probability of failure should consider many possibilities: failure in each mode, failure in pairs, failure in triplets, etc. The calculation of the total probability becomes quite involved, unless the system has just two modes. In this case, the total probability is given by

$$P_T = P(1) + P(2) - P(1,2) \quad (11.36)$$

in which $P(1)$ and $P(2)$ are the probabilities for the individual modes, and $P(1,2)$ is the probability of both modes occurring simultaneously. In the general case, however, it is possible to calculate bounds for the total probability. For example, it is easily shown that P_T must be lower than the sum of all the individual mode probabilities but greater than the largest individual. These bounds depend only upon probabilities for the individual modes but they may be quite wide. A narrower set of bounds has been developed by Ditlevsen (Madsen *et al.*, 1986), but they require calculation of joint probabilities as well as the individual ones. Usually, the Ditlevsen bounds are sufficiently narrow as to allow using their average for the estimate of P_T. The joint probabilities depend upon the correlation between modes if the corresponding G functions share some variables: for example, since bending and shear failure of a beam share the variable for the applied load, the two modes are correlated. That is, a high load can be produce simultaneous failure in bending *and* shear. Madsen *et al.* (1986) or Ditlevsen (1996) present details on the estimation of joint probabilities and correlations between failure modes.

Problems of multiple load combinations have also been studied, particularly the coincidence problem of load sequences with different arrival statistics over time (Wen, 1990).

11.4 CASE EXAMPLE NO. 1: BENDING FAILURE OF A GLUED-LAMINATED BEAM

To illustrate a *reliability assessment problem*, consider the laminated beam shown in Figure 11.4. It is simply supported, with a span L, and carries a distributed permanent load d and a concentrated, live load Q at midspan. The laminations have different moduli of elasticity and different tensile strength. Two failure modes will be considered: (1) failure of the bottom lamination; and (2) failure of the second from the bottom lamination. Of course, the two bottom laminations could fail together. Figure 11.4 shows end joints (fingerjoints) within the laminations. Although this would be the normal practice, for this example fingerjoints are not considered. Their consideration would only increase the number of random variables and the number of failure modes (including failure of the fingerjoints themselves). The example considers $N = 5$ laminations.

The analysis of stresses is carried out by assuming that plane sections remain plane and at 90° to the deformed beam axis (that is, no shear deformations are taken into account). The theory of

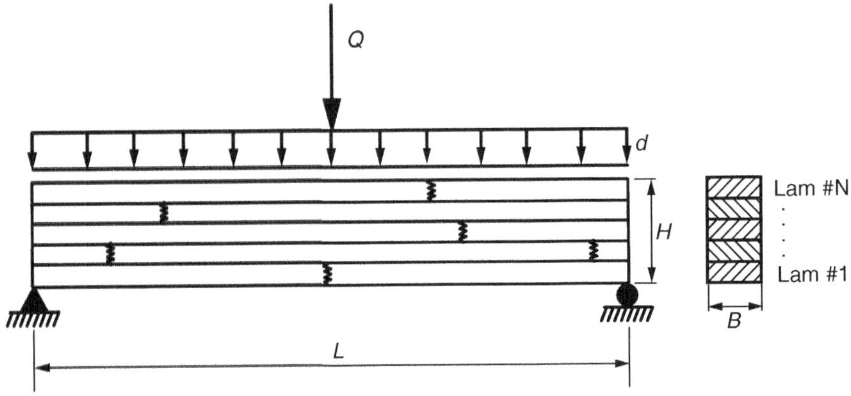

Figure 11.4 Laminated beam under a permanent distributed load and a concentrated live load

composite beams is then applied, requiring first the calculation of the centre of gravity of the composite section, and then the composite moment of inertia. Thus, if z is the distance from a point in the cross-section to the bottom of the beam, the position of the centre of gravity is first obtained from

$$\bar{z} = \frac{\int_A E z \, dA}{\int_A E \, dA} \tag{11.37}$$

and the composite moment of inertia is then calculated from

$$EI = \int_A E(z - \bar{z})^2 \, dA \tag{11.38}$$

Finally, the stresses at a distance z from the bottom are

$$\sigma = M \frac{E(\bar{z} - z)}{(EI)} \tag{11.39}$$

Two performance functions are then written:

Mode 1: for failure of the bottom lamination:

$$G_1(x) = f_1 - \frac{E(\bar{z} - z_1)}{(EI)}(dL^2/8 + QL/4) \tag{11.40}$$

Mode 2: for failure of the second, next to the bottom lamination:

$$G_2(x) = f_2 - \frac{E(\bar{z} - z_2)}{(EI)}(dL^2/8 + QL/4) \tag{11.41}$$

Variables Definition

X(1) = modulus of elasticity E of lamination 1 (bottom)
X(2) = modulus of elasticity E for lamination 2
X(3) = modulus of elasticity E for lamination 3
X(4) = modulus of elasticity E for lamination 4
X(5) = modulus of elasticity E for lamination 5 (top)
X(6) = permanent load d
X(7) = concentrated live load Q
X(8) = strength of lamination 1, f_1
X(9) = strength of lamination 2, f_2

The beam span is $L = 6.0$ m, its width is $B = 0.20$ m and the lamination thickness is $t = 0.040$ m.

Table 11.1 shows the corresponding statistical information and the assumed probability distributions. It is assumed that the modulus of elasticity and the strength of a lamination are correlated. In this case, X(1) is correlated with X(8), and X(2) with X(9). As a quality control measure, a minimum for the modulus X(1) may be enforced.

All computer packages require the definition of the performance functions, as per Equations (11.40) and (11.41), and the probabilistic description of the intervening random variables. As an example, the problem was run with the program RELAN developed at the University of British Columbia.

For a case in which no correlations and no lower bound for the E of the bottom lamination

Table 11.1 Statistics and distribution for variables, Case Example 1

Variable Number	Distribution Type	Mean value	Std. Deviation
1	Lognormal	$10\,000 \times 10^3$ kN/m^2	1500×10^3 kN/m^2
2	Lognormal	$10\,000 \times 10^3$ kN/m^2	1500×10^3 kN/m^2
3	Lognormal	$10\,000 \times 10^3$ kN/m^2	1500×10^3 kN/m^2
4	Lognormal	$10\,000 \times 10^3$ kN/m^2	1500×10^3 kN/m^2
5	Lognormal	$10\,000 \times 10^3$ kN/m^2	1500×10^3 kN/m^2
6	Normal	2.0 kN/m	0.10 kN/m
7	Extreme Type I	8.0 kN	2.0 kN
8	Weibull, 2-p	35.0×10^3 kN/m^2	8.75×10^3 kN/m^2
9	Weibull, 2-p	35.0×10^3 kN/m^2	8.75×10^3 kN/m^2

Correlations: $\rho_{1,8} = 0.70$, $\rho_{2,9} = 0.70$.
Minimum E, bottom lamination = 9000×10^3 kN/m^2.

are considered, the output can be summarised as follows:

Mode 1: FORM $\beta_1 = 2.061 \quad P_f = 0.01964$

Most important variable = X(8)
Second most important = X(7)
Third most important = X(1)

Importance Sampling ($N = 1000$) $\beta = 2.004$ or $P_f = 0.02252$

Mode 2: FORM $\beta_2 = 2.833 \quad P_f = 0.00231$

Most important variable = X(9)
Second most important = X(7)
Third most important = X(2)

Importance Sampling ($N = 1000$) $\beta = 2.817$ or $P_f = 0.00242$

Series System Reliability Bounds = 2.016–2.016 (Ditlevsen)

This shows that, even for the highly nonlinear performance functions of this problem, the results from FORM are quite similar to those from the Importance Sampling simulation around the design points P. The number of iterations for the calculation of β (Hasofer–Lind algorithm) was three. The importance of each variable is measured by the sensitivity coefficients corresponding to the components of a unit vector from O to P. The results also show the narrowness of the Ditlevsen bounds, when failure of the beam is deemed to have happened when the beam had failed in at least one of the two modes. The system reliability index is lower than the lowest of the individual modes, reflecting the possibility of failure in the other, less likely modes.

Table 11.2 Results of reliability analysis for a glued-laminated beam

Reliability Index β	Case 1	Case 2	Case 3
β_1	2.061	2.256	2.133
β_2	2.833	3.193	3.261
β series system	2.016	2.235	2.120

Table 11.2 shows a summary of results for three cases. In Case 1, no correlations are assumed and no lower bounds are imposed. In Case 2, a correlation coefficient 0.7 is assumed between E and strength of the two bottom laminations. In Case 3, in addition to the correlations, a lower bound is imposed on the modulus of elasticity for the bottom lamination.

Case 2 shows that taking into account the correlation between modulus of elasticity and strength produces a higher reliability. Since this correlation does in fact occur in wood products, not considering it leads to a conservative conclusion. It is of interest to see the effect of a proposed quality control measure. Suppose that the modulus of elasticity of the bottom lamination is enforced to be >9000 MPa. When this lower bound is enforced,

Case 3, the reliability actually decreases for the bottom lamination but increases for the second lamination, with an overall decrease for system reliability in comparison to Case 2. This result can be explained as follows: increasing the modulus of elasticity of the bottom lamination attracts more stress to it, while the correlation between strength and E is not strong enough to guarantee a sufficient increase in strength to support the increase in stress. As the second lamination is relieved of stress, its reliability increases. Overall, it would appear that there is no point in implementing this quality control measure, since it relies on a correlation between strength and E which, at 0.70, it is not sufficiently strong. Investigations of this kind are only feasible by the introduction of probabilistic methods and reliability analysis. It can also be concluded that these methods provide a framework for an optimisation of the resource and costs of production.

11.5 CASE EXAMPLE NO. 2: PERFORMANCE-BASED DESIGN OF A TIMBER BEAM IN BENDING (NORMAL AND FIRE CONDITIONS)

To illustrate a problem of *performance-based design*, consider the timber beam shown in Figure 11.5, with cross-sectional dimensions B and H, and a span L. It carries loads P and Q, both uniformly distributed but with different statistical properties. P is a dead load, with low variability, Q is a live load with higher variability. The bending strength of the beam is f and the modulus of elasticity is E.

The performance requirements under normal circumstances are: (1) the beam must carry the loads without bending failure; and (2) it should not deflect more than 1/180 of the span. In a fire, the beam should not fail before 60 minutes. These are three *performance criteria*. In performance-based design, the reliability levels with which those criteria are to be met must also be specified.

For the fire situation, it is necessary to know the rate at which the wood burns, decreasing the dimensions of the beam by an amount e, as shown in Figure 11.5. Of course, the relationship between e and time might be complicated, depending on the environmental conditions. For this example, however, it is sufficient to assume that the burning rate is a constant α, so that after time T the amount burnt is $e = \alpha T$.

The three performance functions are written as follows:

1. Bending failure under normal conditions:

$$G_1 = f - \frac{(Q+P)L^2}{8}\frac{1}{(BH^3/12)}\left(\frac{H}{2}\right) \tag{11.42}$$

2. Deflection under normal conditions:

$$G_2 = \frac{L}{180} - \frac{5(Q+P)L^4}{384}\frac{1}{E(BH^3/12)} \tag{11.43}$$

3. Bending failure under fire conditions:

$$G_3 = f - \frac{(Q+P)L^2}{8}\frac{1}{((B-2e)(H-2e)^3/12)} \times \left(\frac{H}{2} - e\right) \tag{11.44}$$

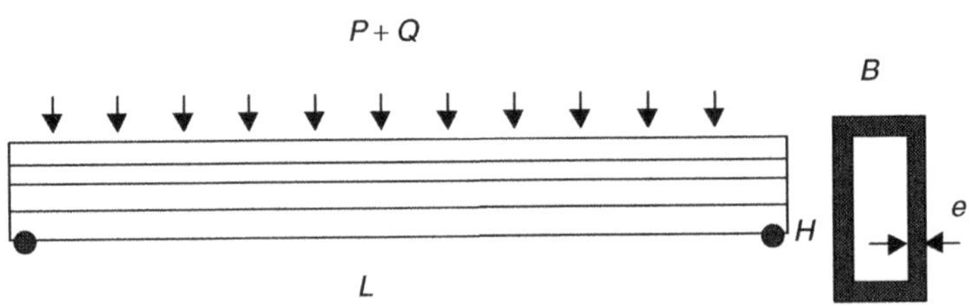

Figure 11.5 Design of a timber beam (normal and fire conditions)

Table 11.3 Statistics and distribution for variables, Case Example 2

Random variable	Mean	Std. Deviation	Distribution
f, bending strength (kN/m^2)	35 000	5250	Lognormal
P (kN/m)	2.0	0.2	Normal
Q (kN/m)	8.0	2.0	Extreme Type I
α (burn rate, mm/minute)	0.5	0.125	Lognormal
E, modulus of elasticity (kN/m^2)	$12\,000 \times 10^3$	1800×10^3	Lognormal

Table 11.4 Optimum values, H and B, Case 2, two strategies

H(m)	B(m)	β_1 Achieved	β_2 Achieved	β_3 Achieved	Optimisation strategy
0.533	0.249	4.32	1.91	2.32	Unconstrained
0.536	0.257	4.47	2.09	2.50	Forcing $\beta_3 = 2.50$
Target β		4.00	2.00	2.50	

Corresponding target reliability levels are:

$\beta_1 = 4.0$ for failure under normal circumstances
$\beta_2 = 2.0$ for deflection under normal circumstances
$\beta_3 = 2.5$ for failure under fire conditions after $T = 60$ minutes of exposure

The problem is to find optimum values H and B for the cross-section that will achieve the specified reliabilities or deviate a minimum from them. This is achieved by minimising the function

$$\Psi = \sum_{j=1}^{NP} (\beta_j^T - \beta_j(B, H))^2 \qquad (11.45)$$

in which NP is the number of performance requirements or reliability constraints, β^T are the corresponding target reliabilities and β_i (B, H) are the achieved reliabilities with dimensions B and H. This is then a reliability-based optimisation problem (Li, 1998, Foschi, 2001), and several computer packages are also available to carry out the calculations. The results presented here were obtained with a performance-based design package, IRELAN, developed at the University of British Columbia.

The problem has five intervening random variables which are identified in Table 11.3.

The dimensions B and H are also considered random (Normal), with a very small coefficient of variation. The design parameters are the *mean values* of B and H. The results are shown in Table 11.4 for a span $L = 10$ m.

The results show that it is not possible to find values of H and B which will achieve exactly the prescribed target reliabilities. Optimum values are obtained which minimize the difference between the targets and the achieved reliabilities. If performance under fire is deemed to be the most important design criterion, and if $\beta_3 = 2.32$ is considered low in relation to the target $\beta_3 = 2.5$, a constrained optimisation can be run to enforce $\beta_3 = 2.5$. Accordingly, as shown in Table 11.4, the reliabilities achieved for the first two performance criteria increase somewhat, with a small increase in the optimum beam dimensions.

11.6 CASE EXAMPLE NO. 3: RELIABILITY EVALUATION OF SYSTEM BEHAVIOUR

Generally, one of the characteristics of wood products is their relatively high variability in comparison to, for example, steel. However, wood products are commonly used as part of redundant systems, with multiple members or components. It is important to consider this system behaviour so as to not unduly penalise the performance of the products (Foschi *et al.*, 1990).

As an example, consider the timber floor structure of Figure 11.6 under a uniformly distributed load. Structurally, this is a plate stiffened with ribs

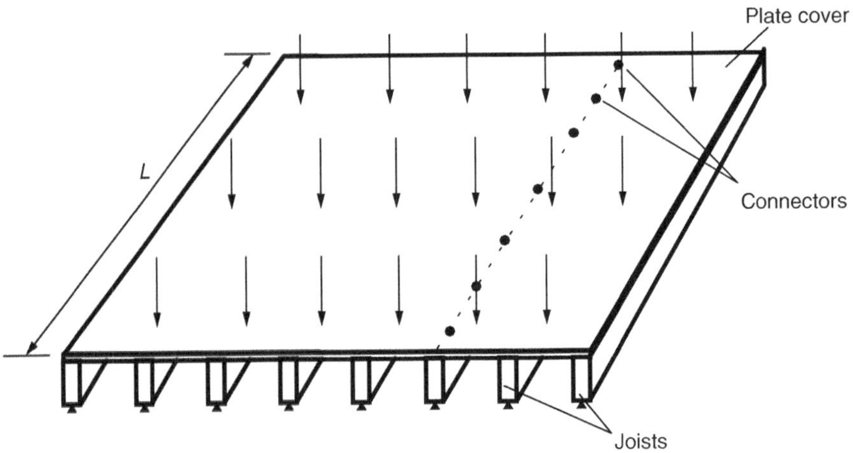

Figure 11.6 Floor system under uniformly distributed load

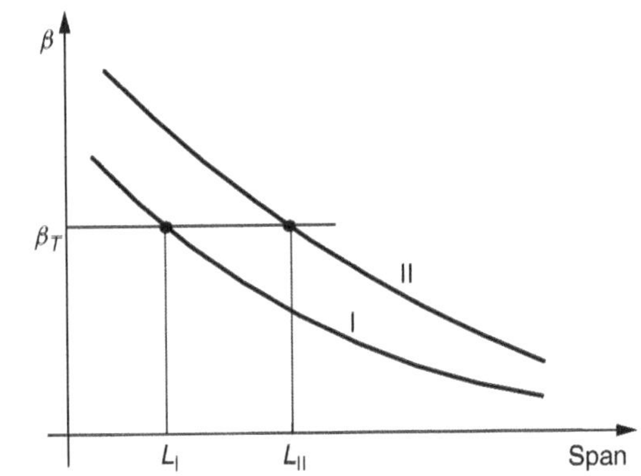

Figure 11.7 Calculation of a system factor, component vs. system reliability

and with non-rigid connectors between the plate and the stiffeners (nails, screws or adhesive). If, using a function $G(x)$ as Equation (11.42), one analyses, for different spans, the bending failure of a stiffener (joist) as a single, detached beam under a contributory area load, one would obtain curve I of Figure 11.7. This curve shows the corresponding variation in reliability index β with span L.

However, if the demand in the performance function for each joist is based on the bending stress calculated with a comprehensive (stiffened plate) structural analysis, taking into account the load-sharing among joists and the cover plate, the corresponding relationship between β and span would be as shown by curve II in Figure 11.7. Thus, the 'system effect' produces a higher reliability for the same span or, conversely, consideration of the system stress would allow a longer span at the same reliability. To obtain curve II it is necessary to define what is meant by system failure, and several options can be followed. For example,

1. After calculating the reliability of each joist using the corresponding system stresses, the reliability of the system could be taken as the minimum of the individual joists' reliabilities;
2. The floor could be considered failed if at least one joist fails. This is a series system assumption, and one of its consequences is a size of floor effect (the reliability of the system will decrease with the number of joists in it).
3. The progressive failure of the system can be studied by simulation, eliminating joists as they fail during the load increase.

It could be argued that, while Option 3 is the more realistic, Option 2 is the most practical from the point of view of the owner. However, because of the floor size effect introduced by Option 2, a more realistic application of this Option should include the spatial variability of the applied loads, particularly for floors covering large areas. The same could be said for Option 3. For floors of standard sizes for residential or small office construction, Option 1 is a simpler and not too conservative assumption over Option 2, resulting in similar curves II.

In a customised design, only curve II would be sought. In a code application, in which a simple beam design equation would probably have been prescribed, a factor accounting for the system effect should be introduced. Such a factor should convert the span calculated with the simple beam approach into the one that would be permitted by the system at the same reliability. The data in Figure 11.7 would then allow the calculation of the system factor corresponding to the target reliability used in the code. Since for distributed loads the demands are proportional to the square of the span, a system factor would be obtained in this case as the ratio of the squares of the spans corresponding, at the same reliability, to curves II and I, $(L_{II}/L_I)^2$.

The system effect could be altered by changing the configuration of the floor: different joist or connector spacing, different properties for the plate cover or the connectors, etc. Thus, the system factor will vary substantially with the configuration, and specific systems may be penalised if a single system factor is used as a conservative approximation for all constructions. Calculations of floor system factors for the Canadian code show a range from 1.4–2.0 (Foschi *et al.*, 1990), with 1.4 adopted as a common denominator. One of the advantages of using a customised, performance-based design, is to avoid such penalties and to give adequate treatment to possibly innovative, engineered solutions.

A similar approach can be followed in the study of any other system. For example, acceptance of a floor may be controlled by serviceability vibration requirements, and this would require the availability of a dynamic floor analysis to quantify the demand and a proper definition of acceptable performance. Another example would be the reliability of a shear wall under lateral forces due to earthquake or wind excitation, when the performance is measured by the amount of damage or deformations produced. The deformation demand has to be calculated with a comprehensive structural analysis of the wall, and the capacity defined in terms of deformation or damage thresholds. In all cases, system factors would be highly configuration-dependent, and the objectives of performance-based design would be more properly achieved by customised design.

11.7 CASE EXAMPLE NO. 4: CONNECTION DESIGN UNDER EARTHQUAKE EXCITATION, APPLICATION OF RESPONSE SURFACES

Consider now an earthquake engineering problem for the inverted pendulum shown in Figure 11.8. The timber column carries an overhanging mass M and it is connected at its base by four dowels and two steel side plates fixed to the ground. This moves with an acceleration $a_G(t)$. The drift, or horizontal displacement at the top of the column, is Δ. Δ_{max} corresponds to the maximum during the earthquake. The column is assumed to be sufficiently rigid, such that most of the displacement is due to the deformation of the connectors. Since the cyclic response of the connectors is highly nonlinear, an appropriate nonlinear dynamic analysis must be used to calculate the time history $\Delta(t)$, taking into account the hysteretic behaviour of the

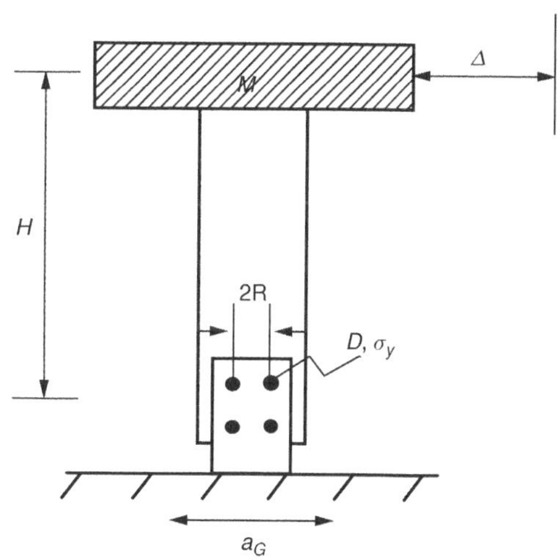

Figure 11.8 Dowel connection, earthquake excitation

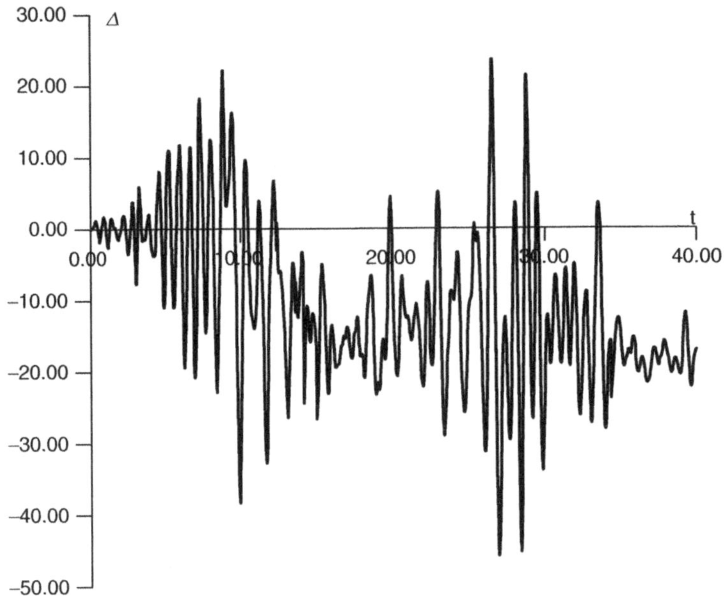

Figure 11.9 Calculated displacement history $\Delta(t)$

dowels. Figure 11.9 shows a typical result for the Landers, California, earthquake of 1992 (Joshua Tree Station).

The nonlinear analysis used here incorporates, at each time step, the calculation of the hysteresis loop for each of the dowels, taking into account their own elasto-plastic response and the nonlinear behaviour of the wood medium around them (Foschi, 2000). The performance criteria are associated with different damage levels, and are expressed in terms of the maximum horizontal drift Δ_{max} and limits H/K, with K being a constant and H the column height. For each limit state, the performance function $G(x)$ can then be written as

$$G(x) = H/K - \Delta_{max}(x) \tag{11.46}$$

Δ_{max} is a function of the random variables associated with the problem. Some of these are linked to the structural properties and others to the uncertainty in the earthquake itself.

Although the formulation of the problem is simple, the evaluation of G for each vector x is time-consuming, and generally makes standard simulation procedures quite inefficient. Further, the shape of G may show peaks and valleys in coincidence with frequencies near resonant conditions, making the FORM algorithm susceptible to slow or difficult convergence. Even if convergence is achieved, the nonlinearity of G may make the approximation of failure probability, based on the reliability index, quite inaccurate. A response surface replacement for the actual G function may be used, but it is difficult to find a simple surface which will represent with confidence all the nonlinearities and oscillations in G, particularly over the entire domain of interest. On the other hand, a different treatment of the response surface may be used.

Five random variables were considered and their statistics are shown in Table 11.5. One of the them is the Peak Ground Acceleration (PGA) during the event, and its cumulative distribution was consistent with a design value of 0.25 g (475 year-return), a coefficient of variation COV = 0.6 and a mean earthquake arrival rate $\nu = 0.1$ events/year. The peak acceleration was expressed as a random factor f times the peak of the historical record. The mass M had a mean value of 800 kg, and a COV = 0.10. The dowel diameter D and the arrangement

Table 11.5 Statistics and distributions for variables, Case Example 4

Variable	Mean	Standard Deviation	Type
f, PGA factor[a]	0.357	0.214	Lognormal
Mass M (kN·sec^2/mm)	0.0008	0.00008	Normal
Radius R (mm)	100.0	0.1	Normal
DowelYield Point σ_Y (kN/mm^2)	0.25	0.025	Lognormal
Dowel Diameter D (mm)	15.0	0.015	Normal

[a]factor applied to the PGA for the historical record of the Landers earthquake

radius R were deterministic, but considered random with a very small coefficient of variation. The yield point for the dowels, σ_Y, was assumed to have a COV = 0.10. A database for Δ_{max} can be generated by dynamic analysis for several combinations of the five variables, augmenting it with zero responses at zero peak ground acceleration, for all values of the remaining variables.

The design parameters are the diameter D, the radius R and the yield point σ_Y. The reliability corresponding to given values for these three parameters can be obtained in the following manner (Foschi *et al.*, 2001): (1) the database for Δ_{max} is used to first fit an approximate response surface using Equation (11.34); (2) this response surface is used with FORM to determine a design point P; (3) the probability of failure is obtained by Importance Sampling simulation around P, calculating the values of $G(x)$ by a local interpolation of the database.

Thus, the database is obtained independently of the reliability calculation, and it is then accessed by local interpolation or to train a neural network for the estimation of $G(x)$ at x not in the database. The results are shown in Table 11.6 in terms of reliability indices β, for several performance level definitions.

Table 11.6 Results, Example Case 4

Drift limit Δ_{max}	FORM	FORM + Importance Sampling
H/30	2.55	2.67
H/60	2.39	2.46
H/80	2.38	2.36
H/100	1.50	2.03
H/200	2.19	1.12
H/250	0.40	0.77

In general, the results for FORM from the fitted, approximate response surface from Equation (11.34) are not satisfactory, particularly for the last three performance levels with higher exceedence probabilities. The results are improved and more consistent when the Importance Sampling simulation is performed around the design point from FORM.

For performance-based design, the problem would be formulated as follows: obtain the optimum vector $\{d\}$ of design parameters,

$$\{d\} = \{\overline{R}, \overline{D}, \overline{\sigma_Y}\} \tag{11.47}$$

in which the components of $\{d\}$ are the mean values of the variables R, D and σ_Y, to meet the following three performance criteria with specified reliabilities:

$$\begin{aligned} G_1(x) &= H/30 - \Delta_{max} \\ G_2(x) &= H/100 - \Delta_{max} \\ G_3(x) &= F_{max} - F \end{aligned} \tag{11.48}$$

G_1 is assumed to represent a collapse situation or of excessive damage; G_2 a serviceability condition or of controlled, repairable damage; and G_3 a collapse situation corresponding to dowel forces F exceeding a maximum F_{max} controlled by wood cracking. In this case, F_{max} was assumed to be 10 kN. Given these three criteria for performance, three reliability levels were imposed: $\beta_1 = 3.0$ for G_1, $\beta_2 = 2.0$ for G_2, and $\beta_3 = 3.0$ for G_3. The reliabilities achieved at the optimum mean parameters $D = 13.2$ mm, $R = 117.5$ mm, and $\sigma_Y = 305$ MPa, were $\beta_1 = 3.01$, $\beta_2 = 1.98$ and $\beta_3 = 3.00$.

Table 11.7 Variable statistics and distributions, Case Example 6

Variable	Distribution	Mean	Coefficient of Variation
Bending Strength, R	Lognormal	30.0 Mpa	0.20
Vertical load, V	Normal	5.0 kN	0.10
Horizontal load, Q	Extreme Type I	22.20 kN	0.18

11.8 CASE EXAMPLE NO. 5: EVALUATING AN EXISTING STRUCTURE

An important application of probabilistic methods is the assessment of existing structures for continuing use under a change in load regime. The difficulty here is that the actual capacity of the structure is not known (it would have to be tested to failure to obtain it). What is known is the fact that the structure has survived for a certain time T under a load regime p. This piece of information can then be used, together with Bayes' Theorem, to assess the probable distribution of capacity. We apply Equation (11.25) for the case when the different sets A_i are different levels of capacity R. What is known is the prior density function for R, $f(R)$, as it was used in the original design or from standard laboratory tests of similar structures. Bayes' Theorem provides a tool to update this density function on the basis of the information that the structure survived the maximum loads that could have occurred during the prior period T:

$$f^*(R/s) = \frac{P(s/R)f(R)}{\int_0^{\infty} P(s/R)f(R)\,dR} = f^*(R) \tag{11.49}$$

where $f^*(R)$ is the updated density function given that the structure has survived, and $P(s/R)$ is the probability that the structure would have survived the period T if the capacity were R.

With the updated density function, the probability of failure under the new load regime, for the next time window T_1, would be calculated as follows:

$$P_f = \int_0^{\infty} P(f/R)f^*(R)\,dR$$

or, from Equation (11.49),

$$P_f = \frac{\int_0^{\infty} P(f/R)P(s/R)f(R)\,dR}{\int_0^{\infty} P(s/R)f(R)\,dR} \tag{11.50}$$

where $P(f/R)$ is the probability of failure under the new load regime on the assumption that the capacity is R. Equation (11.50) can be readily implemented by fixing several values of R within the corresponding range and, for each one, calculating the probability of prior survival and the probability of future failure. The numerator of Equation (11.50) is the average of the product $P(f/R)P(s/R)$, while the denominator is the average of the probabilities $P(s/R)$.

The density function for R can also be further upgraded taking into account information from non-destructive testing or proof loading.

11.9 CASE EXAMPLE NO. 6: CALIBRATION OF A CODIFIED DESIGN EQUATION

Traditional timber engineering design procedures have depended upon codified approaches. In these, *design equations* are provided by Codes and the engineer uses them to calculate required geometrical parameters like cross-section dimensions, allowable spans, etc. A design equation usually contains a number of *design factors* (some for loads and some for capacities), applied to *nominal* or *design values* of the variables, as in the following example:

$$\alpha_D D_N + \alpha_Q Q_N = R_N S/\gamma \tag{11.51}$$

in which D_N = effect of the nominal or design dead (or permanent) load D, α_D = load factor for dead loads, Q_N = effect of the nominal or design live load Q, α_Q = load factor for live loads, γ = resistance factor, R_N = nominal or design resistance, and S = Geometric parameter (e.g. cross-sectional modulus for bending).

The nominal dead loads are normally chosen equal to their mean values. Design live loads, on the other hand, are chosen in correspondence with a return period (e.g. 30-year snow, 475-year earthquake, etc.). The nominal resistance is normally chosen equal to a lower percentile of the strength distribution (e.g. for wood properties, normally the fifth-percentile). The load and resistance factors would normally be *calibrated* so that, when using the code equation, the resulting parameter S will ensure a desired target reliability. In each case, the reliability achieved can be evaluated by the direct procedures described earlier in this chapter. Of course, the target cannot be achieved in all situations with just a few factors in the design equation and, as a result, the factors must be calibrated by optimisation, minimising the difference between achieved and target reliabilities over a large number of calibration design conditions.

Although a code design procedure is simple and can be applied without resorting to software, a major disadvantage of the approach is that the actual reliability achieved could vary across situations. For example, code prescriptions for wood pole power transmission structures could have been calibrated using wind and icing conditions for five different locations across North America and, for those calibration sites, the design equation would result in reliabilities fairly close to the target. There is no guarantee, however, that the reliability achieved for other sites would be close to the desired level. The calibration would only produce a code with an applicability range depending on the situations considered in the calibration exercise.

As an illustration, let us consider a bending design equation for a wood pole problem as shown in Figure 11.10:

$$\alpha_V V_N e + \alpha_Q Q_N h = R_N S/\gamma \qquad (11.52)$$

Q is an horizontal load produced by wind on iced conductors, and V is a permanent vertical load from an attached transformer. R is the bending strength of the wood pole.

The statistics for a calibration site are those from Table 11.7. The nominal or design values are taken to be:

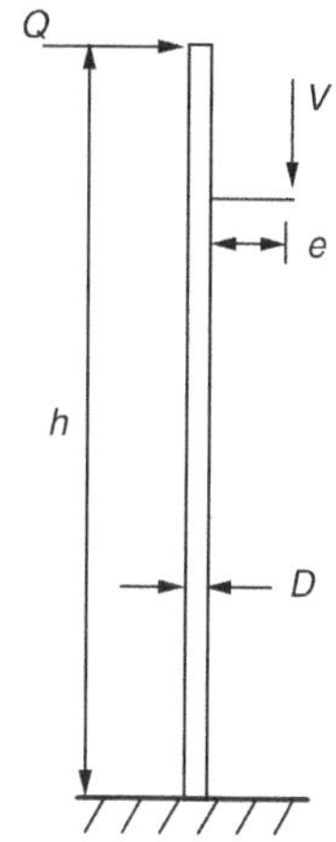

Figure 11.10 Wood pole for power transmission

Vertical load, V_N = Mean value of $V = 5.0$ kN;
Horizontal load, Q_N = 50-year return load (annual exceedence 1/50) $= Q_{50} = 20.39$ kN
Bending strength, R_N = fifth-percentile of $R =$ 21.24 MPa

To calibrate the factors, the parameter S obtained from the design Equation (11.52) is introduced into the performance function G:

$$G = RS - (Qh + Ve) \qquad (11.53)$$

In this manner, the function G now becomes a function of the load and resistance factors:

$$G = R/R_N - \left(\frac{1}{\gamma(\alpha_V V_N e + \alpha_Q Q_N)}\right)(Ve + Qh) \qquad (11.54)$$

The reliability associated with the probability of $G < 0$ can now be calculated for different combinations of the factors. This is done by the direct methods described earlier in this chapter. Table 11.8 shows the achieved reliability for different choices for the factors when the pole height is $h = 8$ m and the eccentricity $e = 0.8$ m.

For a *target reliability index* $\beta = 3.0$, Table 11.8 shows that $\alpha_V = 1.20$, $\alpha_Q = 1.60$ and $\gamma = 1.1$ can be used in the design equation to achieve quite closely the target for a design situation with the statistics of Table 11.7. However, if the statistics of the horizontal load Q are changed to those of a different site, not used in the previous calibration,

Table 11.8 Load and resistance factors, and achieved reliability

α_v	α_Q	γ	Reliability achieved, β
1.20	1.50	1.10	2.783
1.25	1.55	1.10	2.891
1.20	1.60	1.10	2.989
1.25	1.65	1.10	3.090
1.25	1.70	1.10	3.184

still having an Extreme I distribution but now with a mean of 30 kN and a coefficient of variation of 0.30, the reliability index achieved with the same factors would now be $\beta = 2.232$ (*substantially lower than the intended target of 3.0*).

This simple example shows the steps in a code calibration procedure, and also points out the shortcomings of a codified, factored approach. To achieve good results, the load factors must be optimised taking into account a sufficiently large number of calibration sites, reasonably covering the range of code applicability.

11.10 CONCLUSIONS

This chapter has presented a brief review of reliability concepts and methods. These have been illustrated with applications to several typical problems in timber engineering: deflections and bending failure of a glued-laminated beam under service or fire conditions; failure of a floor system with multiple joists; design of a doweled connection under earthquake excitation; assessment of an existing structure for reliability under a change of service conditions; and finally, the calibration of a factored code equation.

As in all aspects of civil engineering, reliability theory and probabilistic methods provide also in timber engineering a realistic approach to problem solving, taking into account the unavoidable uncertainties in the intervening variables. In timber engineering, reliability theory also offers, in particular, a sound framework for the study of quality control implementation, material mixing and optimisation. The vast potential of the approach, however, is fully exploited when problems are solved on a one-to-one basis, customising the solution. Although the techniques could also be used to calibrate codified approaches to a target reliability, this objective cannot be achieved for all design situations, given the simplified nature of the code equations and the few factors or degrees of freedom involved. The aim of performance-based design can be achieved through a customised approach, a task facilitated by the availability of sophisticated structural and reliability analysis software, coupled with ever increasing computer power.

REFERENCES

Ang A.H.S and Tang W. (1984a) *Probability Concepts in Engineering Planning and Design. Volume I: Basic Principles*. Bookmasters Inc., Mansfield, Ohio.

Ang A.H.S and Tang W. (1984b) *Probability Concepts in Engineering Planning and Design. Volume II: Decision, Risk and Reliability*. Bookmasters Inc., Mansfield, Ohio.

Augusti G., Baratta A. and Casciati F. (1984) *Probabilistic Methods in Structural Engineering*. Chapman & Hall.

Benjamin J. and Cornell A. (1970) *Probability, Statistics and Decision for Civil Engineers*. McGraw-Hill, New York.

Bucher C.G. and Bourgund U. (1990) A fast and efficient response surface approach for structural reliability problems. *Structural Safety*, **7**, 57–66.

Der Kiureghian A. and Liu P. (1986) Structural reliability under incomplete probability information. *Journal of Engineering Mechanics*, **112**(1).

Ditlevsen O. (1981) Principle of normal tail approximation. *Journal of Engineering Mechanics*, **107**(6), 1191–1208.

Ditlevsen O. and Madsen H.O. (1996) *Structural Reliability Methods*. Wiley, Chichester.

Faravelli L. (1989) Response surface approach for reliability analysis. *Journal of Engineering Mechanics*, **115**, 2763–2781.

Foschi R.O., Folz B. and Yao F. (1990) Reliability-Based Design of Wood Structures. Structural Research Series Report No. 34, Department of Civil Engineering, University of British Columbia, Vancouver, Canada.

Foschi R.O., Li H., Folz B., Yao F. and Baldwin J. (2000) RELAN: A General Software Package for Reliability Analysis. Department of Civil Engineering, University of British Columbia, Vancouver, Canada.

Foschi R.O. (2000) Modeling the hysteretic response of mechanical connections for wood structures. *Proceedings World Timber Engineering Conference*, Whistler, B.C. Canada.

Foschi R.O. and Li H. (2001) A methodology for performance-based design in earthquake engineering. *Proceedings 8th International Conference on Structural Safety and Reliability, ICOSSAR'01*, Newport Beach, CA.

Hasofer A.M. and Lind N.C. (1974) Exact and invariant second-moment code format. *Journal of Engineering Mechanics*, **100**(1), 111–121.

Li H. and Foschi R.O. (1998) An inverse reliability method and applications. *Structural Safety*, **20**, 257–270.

Liu Y.W. and Moses F. (1994) A sequential response surface method and its application in the reliability analysis of aircraft structural systems. *Structural Safety*, **19**, 3–19.

Madsen H.O., Krenk S. and Lind N.C. (1986) *Methods of Structural Safety*. Prentice-Hall, Eaglewood Cliffs, NJ.

Melchers R. (1987) *Structural Reliability: Analysis and Prediction*. Ellis Horwood, Chichester.

Rackwitz R. and Fiessler B. (1978) Structural reliability under combined random load sequence. *Computers and Structures*, **9**(5), 489–494.

Schueller G.I., Bucher C.G. and Bourgund U. (1989) On efficient computational schemes to calculate structural failure probabilities. *Probabilistic Engineering Mechanics*, **4**(1), 10–18.

Thoft-Christensen P. and Baker M. (1982) *Structural Reliability Theory and Its Applications*. Springer-Verlag.

Thoft-Christensen P. and Murotsu Y. (1986) *Application of Structural Systems Reliability Theory*. Springer-Verlag.

Wen Y.K. (1990) *Structural Load Modeling and Combination for Performance and Safety Evaluation*. Elsevier.

12

Design of Structures Based on Glulam, LVL and Other Solid Timber Products

Hans J. Larsen

12.1 INTRODUCTION

12.1.1 General

For the design of structural elements, glulam is normally treated as a homogenous, orthotropic, elastic material with one main axis in the fibre direction and with the same properties in all directions perpendicular to this axis. How to derive the properties from those of the laminations is treated in Chapter 5. The special problems in the design of glulam are then related to the special shapes that are possible or more common for glulam than for solid timber, e.g. curved or tapered members. Furthermore, the volumes of glulam structures are often large: size/volume effects and instability problems may therefore be more severe. These problems are covered in the following.

In some cases, it is necessary or advantageous to carry out a more detailed analysis. Finite element methods may be very effective, either 2D or 3D methods making it possible to take into account the influence of the year ring pattern. This topic is dealt with in Section 12.1.6.

Laminated Veneer Lumber (LVL) with all the fibres in the same direction, behaves in principle as solid timber: the strength and stiffness parameters and the magnitude of the size effect may differ, but the design equations are the same. Therefore,

LVL will not be dealt with separately. The same applies to parallam and similar products.

LVLcross, where the fibre direction of some of the veneers is perpendicular to the main direction, behaves in the same way as plywood.

12.1.2 Assumptions

Unless otherwise stated, members with rectangular cross-sections are regarded with axes as shown in Figure 12.1. In cases where it is convenient to distinguish between the axes, the z-coordinates are in the strong direction of the member, and M_y corresponds to the bending moment about the strong axis.

12.1.3 Notation

The notation follows that of Draft Eurocode 5 (2000). Frequently used symbols are listed below; other symbols are defined where they are used.

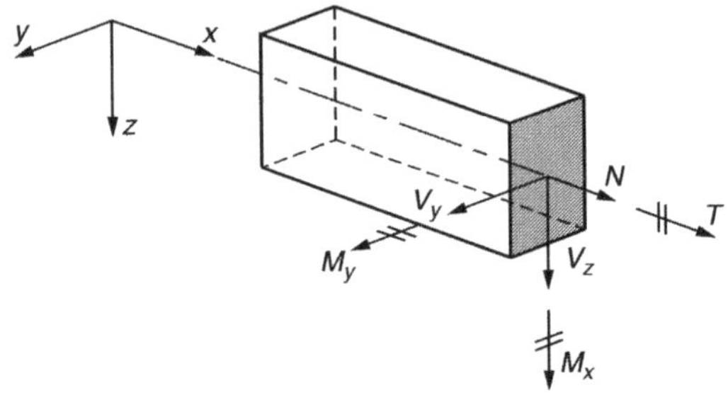

Figure 12.1 Axes and positive directions of stress resultants

12.1.4 Stress-strain Curve

According to Blass (1988) and Glos (1978), the stress-strain curve under axial load is as shown in Figure 12.2. In tension there is a linear relationship. In compression the relation is described by the initial modulus of elasticity $E = \tan\alpha$, the

A	Area of cross-section
E	Modulus of Elasticity
F	Force
G	Shear modulus
I	Second moment of area
M	Moment
N	Axial force
V	Volume
W	Section modulus
b	Width
f	Strength
h	Depth, height
i	Radius of inertia
k	Factor, coefficient
l	Length
r	Radius
t	Thickness
u	Deflection
α	Angle
γ	Angle, Partial coefficient
ε	Strain
λ	Slenderness ratio
ρ	Density (in kg/m^3)
σ	Stress
τ	Shear stress

Subscripts:

c	Compression
crit	Critical
d	Design
def	Deformation
dis	Distribution
ef	Effective
k	Characteristic
m	Bending, material partial safety factor
mod	Modification
rel	Relative
t	Tension
ult	Ultimate
v	Shear
vol	Volume
x, y, z	Directions of axes
y	Yielding

0, α and 90 denotes the angle (°) between the force/stress-direction and the grain direction, i.e. 0 is parallel and 90 perpendicular to grain.

The fibre direction is marked with ➡

compression strength f_c, the asymptotic final compression stress $f_{c,y}$, the strain ε_c at maximum stress and the ultimate strain ε_u. The following empirical relation is assumed:

$$\sigma = \frac{E\varepsilon + k_1(\varepsilon/\varepsilon_c)^7}{1 + k_2\varepsilon + \dfrac{k_1(\varepsilon/\varepsilon_c)^7}{f_{c,y}}} \quad (12.1)$$

where

$$k_1 = \frac{f_{c,y}}{6(1 - (f_{c,y}/f_c))} \quad (12.2)$$

$$k_2 = \frac{E}{f_c} - \frac{7}{6\varepsilon_c} \quad (12.3)$$

Typical values for the parameters are:

$$f_{c,y}/f_c \sim 0.8 \quad \varepsilon_c = 0.8\text{–}1.2\% \quad \varepsilon_u \sim 3\varepsilon_c \quad (12.4)$$

12.1.5 Axial Load and Bending

The behaviour for axial load in combination with a moment about the y-axis can be determined by equilibrium conditions. The procedure can be illustrated for the simplified linear elastic–plastic stress-strain relation shown in Figure 12.2. It is

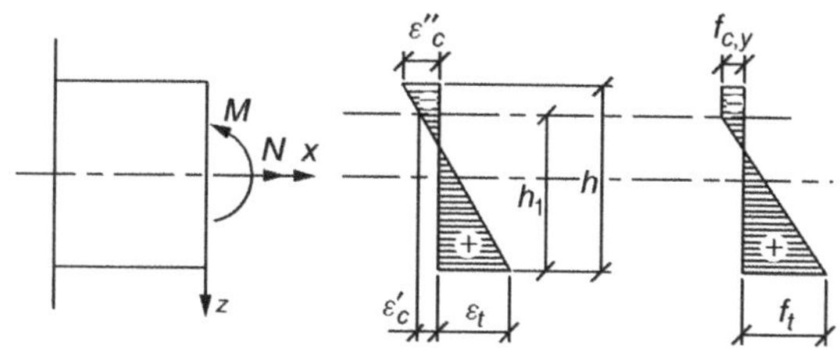

Figure 12.3 Strain and stress distribution at failure for a rectangular cross-section (width = b) for an axial load N and a bending moment M

assumed that at failure, the strain and stress variation over the depth is as shown in Figure 12.3.

Axial equilibrium gives:

$$N = b\left(-hf_{c,y} + \tfrac{1}{2}h_1(f_{c,y} + f_t)\right) \quad (12.5)$$

from which

$$h_1 = 2h\frac{N/(bh) + f_{c,y}}{f_{c,y} + f_t} \quad (12.6)$$

Moment equilibrium taken at $z = 0.5h$ gives

$$M - N\frac{h}{2} = b\left(\frac{1}{2}f_{c,y}h^2 - \frac{1}{6}h_1^2(f_{c,y} + f_t)\right) \quad (12.7)$$

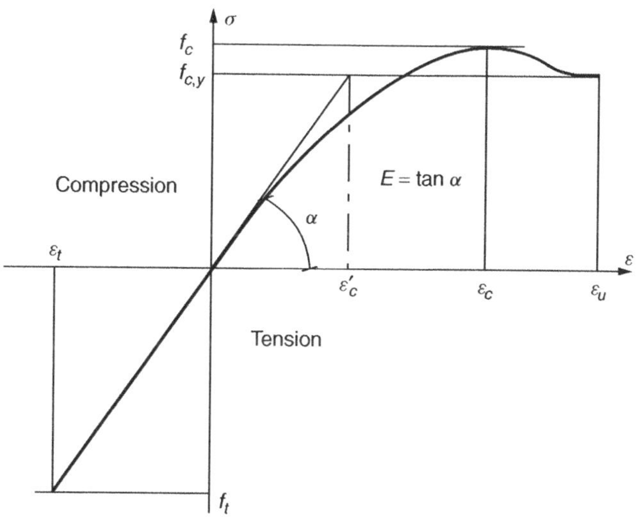

Figure 12.2 Stress-strain relation according to Blass (1988). A simplified linear elastic plastic stress-strain curve is shown with thin line

The permissible combinations of N and M are found by inserting (12.5) into (12.7):

$$M = \frac{1}{2}Nh - b\left(\frac{2}{3}\frac{(N/b + hf_{c,y})^2}{f_{c,y} + f_t} - \frac{1}{2}f_{c,y}h^2\right) \quad (12.8)$$

The moment may be expressed as a formal bending stress σ_m:

$$\sigma_m = \frac{M}{W} = \frac{6M}{bh^2} = 3(\sigma_N + f_{c,y}) - 4\frac{(\sigma_N + f_{c,y})^2}{f_{c,y} + f_t} \quad (12.9)$$

where $\sigma_N = N/A$.

The following relation between the ultimate strains in tension and compression is found from similar triangles:

$$\frac{h_1}{h} = \frac{\varepsilon'_c + \varepsilon_t}{\varepsilon''_c + \varepsilon_t} \quad \text{and} \quad \frac{\varepsilon'_c}{\varepsilon_t} = \frac{f_{c,y}}{f_t} \quad (12.10)$$

Inserting (12.6) into (12.10) gives:

$$\varepsilon''_c = \varepsilon_t\left(\frac{(1 + f_{c,y}/f_t)^2}{2(\sigma_N/f_t + f_{c,y}/f_t)} - 1\right) \quad (12.11)$$

Expressions (12.7), (12.8) and (12.11) are only valid for $h_1 \leq h$. For larger tensile forces, the conditions are elastic.

For pure bending, the following relation applies:

$$M = \frac{bh^2}{6}f_{c,y}\left(3 - \frac{4f_{c,y}}{f_{c,y} + f_t}\right) \quad (12.12)$$

$$\sigma_m = f_{c,y}\left(3 - \frac{4f_{c,y}}{f_{c,y} + f_t}\right) \quad (12.13)$$

$$\varepsilon_c = \frac{\varepsilon_t}{2}\left(\frac{f_t}{f_{c,y}} + \frac{f_{c,y}}{f_t}\right) \quad (12.14)$$

Corresponding values of σ_m and σ_N at failure are shown in Figure 12.4. Parts of the curves are shown dotted. They correspond to $\varepsilon_{c''} > 3\varepsilon_t$, which is hardly likely. The failure mode depends upon the magnitude of the axial load. For large positive values (tension), the conditions are linear elastic until failure. For small values (the curved parts), the fibres in the compression side will initially 'yield' until final tensile failure. For large values of f_t/f_c, i.e. for clear wood, the moment capacity is reduced when superimposed by a compression load. For small values of f_t/f_c, the moment capacity is increased by a compression stress, because the compression action relieves the tension side.

The tensile strength is determined from tests in which the axial force acts in the geometrical centre in the end cross-sections. Knots and density

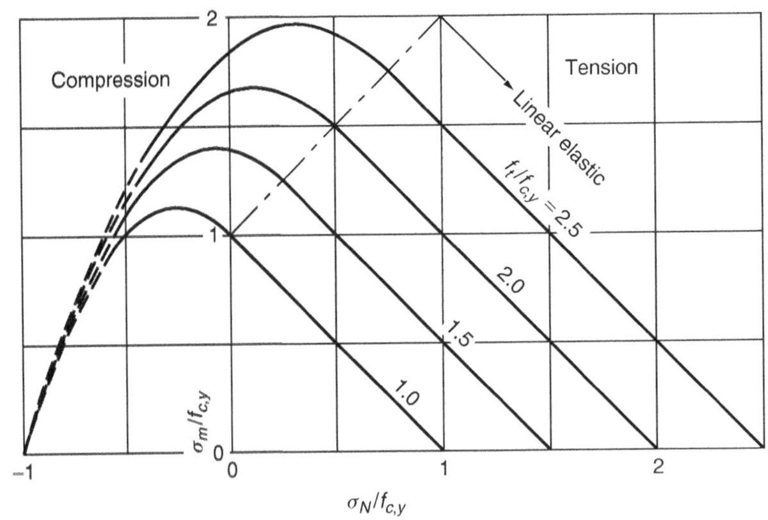

Figure 12.4 Corresponding values of axial stresses and bending stresses

variations along the test piece and across the cross-sections result in bending stresses that can be rather high. In a beam cross-section, these secondary effects are prevented and further stresses can be redistributed when failure starts. This means that a higher tensile strength than that used in the design of tension members should be inserted in the expressions derived above. A value of $f_t \sim 1.5 f_m$ gives reasonable results.

For the cases where the bending stresses in compression are so high that the fibres yield, the shear stress distribution will also be influenced (see Figure 12.5). The bending stresses are almost constant in the plastic zone, and the shear stresses are therefore zero. The shear force has to be taken on the reduced depth, which can be calculated to $(3 - \sigma_m / f_{c,y}) h/2$, i.e.:

$$\tau_{\max} = \frac{2 f_{c,y}}{3 f_{c,y} - \sigma_m} \frac{1.5V}{bh} \tag{12.15}$$

12.1.6 Finite Element Calculations

Most of the design equations given in the following aim at 'manual calculations'. They are usually on the safe side, and it may be advantageous to make a more detailed analysis of the stress distribution. This may be the case for a special type of structure, or for economically important members in a structure. An important tool for such analyses is finite element methods – either 2D methods, where the members are regarded as plane structures, or 3D methods, where the cylindrical, orthotropic structure is taken into account. With the increase of computer power, 3D methods are becoming more and more common and their use may lead to quite surprising results because of the difference in stiffness in the radial and tangential directions, the stiffness in the former being the bigger. Two examples will be mentioned: the first is the stress distribution perpendicular to grain in a curved beam; the second is the stress distribution in a glulam block loaded in compression perpendicular to grain.

In Dahlblom and Ormarsson (1993), the stress distribution in a curved beam is calculated taking into account the different wood properties in the radial and tangential directions. The results are shown in Figure 12.6. For a 22° growth ring slope at the surface, the maximum stress perpendicular to grain is about 30% higher than the average stress. For a 45° slope, the increase is about 70%. An example on moisture-induced stresses in curved beams is given in Chapter 9.

The other example is the results of tests to determine the compression strength of glulam perpendicular to grain (see Figure 12.7) (Damkilde *et al.*, 1999). It was expected that failure would take place for a uniform stress perpendicular to the glue lines of about 6 MPa, which is the value assumed in most codes. Instead, failure occurred as tensile failure perpendicular to the grain: a crack started in the origo and spread along the x-axis. The compression strength was only 3 MPa. A detailed analysis revealed that the compression stress was not uniform across the cross-section (σ_x in the section $y = 0$ is about 1.5 the average stress, σ_c). More importantly, the inclination of the growth rings gave tension stresses (σ_y) over and compression stresses under the glue lines, the maximum tensile value being about 0.5 σ_c. As a

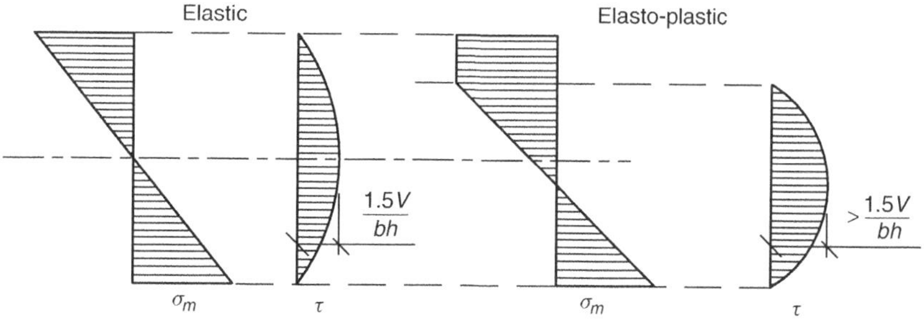

Figure 12.5 The stress variations over the depth for a rectangular cross-section with moment and a shear force V

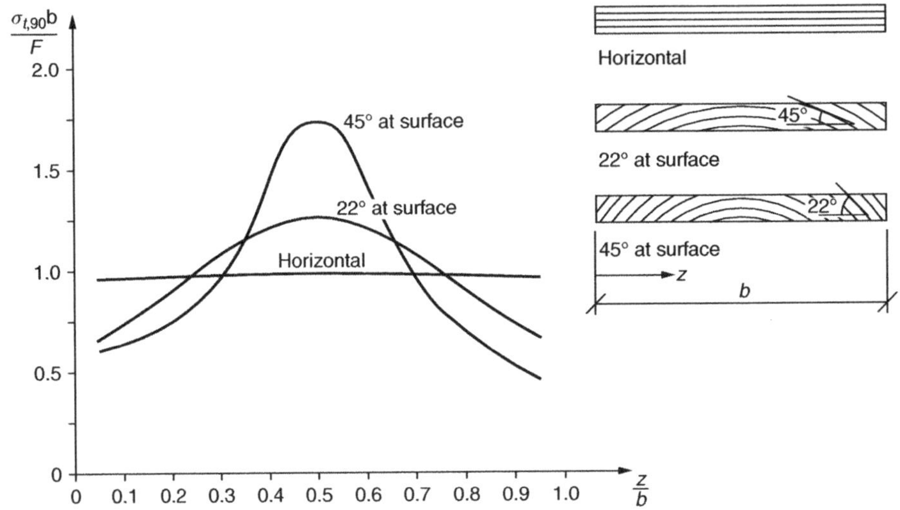

Figure 12.6 Tensile stresses perpendicular to grain in a curved glulam beam with constant bending moment

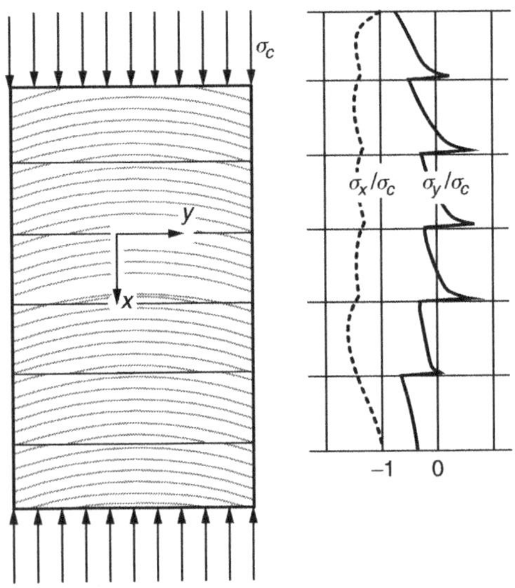

Figure 12.7 Compression test stresses in the section at $y = 0$

consequence, code values were halved, which has created many problems.

Typical stiffness parameters for glulam L40 are:

$$\begin{array}{lll} E_L = 14\,000 \text{ MPa} & E_R = 800 \text{ MPa} & E_T = 500 \text{ MPa} \\ G_{LR} = 800 \text{ MPa} & G_{LT} = 500 \text{ MPa} & G_{RT} = 60 \text{ MPa} \\ \nu_{LR} = 0.02 & \nu_{LT} = 0.02 & \nu_{RT} = 0.45 \end{array} \quad (12.16)$$

For a 2D analysis, the normal failure criteria and design values apply. For a 3D analysis, higher values may be applied, since many of the effects that reduce the formal strength have been taken into consideration. This may be of particular importance for tension perpendicular to grain, where values 50% higher than normal may be used. For failure analysis at a point, the tensile strength parallel to grain may be taken as $f_{t,0} = 1.5 f_m$ (cf. Section 12.1.5).

12.2 FAILURE CRITERIA AND VOLUME EFFECTS

12.2.1 Failure Criteria

For cases other than the simple ones – unidirectional stresses in the main directions or pure shear – it is necessary to have failure criteria. Various criteria, originally pioneered by aircraft building, have been proposed. Recently, failure models for orthotropic materials have attracted renewed interest in connection with the development of modern fibre composite materials.

The simplest failure criterion is the *maximum stress criterion*, where the stresses in the main direction and the shear stresses are limited to the corresponding strength values, i.e. the interaction between the stresses is – often rather on the unsafe side – disregarded. For tension and shear, for example:

$$\sigma_{t,0} \le f_{t,0} \quad \sigma_{t,90} \le f_{t,90} \quad \tau \le f_v \qquad (12.17)$$

One of the most common failure criteria for isotropic materials is the von Mises yield criterion, and most of the proposals found in the literature take this as their point of departure, by postulating that failure takes place when:

$$k_1\sigma_0^2 + 2k_2\sigma_0\sigma_{90} + k_3\sigma_{90}^2 + k_4\tau^2 + k_5\sigma_0 + k_6\sigma_{90} = 1 \qquad (12.18)$$

Equation (12.18) contains six parameters, five of which can be found from simple uniaxial tension or compression tests or shear tests. The parameter k_2 can only be determined from complicated combined stress tests, and is often treated as an empirical constant. Inserting the basic strength values, the failure criterion becomes:

$$\frac{\sigma_0^2}{f_{t,0}f_{c,0}} + 2k_2\sigma_0\sigma_{90} + \frac{\sigma_{90}^2}{f_{t,90}f_{c,90}} + \frac{\tau^2}{f_v^2} + \left(\frac{1}{f_{t,0}} - \frac{1}{f_{c,0}}\right)\sigma_0 + \left(\frac{1}{f_{t,90}} - \frac{1}{f_{c,90}}\right)\sigma_{90} = 1 \qquad (12.19)$$

σ_0 and σ_{90} are the stresses parallel and perpendicular to grain (tension is positive), τ are the shear stresses, and f the corresponding strength values. Tsai (2000) proposed writing the interaction constant k_2 as

$$k_2 = k(f_{t,0}f_{t,90}f_{c,0}f_{c,90})^{-0.5} \qquad (12.20)$$

where $-1 \le k \le 0$ for all materials. Figure 12.8 shows the failure envelope for $\tau = 0$ according to (12.19)–(12.20) – often denoted the *Tsai–Wu failure criterion* – for $f_{c,0} = 0.67f_{t,0}$, $f_{t,90} = 0.025f_{t,0}$, $f_{c,90} = 0.10f_{t,0}$ and $k = 0$ or $k = -0.5$.

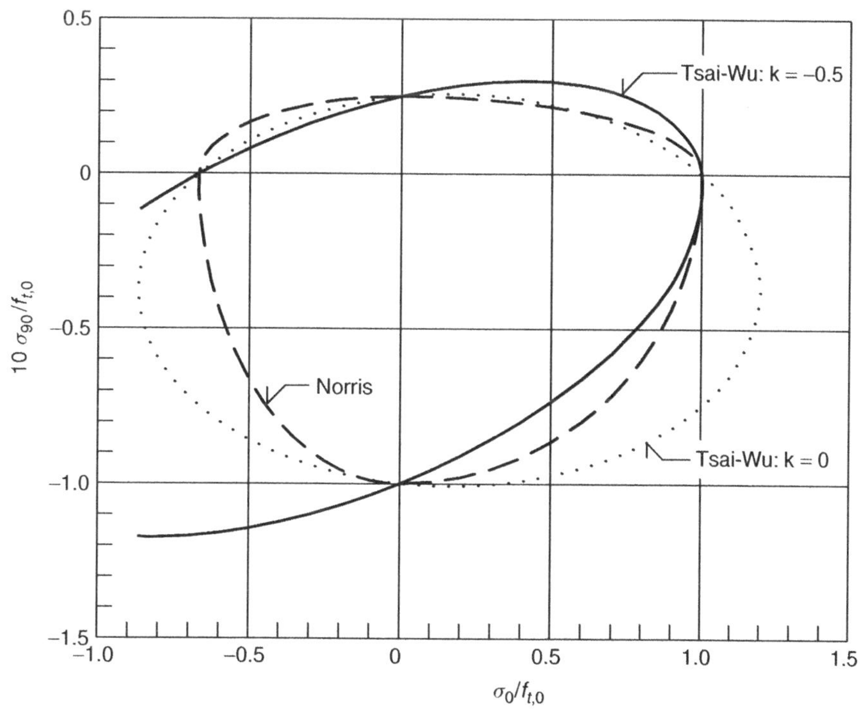

Figure 12.8 Failure envelopes

A simple semi-empirical failure condition is the quadric expression proposed in Norris (1962):

$$\left[\left(\frac{\sigma_0}{f_0}\right)^2+\left(\frac{\sigma_{90}}{f_{90}}\right)^2+\left(\frac{\tau}{f_v}\right)^2\right]^{\frac{1}{2}} \le 1 \quad (12.21)$$

The Norris criterion – composed of four ellipses – is also shown in Figure 12.8. The difference between Norris and Tsai–Wu with $k = 0$ is small, and since there are very few tests with combined stresses, it is recommend that the Norris criterion be used. A more detailed treatment of the Norris criterion is given in Chapter 7.

In many codes, a very simple criterion – the Hankinson expression (Hankinson, 1921) – is used for the strength f_α at an angle α to the grain (see Figure 12.9). For tension, it is assumed that the strength is:

$$f_{t,\alpha} = \frac{f_{t,0}f_{t,90}}{f_{t,0}\sin^2\alpha + f_{t,90}\cos^2\alpha} \quad (12.22)$$

For compression, the tensile strengths are replaced by the corresponding compression strengths. The *Hankinson criterion* is generally on the unsafe side, especially at small angles to the grain – the most common situation in practice. Reference is also made to Chapter 7.

12.2.2 Volume Effects

It is often claimed that the strength of timber depends upon the size of the stressed volume: deep beams are weaker than shallow ones, wide and long tension members are weaker than narrow and short ones, etc. Whether this is really the case is much discussed; for instance, many maintain that the effect for ordinary timber beams is due to cutting and grading practices. There is, however, no doubt that the effect is real for tension perpendicular to the grain and other brittle failures. The effect may be explained by the size effect always inherent in fracture mechanics problems, or by the so-called Weibull theory, which will be described briefly in the following. For a more general treatment, reference is made to Gustafsson (2001). Volume effects in beams (effects of length and load configuration) are treated in Chapter 4.

The Weibull theory (Weibull, 1939), also called the weakest link theory, is based on the following assumptions:

- The stressed volume, V, consists of a large number of basic volumes, each containing a weak spot.
- The probability of failure is the same for all basic volumes.
- Failure in one basic volume will lead to brittle failure of the total volume.

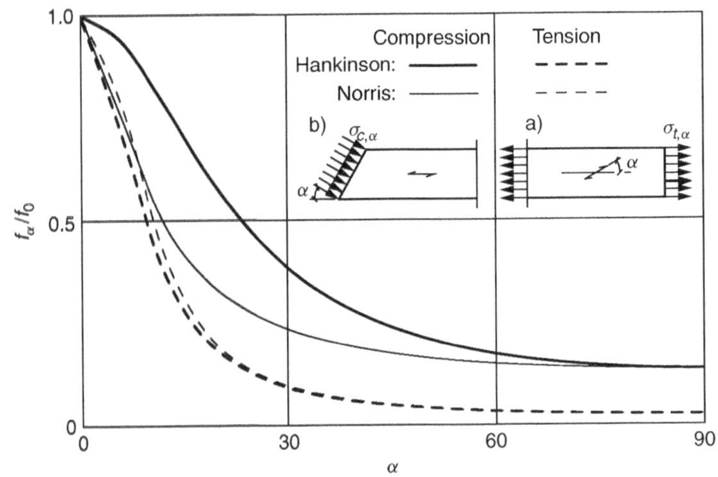

Figure 12.9 Relative unidirectional strength at an angle to the grain according to Norris and Hankinson

The probability of failure is then:

$$P_f(\sigma) = 1 - \exp\left(-V\left(\frac{\sigma - \sigma_o}{m}\right)^k\right) \quad (12.23)$$

This model is based on the three-parameter Weibull distribution. σ is the tensile stress perpendicular to the grain, and k and m are material parameters. k may be found from the mean and the coefficient of variation for the tensile strength, and is about 3–6. In many codes, a value of $1/k = 0.2$ is assumed. The threshold stress (the minimum failure strength) σ_0 can often with good approximation be taken equal to 0, resulting in the two-parameter Weibull distribution:

$$P_f(\sigma) = 1 - \exp\left(-V\left(\frac{\sigma}{m}\right)^k\right) \quad (12.24)$$

From Equation (12.24), it can be seen that two uniformly stressed volumes, V_1 and V_2, have the same probability of failure if the stresses σ_1 and σ_2 fulfil the condition:

$$\sigma_1/\sigma_2 = (V_2/V_1)^{1/k} \quad (12.25)$$

The probability of failure of a volume V with a varying stress $\sigma(x, y, z)$ with the maximum value $\sigma_{\max}$ is the same as that of a volume

$$V^* = \int \frac{\sigma(x, y, z)^k}{\sigma_{\max}} dV \quad (12.26)$$

The ratio $(V/V^*)^{1/k}$ is called the distribution factor, k_{dis}, and is convenient, for instance, for calculating the tension perpendicular to grain strength of a volume V for different load or stress configurations:

$$f_{t,90,d}^V = k_{vol} k_{dis} f_{t,90,d} \quad (12.27)$$

where $f_{t,90,d}$ refers to a reference volume V_0 under uniform stress and $k_{vol} = (V_0/V)^{1/k}$. Examples of k_{dis} may be found in Colling (1986). As an example, for $k = 5$, $k_{dis} = 1.22$ for the parabolic variation of the tension stresses perpendicular to the grain found in a curved beam with constant moment.

Eurocode 5 also gives k_{dis} for cambered and pitched cambered beams, but the values are very crude, and in most cases a determination according to Equation (12.26) will prove advantageous.

12.2.3 Fracture Mechanics

At notches and holes in beams, for example, the stresses become theoretically infinite, and failure is often very sudden and brittle and not preceded by large deformations or other visible warning. In these cases, it is meaningless to try to determine the load-carrying capacity by means of conventional stress criteria. Instead, fracture mechanics may offer a solution (see Chapter 7).

12.3 CURVED BEAMS

12.3.1 Introduction

Curved beams have lower strength than straight ones because of bending prestresses in the laminations. Furthermore, there are stresses perpendicular to grain and the stress variation over depth is not linear, even when assuming that plane cross-sections perpendicular to the axis remain plane (Bernouilli beam). Lekhnitskii (1968) gives expressions from which the edge stresses parallel to grain, σ_i, σ_o, and the maximum stress perpendicular to grain, $\max \sigma_{t,90}$, may be found, taking the orthotropy into account. Approximate solutions are derived in Sections 12.3.3 and 12.3.4.

12.3.2 Prestresses

High eigenstresses are built into the laminations during production. Their magnitude is

$$\sigma_m = Et/(2r + t) \quad (12.28)$$

where t is the thickness of the laminations and r is the minimum radius. For glulam laminations, $E/f_m \sim 300$. Normally, r/t varies between 125 and 250, i.e. σ_m varies between $0.6f_m$ and $1.2f_m$. One would expect some laminations to break during production and the strength of the curved member to be reduced considerably. However, according to Wilson (1939) and Hudson (1960), that is not the case. The bending strength is only reduced marginally. The explanation may be creep, especially during production where there is some moisture from the glue. Based on the tests in these references, the following reduction factor to the bending strength of straight beams seems appropriate for $r/t < 240$:

$$k_{curvature} = 0.76 + 0.001r/t \qquad (12.29)$$

For $r/t = 150$, where the eigenstress is $1.2f_m$, k is only about 0.9.

12.3.3 Bending Stress Distribution

Because of the varying length of the arcs, the stress variation with depth is as shown in Figure 12.10. The mean radius is changed from r_{mean} to ρ and the neutral axis is shifted downwards. The initial arc lengths are $s = s_0\ (1 - z/r_{mean})$. The arc lengths are increased by ds_0 and ds:

$$s + ds = (s_0 + ds_0)(1 - z/\rho) \qquad (12.30)$$

$$1 + \frac{ds}{s} = 1 + \varepsilon_z = \left(1 + \frac{ds_0}{s_0}\right)\frac{1 - z/\rho}{1 - z/r_{mean}}$$

$$= (1 + \varepsilon_0)\frac{1 - z/\rho}{1 - z/r_{mean}} \qquad (12.31)$$

By rearranging and disregarding second order terms in ε,

$$\varepsilon_z = \varepsilon_0 + k\frac{z}{1 - z/r_{mean}} \qquad (12.32)$$

$$\sigma = E \cdot \varepsilon_z = \sigma_0 + Ek\frac{z}{1 - z/r_{mean}} \qquad (12.33)$$

where

$$k = 1/r_{mean} - 1/\rho \qquad (12.34)$$

The equilibrium conditions $\int \sigma b\, dz = 0$ and $\int \sigma bz\, dz = M$, together with Equation (12.33), lead to:

$$\sigma(z) = \frac{M}{bhr_{mean}} + \frac{M}{I_{ef}}\frac{z}{1 + z/r_{mean}} \qquad (12.35)$$

where an effective second order moment I_{ef} has been introduced:

$$I_{ef} = \int \frac{z^2 b}{1 - z/r_{mean}} dz \qquad (12.36)$$

If there is also an axial force F, Equation (12.36) is changed to

$$\sigma(z) = \frac{N + M/r_{mean}}{bh} + \frac{M}{I_{ef}}\frac{z}{1 + z/r_{mean}} \qquad (12.37)$$

Equilibrium requires stresses perpendicular to the grain. The maximum force perpendicular

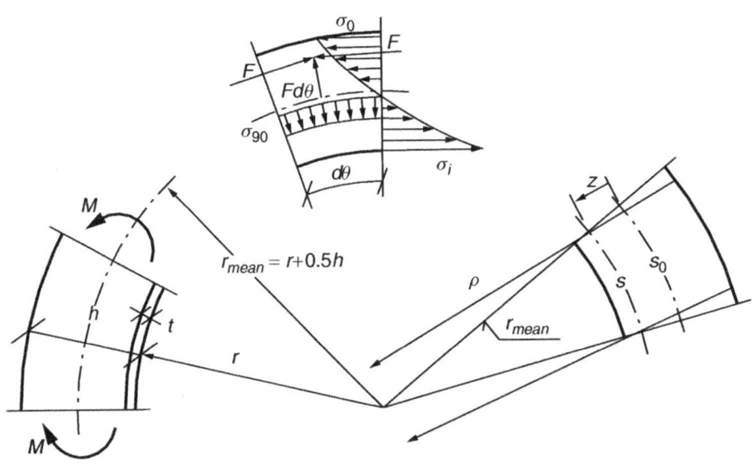

Figure 12.10 Plane curved beam with constant moment M. The cross-section is rectangular with thickness b. The inner radius is r and the mean radius is r_{mean}, from which z is measured

to the grain acts in the neutral axis for pure bending, i.e. for $z = r_{mean} I_{ef}/(I_{ef} + bhr_{mean}^2)$. A good approximation illustrating the problem can be found by approximating the moment by two forces $F = 1.5M/h$ (i.e. the influence of the curvature on the stress distribution is disregarded). Equilibrium of the small segment $d\theta$ gives

$$\sigma_{90} b r_{mean} d\theta = F d\theta$$

$$\sigma_{90} = 1.5\ M/(bhr_{mean})$$

12.3.4 Failure Criteria

The bending stresses shall satisfy the following condition:

$$\sigma_m \leq k_{curvature} f_m \tag{12.38}$$

where $k_{curvature}$ (see Equation (12.29)) takes into account the strength reduction due to the curvature of the innermost fibre. The stresses perpendicular to grain shall satisfy the following condition:

$$\sigma_{t,90} \leq k_{dis}(V_{ref}/V)^{1/k} f_{t,90} \tag{12.39}$$

For $k = 5$ and with a reference volume $V_{ref} = 0.01\ m^3$, Equation (12.39) becomes with $k_{dis} = 1.22$, see (12.27):

$$\sigma_{t,90} \leq 1.22 \cdot 0.01^{0.2} V^{-0.2} f_{t,90} = 0.49 V^{-0.2} f_{t,90} \tag{12.40}$$

V is the volume of the curved part in m^3.

12.4 TAPERED, PITCHED AND CAMBERED BEAMS

12.4.1 Tapered Beams

It is often advantageous to vary the depth of beams to obtain an inclined roof or to adjust the depth to the moments or shear forces. The depth variation influences the stress distribution in a beam (see Section B in Figure 12.11). A bending moment will, in addition to the normal stresses in the beam direction, result in stresses perpendicular to grain and in shear stresses: the stresses at the top must necessarily be parallel to the surface, and therefore have a component at an angle α to the grain/beam direction. The stresses perpendicular to the grain reduce the strength in accordance with Section 12.2.1.

According to Riberholt (1979), the stresses parallel to surfaces can be written as:

$$\sigma_{surface} = k_\sigma \cdot 6M/(bh^2) \tag{12.41}$$

As an approximation, the following values are found:

- at the surface where there is an angle α to the grain:

$$k_\sigma = 1 - 4.4 \tan^2 \alpha \tag{12.42}$$

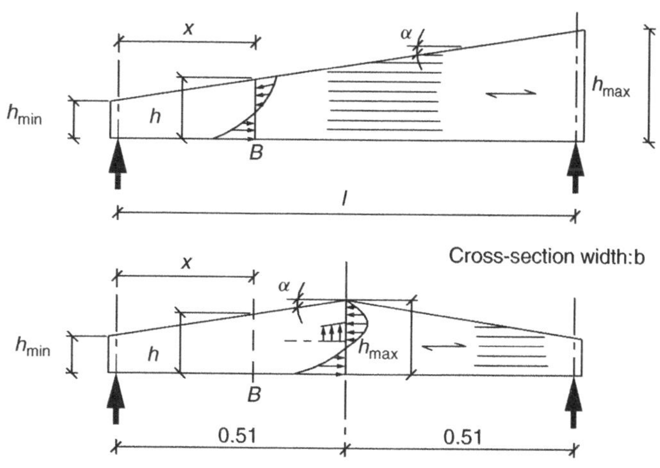

Figure 12.11 Stress distributions in a tapered beam and in a pitched beam. The stresses in section B in the pitched beam are similar to those in the cambered beam

- at the surface where $\alpha = 0$:

$$k_\sigma = 1 + 3.7 \tan^2 \alpha \quad (12.43)$$

At the surface with an angle to the grain, the stresses relative to the principal axes of the wood are:

$$\sigma_0 = \sigma_{surface} \cos^2 \alpha \quad \sigma_{90} = \sigma_{surface} \sin^2 \alpha$$

$$\tau = \sigma_{surface} \cos \alpha \sin \alpha \quad (12.44)$$

These stresses should be inserted in the relevant failure criterion.

12.4.2 Pitched and Pitched Cambered Beams

In a pitched beam (see Figures 12.11 and 12.12) the stresses in the shaded zones are heavily influenced by the pitch. The bending stresses at the upper surface at the apex are zero and the bending stresses at the lower surface are increased. Furthermore, there will be stresses perpendicular to the grain in the whole volume. The biggest stresses can be expressed as

$$\sigma_{m,i} = k_l \frac{6M_{ap}}{bh_{ap}^2} \quad (12.45)$$

$$\sigma_{t,90} = k_r \frac{6M_{ap}}{bh_{ap}^2} \quad (12.46)$$

Analytical expressions for the factors k_l and k_r are given in Figures 12.13 and 12.14. M_{ap} is the bending moment in the cross-section through the apex and α, h_{ap} and r are as defined in Figure 12.12. The factors are determined by a finite element analysis (see Blumer (1979) and Riberholt (1979)). The bending stress is limited to $k_{curvature} f_m$, where $k_{curvature}$ is the reduction

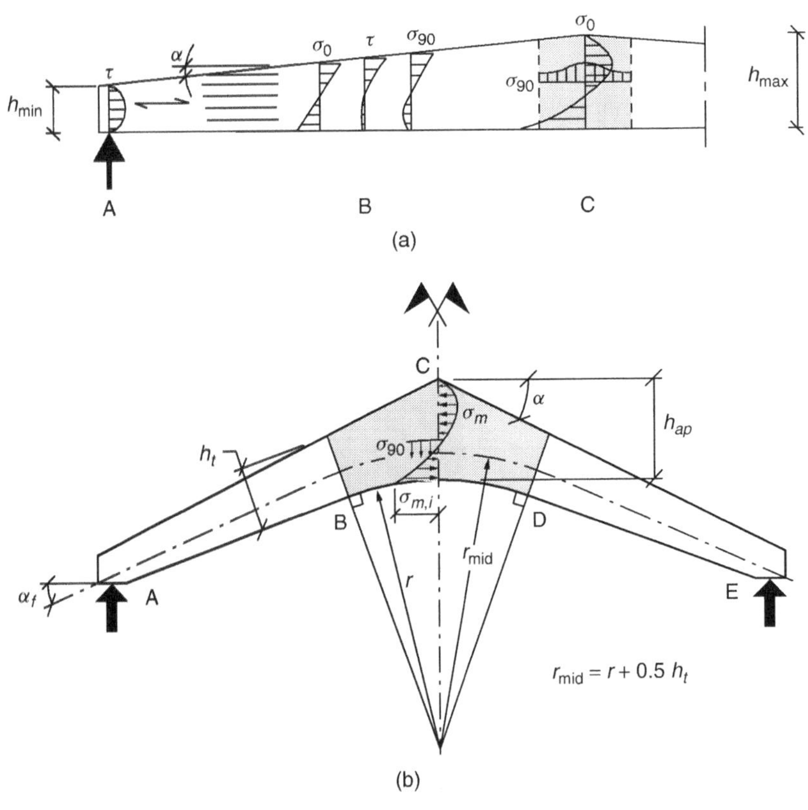

Figure 12.12 Pitched beam (a), and pitched cambered beam (b)

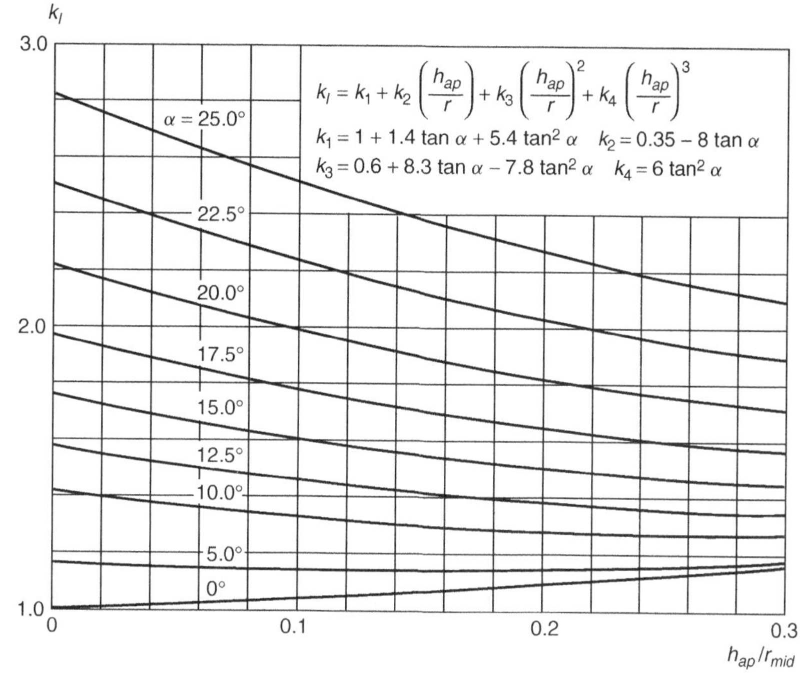

Figure 12.13 k_l for pitched cambered beam

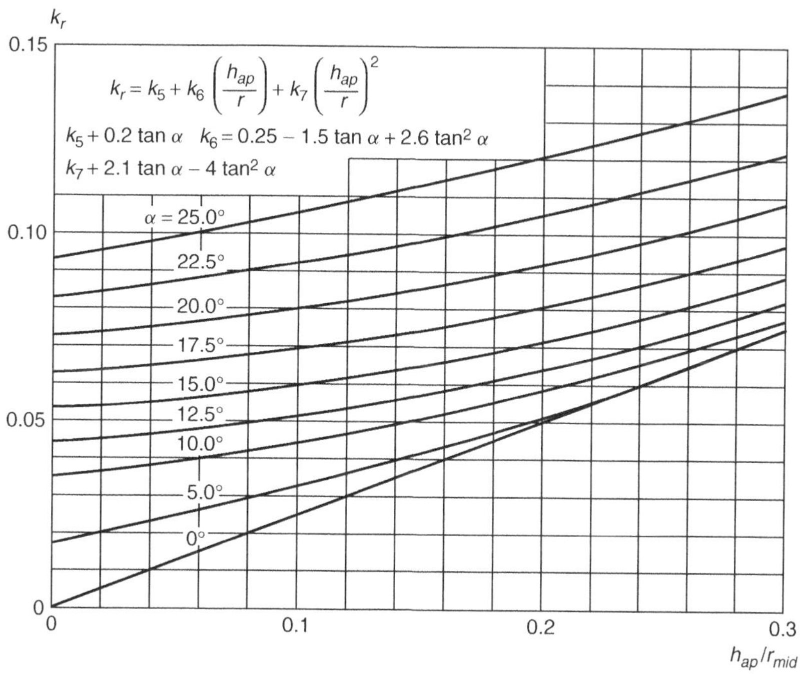

Figure 12.14 k_r for pitched cambered beam

factor due to the curvature of the inner lamination (see Section 12.3.2). The stresses perpendicular to the grain are limited to $k_{dis}(V_{ref}/V)^{0.2} f_{t,90}$. $V_{ref} = 0.01$ m^3 is a reference volume, and V is the volume of the shaded areas. According to Draft Eurocode 5 (2000), $k_{dis} = 1.7$. For the straight pitched beam, $k_{dis} = 1.4$ with a volume bh_{ap}^2, where b is the width of the beam. Both values are rather crude and, as mentioned in Section 12.2.2, a more detailed analysis should be done for important structures.

12.5 END-NOTCHED BEAMS AND BEAMS WITH HOLES

12.5.1 End-notched Beams

According to Gustafsson (1988) (See also Chapter 7), for cases a) and b) in Figure 12.15:

$$V_{cr} = bh_{ef}\sqrt{\frac{G_f E}{h}} \times \left(0.6(\alpha - \alpha^2)\frac{E}{G} + \beta\sqrt{6\left(\frac{1}{\alpha} - \alpha^2\right)}\right) \tag{12.47}$$

where $\alpha = h_{ef}/h$ and β is defined in Figure 12.15. G_f is the perpendicular to the grain fracture energy. With $E/G \sim 16$ and assuming that both the shear strength f_v and G_f are proportional to G, the strength criterion can be formulated as a usual strength requirement:

$$\tau = \frac{3}{2}\frac{V}{bh_{ef}} \leq k_v f_v \tag{12.48}$$

$$k_v = \min\begin{cases} 1 \\ k_n\left(1 + \dfrac{1.1 i^{1.5}}{\sqrt{h}}\right)\dfrac{1}{\sqrt{h}} \\ \quad \times\left(\sqrt{\alpha - \alpha^2} + 0.8\beta\sqrt{\dfrac{1}{\alpha} - \alpha^2}\right)^{-1} \end{cases} \tag{12.49}$$

where $k_n = 6.5$. The reason why there are two conditions in Equation (12.4.9) is that, for small values of the notched depth, the shear strength, and not cracking, is decisive. The term with i takes into account the fact that the load-carrying capacity is increased by the taper (see Riberholt *et al.*, 1991). For case c) $k_v = 1$.

12.6 INSTABILITY

12.6.1 Centrally Loaded Columns

It is assumed that the column deflects in the direction of one of the main axes. The corresponding second order moment is I and the radius of gyration is i. The free column length is l and the geometrical slenderness ratio $\lambda = l/i$. The compression stress is $\sigma_c = F/A$, where F is the axial load. The column strength is expressed as $k_c f_c$. The relative slenderness ratio is defined by

$$\lambda_{rel} = \lambda\sqrt{\frac{f_c}{\pi^2 E}} \tag{12.50}$$

The compression strength parallel to grain $f_c = f_{c,0}$ is determined from tests with $\lambda_{rel} \sim 0.3$ and, therefore, $k_c = 1$ for $\lambda_{rel} \leq 0.3$. For slender columns, the load-carrying capacity is always with a good approximation the Euler load, i.e.

$$k_c = 1/\lambda_{rel}^2 \tag{12.51}$$

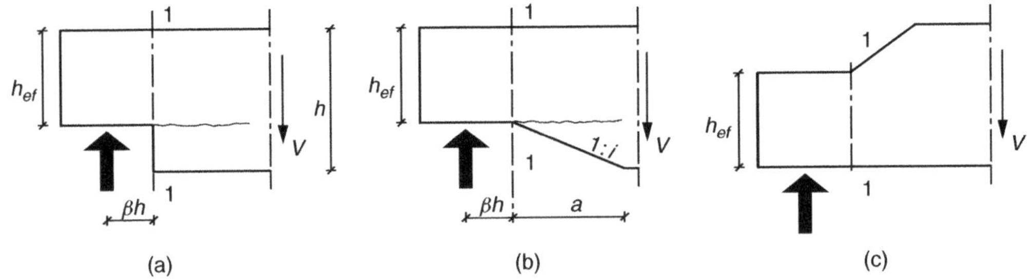

Figure 12.15 Geometry of end-notched beams

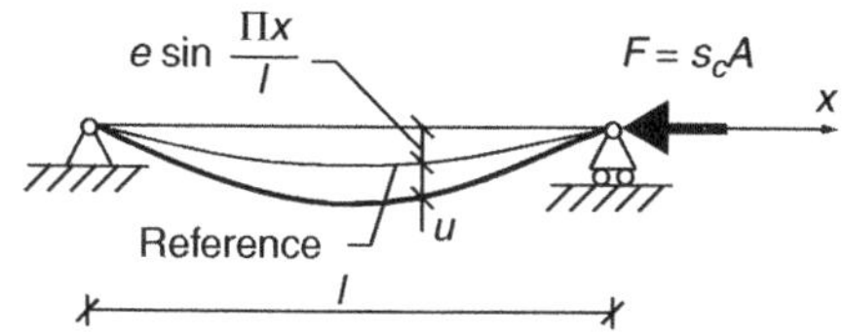

Figure 12.16 Simply supported centrally loaded column with an initial deviation from straightness corresponding to e (reference condition)

To demonstrate how k_c is determined, a very simple case is initially investigated. It is assumed that the column initially has a sinusoidal deviation from straightness with a midspan value e (see Figure 12.16).

From the differential equation for the column:

$$EI\frac{d^2u}{dx^2} = -M = -F\left(u + e\sin\frac{\pi x}{l}\right) \quad (12.52)$$

the resulting midspan deviation is found:

$$e + u_{mid} = e\frac{1}{1 - \lambda_{rel}^2\sigma_c/f_c} \quad (12.53)$$

The stresses in the middle are σ_c and $\sigma_m = F(e + u_{mid})/W$. Inserting these stresses in the simple failure criterion $\sigma_c/f_c + \sigma_m/f_m = 1$ results in:

$$\frac{\sigma_c}{f_c} + \frac{\sigma_c A}{W f_m}e\frac{1}{1 - \lambda_{rel}^2\sigma_c/f_c} \leq 1 \quad (12.54)$$

The equal sign corresponds to $\sigma_c/f_c = k_c$, i.e. k_c can be found from:

$$k_c\left(1 + \frac{eA}{W}\frac{f_c}{f_m}\frac{1}{1 - \lambda_{rel}^2 k_c}\right) = 1 \quad (12.55)$$

The column formula in Draft Eurocode 5 (2000) is, in accordance with Blass (1988), derived along the same lines, but the parabolic stress-strain curve from Figure 12.3 is used. In this case, it is not possible to give a closed solution, but it is necessary to use an iterative numerical method:

- An initial deviation from straightness is estimated. For a simply supported column, a sinusoidal deviation from straightness is appropriate (see Figure 12.16). For other conditions, the deflection should correspond to the expected deflection curve.
- An axial load is estimated and the initial moment distribution is calculated.
- The stiffness in all cross-sections is calculated (the stiffness decreases with increasing axial load and moment).
- The deflections are calculated and a new moment distribution is found, resulting in new stiffness and moments, etc.
- If the deflections diverge, the critical load is exceeded and a new, lower axial load has to be estimated.
- If the deflections converge, it must be checked that the load-carrying capacity is not exceeded. The capacity is exceeded if
  - f_c (or ε_c) is exceeded,
  - the capacity of the compression zone is exceeded,
  - the tensile strength is exceeded.
- If the load-carrying capacity is exceeded/not exceeded, a new trial is made with a lower/higher axial load.
- The calculations are repeated until equilibrium is obtained with an axial load corresponding to the capacity of the cross-sections.

The calculations can be performed using deterministic design values or the parameters can be assumed to be stochastic variables. The following expressions given in Draft Eurocode 5 (2000) are based on curve fitting to results from a Monte Carlo simulation (see Blass, 1988):

$$k_c = \begin{cases} 1 & \text{for } \lambda_{rel} \leq 0.3 \\ \dfrac{1}{k + \sqrt{k^2 - \lambda_{rel}^2}} & \text{for } \lambda_{rel} > 0.3 \end{cases} \quad (12.56)$$

$$k = 0.5(1 + \beta_c(\lambda_{rel} - 0.3) + \lambda_{rel}^2) \quad (12.57)$$

For structural timber, an initial deviation from straightness of $e = 0.005l$ has been assumed, and for glulam, $e = 0.003l$.

12.6.2 Laterally Loaded Columns

A simply supported column is considered (see Figure 12.17). In addition to the axial force, there

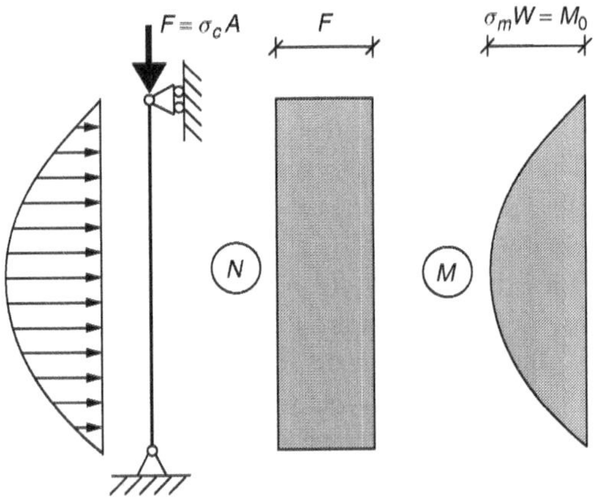

Figure 12.17 Laterally loaded column

is a lateral load giving a bending moment M_0 and a corresponding bending stress σ_m in the middle of the undeformed column (note that σ_m has another meaning than in relation to Equation (12.54)). This case is in principle treated as the centrally loaded column, the only difference being that the midpoint deflection is formally $(e + M_0/F)$.

The results of calculations for typical material parameters are given in Figure 12.18, showing acceptable combinations of σ_c and σ_m. The values on the ordinate correspond to centrally loaded columns, and are thus k_c.

Codes normally give the following simple interaction formula:

$$\sigma_c/(k_c f_c) + \sigma_m/f_m \leq 1 \tag{12.58}$$

For very slender columns, Equation (12.58) is a good approximation but generally on the safe side – in most practical cases very much so – and a more general calculation can lead to considerable savings.

For the case with bending about two axes, no verified design method exists, but it is assumed that structures are safe if both of the following interaction formulae are satisfied:

$$\frac{\sigma_c}{k_{c,y} f_c} + \frac{\sigma_{m,y}}{f_{m,y}} + k\frac{\sigma_{m,z}}{f_{m,z}} \leq 1 \tag{12.59}$$

$$\frac{\sigma_c}{k_{c,z} f_c} + k\frac{\sigma_{m,y}}{f_{m,y}} + \frac{\sigma_{m,z}}{f_{m,z}} \leq 1$$

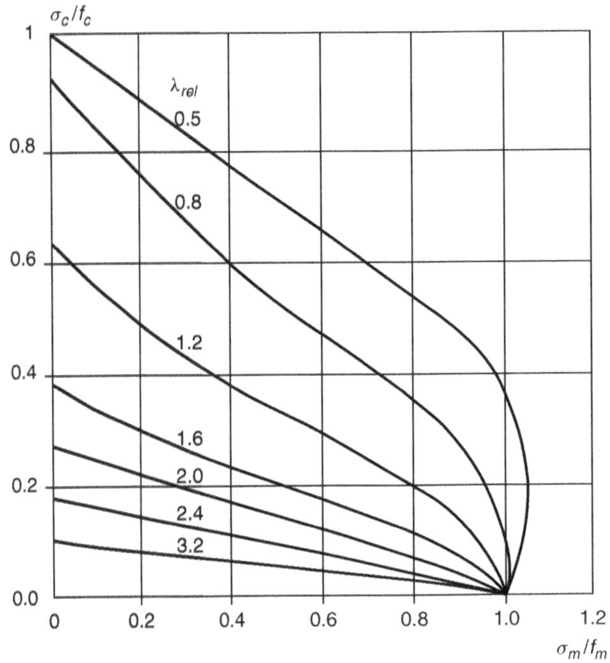

Figure 12.18 Acceptable combinations of σ_m/f_m and σ_c/f_c

$k_{c,y}$ is the column factor for bending about the y-axis (deflection in the z-direction) and $k_{c,z}$ is the factor for deflection in the y-direction. Account is taken of the fact that the bending strengths may be different due to size factors. k is a reduction factor (form factor). In Eurocode 5, $k = 0.7$.

12.6.3 Lateral Instability of Beams

Slender beams – beams that are long and deep relative to the width – fail in lateral instability, i.e. in torsion and lateral deflection of the compression flange (see Figure 12.19). The critical moment can (Timoshenko and Gere, 1961) be expressed as shown in Equation (12.60). Slightly more complicated expressions may be found in the literature that take into consideration secondary effects (e.g. warping) that are unimportant for timber.

$$M_{y,cr} = M_{cr} = \frac{\pi}{l_{ef}}\sqrt{EI_z GI_{tor}} \qquad (12.60)$$

where EI_z is the bending stiffness about the z-axis (the weak direction), GI_{tor} is the torsional stiffness (I_{tor} is the torsional second order moment, not the polar moment, see below), and l_{ef} is the effective length. l_{ef} depends upon the geometry, the load configuration and the support conditions. Examples are shown in Figure 12.20.

The reduction factor due to instability is:

$$k_m = \sigma_{cr}/f_m = M_{cr}/(Wf_m) = 1/\lambda_m^2 \qquad (12.61)$$

where the slenderness ratio λ_m has been introduced in analogy with Equation (12.50). W is the

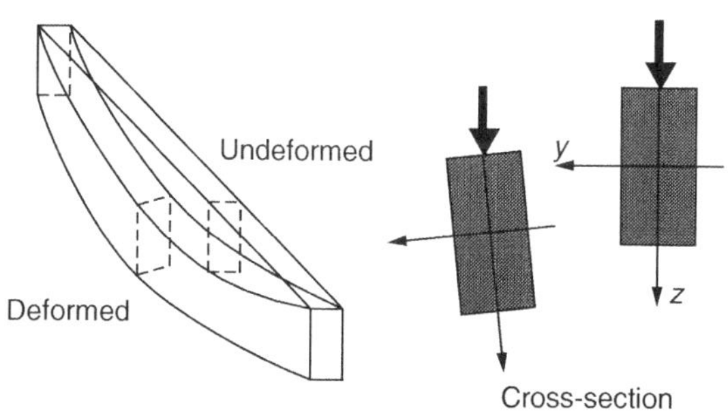

Figure 12.19 Deformations by lateral instability

section modulus. Inserting the values for a rectangular cross-section – $W = bh^2/6$, $I_z = b^3h/12$ and $I_{tor} \sim hb^3(1 - 0.63b/h)/3$ – and utilising the fact that b/h is small for beams where lateral instability occurs:

$$\frac{W}{\sqrt{I_z I_{tor}}} \sim \frac{\frac{1}{6}bh^2}{\sqrt{\frac{1}{12}b^3h\frac{1}{3}b^3h(1-0.63b/h)}} = \frac{h}{b^2}\Big/\sqrt{1-0.63b/h} \sim \frac{h}{b^2} \qquad (12.62)$$

The slenderness ratio then becomes:

$$\lambda_m^2 = \sqrt{\frac{l_{ef}h}{\pi b^2}\frac{f_m}{E}\sqrt{\frac{E}{G}}} \qquad (12.63)$$

For glulam, the mean values are $E/f_m \sim 275$, and $E/G \sim 15$, i.e.

$$\lambda_m \sim 0.067\frac{\sqrt{l_{ef}h}}{b} \qquad (12.64)$$

Equation (12.61) only applies in the linear elastic range, i.e. for $\sigma_m/f_m < 0.5$ corresponding to $\lambda_m > 1.4$. The bending strength is found from tests with $h/b \sim 3$ and $l_e/h \sim 15$, i.e. with $\lambda_m \sim 0.75$, and in this case, lateral buckling is included, i.e. $k_m = 1$. In the interval between 0.75 and 1.4, linear interpolation is normally used in codes, resulting in Figure 12.21.

The expressions are correct for equal end moments and for isotropic materials. For other cases, the effective lengths given in Figure 12.20 may be used. However, the results are only approximations, and often rather on the safe side. For important structures it is therefore recommended that a more correct finite element analysis taking into account the anisotropy be performed. A very effective calculation model is described in Eggen (2000), which also contains examples on the critical load.

12.6.4 Combined Column Effect and Lateral Instability

The expressions in Section 12.6.2 are only valid if lateral instability is prevented. This is the case for

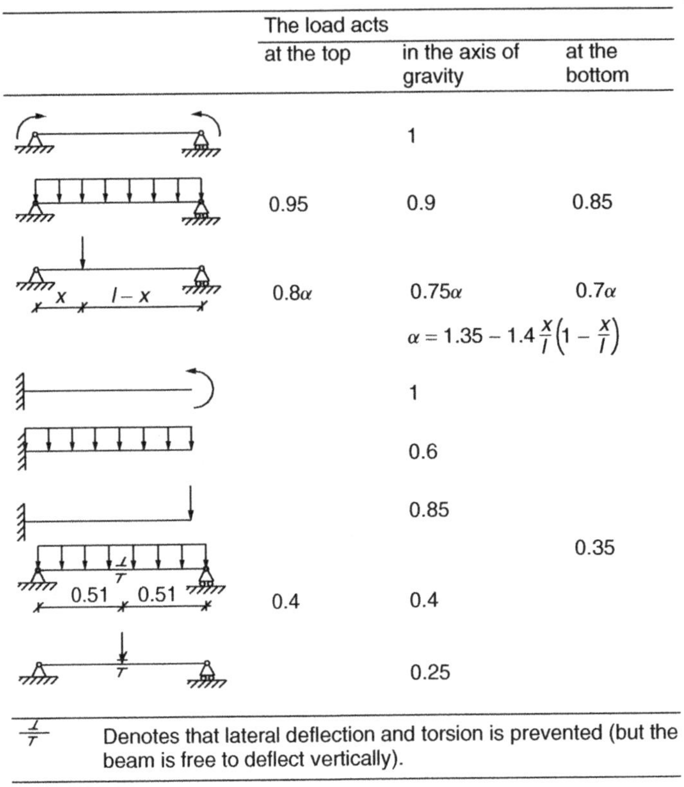

Figure 12.20 Examples on l_e/l

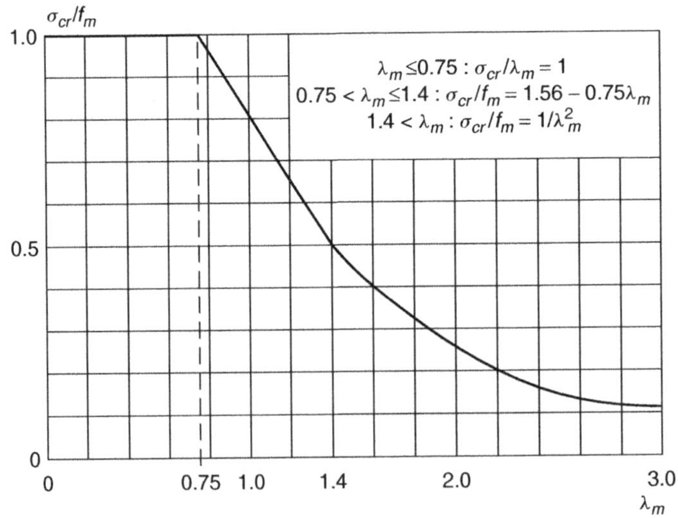

Figure 12.21 The critical bending stress as a function of the slenderness ratio in bending

bending about the weak axis. For bending about the strong axis, it is a condition that bending about the weak axis is prevented, e.g. by holding the compression flange effectively in place. There is no theoretical solution for the general case with a beam free to deflect and with an axial compression load. Eggen has developed a very effective finite element program to solve the general problem of beam/column instability (Eggen, 2000), and has investigated the interaction between the two types of failure. On the basis of these, he suggests, in addition to the column interaction formulae (12.59), that, for bending about the strong axis, it be verified that:

$$\sigma_c/(k_c f_c) + (\sigma_m/\sigma_{cr})^n \leq 1 \qquad (12.65)$$

where σ_{cr} is defined in (12.61).

A suitable value seems to be $n = 1.75 - 2.0$. A comparison between (12.59) and (12.65) is shown in Figure 12.22. It will be seen that even small axial forces have a serious influence on the buckling moment.

12.6.5 Safety Elements

For non-slender columns and beams, the load-carrying capacity depends upon the material strengths – compression and bending strength – and design values will be used, i.e. characteristic five-percentiles reduced by the partial coefficient and load duration factors. For slender members, the load-carrying capacity depends upon stiffness values (see Equations (12.59) and (12.65)). The coefficient of variation is about the same for strength and stiffness values, and it is obvious that the modulus of elasticity and the shear modulus to be used are the five-percentiles reduced by the normal partial coefficient. It is normally assumed that the strength development with time is the same for the two situations, and that the k_{mod} for timber can also be used for the stiffness values. It is doubtful, however, whether the use of k_{mod} is sufficient. The design equations are based on a second-order method, where the internal forces are calculated in the deformed state, and the reduction of stiffness due to creep may be much larger than corresponding to k_{mod}. It is

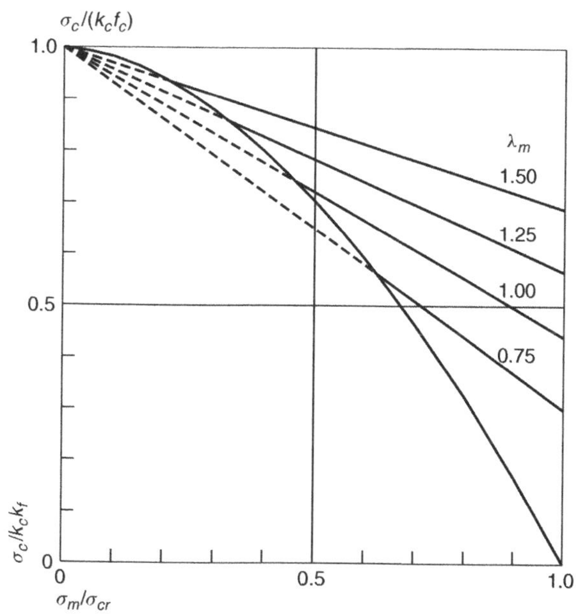

Figure 12.22 Comparison between the requirements (12.59) (straight lines) and (12.65) (curve) with $n = 1.75$. The dotted lines show where (12.59) is on the safe side

therefore recommended that the lower value of $1/(1 + k_{creep})$ and k_{mod} be used.

REFERENCES

Blass H.J. (1988) Traglastberechnung von Drückstäben aus Brettschichtholz. *Bauingenieur*, **63**, 245–251.

Blumer H. (1979) Spannungsberechnung an isotropen Kreisbogenscheiben und Satteldachträgern mit konstanter Dicke. Lehrstuhl für Ingenieurholzbau und Baukonstruktionen. Universität Karlsruhe, Germany.

Colling F. (1986) Einfluss des Volumens und der Spannungsverteilung auf die Festigkeit eines Rechteckträgers. *Holz als Roh- und Werkstoff*, **121**, 179–183.

Dahlblom O. and Ormarsson S. (1993) Analysis of stress distribution. In: Dahlblom *et al.*, *Fibre Reinforcement of Glulam*, Report TVSM–7083, Division of Structural Mechanics, Lund Institute of Technology, Lund, Sweden.

Damkilde L., Hoffmeyer P. and Pedersen N.P. (1998) Compression strength perpendicular to grain of structural timber and glulam. CIB W18, Paper CIB W18/31-6-4, Savonlinna, Finland.

Draft Eurocode 5 (2000) Draft prEN 1995-1-1, Design of timber structures. Part 1.1, General rules and rules for buildings.

Eggen T.E. (2000) Buckling and geometrical nonlinear beam-type analyses of timber structures. Dissertation, Department of Structural Engineering, Norwegian University of Science and Technology, Trondheim.

Glos P. (1978) Zur Bestimmung des Festigkeitsverhalten von Brettschichtholz bei Druch-beanspruchung aus Werkstoff- und Einwirkungskenngrössen. Dissertation, TU München.

Gustafsson P.J. (1988) A study of Strength of Notched Beams. CIB, Working Commission W18, Paper W18/21-10-1.

Hankinson R.L. (1921) Investigation of crushing strength of spruce at varying angles of grain. *Air Service Information Circular*, **III**(259).

Hudson M. (1960) The effect of initial bending on the strength of curved laminated beams. *Wood*, **25**.

Norris C.B. (1962) Strength of Orthotropic Materials subjected to Combined stresses. Report 1816, Forest Products Laboratory, Madison, WI.

Riberholt H. (1979) Tapered Timber Beams. CIB, Working Commission W18, Paper W18/11-10-1.

Riberholt H. Enquist B. Gustafsson P.J. and Jensen R.B. (1991) Timber beams notched at the support. Report TVSM-7071, Lund University, Sweden.

Tsai S.W. (2000) Comparison of various failure criteria of orthotropic materials. *Proceedings of the International Conf. on Wood and Wood Fibre Composites*, Stuttgart, Germany, pp. 573–584.

Wilson T.R.C. (1939) The glued laminated wooden arch. U.S. Dept. for Agriculture, Technical Bulletin No. 691.

13

Short- and Long-term Deformations of Timber Structures

Annika Mårtensson

13.1 INTRODUCTION

Deflections and movements in a building must be limited to avoid damage and other undesirable effects in service. Lightweight structures like timber are especially prone to exhibit deformations. Several factors influence deformations in timber structures. Since wood is an anisotropic material, different loading modes will lead to different types of response. One of the most important factors affecting the deformation is the combined effect of loading and moisture changes.

Deformations may be unfavourable for the utilization of the building in many ways. Examples are an insufficient slope of roofs, and problems with opening doors and windows. In special buildings such as gymnasiums, or facilities with strong requirements of surface planeness (e.g. due to sensitive equipment), deformations may also lead to problems. In the latter case, it is important that the design is made in co-operation with the equipment manufacturer.

Deformation limitations should also be applied to buildings and their components so that non-structural elements are not introduced into the load path. If there is insufficient separation between deflecting primary structural elements and non-structural components, the load may be transferred through the non-structural element. Such non-structural elements include walls, windows, ceilings, floor and roof coverings, facades, lifts, etc. The resulting damage may include cracking and buckling of some elements of moderate stiffness and integrity, or tearing and folding of more flexible items such as coverings and linings. Whereas these forms of failure in general do not affect the stability of the structure, they may render the building unserviceable. Limitations to prevent damage to non-structural elements should, in many cases, be determined on the basis of variable loads, with respect to the higher values that these may attain during shorter periods.

Excessive deformations do not look attractive to the public, and may lead to a feeling of insecurity. These observations are subjective, and although the deformations will not endanger the structure or shorten its lifetime, it is appropriate to limit deformation to ensure that confidence is maintained. Such limits especially apply to long-term deformation, and not so much to shorter, but recoverable deformations that may occur due to high load peaks.

Requirements on the limitations of deformation will in many cases be the decisive criterion in the design process. This means it is important that the magnitude of the deformations can be predicted with adequate accuracy, and that the requirements on the magnitude are correctly formulated. This is a problem in the design process today, since the knowledge is lacking with regards to both requirements and prediction methods. In the following, a description of knowledge concerning deformations in timber structures will be given.

This chapter mainly discusses the behaviour in the serviceability state, i.e. the stress levels will be moderate.

13.2 STATE OF THE ART

Deformations occur in practically all timber structures, but they are of more interest in some cases than in others. In Figure 13.1, some typical situations where deformations may lead to problems are indicated. Beams subjected to loading will deflect under load, a deflection that is also affected by current moisture content as well as moisture content variations. This deflection can be of such a magnitude that it severely affects the behaviour of the individual structural element, and also the behaviour of the surrounding elements.

In many structural systems based on timber or other wooden products, wood will be subjected to

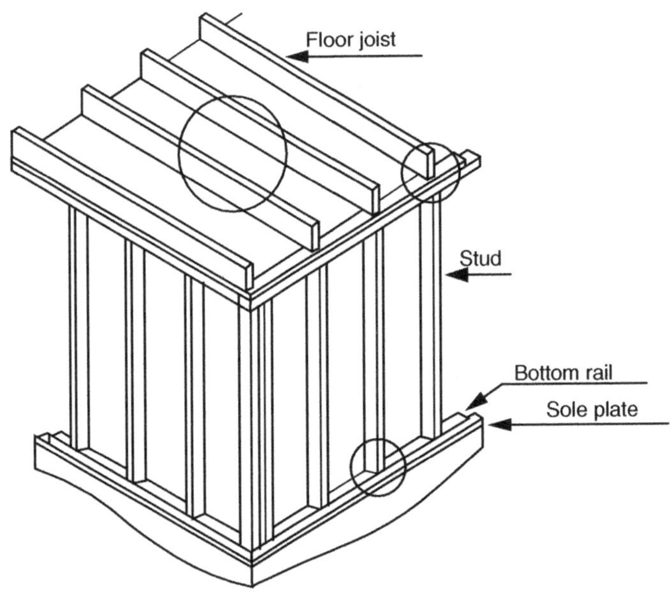

Figure 13.1 Timber frame structure. Positions where significant deformations may occur are indicated by circles

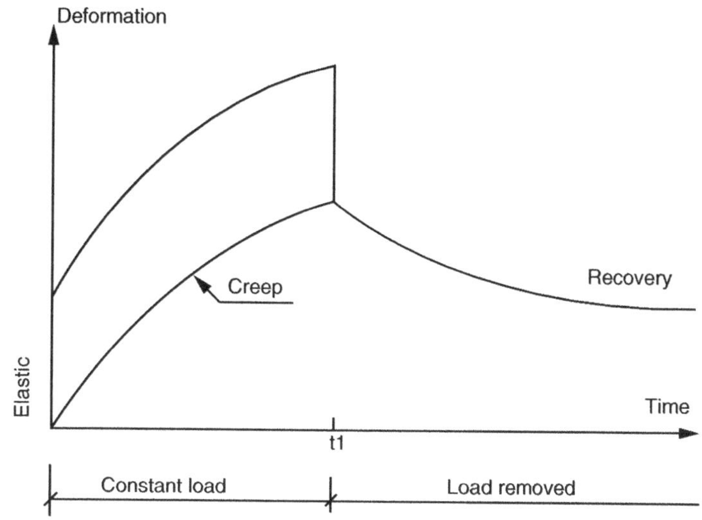

Figure 13.2 The components of the deformation of wood

compression perpendicular to grain. Since wood has low stiffness in this direction and exhibits large moisture-induced movements, compression will lead to large deformations. When this occurs in a connection between floor and wall structures, it may lead to undesired settlements in the buildings.

Platform frame constructions have walls which are one storey high with the floor joists resting on top of the wall (see Figure 13.1). Each storey is built up from a floor platform, which also serves as the ceiling of the storey below it. Deformations in joints (see Figure 13.1) where structural wood elements are subject to compressive loads perpendicular to grain can then lead to large vertical movements. In a multi-storey wood frame building, large compressive loads may occur in the lower floors and at the foundation level. High loads will also occur if the room layout varies from floor to floor and a concrete topping is used to limit noise.

The deformations due to load are further compounded by wood shrinkage perpendicular to grain, which is far more severe than shrinkage parallel to the grain, which is negligible. In the case of a 38 × 235 mm dropped header, or joist at floor level, Thorson (1989) has observed real field values for vertical movements of 13 mm. These field values represent the combined effect of related shrinkage and deformation under load.

Under service conditions, timber often has to withstand an imposed load for many years. In Figure 13.2 a generalised interpretation of the variation of deformations is given. On application of a load at time zero, an instantaneous (and reversible) deformation occurs which represents elastic behaviour. On maintaining the load to time t_1 the deformation increases, though the rate of increase is continually decreasing; this increase in deformation with time is termed 'creep'. On removal of the load at time t_1, an instantaneous reduction in deformation occurs, which is approximately equal in magnitude to the initial elastic deformation. With time, the remaining deformation will decrease at a decreasing rate.

Moisture variations will influence the deformation curve. As discussed later, moisture variations will significantly increase the deformations. This effect is commonly referred to as 'mechano-sorptive behaviour'.

The deformations can then be divided into elastic deformations, viscoelastic deformations and mechano-sorptive deformations. It can also be added that at higher stress levels, the material exhibits plastic behaviour, but this effect will not be discussed here.

13.2.1 Elastic Deformation

When a sample of timber is loaded in tension, compression or bending, the instantaneous deformations obtained with increasing load are approximately proportional to the values of the applied load. For all loading directions a point on the load-deflection curve known as the limit of proportionality can be identified, below which the relationship between load and deformation is linear, and above which nonlinearity occurs. In Figure 13.3 typical load-deformation curves for tension and compression are presented. As can be seen, there is a difference between these two loading modes. The linearity is more valid in tension than in compression.

At the lower levels of loading where the straight-line relationship appears to be valid, the material is said to be linearly elastic, and the linearity is described with the elastic modulus E. The value of the elastic modulus is often determined at a certain moisture level, and if the moisture content at loading differs from the reference state the value of the elastic modulus is changed. The higher the moisture content the smaller the elastic modulus, hence larger deformations occur. It is therefore important to have an accurate estimation of the moisture content in the timber.

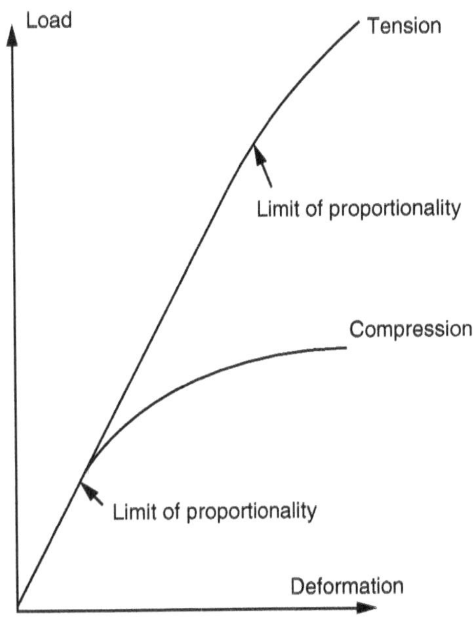

Figure 13.3 Load-deformation curves for timber stressed in tension and compression. The limit of proportionality is indicated

In the direction perpendicular to grain the elastic modulus is much smaller than along the grain, thus a load will induce much larger deformations in that direction. For this special loading case, there are some aspects that have to be specially recognised and they will be discussed in the following. To determine the load-deformation curve tests have to be carried out, and in the case of compression perpendicular to grain, a test set-up like that in Figure 13.4 could be used.

If the loaded length covers the whole surface of the beam, i.e. $L_e = 0$, the deformation Δ would be given by

$$\Delta = \frac{\sigma_{90} d}{E_{90}} \tag{13.1}$$

where d is the beam depth, E_{90} is the elastic modulus perpendicular to grain, and the stress perpendicular to grain is given by

$$\sigma_{90} = \frac{P}{bL} \tag{13.2}$$

where b is the beam width, L is the length of the loaded surface and P is the applied load. Experiments have shown that in the case where the load does not cover the entire surface, the deformation is less than that determined by Equations (13.1) and (13.2).

Figure 13.5 illustrates an example result from tests made on test specimens designed from 45 × 95 mm dimensioned timber from Norway spruce (*Picea Abies*) (see Frater and Thelandersson, 1996). The moisture content in the specimens was between 10% and 13%. The specimen configurations were designed to examine the bearing performance between studs, plates and joists. In Figure 13.5, the load-deformation curve for a test of type SPP (one stud resting on two plates) is shown.

Other very comprehensive test series of the behaviour of wood subjected to compressive stresses perpendicular to grain have been performed by Backsell (1966) and Gehri (1997), and a number of parameters were investigated in

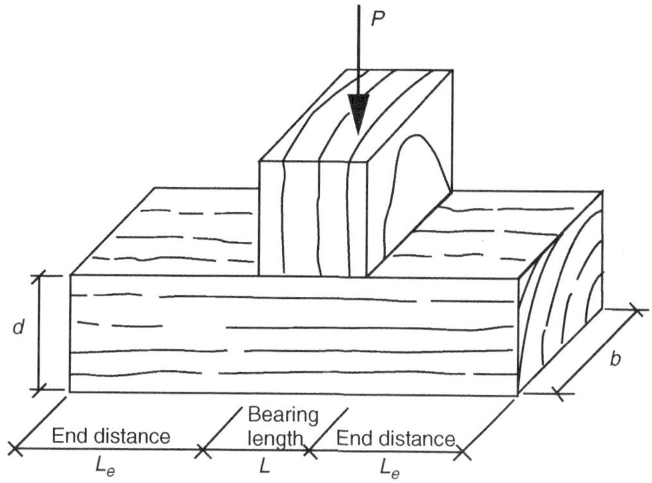

Figure 13.4 Test set-up

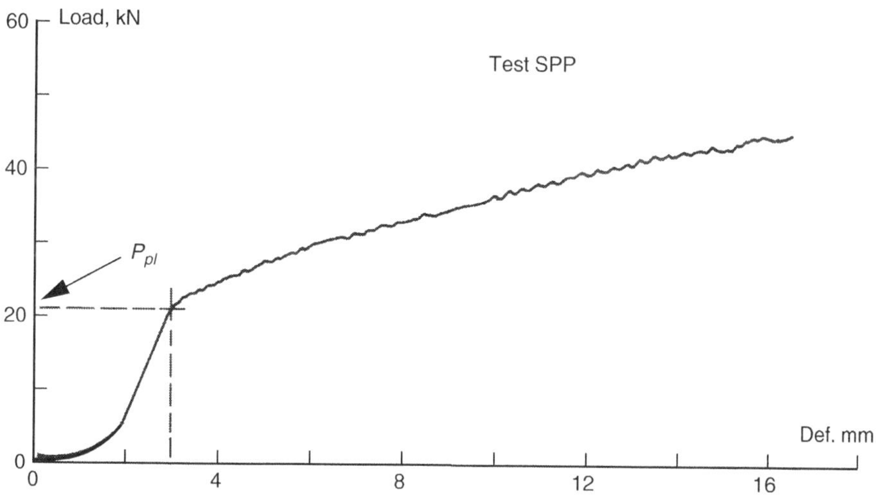

Figure 13.5 Load deformation curve for test SPP

those studies, where test set-ups similar to that in Figure 13.4 were used.

Conclusions that can be drawn from short-term tests are:

- The load-deformation relations show a distinct change in slope at a certain stress level, which is defined as the proportional limit.
- The initial stiffness increases significantly with decreasing bearing length (with $L_e > 0$).
- There is also a tendency that a smaller bearing length leads to higher stress at the proportional limit.
- The initial stiffness decreases strongly with increasing moisture content, but increases with wood density.
- When two members loaded perpendicular to grain are in contact at the bearing area the deformations are significantly higher compared

to the normal case when one of the members is loaded parallel to the grain.

The conclusion from tests is that Equations (13.1) and (13.2) will not give accurate values of the deformation for situations when $L_e > 0$.

13.2.2 Viscoelastic Deformation

Viscoelasticity infers that the behaviour of the material is time-dependent; at any instant in time under load its performance will be a function of its past history. In Figure 13.2 this time dependence was described as creep and recovery (the reduction with time in deformation after unloading). For beams the creep may lead to failure at high load levels, but for normal loading conditions the increase in deflection will be relatively moderate. No moisture variations are included in a viscoelastic description, but the creep will be larger the higher the moisture content.

If the load varies with time this will affect the deformation behaviour. When the load increases the deflection increases and the creep increases due to a higher load level. As long as the load increase is moderate, the effect of an increase in load can be described by linear superposition of the effect of all load increments. If the load is decreased there will be an immediate spring back in deflection, i.e. the elastic response. Besides this, creep recovery will take place, i.e. a part of the creep deflection achieved earlier will vanish with time. If some of the load is still applied, further creep will, however, occur.

Even if constant climate condition is very rare in practice, it is important to have knowledge of the behaviour under these circumstances. A number of experiments have been carried out, both to achieve knowledge of the creep effects but also with the aim of studying the effect of load duration on strength, see Chapter 8. In the latter case, the load levels are normally rather high, since the failure is of interest. For moderate load levels and for structural size timber, relative deflection after one year from loading varies between 0.3 and 0.4 (Ranta-Maunus, 1991). Relative deflection is defined as the creep deflection divided by initial elastic deflection.

13.2.3 Mechano-Sorptive Deformation

Wood under load, when subjected to moisture content changes, exhibits much greater deformations than under constant humidity conditions. This phenomenon is often referred to as the mechano-sorptive effect, and has been known since the late 1950s and early 1960s (Perkitny, 1960. Armstrong and Christensen, 1961. Armstrong and Kingston, 1962). Extensive work has been performed by a number of researchers during the last thirty years to clarify the phenomenon, both its nature and its origin. A lot of experimental data is now available in the literature, but the results from all this work are sometimes more confusing than clarifying. The results are sometimes contradictory, and new effects seem to be found all the time. A number of factors influence the mechano-sorptive behaviour. Primarily the nature of the wood itself affects behaviour. Different species behave differently, within the tree trunk different parts can behave differently, the orthotropy of wood and the heterogeneous structure with defects and anomalies makes it difficult to predict and model the mechano-sorptive response. The apparent differences between different loading modes, including recovery effects, also complicate the description. It is clear that one model cannot cope with all situations, and it is necessary to restrain a study to some specific problems and to perform special tests for these problems.

Testing is the only way to achieve basic information about the mechano-sorptive effects. It is, however, a very tedious task and it can often be difficult to analyse the results. Most long-term tests have been performed in bending on small clear specimens. A rather large number of tests have also been performed in tension and compression parallel to grain. When analysing test results and comparing test results it is important to distinguish between tests performed on small clear specimens, which excludes most effects of knots and other anomalies, and tests performed on full size timber.

One problem with experimental studies of mechano-sorption is that a mixture of several effects is measured. The different effects then have to be identified and quantified. To quantify mechano-sorption, some type of model is needed, and different models define the effect in somewhat different ways.

Descriptions of mechano-sorptive effects have been given by, for example, Grossman (1976), Hoffmeyer (1990), Hunt (1991) and Mårtensson (1992). The behaviour at high stress levels becomes nonlinear with stress, which makes the evaluation more difficult. The stress level where mechano-sorptive behaviour becomes nonlinear is hard to define with certainty, but in compression the non-linearity starts at about 10–20% of ultimate stress (Hunt, 1989), and in tension and bending at about 20–30% of ultimate stress. The working stresses in normal structures are in most cases lower than these values. Since the main aim here is to deal with serviceability problems, it is considered to be correct to assume a linear behaviour.

The response in bending is related to the response in the compression and tension zones in the beam. In Figure 13.6 the response for a beam subjected to constant load and simultaneous moisture variations is shown. Initially an elastic deflection takes place, δ_i. If the humidity as well as the load thereafter are kept constant, pure creep takes place, δ_c. If the humidity is instead varied, the deflection curve will oscillate. Since free shrinkage-swelling does not influence the curvature of a beam, the oscillations represent the difference in response between the compressive and tensile sides. In this case, the mechano-sorptive deflection is defined as $\delta_{ms} = \delta_t - \delta_c - \delta_i$, since no reduction for free shrinkage-swelling is needed. This makes it easier to estimate the magnitude of mechano-sorption from bending tests than from uniaxial tests. The disadvantage of the bending tests, however, is that it is difficult to distinguish the influences from the two different zones in compression and tension.

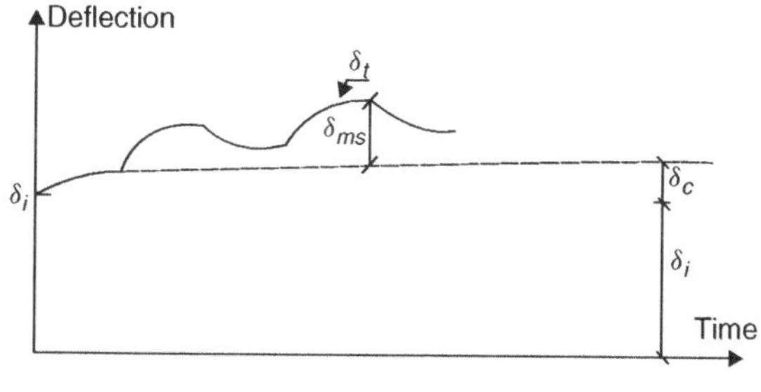

Figure 13.6 Principal sketch of bending deflection in a wooden beam

Some important conclusions regarding moisture variations and their effect on deformations are:

- In bending of small clear specimens the first adsorption and all desorption periods cause increases in deflection, while later adsorptions tend to decrease the deflection (Figure 13.7) (Hearmon and Paton, 1964). This is applicable in cases where the moisture content changes take place within previously attained moisture content limits.
- The effect of moisture content change on the deformation is in general much greater than the effect of time.
- When moisture cycling takes place between two fixed limit values, the increase in bending deflection within one cycle tends to decrease, i.e. the deflection goes towards a limiting value (Hearmon and Paton, 1964. Mohager, 1987; Hunt and Shelton, 1988) (see also Figure 13.7). The same behaviour has been found in uniaxial tests, i.e. the strain approaches a limiting value.
- Mechano-sorption also gives rise to increased deflection for structural timber (Bethe, 1969. Meierhofer and Sell, 1979. Hoyle *et al.*, 1986. Mohager, 1987. Leivo, 1988). Mohager (1987), for instance, has shown that the relative deflection of full size beams (44 × 94 mm, length 4 m) is as large as the relative deflection of small beams (10 × 10 mm, length 300 mm).
- After unloading, an instantaneous elastic recovery takes place. The magnitude of this recovery is equal to or greater than the initial elastic response (Gibson, 1965. Mohager, 1987) (see Figure 13.8). The recovery of wood after unloading is similar to that for normal creep, but is accelerated by moisture cycling (Christensen, 1962. Arima and Grossman, 1978), and is larger during adsorption than during desorption (Armstrong and Christensen, 1961, Armstrong and Kingston, 1962. Ranta-Maunus, 1975. Mohager, 1987). If the recovery is total, i.e. if the accumulated mechano-sorptive strain has vanished totally, a new loading period with moisture

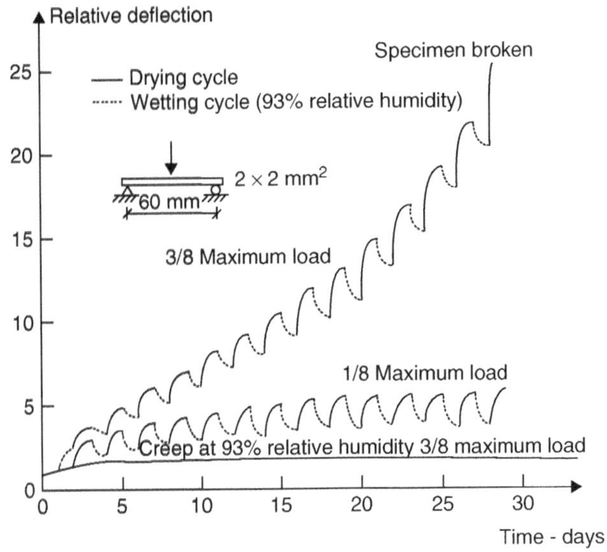

Figure 13.7 Small clear wooden beam subjected to moisture cycling under load (Hearmon and Paton, 1964)

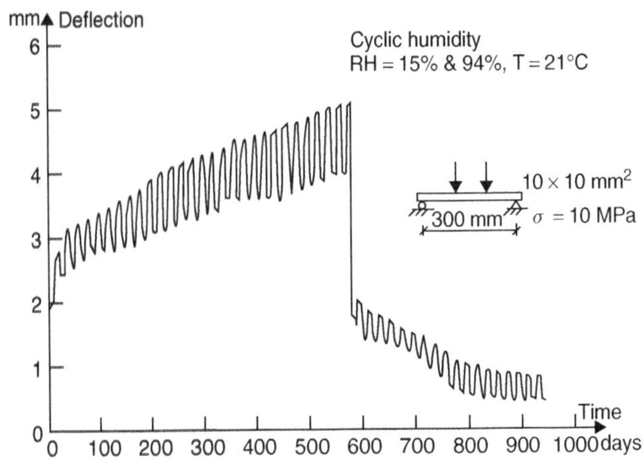

Figure 13.8 Small clear wooden beam subjected to moisture cycling under load. After a number of moisture cycles the load is removed and recovery takes place. The recovery is increased by further moisture cycling (Mohager, 1987)

content changes shows the same behaviour as the previous one (Christensen, 1962).

- The mechano-sorptive effect has also been found in wood-based materials such as particle boards (Armstrong and Grossman, 1972. Gressel, 1984. Dinwoodie *et al.*, 1990) and fibre boards (Sauer and Haygreen, 1968. Mårtensson, 1988a, 1988b. Mårtensson and Thelandersson, 1990). The behaviour differs, however, in a fundamental way between wood and wood-based materials. The typical fluctuations in deflection that occur in bending under moisture variations for wood are not as pronounced for board materials (Dinwoodie *et al.*, 1990). The increase in total deflection is, however, more significant for board materials than for wood. Large differences in behaviour also materials, depending upon the size of the chips or fibres in the material, and on the type of components added to the boards.

- Chemical treatments such as acetylation and formaldehyde cross-linking (Norimoto *et al.*, 1987) have been shown to reduce the mechano-sorptive effect. Acetylation means that the hydroxyl groups in the wood react with acetic anhydride, with the result that the moisture uptake is reduced (Feist *et al.*, 1991).

One experimental study presented by Gowda *et al.* (1996) and Ranta-Maunus and Kortesmaa (2000) presents results from long-term tests carried out on structural size timber, glulam, LVL and I-beams under low stress levels. Some of the beams were treated either with surface coatings or with impregnation. The results from these tests are very interesting, since the tests were carried out during a long period of time, and since they also show the effect of different treatments. The tests were carried out in Finland under natural sheltered conditions. An example of the humidity and temperature variations is given in Figure 13.9. Figures 13.10 and 13.11 show the relative creep, i.e. total deflection divided by initial deflection. In Figure 13.10 results from different types of beams are presented. The stress levels are similar, and it is obvious that spruce and glulam beams exhibit less creep than the LVL beams and the I-beams. A significant part of the creep takes place during the first six months. For the glulam and spruce beams it will take at least ten years to double the six months creep deflection. For the LVL beams and the I-beams the rate of increase is larger, but for all cases the creep rate is low after three years.

In Figure 13.11 the effect of different treatments is presented. The effect of treatment is clear when comparison is made with a non-treated beam. The beam impregnated with creosote exhibits very little creep, which is natural since the impregnation prevents the moisture variations in the beam. The surface coatings are less effective and the effectiveness depends upon the type of coating.

Figure 13.12 illustrates examples of the specimens and the testing rig used for long-term tests (Jordow and Enockson, 1996), where a wood plate(s) was loaded perpendicular to grain. Both the wood on the top of the plate and below the plate was loaded parallel to the fibre direction (studs). The dimensions of the plate (denoted by ‘p’) and stud (denoted by ‘s’) were 45 × 120 mm. A constant load was applied to the specimens, producing a constant stress level in all tests of 2.6 MPa. The stress level was determined so that it

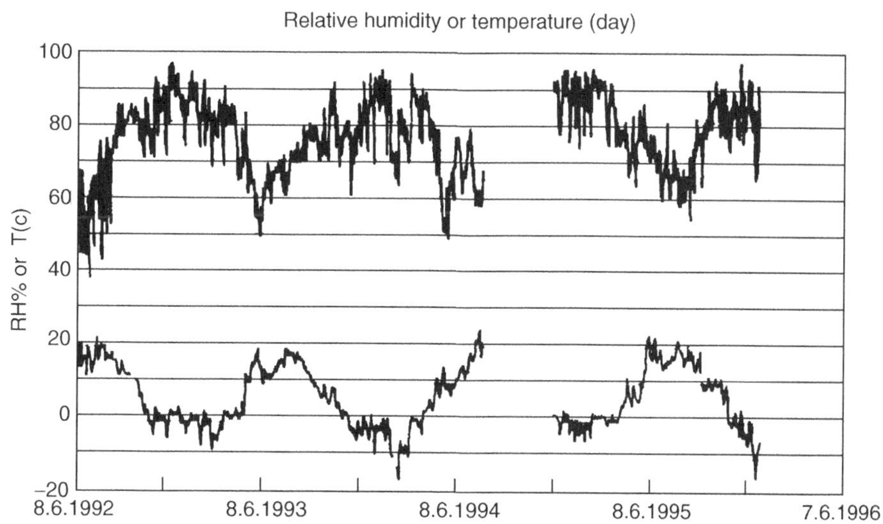

Figure 13.9 Changes in temperature and relative humidity versus time (Gowda *et al.*, 1996)

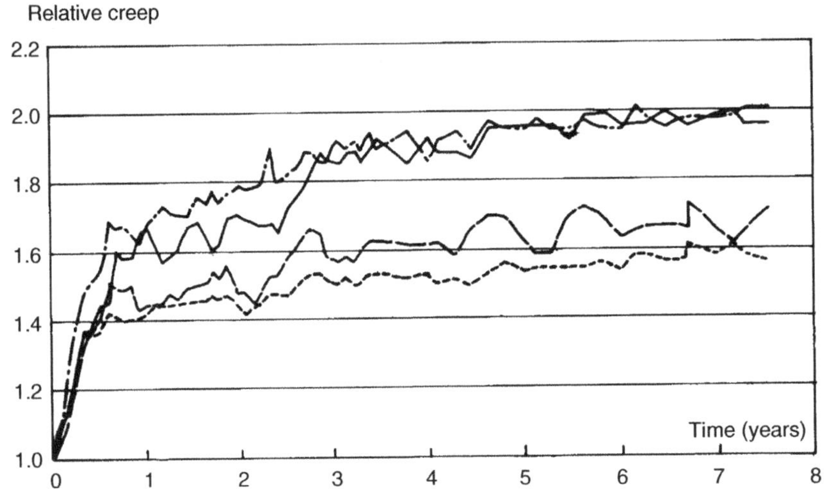

Figure 13.10 Average relative creep in sheltered environment.———; LVL:– – – –; glulam:- - - - -; spruce:-·-·-·-·-; I-beam. Stress level 2 MPa (Ranta-Maunus and Kortesmaa, 2000)

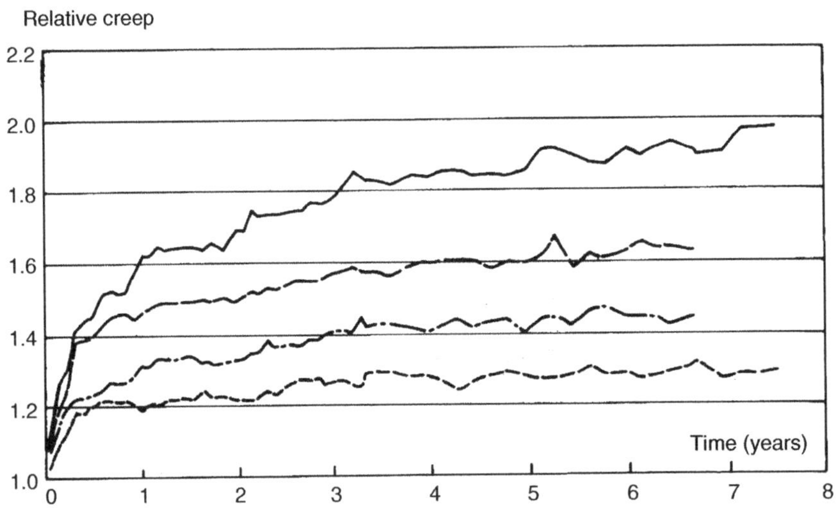

Figure 13.11 Average relative creep in sheltered environment for different surface coatings and treatments (pine, 7 MPa)———; non-treated:– – – –; emulsion paint:- - - - -; creosoted:-·-·-·-·-; alkyd paint (Ranta-Maunus and Kortesmaa, 2000)

corresponded to a long-term load for a four storey timber frame building. Thus, it is not determined in relation to any ultimate limit state, but to a serviceability state that gives a load which would be applied on a structure for a long time period without regard to any load peaks. The tests were carried out on timber conditioned in 65% RH and 20 °C before loading. During the experiments, the relative humidity changed from 60% to 30% and the temperature varied between 18 °C and 20 °C. Figure 13.13 presents the load-deformation curves for each specimen type. There is a significant effect of creep and mechano-sorptive deformation in the tests, and already after two weeks the deformation has increased to twice the initial deformation. During that time the shrinkage was about 1/3 of the additional deformation increase after the initial deformation. Tests have also shown that there is a significant difference between tests carried out in constant climate conditions and varying climate conditions, i.e. the increase in deformation with time is less if the climate is constant compared to the case when the climate varies (Jordow and Enockson, 1996).

A number of researchers have given hypothetical explanations of the mechano-sorptive effects, but none has been able to explain all the phenomena observed in tests. There are probably a number of factors involved both on the molecular, microstructural and macrostructural levels.

There is only a limited amount of data concerning the effects of varying load under varying climate conditions. In some cases, experiments have been performed with varying load, but then mainly with increasing load, with the aim of studying the duration of load effects and not the deformations. It is, however, reasonable to assume that the effects of varying load can be predicted by means of superposition, at least under service load conditions.

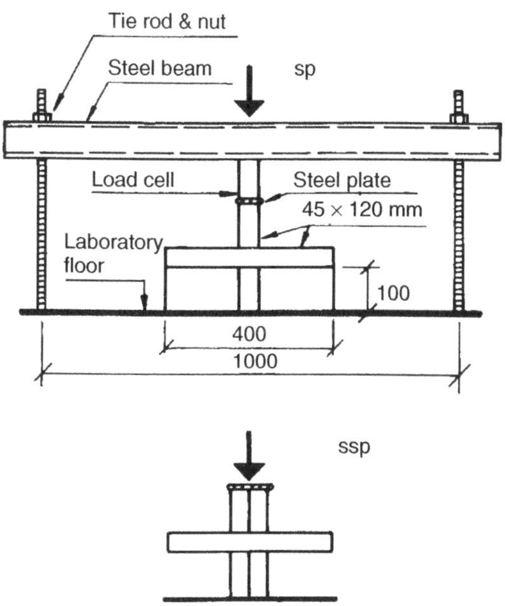

Figure 13.12 Test specimens and test set-up for long-term tests

One important fact to recognise is that moisture variations in some cases lead to a more efficient recovery process. It has been shown by Mårtensson (1992) and Toratti (1992) that the effects of moisture variations are generally smaller for variable loads than for permanent loads. The creep induced by moisture variations for high loads acting in short intervals separated by longer periods of low loads is smaller than if the high load is acting during an unbroken time period equal to the sum of the short intervals. This is due to the fact that, during periods with low load, the deformations induced earlier by the load peaks are diminished due to recovery.

13.2.4 Constitutive Models

Based on results from tests like those presented in previous sections, constitutive models can be formulated to describe the response of wood under stress and varying moisture conditions. Only the basic ideas behind such models will be presented here. In general, the majority of all models proposed in the literature can be formulated as

$$\dot{\varepsilon} = \dot{\varepsilon}(\sigma, \dot{\sigma}, w, \dot{w}, A) \tag{13.3}$$

where $\dot{\varepsilon}$ is the total strain rate, and the dot denotes time derivative. The strain rate is considered as a function of stress σ, stress rate $\dot{\sigma}$, moisture content w, rate of change in moisture content $\dot{w}$ and some strain limit value A. The formulation may be given in three-dimensional form for the most complex cases, but in a number of cases a one-dimensional formulation is sufficient.

When using this formulation one has to be able to predict the moisture content distribution within the timber. This is especially important when studying the deformation of beams in structural sizes. In pure compression or tension it can be enough to use the average moisture content and the variations in moisture content with time.

Equation (13.3) is very often written as

$$\dot{\varepsilon} = \dot{\varepsilon}_e + \dot{\varepsilon}_s + \dot{\varepsilon}_c + \dot{\varepsilon}_{ms} \tag{13.4}$$

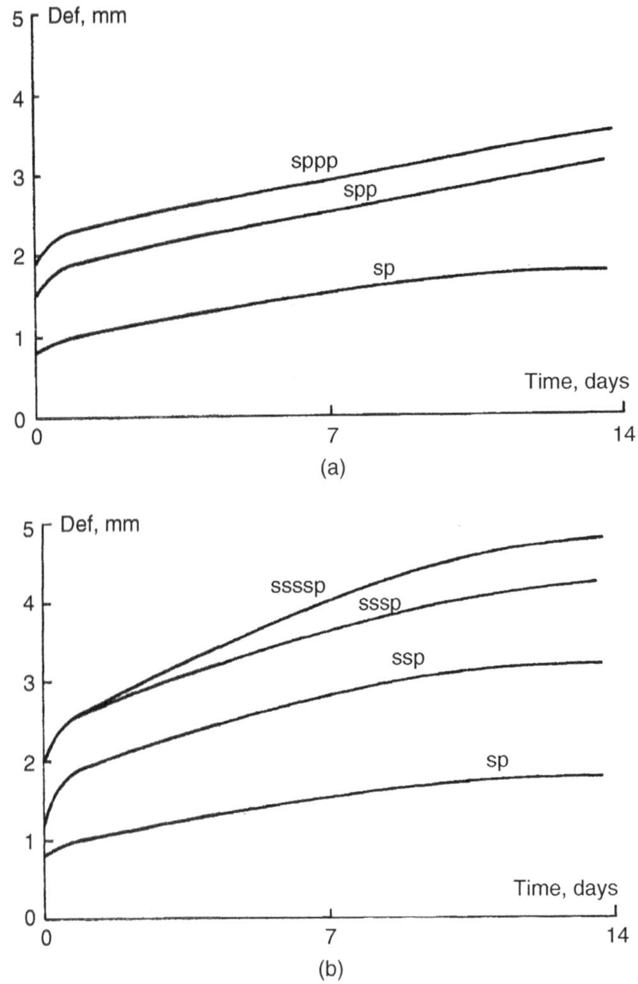

Figure 13.13 Test results for long-term tests. (a) Different number of plates. b) different number of studs

where $\dot{\varepsilon}_e$ is the elastic strain rate described by

$$\dot{\varepsilon}_e = \frac{\dot{\sigma}}{E(w)} \tag{13.5}$$

$\dot{\varepsilon}_s$ is the free shrinkage-swelling strain rate described by

$$\dot{\varepsilon}_s = \alpha \dot{w} \tag{13.6}$$

where α is the shrinkage–swelling coefficient.

$\dot{\varepsilon}_c$ is the creep strain rate. This strain rate can be described with any suitable creep formulation. At normal room temperature and in the longitudinal direction, the creep will be fairly small and can therefore be neglected in a number of practical cases. Perpendicular to grain, however, the creep is much more significant.

$\dot{\varepsilon}_{ms}$ is the mechano-sorptive strain rate, described by

$$\dot{\varepsilon}_{ms} = \dot{\varepsilon}_{ms}(\sigma, w, \dot{w}, A) \tag{13.7}$$

where A is a measure of the accumulated strain. This accumulated strain has been described in a number of different ways (Mårtensson, 1992, Toratti, 1992). Some researchers claim that A may be zero, that is, the material has no memory at all.

13.3 PREDICTION AND LIMITATION OF DEFORMATIONS

A detailed description of deformations in wood under load taking into account the effects of climate conditions is very complex. In the design of timber structures the prediction of deformations must be simplified so it can be handled in an efficient way. Methods for the design of timber structures in the serviceability limit state will be discussed in this section. Examples will be given for bending and compression perpendicular to grain.

13.3.1 General Formulation

In Section 13.2.4, a model of the strain rate was given. In accordance with that formulation, the total strain can be described as

$$\varepsilon = \varepsilon_e + \varepsilon_s + \varepsilon_c + \varepsilon_{ms} \tag{13.8}$$

where the elastic strain is given by

$$\varepsilon_e = \frac{\sigma}{E} \tag{13.9}$$

The free shrinkage ε_s is defined by Equation (13.6). The two parts creep ε_c and mechano-sorptive strain ε_{ms} may be described individually, but for practical purposes they are considered as one component. This strain component, i.e. the sum of viscoelastic and mechano-sorptive strain, is often defined as

$$\varepsilon_{c,ms} = k_{\text{def}}\varepsilon_e \tag{13.10}$$

where k_{def} is a creep-mechano-sorptive coefficient and ε_e is the elastic strain. This gives the following formulation for the total strain:

$$\varepsilon = \frac{\sigma}{E}(1 + k_{\text{def}}) + \varepsilon_s \tag{13.11}$$

where the stress is assumed to be constant. In some cases, the free shrinkage is negligible, for instance when deformations of a beam subjected to bending is studied, i.e. when the primary loading direction is along the grain. In cases when wood is subjected to loading perpendicular to grain, the shrinkage is often significant.

When calculating load-induced deformations, an effective elastic modulus E_{eff} can be defined as

$$E_{\text{eff}} = \frac{E}{1 + k_{\text{def}}} \tag{13.12}$$

where E is a representative value of the elastic modulus.

13.3.2 Design Principles for Deformation Control

The loads given in the codes are commonly defined as permanent loads and variable loads, where the latter can be imposed loads, snow loads and wind loads. The fact that variable loads often dominate in timber structures means that the deflection will fluctuate to a great extent during the lifetime of the structure. This must be considered in the choice of design criteria for deformations.

The discussion below relates to beam bending, but the same approach may be used for other loading modes.

The total deflection achieved after a long time may be calculated as

$$\delta_{\text{total}} = \sum (\delta_{\text{inst},i} + \delta_{cr,i}) \tag{13.13}$$

where $\delta_{\text{inst},i}$ is the instantaneous deflection corresponding to each load component i and $\delta_{cr,i}$ is the creep deflection calculated for each load component i, i.e.

$$\delta_{cr,i} = \delta_{\text{inst},i} \cdot k_{\text{def}} \tag{13.14}$$

where k_{def} is the creep factor (see also Equation (13.10)). For load that occurs during very short time intervals, the creep effect is small, i.e. the value on k_{def} should be close to unity. Equation (13.13) can now be rewritten and used for a criterion used in design

$$\delta_{\text{total}} = \sum \delta_{\text{inst},i}(1 + k_{\text{def},i}) \leq \delta_{\text{lim}} \tag{13.15}$$

where the limiting value of deflection δ_{lim} generally depends upon an acceptable level of deflection.

In Equation (13.15) the deflection for each load component is calculated, which can be a tedious and unnecessarily complicated way to perform the calculations. Instead the following approach

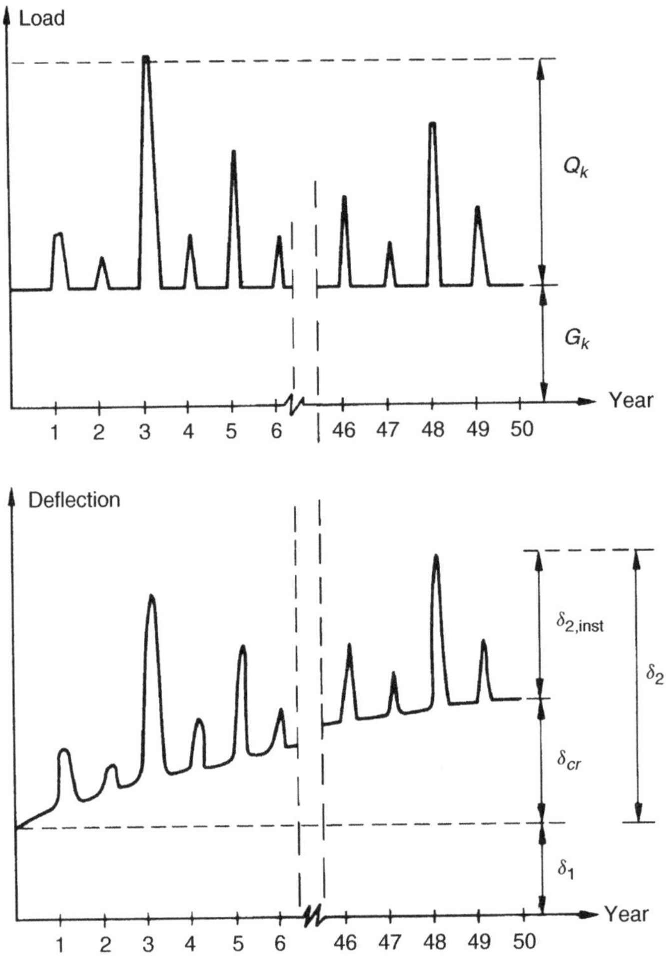

Figure 13.14 Time variation in principle for deflection (lower figure) of a beam with permanent and variable loads, according to the upper figure

may be used (see Figure 13.14) that illustrates in principle the deflection behaviour of a beam loaded with permanent load, G, and variable load, Q, e.g. see Mårtensson and Thelandersson (1992).

The total deflection can for simplicity be subdivided into one part δ_1 due to permanent load immediately after loading, and one part δ_2 which is variable during the lifetime of the structure. The variable part δ_2 consists of a reversible portion $\delta_{2,\text{inst}}$ corresponding to the immediate response to fluctuations of variable load, and a continuously increasing portion δ_{cr}, which for all practical purposes may be considered as irreversible.

The component δ_2 varies with time and is given by

$$\delta_2 = \delta_{2,\text{inst}} + \delta_{cr} \tag{13.16}$$

where $\delta_{2,\text{inst}}$ is instantaneous deflection due to variable loads, and δ_{cr} is the total creep deflection due to permanent and variable loads. Here, $\delta_{2,\text{inst}}$ can be conceived as a temporary, reversible part, and δ_{cr} as a permanent, non-reversible part of δ_2.

The maximum creep deflection $\delta_{cr,\max}$ developed during the lifetime depends upon the permanent load as well as on the variable load. With this approach the most extreme short-term values of the load should not be used in the calculation of the creep, since this would overestimate the time effects of the variable load. Instead, a time averaged value of the variable loads together with the permanent load can be used. The sum of these will in the following be referred to as the quasi-permanent load.

In the calculation of $\delta_{cr,\max}$ the increase in deflection due to moisture variations must also be included.

The following criterion may be used to control the deflections:

$$\delta_{\max} = \delta_1 + \delta_{2,\text{inst}} + \delta_{cr,\max} \leq \delta_{\text{lim}} \qquad (13.17)$$

where δ_1 is the instantaneous deflection due to permanent loads, $\delta_{2,\text{inst}}$ is the instantaneous deflection due to variable loads, and $\delta_{cr,\max}$ is the maximum creep during the building lifetime given by

$$\delta_{cr,\max} = \delta_{qp} k_{\text{def}} \qquad (13.18)$$

where k_{def} is a creep coefficient (see also Equation (13.10)) and δ_{qp} is the instantaneous deflection calculated on the basis of the value on the quasi-permanent load. The criterion given by Equation (13.17) limits the maximum deflection occurring in a timber structure during its lifetime. A typical case when it is of interest to use this criterion is when excessive deformations may cause permanent damage to partitions, to members attached to or in contact with the member considered and to fixtures and finishes. In this case, the risk of passage of the deflection limit should be kept small.

From the point of view of appearance and general utility, it may be desirable to avoid excessive deflections which are permanent or occur during long periods. Occasional passages of the deformation limit may, however, be acceptable if they are reversible and occur only during short periods. An appropriate deflection criterion for this case is

$$\delta_{qp,\max} = \delta_{qp}(1 + k_{\text{def}}) \leq \delta_{\text{acc}} \qquad (13.19)$$

where δ_{acc} can be seen as the acceptable deflection limit with respect to appearance and general utility.

The effect of different climate conditions can be accounted for by making the coefficient k_{def} dependent upon climate conditions. High relative humidities lead to higher values on k_{def} than lower humidity values. It is also important to consider humidity variations, i.e. to consider that variations lead to larger deformations. In most cases, changes in humidity are more severe than a constantly high humidity. The values of k_{def} should therefore depend both upon the expected maximum value of the relative humidity and on expected changes in it.

Long-term Beam Deflections

Values of the creep factor k_{def} for beam deflections taken from experiments will be given here. The experiments discussed in the following were performed with constant load levels corresponding to what could be referred to as a quasi-permanent service load.

Experimental data from tests made on full size timber and glulam beams are available to some extent (e.g. see Ranta-Maunus, 1991, Ranta-Maunus and Kortesmaa, 2000, and Taylor *et al.*, 1991). Theoretical studies such as Mårtensson (1992) and Toratti (1992) can also be used to obtain qualitative information concerning long-term deflections.

According to Ranta-Maunus (1991), the basic creep factor for timber and glulam after one year is 0.3–0.4. These figures were obtained from tests on full size beams in a nearly constant climate, with different relative humidities in different test series varying from 35% to 90%. The amount of creep increases to some extent with the relative humidity level, but this influence is generally very small.

Only a few tests are available for naturally varying conditions. Some test series (see Ranta-Maunus, 1991, and Taylor *et al.*, 1991) have been performed in outdoor environments with weather protection. Under these conditions, one year creep factors between 0.3 and 1.4 have been found by various investigators. It was found by Taylor *et al.* (1991) that the one year creep was much lower for beams with varnish ($k_{\text{def}} = 0.6$) than for unvarnished beams ($k_{\text{def}} = 1.4$). Ranta-Maunus and Kortesmaa (2000) showed that a value of the ten year creep factor of 0.67 is valid for a sheltered outdoor climate for non-treated spruce and glulam beams, while it is higher for Kerto-LVL and I-beams. This investigation also shows the significant effect of treatment (see Section 13.2.3).

Experimental data for different types of board materials is available (Dinwoodie and Bonfield, 1994, Thelandersson *et al.*, 1993). Since these types of materials vary considerable with respect to their composition and material behaviour, the scatter in measured creep is significant. The long-term deflection for board materials is generally larger than for timber and glulam, except for plywood where the creep in some cases is smaller than in timber. Processed board materials tend to be more sensitive to the moisture content level than to the variation of the moisture content.

Values of k_{def} representative for typical building lifetimes may be estimated on the basis of these types of data. Various effects may have an influence on the long-term deflection, and can lead to changes in the value of the creep factor. For instance, large size glulam beams are less sensitive to moisture variations than small size members. Surface coatings of paint or varnish also tend to reduce the moisture effects, provided that the coating is durable. Using green timber during construction may lead to rather large deflections in the initial phase of structural use.

For wooden structures used in climate separating structures, moisture gradients in service may increase deflection considerably. An example of such a structure is a crawl space floor where higher moisture content at the bottom side than on the top side will create an additional deflection unless special measures are taken during installation of the floor (Mårtensson, 1992). In this case, the creep factor should be magnified. For a roof structure with higher moisture content at the top side the moisture gradient will have the opposite effect.

Compression Perpendicular to Grain

There are two main effects that have to be considered in calculation of deformation perpendicular to grain. Experiments show that the shorter the bearing length is, the less deformation is achieved for the same stress level, which can be seen as an increase in stiffness with decreasing bearing length. According to Madsen (1992), this increase can be attributed to shear forces becoming more and more dominating as the bearing plate length is shortened.

The other effect is the distribution of stress in the member loaded perpendicular to grain. For example, consider the case shown in Figure 13.4. Directly under the contact area between the members the stress will be equal to the value given by Equation (13.2). If it is assumed that the beam is resting on the whole bottom surface, the stress would be less at that boundary than at the top boundary, since the loaded area is larger there. This would mean that the deformation over the beam depth is not as large as a calculation based on Equations (13.1) and (13.2) would indicate.

A simple way to determine the elastic deformation Δ_e would be to use the relation

$$\Delta_e = \frac{P}{AE_{90}}d \qquad (13.20)$$

where P is the load, A is the bearing area (bL) and d is the height of the wood subjected to compression perpendicular to grain. This will give too high deformations, however. The deformations observed in tests for moderate stresses are generally significantly lower than those predicted from Equation (13.20) (Backsell, 1966, Madsen *et al.*, 1982, Frater and Thelandersson, 1996, Jordow and Enockson 1996, Gehri, 1997).

The actual deformation in bearing depends in a more complicated manner upon the geometrical factors. The above-mentioned tests show that the deformation decreases with decreasing bearing length for the same nominal stress. There is also a nonlinearity in the dependence of height, i.e. if the height d is doubled the deformation will not be doubled.

These effects can be accounted for approximately by replacing A in Equation (13.20) with an effective area A^* which is greater than the actual bearing area A. The following relation has been proposed for A^* (Madsen *et al.*, 1982, Enockson and Mårtensson, 1996):

$$A^* = A\left(1 + \mu\frac{d}{L}\right) \quad \text{for } L_e \geq L_{e1} \qquad (13.21)$$

where L is the bearing length and μ is a factor that considers the effect of the height d. Equation (13.21) implies that the effective area over which the load is assumed to be distributed increases with the ratio d/L. The relation is valid for cases where the end distance L_e is greater than a certain limit L_{e1}.

Tests and finite element calculations made by Enockson and Mårtensson (1996) indicate that a value of $\mu \approx 0.3$ would be correct. The limit L_{e1} can based on finite element calculations be set to $0.5d$–$0.8d$. L_{e1} could also be defined as a fixed distance in mm.

In the absence of experimental results for loading near the end ($L_e < L_{e1}$), the value $A^* = A(1 + \frac{1}{2}\mu\frac{d}{L})$ could be used for the case $L_e = 0$. For intermediate values ($0 < L_e < L_{e1}$) linear interpolation could be used. Madsen *et al.* (1982) suggests another – but not so simple – method to deal with bearing near the ends of the member.

Thus, the elastic-bearing deformation can be calculated from

$$\Delta_e = \frac{P}{A^* E_{90}} d \qquad (13.22)$$

E_{90} used in Equation (13.22) should be the mean value given by codes for serviceability design.

As stated earlier, the time-dependent deformation depends upon load level, time passed from loading and the moisture variations during this time. In Section 13.3.1, it was said that a reduction of the elastic modulus could be used to determine the effects of creep and mechano-sorption, i.e. the elastic modulus in Equation (13.22) is substituted with the effective elastic modulus given by Equation (13.12). In most cases, no value is defined in codes for the coefficient k_{def} to be used for loading perpendicular to grain. A reasonable value of this coefficient is 3 based on a prediction of the effects during the first year after erection of the building, and 7–10 for longer time intervals (Enockson and Mårtensson, 1996). These values are valid for an indoor climate and for quasi-permanent load. Pure shrinkage is not included in these figures.

In the case of deformations perpendicular to grain, the shrinkage may be of interest. This deformation component can be calculated from the relation

$$\Delta_s = \alpha \cdot \Delta u \cdot d \qquad (13.23)$$

where α is the shrinkage coefficient perpendicular to grain and Δu is the change in moisture content. The initial moisture content should correspond to the moisture content in the timber when it is delivered to the building site, provided that the timber thereafter is protected from direct water exposure. The final in-service moisture content of protected timber elements in an indoor environment is usually between 8% and 12% in most structures. For a rough estimate an in-service value of 10% may be used. The shrinkage coefficient perpendicular to grain for structural softwood is of the order 0.002 per 1% change of MC.

13.4 AVOIDANCE OF PROBLEMS WITH DEFORMATIONS

In practice it is impossible to avoid deformations when using timber as a structural material. It is therefore of interest to find design solutions, which reduce deformations. For beams it is important to reduce the humidity variations. This means that one has to be observant on the climate conditions in the building where the timber is to be used. One especially important aspect of the climate is if the timber structure is situated in a structure where the climate is different on different sides of the structure, as in the case with a climate separation structure.

For joints subjected to compression it is important to avoid compression perpendicular to grain over larger depths, since this is the most sensitive direction both with regard to load and to the effects of moisture variations. In Figure 13.15 two examples of joints are presented where only a minor part of the timber is subjected to compression perpendicular to grain. There may also be problems if a different type of material is used in the building. If, for instance, a beam is situated on

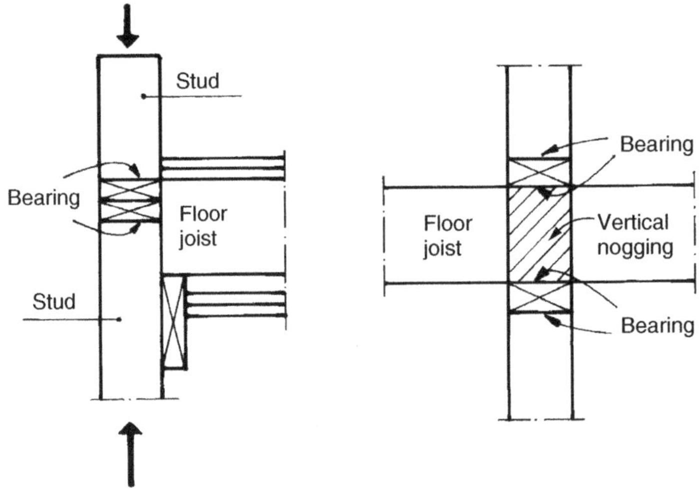

Figure 13.15 Examples of joints that can be used to reduce the problems with compression perpendicular to grain

a concrete wall at one end and on a timber structure at the other, this often leads to inclination of the beam. In this case, one may introduce a wooden piece between the concrete and the beam so that deformations of the same magnitude will occur at both ends.

The use of 'Balloon Framing' is another way to avoid deformations due to both shrinkage and crushing deformations from gravity loads. In this form of construction, the external wall studs are full height from the sole plate to the roof. The disadvantage of this method is an economic one, since the use of prefabricated wall panels would not be possible due to the size and weight of multi-storey high wall panels, and the associated erection problems.

Another method to avoid the negative effect of moisture variations in timber structures, and thereby decreasing the deformations, is to use some sort of coating that makes it more difficult for the moisture to move into the timber. This is not practical for most cases, since it is laborious and uneconomical. In some special situations, however, it may be very good to use a coating. Impregnation is another method that may be used, but as in the case with coating, it is not especially practical. For both these methods, there is also the problem with time and decay. The maintenance of the treatment has to be efficient, otherwise problems will arise with time.

13.5 CONCLUDING REMARKS

It is in practice not possible to avoid the occurrence of deformations in timber structures completely. With adequate knowledge, however, the problems that deformations may give in buildings can be reduced, and in many cases avoided. To reduce the negative effects, knowledge about the behaviour of wood with regard to deformations due to external loading and moisture variations have to be adopted in the design process. The two main problem areas with regard to deformations are beam deflection and compression perpendicular to grain. In both cases, moisture variations have a significant influence on the behaviour.

Building structures must be designed for strength and serviceability. In general, building control documents concentrate primarily on strength requirements. Much effort has been made to refine the design methods used in the ultimate limit state. However, serviceability requirements such as the limitation of deflection are often decisive for the dimensions of horizontal members of glulam or timber. This is especially true for higher strength classes. The deflection criteria and the associated design principles recommended should therefore be based on rational and practical arguments.

Design with respect to bearing stress is classified as an ultimate limit state in many codes. It would be more rational to consider bearing behaviour as a serviceability problem. The actual problem in bearing is most often to limit the deformations. With respect to serviceability, the actual deformation that may occur in the connection should be estimated. To determine the deformation, an effective stress reflecting the effects of bearing length and beam depth can be used in the calculation.

One of the most important aspects of serviceability design is the question: what is the acceptable serviceability level? To the author's knowledge, very limited information is available to help the designer answer this question. Better knowledge about acceptable deformations in different situations and application of this knowledge in practice is one of the most important factors in a rational and cost-effective design of timber structures.

REFERENCES

Arima T. and Grossman P.U.A. (1978) Recovery of wood after mechano-sorptive deformation. *J. Inst. Wood Sci.* **8**, 47–52.

Armstrong L.D. and Christensen G.N. (1961) Influence of moisture changes on deformation of wood under stress. *Nature*, 4791, 869–870.

Armstrong L.D. and Grossman P.U.A. (1972) The behaviour of particle board and hardboard beams during moisture cycling. *Wood Sci. Technol.* **6**, 128–137.

Armstrong L.D. and Kingston R.S.T. (1962) The effect of moisture content changes on the deformation of wood under stress.

Backsell G. (1966) Experimental Investigations into Deformations Resulting from Stresses Perpendicular to Grain in Swedish Whitewood and Redwood in Respect of the Dimensioning of Concrete Formwork. Byggforskningen, Report 12:1966, Stockholm, Sweden.

Bethe E. (1969) Strength properties of construction wood stored under changing climates and mechanical load (in German). *Holz als Roh- und Werkstoff*, **27**(8), 291–303.

Christensen G.N. (1962) The use of small specimens for studying the effect of moisture content changes on the deformation of wood under load. *Aust. J. Appl. Sci.* **13**(4), 242–257.

Dinwoodie J.M., Higgins J.A., Paxton B.H. and Robson D.J. (1990) Creep research on particleboard – 15 years work at the UK Building Research Establishment. *Holz als Roh- und Werkstoff*, **48**, 5–10.

Dinwoodie J.M. and Bonfield P.W. (1994) Rheological behaviour of wood based panels. An overview of current and recent European research. *Duration of Load Workshop*, Toronto.

Enockson P. and Mårtensson A. (1996) Deformations in timber buildings (in Swedish). Report TVBK-7054, Department of Structural Engineering, Lund University, Lund, Sweden.

Feist W.C., Rowell R.M. and Ellis W.D. (1991) Moisture sorption and accelerated weathering of acetylated and methacrylated aspen. *Wood and Fiber Science*, **23**(1), 128–136.

Frater G.S. and Thelandersson S. (1996) Bearing Performance of Dimensioned Lumber in a Platform Frame An Experimental Study. Department of Structural Engineering, Lund University, Lund, Sweden.

Gehri E. (1997) Timber in compression perpendicular to the grain. *Proc. Conference of IUFRO S5.02*, Timber Engineering, pp. 355–374.

Gibson E.J. (1965) Creep of wood: Role of water and effect of a changing moisture content. *Nature*, **206**(4980), 213–215.

Gowda S., Kortesmaa M. and Ranta-Maunus A. (1996) Long term creep tests on timber beams in heated and non-heated environments. VTT Publications.

Gressel P. (1984) Influence of sorption behaviour on particleboard properties (in German). *Holz als Roh- und Werkstoff*, **42**, 393–398.

Grossman P.U.A. (1976) Requirements for a model that exhibits mechano-sorptive behaviour. *Wood Sci. Technol.* **10**, 163–168.

Hearmon R.F.S. and Paton J.M. (1964) Moisture content changes and creep of wood. *For. Prod. J.*, 8, 357–359.

Hoffmeyer P. (1990) Failure of wood as influenced by moisture and duration of load. Doctoral dissertation, State University of New York, College of Environmental Science and Forestry, Syracuse, New York.

Hoyle R.J., Itani R.Y. and Eckard J.J. (1986) Creep of Douglas-fir beams due to cyclic humidity fluctuations. *Wood Fiber Sci.* **18**(3), 468–477.

Hunt D.G. (1989) Linearity and non-linearity in mechano-sorptive creep of softwood in compression and bending. *Wood Sci. Technol.* **23**, 323–333.

Hunt D.G. (1991) Present knowledge of mechano-sorptive creep of wood. *RILEM TC 112*.

Hunt D.G. and Shelton C.F. (1988) Longitudinal moisture-shrinkage coefficients of softwood at the mechano-sorptive creep limit. *Wood Sci. Technol.* **22**.

Jordow N. and Enockson P. (1996) Moisture Movements and Deformations in Timber Framed Buildings (in Swedish). Report TVBK-5079, Department of Structural Engineering, Lund University, Lund, Sweden.

Leivo M. (1988) Effect of moisture changes on the deformations of wooden beams and nail plate joints. *IUFRO Timber Engineering Meeting*, Finland.

Madsen B., Hooley R.F. and Hall C.P. (1982) A design method for bearing stresses in wood. *Canadian J. Civil Eng.* **9**, 338–349.

Madsen B. (1992) Structural Behaviour of Timber. Timber Engineering Ltd., Vancouver, British Columbia, Canada.

Meierhofer U. and Sell J. (1979) Physical phenomena in timber construction elements exposed to weathering. Part 2: Structural timber elements exposed to outdoor weathering under shelter (in German). *Holz als Roh- und Werkstoff*, **37**, 227–234.

Mohager S. (1987) Studies of creep in wood (in Swedish). The Royal Institute of Technology, TRITA-BYMA 1987:8, Stockholm, Sweden.

Mårtensson A. (1988a) Interaction between moisture and stress in wooden materials. Report TVSM-3011, Lund Institute of Technology, Lund, Sweden.

Mårtensson A. (1988b) Tensile behaviour of hardboard under combined mechanical and moisture loading. *Wood Sci. Technol.* **22**, 129–142.

Mårtensson A. and Thelandersson S. (1990) Effect of moisture and mechanical loading on wooden materials. *Wood Sci. Technol.* **24**, 247–261.

Mårtensson A. (1992) Mechanical behaviour of wood exposed to humidity variations. Department of Structural Engineering, Lund Institute of Technology, Report TVBK-1006.

Mårtensson A. and Thelandersson S. (1992) Control of deflections in timber structures with reference to Eurocode 5. CIB-W18A/25-102-2, Åhus.

Norimoto M., Gril J., Minato K., Okamura K., Mukudai J. and Rowell R.M. (1987) Suppression of creep of wood under humidity change through chemical modification (in Japanese). *Mokuzai Kogyo*, **42**, 504–508.

Perkitny T. (1960) Pressure variations in differently pre-pressed and rigidly clamped wood specimens (in German). *Holz als Roh- und Werkstoff*, **18**(6), 200–210.

Ranta-Maunus A. (1975) The viscoelasticity of wood at varying moisture content. *Wood Sci. Technol.* **9**, 189–205.

Ranta-Maunus A. (1991) Collection of creep data of timber. CIB-W18A/24-9-2, Oxford.

Ranta-Maunus A. and Kortesmaa M. (2000) Creep of timber during eight years in natural environments. *WCTE2000 Conference*, Whistler, CA.

Sauer D.J. and Haygreen J.G. (1968) Effects of sorption on the flexural creep behaviour of hardboard. *For. Prod. J.* **18**(10), 57–63.

Taylor G.D., West D.J. and Hilson B.O. (1991) Creep of glued laminated timber under conditions of varying humidity. *Int. Timber Eng. Conference*, London, UK.

Toratti T. (1992) Creep of timber beams in a variable environment. Laboratory of Structural Engineering and Building Physics, Helsinki University of Technology, Report 31.

Thorson B.R. (1989) Structural Design of Multi-storey Timber Buildings in Canada. *Proceedings Pacific Timber Engineering Conference*.

14

Vibrations of Timber Floors: Serviceability Aspects

Ian Smith

14.1 INTRODUCTION

Except under unusual circumstances, floor motion (vibration) is a serviceability issue, and therefore this chapter deals only with the dynamic behaviour of floors under 'normal use' conditions. Here normal use implies those loads that will be encountered regularly when a structure is being used for its intended purpose. Vibration serviceability problems are categorised as those due to continuous vibration, or those due to transient vibration. Continuous vibration in floors is caused by periodic forces that arise from repetitive human activities such as dancing, or from the use of machinery. The amplitude of resultant motion is considerably increased if a periodic force is synchronised with a structural frequency, leading to the condition known as resonance. Transient vibration results from unsustained loads such as impacts from people walking or dropped objects. Such motion decays in proportion to the available damping in the structural system.

Historically, engineers have been most interested in vibration serviceability problems related to wind-induced motion of tall structures. It is principally horizontal motion that discomforts human occupants of tall buildings, while principally vertical motion is what discomforts them as users of floors (or footbridges). Irrespective of the type of situation, vibrations must be kept within limits that ensure the occupant's comfort, proper operation of equipment, and integrity of the structure itself. As a topic of scientific study, the serviceability of floors is less mature than the serviceability of tall buildings. Largely, this is because traditional construction practices do not result in large numbers of 'problem floors'. Problem floors are essentially by-products of modern endeavours to built floors in ways that maximise their strength-to-weight ratio. Timber floors are almost always designed only against the action of static loads, despite everyday evidence that dynamic loads are also of great importance. Surely, nobody who has ever walked across an excessively 'bouncy' floor would doubt the importance of vibrational serviceability.

Serviceability issues in general are becoming increasingly important due to the tendency towards making buildings more material efficient and more lightweight, and as a consequence, more flexible and lightly damped. Field surveys indicate that there is increasing end-user dissatisfaction with timber floors, even if they are properly designed and constructed (Chui and Smith, 1988, Hu *et al.*, 2001). Designs are currently based on verification of serviceability conditions, according to which deterministic structural deformations must comply with admissible limits. Admissible deformation limits are primarily experience-based, and intended to avoid unwanted consequences such as excessive deflection, excessive acceleration or cracking (Holicky and Ostlund, 1993). An unavoidable conclusion is that current design practices are not adequate as far as floor serviceability is concerned. As discussed below, there has been considerable improvement in understanding floor vibration problems over the last few years, but as yet that understanding is not complete.

Design approaches for buildings, and therefore floors, are often segregated on the basis of building size. Most countries allow 'small buildings' (i.e. those with a relatively small footprint and of limited height) to be built according to prescriptive rules laid down by the local civil administration. Nawar (1988) argues that there is bias towards disproportionately high investment in improved serviceability because small buildings are mostly owner-occupied and represent a high proportion of the financial assets of owners. Also, involvement of a broad spectrum of government agencies, professionals and tradespeople in the control, design and construction of small buildings militates again adoption of solutions that are prone to serviceability problems. Small buildings tend to be highly structurally redundant with complex construction details, which makes it extremely difficult to analyse the behaviour of whole systems, sub-systems or components. Design models are simplistic and solutions rarely optimised. Large buildings are often owned as an investment or fulfil purely functional purposes, which lessens or eliminates bias towards disproportionably high investment in improved serviceability. A relatively narrow spectrum of professionals is involved in the control, design and construction of large buildings, and this promotes optimal engineering design solutions. Large buildings tend not to be highly structurally redundant, and have simple and repetitive construction details. This makes it relatively straightforward to analyse behaviour of whole systems, sub-systems or components in a realistic fashion.

There are two dominant types of timber construction: light-frame construction and framed/post-and-beam construction. Light-frame construction has repetitive lumber (small dimension timber) members, or components such as trussed-rafters made from lumber, that are sheathed with panel products made from wood or other materials. A characteristic is that the sheathing helps distribute load laterally between lumber members, resulting in 'load sharing'. Although floor plates traditionally have lumber joists as the stiffening ribs, it is now common practice to use substitute products such as wood I-joists or open-web joists. Framed construction has, as the name implies, a primary structural framework. Timber members in frameworks are normally large sawn timbers or glued-laminated timbers (glulam). Joints in timber frames are usually simple and approximate to pins. Robustness and stability of framed timber systems is generally provide by triangulation of members and/or infill panels of wood or other materials. Timber floors in framed buildings are usually, but not always, infill rib-stiffened plates. Ribs are lumber, glulam, wood I-beams or other 'materials'. The plate layer can be of timber boards, plywood or other materials and is often topped by a non-structural layer of concrete, gypsum-based or other insulating material and/or decorative cum wearing layers. In other cases, timber floors are incorporated in what are essentially masonry structures. Again, the floors are usually constructed as rib-stiffened plates. Ends of ribs are either built into the masonry or face mounted to it using joist hangers. It is clear that the overall building system influences the response characteristics of the floors, and whether or not their behaviour should be considered in isolation from the rest of the system.

Any one person can have varying expectations of different floors on which he/she walks, e.g. floors in their home versus floors in their place of work. When a single human stands on a small lightweight floor (e.g. timber joisted floors in residential construction), he/she is a significant part of the system mass and he/she is a heavily damped component. Therefore, the presence of one or more humans can greatly influence the structural response of a floor. By contrast, for large heavy floors single humans have relatively little influence on either the mass or damping of the system. Therefore, any attempt to characterise the effects of human activities on the behaviour of small lightweight floor systems is doomed to failure if it does not consider the humans. Clearly, assessments of vibrational behaviour of floors should differ depending upon the specifics of the case at hand.

From the above it will be appreciated that the vibration of timber floors is a complex matter, and worthy of any structural designer's attention. The rest of this chapter reviews the nature of vibrational serviceability problems in both lightweight and heavyweight floors, and summarises background scientific studies related to this. Avoidance of vibration serviceability problems in timber floors and their analysis are also addressed, but in lesser detail.

14.2 CONCEPTS OF HOW SYSTEMS VIBRATE

Vibration in any building structure reflects the dynamic loads applied to it, the dynamic system, the dynamic stiffness (usually flexural stiffness) and damping. Although floors and other structural systems have multiple degrees of freedom, the essential nature of their vibration response, and how to control it can be understood from analysis of a Single Degree of Freedom (SDF) system.

Consider simple harmonic motion of a concentrated mass vibrating on a linear elastic spring of zero mass (Figure 14.1(a)). Applying Newton's second law of motion, the equilibrium equation for simple harmonic motion is

$$m\ddot{u} + ku = 0 \quad \text{or} \quad \ddot{u} + \omega_n^2 u = 0 \qquad (14.1)$$

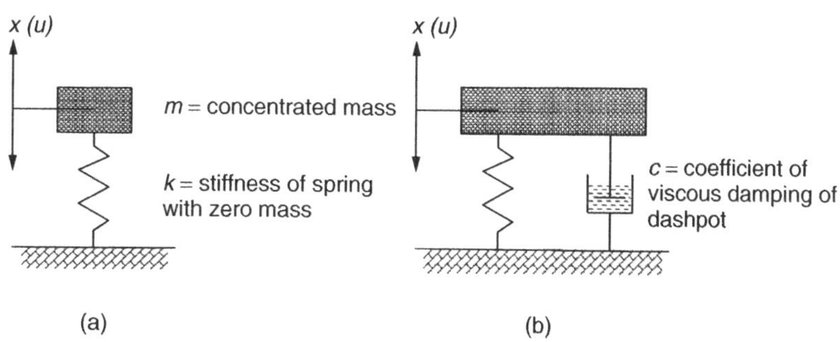

Figure 14.1 Single degree of freedom systems: (a) without damping, (b) with damping

where $\omega_n = \sqrt{k/m}$ = angular natural frequency (radians/s). This leads to the well known SDF equation for the natural frequency of vibration for simple harmonic motion (Hz):

$$f_n = \frac{\omega_n}{2\pi} = \frac{1}{2\pi}\sqrt{\frac{k}{m}} \tag{14.2}$$

The importance of the stiffness to mass ratio, k/m, to vibration of a system is immediately apparent. The period of simple harmonic motion $T = 1/f_n$ (seconds) (Figure 14.2(a)). Adding a linear viscous dashpot to the system (Figure 14.1(b)) the equilibrium equation for damped free vibration becomes

$$m\ddot{u} + c\dot{u} + ku = 0 \quad \text{or} \quad \ddot{u} + \frac{c}{m}\dot{u} + \omega_n^2 u = 0 \tag{14.3}$$

The solution for Equation (14.3) is

$$u = A_1 e^{p_1 t} + A_2 e^{p_2 t} \tag{14.4}$$

where $p_1 = -\frac{c}{2m} + \sqrt{\left(\frac{c}{2m}\right)^2 - \frac{k}{m}}$ and $p_2 = -\frac{c}{2m} - \sqrt{\left(\frac{c}{2m}\right)^2 - \frac{k}{m}}$, and A_1 and A_2 are determined from the initial conditions for the system. Noting the form of the terms within the square-root portions of p_1 and p_2, it is apparent that there are three cases associated with the amount of damping in the system:

Case 1 $(c/2m)^2 > k/m$, which is known as an overdamped (non-oscillatory) condition. The system is too damped to vibrate, i.e. damping is enough to prevent rebound in the system.

Case 2 $(c/2m)^2 < k/m$, which is known as an underdamped (oscillatory) condition.

Case 3 $(c/2m)^2 = k/m$, which is known as the critical damping condition. It represents the boundary between oscillatory and non-oscillatory motion.

Structural systems are rarely heavily damped, and therefore Case 2 (underdamped) systems are the norm. Making the substitution $\omega_n^2 = k/m$ allows definition of the critical damping coefficient $c_c = 2m\omega_n$. Knowing c_c permits the degree of damping in any system to be expressed in

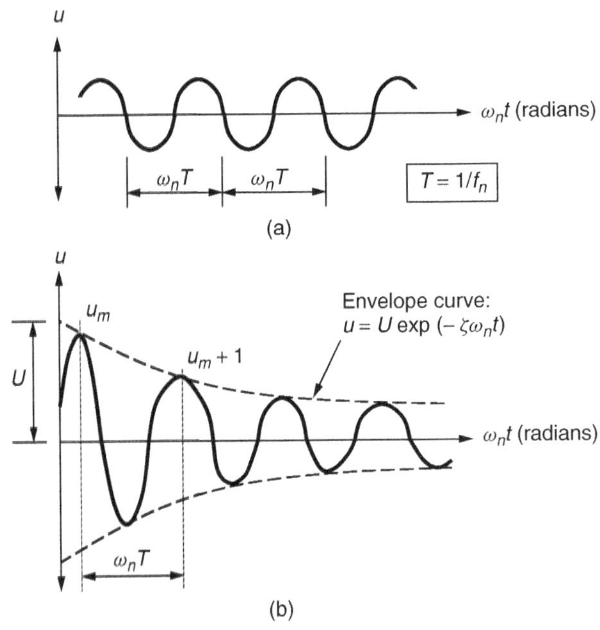

Figure 14.2 Time history responses for single degree of freedom systems: (a) free vibration without damping, (b) free vibration with damping

normalised form as the viscous damping ratio:

$$\zeta = \frac{c}{c_c} = \frac{c}{2m\omega_n} \quad (14.5)$$

Thus, for structural problems, the motion of the mass relative to its equilibrium position is (Figure 14.2(b))

$$u = Ue^{-\zeta\omega_n t}\sin(\sqrt{1-\zeta^2}\omega_n t + \Phi) \quad (14.6)$$

U and Φ are determined from the initial conditions for the system. The angular natural frequency for free damped oscillation is $\omega_{nd} = \omega_n\sqrt{1-\zeta^2}$. Decay in the amplitude of the vibration is proportional to the logarithmic decrement that can be calculated from the heights of successive peaks in the displacement time history (Figure 14.2(b))

$$\delta = \frac{1}{n}\log_e \frac{u_m}{u_{m+n}} \approx 2\pi\zeta \quad (14.7)$$

where n = number of cycles separating the two observations of vibration amplitude used in the calculation. As shown in Equation (14.7), there is an approximate relationship between δ and ζ (Chui and Smith, 1989).

If a harmonic force of $F\cos\omega t$ (forcing frequency $f = \omega/2\pi$) is applied to the system, the equilibrium equation for forced damped vibration is obtained:

$$m\ddot{u} + c\dot{u} + ku = F\cos\omega t \quad (14.8)$$

The amplitude of vibration under the harmonic force is

$$U = \frac{F}{\sqrt{(k-m\omega^2)^2 + (c\omega)^2}} \quad (14.9)$$

Manipulation of Equation (14.9) leads to the dynamic response factor ρ, which is the ratio of the vibration amplitude to the static displacement ($U_{\text{static}} = F/k$):

$$\rho = \frac{U}{U_{\text{static}}} = \frac{1}{\sqrt{(1-\beta_1^2)^2 + (2\zeta\beta_1)^2}} \quad (14.10)$$

where $\beta_1 = f/f_o = f/f_1$ = ratio of the forcing frequency to the fundamental natural frequency of the system. Figure 14.3 shows the relationship between ρ and $f_o/f (= 1/\beta_1)$. It can be seen that the amplitude of movements is greatest when $f = f_o$, which corresponds to the resonance condition. Also, it is apparent that the amount of damping available in the system can strongly influence the magnitude of the dynamic response factor close to resonance. Figure 14.3 indicates the response of systems with ζ values of 0.0, 0.06 and 0.12, which reflects values thought typical for occupied building structures (NRC 1995). Section 14.5.4 discusses factors influencing damping in timber floors.

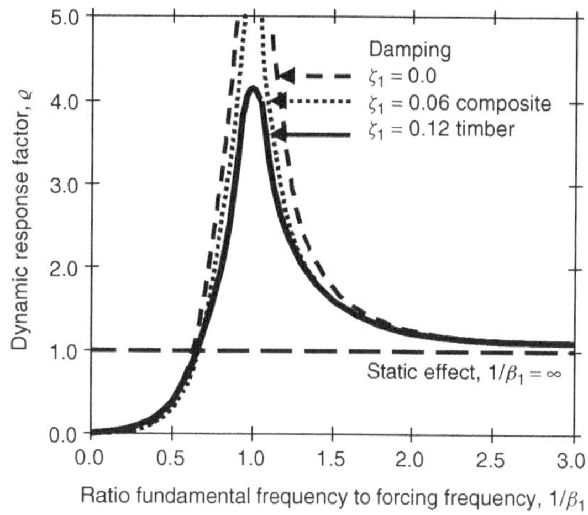

Figure 14.3 Dynamic response factor

With only the above rudimentary understanding of vibration theory, it is possible to understand how floors vibrate and avoid serious error in their design. Apart from the importance of damping, it can be seen from Figure 14.3 that if the natural frequency is much less than the forcing frequency, the dynamic response factor is small, and in the extreme goes to zero. Problematic vibrations tend to be induced by the low frequency content of dynamic loads. For example, it is the low frequency component of footfall impacts that are of concern (see Section 14.4). Also from Figure 14.3, it is apparent that when the natural frequency is much higher than the forcing frequency, the dynamic response factor approaches unity. This suggests ensuring that systems have relatively high natural frequencies as a practical strategy for minimising vibration problems. Certainly this works, but is often only achieved by the addition of extra material or reduced spans. It is thus not surprising that reports of vibration problems in structures have increased in proportion to efforts to reduce the amount of material, which has usually reduced stiffness to mass ratios. It is normally easier to increase the strength-to-weight ratio than to increase the stiffness to mass ratio of construction materials. Gains in ultimate capacities of components are often offset by the appearance of hitherto unknown serviceability problems. Although the situation is more complicated in real structural systems because both the excitation and response have extended and overlapping frequency ranges, solution strategies are essentially as just indicated.

It follows from the above that the primary strategy for avoiding vibration problems in floors and other structural systems is frequency tuning. The most practical strategy is to ensure that the several lowest and most energetic structural frequencies are higher than the excitation frequency and its first one or two harmonics (see Section 14.6). This is known as high frequency tuning. The alternative of low frequency tuning (several lowest structural frequencies are lower than the lowest excitation frequency) is not practical with floors for two reasons. First, loading frequencies associated with human activities can be quite low (Table 14.1), and thus structural frequencies would have to be about 1 Hz or less to avoid resonance. Secondly, human users of floors are particularly sensitive to vibrations in the frequency range 4–8 Hz (see Section 14.3). As it is not usually possible to build floors that have very low structural frequencies and which also meet other design parameters, low frequency tuning is effectively precluded. Low frequency tuning is a more usual strategy for avoiding serviceability vibration problems in tall buildings due to seismic loads, because structural frequencies for horizontal modes can be low. The challenge to designers of floors is to attain high frequency tuning economically.

Table 14.1 Common ranges of activity rates (Rainer *et al.*, 1988)

Activity	Common activity rate (steps per second)
Walking	1.7–2.3
Running	2.0–3.5
Jumping	1.8–3.0
Exercising	2.0–3.0
Dancing	1.9–3.3

Whenever feasible, engineers conceptualise systems as simple, usually rectangular, plan arrangements with idealised supports. Joisted floors are thought of as having either support along the two edges containing joist ends, or along all edges. The ‘all edges supported’ case is often not very realistic, but facilitates calculations (see Section 14.7). What should never be done without due consideration of all factors is to presume floor size equals the room size, but unfortunately this seems to be done quite often. Definition of the dynamic system for any floor is often not simple. Plan geometry, including openings for stairs and services, needs to be known, as does the nature of the structural system for the building as a whole. Characteristics of both supporting and supported components will influence the response of a floor, and determine any continuity between bays of the floor. Physical and mechanical properties of structural and non-structural components effect modal stiffness, modal mass and modal damping characteristics of a floor system. Any evaluation of a floor needs to be related to its intended use and owner/user expectations regarding its serviceability.

The situation portrayed in Figure 14.4 serves to illustrate several issues associated with defining floor systems. For the system shown, the first question is whether the floor system should be treated as two separate spans or one non-symmetric two-span arrangement? Ends of timber joists from the spans overlap at the intermediate support and would be nailed together to form a semi-rigid moment connection. The plywood sheathing, which would be nailed to the joists, also provides some continuity between the spans. The system would probably have enough self-weight and imposed load to prevent joists bouncing on the support (presuming they were not anchored to the masonry wall/supporting beam). Clearly, there is no simple answer, but the system would likely have

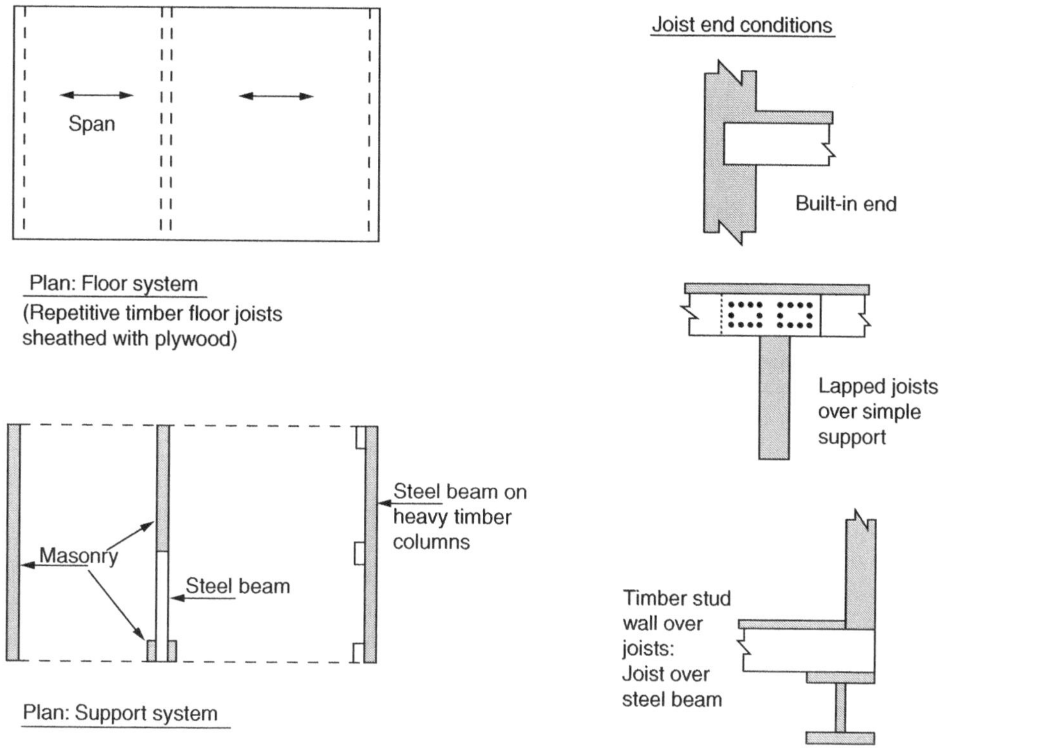

Figure 14.4 Example: Definition of the structural system

behaviour closest to a non-symmetric two span arrangement. The second question is what fixity should be assumed at the left-hand and right-hand end supports? Again, the answer is not straightforward. For spans possible with timber, joists are only a few metres and axial shortening of joists is negligible, so assumptions about fixity in the horizontal direction are not important. Only vertical dynamic equilibrium is of interest, unless there is an unusually complex situation requiring three-dimensional analysis. It is never possible to attain full rotational fixity at the end of a joist, even if it is built into masonry, because timber has low modulus of elasticity perpendicular to grain. Axial shortening in the masonry wall would almost certainly be negligible, and the possibility of vertical movement at the ends could be neglected. The left-hand support would behave closely as though it was fixed vertically and rotation was semi-rigidly restrained. For the right-hand support aspects to be considered are the extent of vertical and rotational restraint. Vertical rigidity of the support would depend mainly upon the stiffness of the steel support beam and the columns beneath it. It would be unlikely that the detail would provide much rotational restraint. The right-hand end support would probably behave like a roller. Here a roller is taken mean an arrangement that prevents both positive and negative movement normal to the joist axis, but does not prevent other movements. A third question that should be addressed for the floor system depicted regards the possibility of significant variability in support conditions along the intermediate and right-hand end supports.

Structural and non-structural walls (partitions) may be placed below or above floors. Architectural considerations may dictate that there is no alignment between walls in different storeys. Whether partitions influence the vibrational serviceability of a floor depends upon the type of partition and the construction detailing. Figure 14.5 illustrates an arrangement where a partition would influence

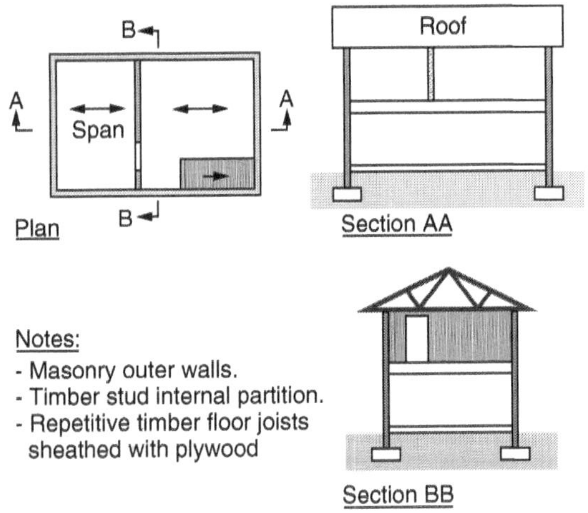

Figure 14.5 Example: Arrangement where a partition (non-structural wall) influences dynamic response of a floor

the dynamic behaviour of a floor. The intent in this example is that the partition is nominally supported by the floor, but in practice has enough rigidity to arch over the floor if it deflects below the neutral position (Figure 14.6). Such a situation is realistic for a timber stud wall sheathed with plasterboard. Upward motion of the floor would be restricted by the partition. Damping would be imparted to the system as the floor contacts the underside of the partition (impact damping). Obviously, the vibration response is not symmetric about the neutral position. Onysko (1988) reported that attachment of non-structural timber stud walls (partitions) to floors by nailing the bottom plate to the surface of the floor can benefit both static stiffness under concentrated load and dynamic response significantly. The extent of any stiffening is hard to quantify and the situation is never simple to resolve in practice.

14.3 HUMAN PERCEPTION OF MOTION

Users of buildings and other engineered structures sense low-frequency motion in three ways:

- Acceleration causes forces on the body that are felt by the balance organs.
- Visual cues (e.g. movement of the structure relative to distant fixed objects, movement of objects resting on or hanging from the structure relative to the observer, reflections off shiny surfaces).
- Audio cues (e.g. creaking or rattling created by motion of the structure).

Surveys for domestic buildings indicate that audio cues are very important to perceptions of how well floors perform. Sound produced by vibrational movement of the building itself or furniture, and rattling of objects in or on furniture, or suspended from the building are particularly annoying (Onysko and Bellosillo, 1978).

Human-induced vibrations in structures are almost always a problem of serviceability, in that they are a source of annoyance to users. In some instances, the person experiencing the motion is also the cause of it, while in other instances it is activities of others that cause annoyance. Thus, the activity of the person experiencing the vibration is important. When a person walks across a floor, he or she will tolerate much larger amplitude vibrations than when sitting quietly resting, reading or writing. Categorisation of human perception, and tolerance, needs to reflect both the activity being undertaken and the relationship between the source and the sensor (person experiencing the vibration). In this respect, the following definitions are used (Ohlsson, 1988):

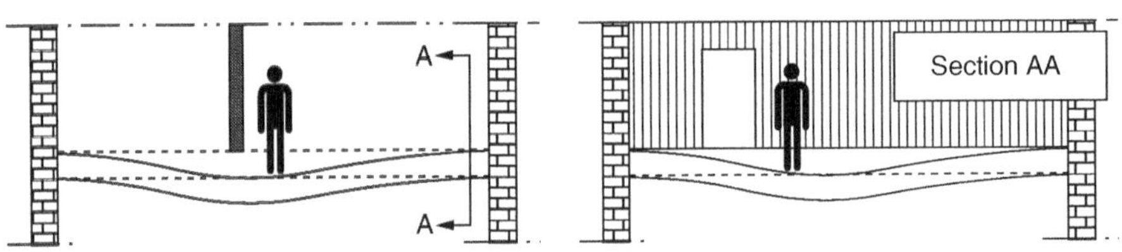

Figure 14.6 'Arching' of a partition over a floor as it moves down

- "*Springiness* of a floor is associated with the sensation of self-generated floor deflection and vibration from a single footstep during the time of contact between foot and floor surface".
- "*Vibrational disturbances* caused by foot-fall on a floor are characterised by perception of floor vibration induced by other persons than the one that is disturbed".

Springiness is a problem usually only associated with lightweight floors or those that are flexible under concentrated load. Such floors are common in light-frame timber construction, and some masonry-wall buildings having timber joisted floors. In terms of the response of a floor system, springiness encompasses static flexibility and impulsive velocity response, while vibrational disturbance encompasses impulsive velocity response and stationary vibration response. Static flexibility relates to cases where the excitation frequency is much lower than the response frequency (forced displacement response is asymptotic to the static response at the peak load level). For example, limiting the static flexibility can account for the localised response that a person might sense when walking across a rib-stiffened plate floor with a flexible plate that flexes between supporting ribs (e.g. floor boards or plywood sheathing over timber joists). This leads to use of the 'static flexibility' as an indirect control on vibration in floors with a 'soft' surface (Ohlsson, 1984).

It is well established that humans are most sensitive to vibrations in the frequency range 4–8 Hz, that being the range of vibration for human internal organs (BSI 1984, ISO 1989), and thus researchers agree that structural frequencies in that range are undesirable. Human sensitivity to low-frequency accelerations is highly variable, and depends upon their activity. At typical building frequencies, the most sensitive of people are able to detect accelerations of 2 milli-g (0.02 m/s^2), while the least sensitive only detect accelerations greater than 20 milli-g (0.20 m/s^2) (Chen and Robertson, 1972). It has been observed that age and sex have only slight influence on human perception (Kanda *et al.*, 1988), but it is important to distinguish between accelerations that can be detected and those that cannot be tolerated (Hansen *et al.*, 1973). Tolerance is related to the return period, e.g. only 2% of occupants living in the top third of a high-rise building will find a 'once in 6 years' root-mean-square (rms) horizontal acceleration of 5 milli-g (0.05 m/s^2) objectionable (Hansen *et al.*, 1973). A ratio of peak to rms acceleration of about 3.5 has been found appropriate based on study of tall buildings (Irwin *et al.*, 1988). Farah (1977) suggested that discomfort is simply proportional to the *vibration power absorbed*. The difficulty with implementing this concept, within floor models, is that it is necessary to incorporate an adequate mechanical analogue of the observer. Foschi and Gupta (1987) took an initial step in that direction in analysis of lumber joisted floors, but their mass-dashpot-spring analog of a person was crude. Crowds stamping, jumping or clapping in unison can produce transient structural oscillations of long duration in large structures such as grandstands, and dance and concert hall floors. Studies of crowd-induced vibrations in floors and temporary grandstands have led to threshold perception levels for steady state transient vibrations (Wiss and Parmelee, 1974, Allen and Rainer, 1976, Murray, 1979, Allen *et al.*, 1985).

With simplified design rules in mind, it has been suggested that human induced vibration can be classified as either 'long span' or 'short and medium span' problems (Ohlsson, 1982). Ohlsson (1988) suggests that "... the disturbance from transient impulsive vibration is in some sense proportional to vibration velocity and inversely proportional to the decay rate. The decay rate with respect to time seems most relevant". Others suggest that disturbance is proportional to acceleration (Chui, 1987). For vibration above 8 Hz rms-velocity or rms-acceleration is used rather than peak velocity to better reflect the waveform. There is, however, no general consensus regarding choice of velocity or acceleration as the 'correct' parameter. This should not be surprising as waveforms are approximately sinusoidal, and the two quantities are therefore related. Based on field evaluations of joisted lumber floors subjected to 'heel drop' impacts, Chui (1987) suggests that 'frequency weighted' rms-acceleration should be limited to 0.45 m/s^2, with the weighting carried out in accordance with British Standard BS6472 (BSI, 1984).

The duration of a disturbance is very important with regard to how people react to it. People can tolerate short duration and heavily damped vibrations of relatively very high amplitude, compared to continuous vibrations. When two activities are being performed on the same floor, the acceleration limit is based on the more sensitive activity. For example, Matthews and Montgomery (1988) proposed that for persons seated in dining areas adjacent to dance floors, the boundary between acceptable and unacceptable acceleration is 0.125 m/s^2 based on tolerance levels for diners. Table 14.2 gives an indication of appropriate limits on acceleration as a function of activity.

Most research studies are predicated on the presumption that deterministic assessments of vibrational serviceability can be made. Contrary to this, there should be explicit recognition of the random nature of functional requirements and structural response (Holicky, 1988). There is inherent variation which causes statistical uncertainty (e.g. in the excitation and properties of structural components) and imprecision and vagueness which causes fuzziness (e.g. effect of workmanship, human tolerance of vibration). Inherent variability can be handled by structural reliability theory. Imprecision and vagueness can be handled using the theory of fuzzy sets (Brown and Yao, 1983, Holicky, 1988, Holicky and Ostlund, 1993). Conceptually at least, it is possible to obtain rational design relationships between physical parameters such as static deflection and peak velocity, and the 'fuzzy probability' of unsatisfactory performance. As yet, this has not been done for timber floors.

Table 14.2 Acceleration limits for continuous vibrations or rhythmic activities (NRC, 1995)

Type of activity/occupancy affected	Acceleration limit (m/s^2)
Office and residential	0.04–0.07
Dancing and dining	0.20
Jumping exercises shared with weight-lifting	0.20
Dining and weight-lifting	0.15–0.25
Live concert or sport event	0.50
Rhythmic activity only	0.40–0.70
Jumping exercises only	0.60

14.4 DYNAMIC LOADS

14.4.1 Human-induced Loads

According to the various load sources and applicable counter-measures, structures affected by human induced vibrations can be grouped as:

- pedestrian structures (footbridges, walkways in shopping malls),
- residential buildings (single occupancy and multiple occupancy),
- office buildings,
- industrial buildings,
- gymnasia and sports halls,
- dance and concert halls without fixed seating,
- concert halls with fixed seating, and
- high-diving platforms.

There are very many variations of rhythmic body movements leading to a large variety of dynamic loads. Activities generating synchronised rhythmic movements such as those due to several or more people dancing or exercising are especially problematic. Several people acting synchronously for 20 seconds or more can lead to approximately periodic loads that produce almost steady state structural vibration. Loads differ in their frequency content and their 'time function', but can be grouped as dynamic loads caused by walking, running, jumping, dancing and clapping (Bachmann, 1988). The most critical loading frequencies of the load-time function are not necessarily the nominal loading frequency. It is necessary, therefore, when representing loads to account for both upper and lower harmonics of the loading function, rather than represent it as just a single harmonic function of the basic loading frequency (e.g. pacing rate).

The simplest way to describe vibration behaviour of any complex system is in terms of modal properties, i.e. modal frequencies, shapes, masses and damping. Once modal properties are known, a transfer function (frequency response function) between force and response can be established, enabling calculation of the response for a given force spectrum. It is therefore desirable to represent dynamic loading as a spectrum in the frequency domain via Fourier decomposition:

$$F(t) = \sum_{n=1}^{N} \left[a_n \cos\left(\frac{2n\pi t}{T}\right) + b_n \sin\left(\frac{2n\pi t}{T}\right) \right] \tag{14.11}$$

where $F(t)$ = load at time t, a_n and b_n are Fourier coefficients, and T = the period. Transient loads are modelled using decaying functions such as:

$$F(t) = A_1 \sin\left(\frac{2\pi t}{T}\right) \exp(-A_2 t) \tag{14.12}$$

where A_1, A_2 and T are fitted parameters. Bachmann and Ammann (1987) give a comprehensive description of human induced vibration serviceability and safety problems.

Footfall Forces

Forces from human motion depend upon many factors including the characteristics of the person(s) involved, the activity being undertaken (e.g. walking, running, jumping), the number of people, whether activities of different people are coordinated, and the characteristics of the floor surface. Annoying vibration of timber floors is commonly associated with walking and running forces. Vertical forces due to an individual foot impact have been measured and characterised (Galbraith and Barton, 1970, Ohlsson, 1982) (Figure 14.7). As can be seen, there are two peaks in a force-time history with the first corresponding to 'heel strike' contact and the second to 'toe-lift off' contact. For 'normal' walking the peak force is about 1.2–1.5 times the static force, and has duration of between 0.5 s and 0.8 s (Eriksson and Ohlsson, 1988). Flexibility of both footwear and floor coverings influences forces. Peak force is much higher, but the duration much shorter, for running than for walking. Assuming a Fourier decomposition, Ohlsson (1982) suggested the approximate relationship: Force intensity = $\sqrt{3} \times 10^2/f$, where f is frequency in Hz. Force intensity has units of N/Hz$^{0.5}$, and is the square root of the power spectral density. The approximation is fairly accurate for $f \leq 50$ Hz. Footstep forces contain significant components at frequencies up to 50 Hz, but are dominated by low frequency components in the range 0–6 Hz. If a structural system contains a low modal frequency (<8 Hz), as can occur in long span floors, the response will be dominated by resonant vibration. As the force is inversely proportional to f, it is apparent that

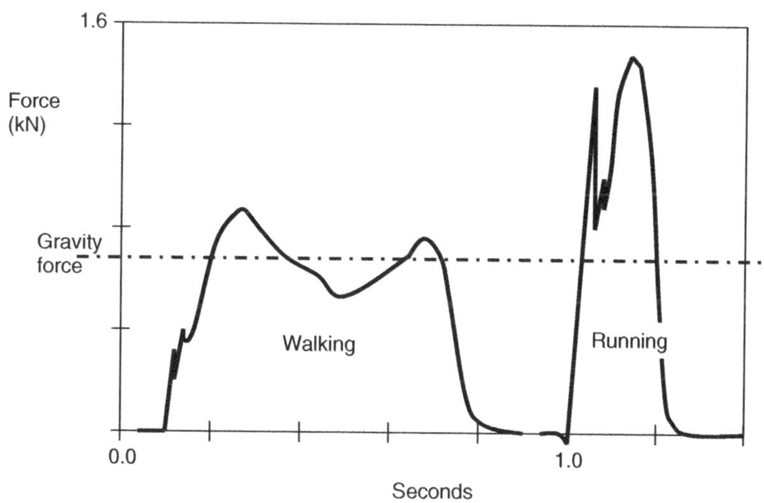

Figure 14.7 Footfall forces (based on Ohlsson, 1982)

the lower modes of structural vibration are most likely to dominate for short to medium span floor systems (cases where modal frequencies lie above 8 Hz). The response of short or medium span floors is therefore a mixture of low-frequency forced vibration and high-frequency resonant vibration. Table 14.1 gives ranges of activity rates (steps per second) associated with human movements.

Stationary force-time histories can be superimposed as illustrated in Figure 14.8 to obtain a 'pulse train'. Whether or not movement across a floor has a significant influence on the forced vibration response of a floor depends upon the size and nature of the structural system and the activity. For example, the effect of forces developed as a single person walks across a floor of moderate span is similar to the effect if that person paces 'on the spot' at mid-span (Hu, 1992). For large floors, zigzag movements such as those produced by ballroom dancers have to be accounted for (Alisjahbana, 2000).

Eriksson and Ohlsson (1988) studied forces generated by a group of people walking on a floor. Focus was on frequencies below 8 Hz, which is relevant to longer span floors and bridges. It was assumed that rms-acceleration governs human comfort in such situations. Footsteps overlap by about 15% at walking rates between 1.4 and 2.5 steps per second. The ratio of the peak force to the static force depends upon the rate of walking. When a group of people walk in a randomised manner across a floor or bridge, it is not reliable to assume the response is similar to when they pace on the spot at mid-span. Eriksson and Ohlsson (1988) deduced dynamic loads developed on a floor by people walking at various speeds and in different manners, by inverse analysis of the acceleration response at mid-span of a simply supported bridge-like floor made from precast concrete elements. Results are specified in terms of spectral density of the load (N^2/ Hz) for a given number or people. Those authors also give a design procedure for predicting rms-acceleration, for which human tolerance limitations are recommended by IS0 Standard 2631-2 (1989).

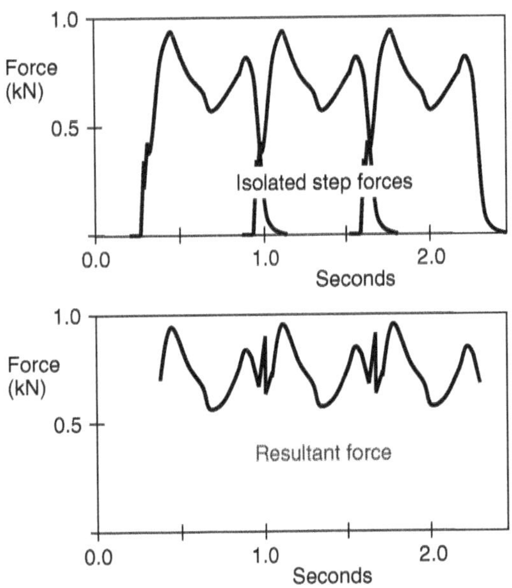

Figure 14.8 Footfall forces from a walking person (based on Ohlsson, 1982)

Crowd Loads

Live loads due to crowds are an important concern for large structures such as stadiums, grandstands, theatres, and dance and concert halls. Problems arise principally if structures have modal frequencies in the range of coherent rhythmic (synchronised) crowd movements. In such cases, loads produce resonant or near-resonant structural vibrations that can be damaging or cause discomfort or be intolerable to members of the crowd (Greimann and Klaiber, 1978, Pernica, 1983). Static occupancy loads associated with crowds can be estimated quite easily, but the dynamic component is activity dependent. Ebrahimpour and Sack (1988) performed measurements to define the dynamic loads generated by *in situ* movements of individuals and groups of people, and to determine parameters influencing 'crowd coherency'. They measured vertical and horizontal forces produced by one, two and four people performing predefined movements. Periodic and transient human loads were proposed. The term 'periodic' refers to the general shape of the load-time function. Loads were applied in the range 2–4 Hz, based on observations that the frequency range for human movements is 0–6 Hz. Members of a crowd react to a prompt event such as beat of music or visual cue. Individual reactions can be characterised by the lag between the prompt and attainment of the peak force (load). Variations in the lag time between people controls whether individuals are, for example, in rhythm with the music (periodic movements) or their reaction time (transient events). Lag times can be transformed into phase lag times. The 'group effect' was found to give coherence to crowd movements. Small groups of people were found to synchronise their movement even for transient activities such as jumping, and phase lag times could not be deduced from behaviour of isolated individuals. Ebrahimpour and Sack used an inverse statistical procedure to deduce phase lag values for groups. It was deduced that phase lags are exponentially distributed, thus movements tend to be synchronised. Group loads were generated and it was found that "In general, crowd movements could be idealised as time-varying point loads. The response of the structure can be obtained by randomly simulating the load histories (*for individuals*), shifting them on the time axis by random time phase lags, and placing loads at the critical location on the structure". Averaging the dynamic loads per person, the periodic load-time history per person is:

$$q(t, M) = \frac{1}{AM}\sum_{j=1}^{M}\sum_{n=1}^{N}[a_n \cos(2n\pi f(t-\phi_j)) + b_n \sin(2n\pi f(t-\phi_j))] \qquad (14.13)$$

where $q(t, M)$ = dynamic load intensity per person (kPa), A = occupied area (e.g. seating area), j = index for the number of people ($j = 1, M$), n = index for Fourier series coefficients ($n = 1, N$), f = fundamental load frequency (Hz), and ϕ_j = jth random phase lag. Ebrahimpour and Sack (1988) report that periodic jumping loads at 2 Hz and 3 Hz asymptotically approach 3.4 kPa. Their simulations imply that this value can be assumed for groups with $M \geq 10$, but some caution should be exercised, as only loads produced by up to four people were measured in experiments.

For design purposes, rhythmic activities can be assumed to generate a sinusoidal component that is additive to the static force (Table 14.3 and Figure 14.9) (see also Pernica, 1983, Allen *et al.*, 1985, 1987, and ATC, 1999). Values of α in the table are based on commonly encountered events with a minimum of 20 participants, and should be increased if there are fewer than 20.

14.4.2 Mechanical Equipment

Powered mechanical equipment usually exerts reciprocating or harmonic loads on floors (e.g. industrial machinery, washing machines, spin dryers). Depending upon the nature of the equipment and the orientation of the axis about which moving parts spin, induced motion can be horizontal, vertical, or a mixture of the two. Ideally, mechanical equipment is isolated from floors used by humans, or those where sensitive equipment is housed. If that cannot be done, attention should be directed to avoiding resonance. Except for some reciprocating machinery, resonance is usually automatically avoided because the excitation frequency is much higher than frequencies of the most energetic structural modes. Audible effects (noise) associated with operation of mechanical equipment heightens

Table 14.3 Estimated loading during rhythmic events (based on NRC, 1995)

Activity	Forcing frequency (Hz)	Weight of participants, w_P (kPa)	Dynamic load factor, α	Dynamic load, αw_P (kPa)
Dancing	1.5–3.0	0.6 (2.5 m^2/couple)	0.5	0.3
Live concert or sports event	1.5–3.0	1.5 (0.5 m^2/person)	0.25	0.4
Jumping exercises				
-first harmonic	2.0–2.75	0.2	1.5	0.3
-second harmonic	4.0–5.5	0.2	0.6	0.12
-third harmonic	6.0–8.2	0.2 (3.5 m^2/person)	0.1	0.02

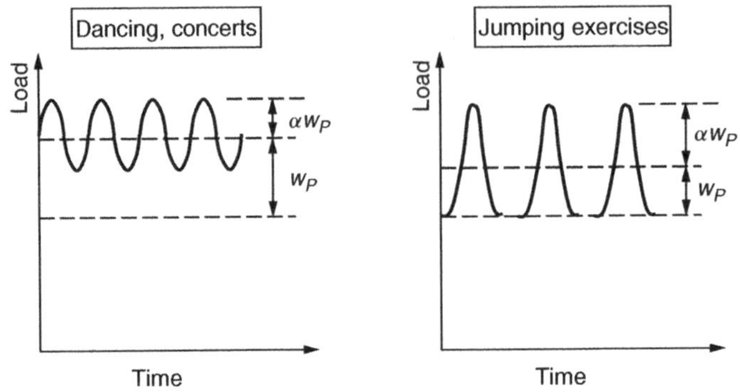

Figure 14.9 Load during rhythmic events (based on NRC, 1995)

human perception of and sensitivity to motion, and so should be minimised.

14.5 FLOOR RESPONSE

14.5.1 Plan Shape and Orthotropy

Most timber floors employ repetitive joists as either primary or secondary structural members. Joists are sheathed (overlain) by a wood or non-wood layer of structural material, the top of which is the floor surface (Figure 14.10). Whole floors with such construction, or floor bay systems, behave as rib-stiffened plates. The plate layer can be either isotropic or orthotropic depending upon the material used. Floors are often stiffened in the across-rib (across-joist) direction by addition of rows of bridging or blocking.

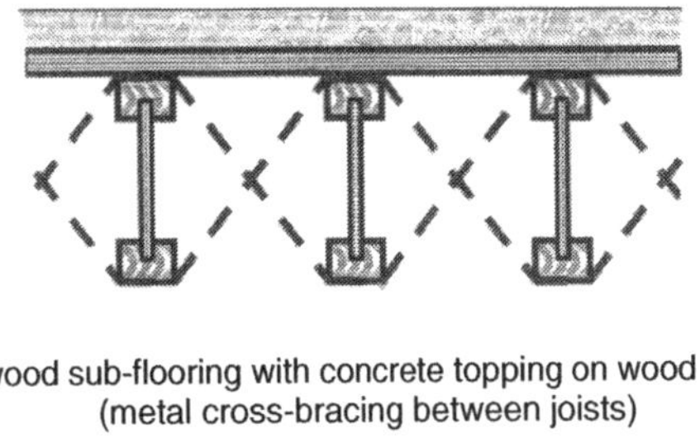

Figure 14.10 Typical joisted floor construction (schematic)

Despite the properties of the plate and the presence of bridging, joisted timber floors are inherently orthotropic. This promotes a tendency towards clustering of the first few modal frequencies, with the second modal frequency often being only 15 –20% higher than the first (Ohlsson, 1982, Smith and Chui, 1988a). Modal clustering has the effect of increasing the amplitude of motion, and thus velocity and acceleration levels experienced by a person or object located on a floor. Because modes are clustered, it is not generally reliable to base vibration assessments of timber floors only on the fundamental mode. The extent to which modes are clustered depends upon parameters such as floor shape, span and width, and flexural rigidity along and across joists. For rectangular plan floors, or floor bays, the first few modes of strongly orthotropic plates exhibit a common 'shape' (usually a single half sine-wave) along the joists, but have different shapes across the joists (Figures 14.11 and 14.12). For example, a bare floor with span 3.6 m, width 4.8 m, 50 × 200 mm softwood joists at 0.6 m on centre, and 22 mm thick chipboard flooring nailed to joists; the first few modal frequencies are $f_1 = 17.2$ Hz, $f_2 = 20.6$ Hz, $f_3 = 25.1$ Hz, $f_4 = 31.5$ Hz and $f_5 = 39.2$ Hz (Chui and Smith, 1988). In this case, the ends of joists were simply supported and edges of

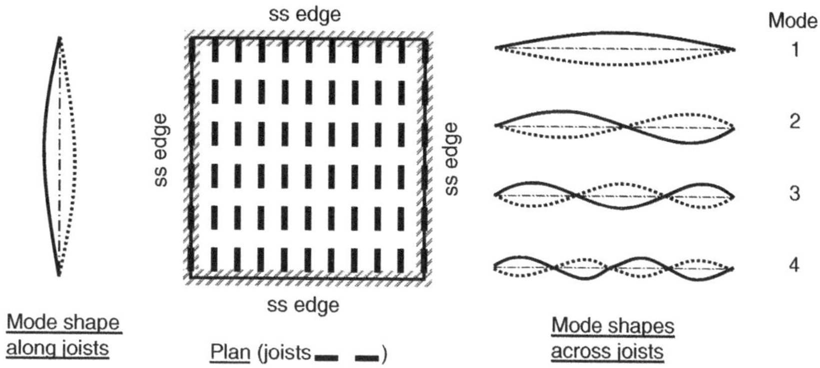

Figure 14.11 Typical mode shapes for a rectangular joisted floor simply supported (ss) on all edges

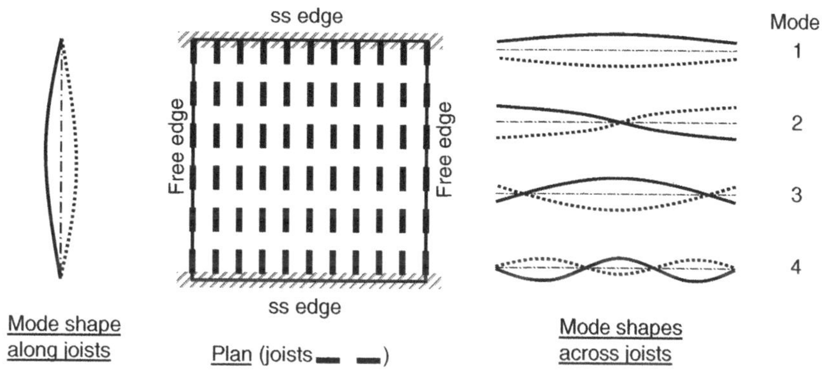

Figure 14.12 Typical mode shapes for a rectangular joisted floor simply supported on two edges

the floor free to move vertically. For isotropic plate floors modal frequencies are well separated, with $f_2 >> f_1$. For a square isotropic plate simply supported on all edges $f_2 \approx 2.5 f_1$ (Equation (14.19)). Any modifications that improve modal separation are inherently good with respect to vibration serviceability of timber floor systems.

14.5.2 Boundary Conditions

Practice varies regarding the number of supported edges on a floor. Two-edge (those containing joist ends), three-edge and four-edge support conditions are all quite common for rectangular joisted floors. Laboratory studies have shown that lumber joisted floors of moderate width can be significantly stiffened when all edges are simply supported, as opposed to just those edges containing joist ends. Although the fundamental natural frequency is very little affected, there is increase in other modal frequencies and static stiffness and reduction in vibration amplitude under impact load (Smith and Chui, 1988a). In light-frame construction joist end conditions approach simple supports. This can be true even for intermediate floors in platform construction in which floors are sandwiched between walls of adjacent storeys (Smith and Chui, 1988a, 1992). An explanation for this is wood's low modulus of elasticity perpendicular to grain, which means that both joists and top and bottom plates of sandwiching stud walls are too deformable to generate more than nominal end fixity (Chui and Smith, 1990).

Continuity of joists across an intermediate support is common in domestic 'all timber' or masonry-timber construction. The nature of the continuity will vary depending upon the building arrangement. Joists may be long enough to be continuous from one span to the next. This often involves adjacent spans of dissimilar length. Alternatively, ends of joists from adjacent spans can be overlap and be nailed together in that region, as illustrated in Figure 14.4. Most often, this will be done to provide integrity to the structure or generate a load path for resisting horizontal loads on buildings. Such connections are semi-rigid and provide some moment continuity. Partial or full moment continuity improves structural stiffness of floor systems. Field assessments of lumber joisted floors in domestic dwellings suggest that the benefit from improved stiffness is lost because of counteracting negative effects associated with transmission of vibrational movement from one span to another (Onysko, 1988), or to other levels in the structure.

14.5.3 Construction Details: Joisted Floors

Construction details have a strong influence on the dynamic behaviour of lightweight timber floors and the acceptability of particular floors to users of buildings. Floors with at least two layers of material in the plate layer (e.g. diagonal board sub-flooring with a hardwood strip overlay) have performed very well in domestic settings. By contrast, floors with a single layer of wood-based sheet material (plywood, chipboard, Oriented Strand Board) can have objectionable dynamic responses. This sensitivity to the make up of the plate layer is not reflected in engineering design practice, provided that individual components satisfy appropriate static strength and stiffness requirements. Thus, it is not enough to simply 'size' components and hope 'all will be well'. Extensive laboratory and *in situ* measurements have been made on full-size lightweight short to medium span floors with lumber joists. Studies typically have assessed how various construction parameters influence static stiffness and transient vibration response.

Reducing the spacing between joists usually has a beneficial influence on floor performance as it increases the stiffness in the along joist direction, and therefore raises the fundamental natural frequency (Smith and Chui, 1988a). However, there is need to exercise caution as there can be a concomitant reduction in the spacing of modal frequencies, and as already mentioned, that has a negative influence on vibration (Smith and Chui, 1988a). Between joists bridging (blocking or cross-bracing) in the interior of the span often has a very beneficial influence on the stiffness in the across joist direction (Figure 14.10) (Smith and Chui, 1988a, Onysko, 1988). Even more effective is bridging combined with tension resistant strapping (thin timber or light gauge steel) tying the arrangement together (Smith and Chui, 1992). Normally the bridging, and any strapping, is fixed to joists with nails. Glued blocking is much more effective than nailed blocking (Smith and Chui, 1992). The bridging stiffens the system in the across joist direction and assists the floor layer to distribute the effect of static concentrated load between joists. If there is significant reduction in the moisture contents of timber components following construction, nailed connections loosen, and bridging can become ineffective (Onysko, 1988). This happens, for example, due to drying of joists installed wetter than the equilibrium moisture condition for the service environment. A recent study by Hu and Tardif (2000) suggests the benefits from applying bridging type components to floors constructed with wood I or open-web joists are less than for floors with lumber joists, but such findings are not conclusive.

There should be at least one row of bridging at mid span of a joisted floor. More rows equally spaced along the span usually leads to better performance under either static or dynamic loads (Chui and Smith, 1988, Onysko, 1988, Smith and Chui, 1992). However, there is minimal benefit from placing rows of bridging closer than about 2 m on centre. When the joists are tall and slender and have low torsional rigidity, bridging and strapping is necessary to prevent torsional movement in joists. If open-web or tall wood I-joists are used, both torsional and warping deformations of the cross-section need to be counteracted by the presence of bridging. Plasterboard ceilings attached directly to the undersides of joists using screws can greatly reduce static deflection under concentrated load (Onysko, 1988). This does not automatically mean that dynamic behaviour is improved, because the mass increases as well as the stiffness and modal frequencies can actually reduce (Chui and Smith, 1988). Onysko (1988) suggested that the flexural stiffness of plasterboard ceilings in the across joists direction is additive to that of the flooring. Most benefit results if a plasterboard ceiling is combined with bridging to stiffen the across-joist behaviour.

Usually there is connectivity between layers in a floor system, i.e. between any wood finishing and the sub-flooring, and between the sub-flooring and the joists. Achieving significant improvement in along joist stiffness depends upon effectiveness of such connections. Sub-flooring is usually connected to timber joists with nails or wood screws placed at between 100 and 200 mm on centre, or a combination of elastomeric adhesive and nailing. Elastomeric adhesives make semi-rigid joints, and from a mechanics point of view the stiffening effect on a connection is additive to stiffness provided by nails or screws. Although composite action is improved, the primary benefit from incorporating elastomeric adhesive is reduced frictional noise (squeaking) generated at interfaces during dynamic movements in floors. Parquet flooring glued to plywood sub-flooring has been found to significantly stiffen the plate layer, with a large reduction in local deflections due to flexing of flooring under concentrated loads between joists. Reductions of more than 20% in static deflection of whole floor systems have been reported when parquet flooring is applied over wood-based sub-flooring (Onysko, 1988). When one or two layers of panel type overlay (e.g. plywood, particleboard) are placed over sub-flooring to create two or three layer flooring, there is no significant composite action between layers in the along joist direction unless each is nailed directly through to the joists. If this is not done and layers are nailed together without regard for the joist locations, the floor stiffness is essentially the sum of flexural rigidity for each layer.

Contemporary practice for many multiple occupancy residential or commercial buildings is to apply a 'normal weight' or lightweight concrete topping (usually 38 mm thick), or gypsum-based topping (usually 19 mm) as an overlay to medium-span floors made with wood I-joists. Such toppings are to improve fire performance, and can incorporate radiant heating systems. They are not reinforced and no special measures are employed to ensure adhesion to wood-based sub-flooring. Available evidence suggests that toppings can significantly enhance static stiffness and modal damping, but reduce modal frequencies as the increased mass more than offsets any gain in stiffness (Hu *et al.*, 1998, Taylor and Hua, 2000). Empirical field and laboratory observations indicate that drying shrinkage can cause concrete toppings to curl at the edges, leading to the flooring and topping layers being unconnected over significant portions of the floor (Hu *et al.*, 1998, Taylor and Hua, 2000). The same problem is not reported to occur with gypsum-based toppings (Taylor and Hua, 2000). It seems unwise to presume composite action involving the topping layer until improved topping materials and better bonding techniques are available. User satisfaction ratings tend to be unaltered or diminish when a floor is modified by application of a topping layer (Taylor and Hua, 2000). Such assessments are subjective, and therefore hard to interpret. They reflect a complex interplay of factors influencing respondents. Although toppings stiffen the plate layer against between joist deflection under concentrated loads and thus reduce springiness, this is offset by increased vibrational disturbance from movements of people other than the observer, i.e. overall motion of the floor becomes a greater factor in ratings.

14.5.4 Damping

Damping is the term commonly used to refer to the energy loss during motion of a system. Energy loss occurs for a number of reasons. Friction between molecules within components is termed material damping and is always present during vibration. Other sources of damping are friction and impact between components of the system, sloshing of liquid, and inertial effects associated with arresting motion of the structure and any imposed objects. The sources of damping can differ between modes of the same structure, so it is quite feasible for damping to differ appreciably between modes (Chui and Smith, 1989). As already illustrated, high modal damping is desirable as it limits the amplitude of movements during forced vibration and reduces the decay time under free vibration. “Some damping values found in the literature are based on free vibration tests and evaluation of logarithmic decrement from multi-frequency decaying vibration. This methodology is not in agreement with the theory of dynamics and usually results in severe overestimation of the floor damping capacity. This is due to the fact that even for a case with equal damping ratio for all modes of vibration, the higher mode components will be damped out more quickly” (Ohlsson, 1988).

End conditions of bending members can greatly affect damp. Chui and Smith (1989) showed that for lumber beams with free-free ends or simply supported ends the viscous damping ratio associated with the fundamental mode is about 1%, which corresponds approximately to the material damping. When the same members had ‘clamped’ ends the viscous damping ratio increased to about 8%. For lumber joisted floors effective viscous damping ratios are in the range of 1–3% depending upon the details of the floor and the mode being considered (Chui, 1987, Ohlsson, 1982). Application of imposed masses (objects) on the surface of a floor can greatly increase the damping, especially if the system is lightweight or small (Chui and Smith, 1988). This reflects inter-component friction and impact, and possibly sloshing, effects. Presence of humans or animals can also significantly increase system damping for such floors. Whether any account should be taken of human presence when assessing performance of a system, depends upon whether the observer is the source of any ‘extra’ damping and upon the criteria employed. As pointed out by Ohlsson (1988), it is, however, not reasonable to count on this extra damping in design calculations, as this damping is nothing else than a sign of vibration absorption in the human body. Humans, except for crowds, will usually not alter either mass or damping of heavyweight floors appreciably.

Because of inherent low self-weight, timber floors often do not exhibit a significant amount of inertial damping. Exceptions can occur with large systems, especially if they have thick concrete topping or support large amounts of imposed mass.

14.6 AVOIDANCE OF VIBRATION PROBLEMS

As succinctly stated by Murray (1988)

> “Although it has been shown that the various criteria differ substantially in the evaluation of a particular floor system, they are useful tools for designers. ..... Control of any of the three fundamental motion parameters, damping, frequency and amplitude (or acceleration) can improve the performance of a floor system. Even a very light flexible floor system will be acceptable if sufficient damping exists. ..... In office/residential environments, the frequency of the floor system must be kept above 3 hertz to avoid resonance with walking. In dancing/exercising environments, it has been suggested (Allen *et al.*, 1985) that the frequency be above 8–9 hertz again to avoid resonance. Further, floor systems having a natural frequency in the 6–8 hertz, without adequate damping, seem to be particularly annoying (possibly because of resonance with body organs). ..... Although amplitude (or acceleration) is an important floor motion parameter, control by modifying characteristics of the floor system is difficult without also changing frequency.”

From a design perspective, frequency tuning is the most controllable and simplest measure for avoiding excessive vibration. A system’s response has to be known in terms of how the modal frequencies relate to the most critical frequencies of the loading function. The objective is to avoid coincidence between loading and response frequencies (see Section 14.2). Most often, the fundamental modal frequency is of greatest concern. Criteria suggested for frequency tuning are (Bachmann, 1988):

- Pedestrian structures: avoidance of the first and second harmonics of the dynamic walking load.

- Office buildings: high-frequency tuning with respect to the third harmonic of dynamic walking loads.
- Gymnasia and sports halls: high-frequency tuning with respect to the second harmonic of dynamic jumping loads.
- Dance and concert halls without fixed seating: high-frequency tuning with respect to the second harmonic of dynamic dancing loads.
- Dance and concert halls with fixed seating: high-frequency tuning with respect to the first harmonic due to dynamic loads produced by people clapping. For lightly damped structures, the second harmonic of dynamic clapping loads should be avoided (or use high-frequency tuning).

These criteria lead to the recommendations in Table 14.4. Their adoption should lead to satisfactory dynamic performance. Although timber structures are not included in the tabulation, it can be seen that there is no great sensitivity to the structural type. Use of recommendations for steel structures will usually be most appropriate. Frequency tuning of timber floors invariably will mean increasing the stiffness to 'improve' the stiffness-to-mass ratio thus raising the lowest few structural frequencies (NRC, 1995, Smith and Chui, 1988a).

Other means of avoiding 'problem' floors are obviously available. A potentially effective approach is to increase the damping (e.g. add artificial damping or tuned mass dampers), but this is normally a complex and/or expensive solution. Attention to details such as provision of adequate blocking or cross-bracing is usually the simplest method of improving floor performance (Smith and Chui, 1992). Design criteria of varying complexity have been proposed to avoid vibration problems in timber floors, but that is outside the scope of this chapter. Readers interested in the status of anti-vibration design methods should refer to recent publications by the Applied Technology Council (1999) and Hu *et al.* (2001).

14.7 PREDICTION OF FLOOR VIBRATION

Engineers typically approximate vibration behaviour of timber floors based on the behaviour of one-dimensional constituent components (e.g. from behaviour of isolated joists). This is, however, not always reliable, as floors are by nature multi-dimensional arrangements. Accurate prediction of their behaviour depends upon stiffness and mass characteristics in at least the two-dimensional plan space. Presumption of a linear-elastic response is reliable in most vibrational serviceability calculations. Any system has an infinite number of modes, but usually only the first several are of concern as they consume the bulk of the energy imparted to a floor by dynamic loads. Modal analysis is commonly used to interpret the behaviour of systems, with the nth mode being characterised by its eigenfrequency/resonance frequency (f_n), modal damping (ζ_n), mode shape ($(\phi_n(x, y)_n$), and modal mass (m_n). Modelling techniques discussed below are capable of predicting modal properties, and time-histories of responses in some cases.

Table 14.4 Recommendations for modal frequencies (Hz) of structures subject to human induced vibration (Bachmann, 1988)

Structural class	Reinforced concrete	Prestressed concrete	Composite steel-concrete	Steel
Pedestrian structures	Avoidance of 1.6–2.4 Hz and 3.5–4.5 Hz.			
Office buildings	>7.5	>8.0	>8.5	>9.0
Gymnasia and sports halls	>7.5	>8.0	>8.5	>9.0
Dance halls and concert halls without fixed seats	>6.5	>7.0	>7.5	>8.0
Dance halls and concert halls with fixed seats	>3.4			

Modal damping is usually assumed to be viscous and is commonly given as a damping ratio (see Section 14.5.4) and estimated from test data for full size floors (typical range $\zeta_n = 0.01-0.03$). Relative modal damping ratio $= c/c_{cr}$, where c is the modal absolute damping and c_{cr} its critical value. In the absence of better information, modal damping can conservatively be taken equal to material damping. Some of the models discussed below are capable of predicting mode shapes. Modal mass is the 'mode shape-weighted' sum of the masses of different parts of a system. Treating a floor as a two-dimensional plate:

$$m_n = \int_0^L \int_0^B g(x, y)\phi(x, y)_n^2 dx\, dy \qquad (14.14)$$

where $g(x, y)$ is the mass per square metre. If the mass is uniform across a floor and it is simply supported along all edges, the modal mass equals one quarter of the floor mass.

As discussed in Sections 14.3 and 14.4, assessment of floors often requires estimation of velocity and acceleration parameters. Models based on numerical methods can accurately predict time-history responses from which velocity and acceleration information can be extracted. Analytical models only allow certain parameters to be estimated for relatively simple systems.

14.7.1 Correlation Studies

Onysko and Bellosillo (1978) make a distinction between different types of correlation studies. Acceptability-based studies are those where general acceptability of a floor is related to a parameter(s) defining its engineering performance. Perception-based studies are those where one aspect of floor performance (e.g. whole body motion) is related to acceptability. Field surveys are the most direct way of finding out whether or not design methods result in products (e.g. cars, buildings) that attain a level of performance acceptable to users.

> "While it is relatively straightforward for a product that is of uniform consistency and to which consumers have equal access, it is not as simple for assessing floor performance where each consumer experiences a different quality of performance and where performance may be different within the same residence. Further, the consumer's reaction to his or her floors is affected by expectations, lifestyle, the placement of furnishings, the demands put on the structure, conditioning, and other factors. None the less, a survey approach can be an effective way to obtain consumer acceptability of floor performance versus studies based solely on human perception of vibration" (Onysko, 1988).

With this philosophy in mind, extensive surveys have been undertaken as a basis for empirical correlations between physical characteristics of timber floor systems and the level of user satisfaction as expressed by occupants of buildings. Figure 14.13 illustrates the principle underlying correlation studies.

Onysko (1982) reported findings from acceptability-based studies on single-family dwellings having lumber joisted floors. Inspections were made in 107 houses to provide data and opinions on 646 floors. There were also informal inspections and interviews for another 186 floors. Data was used to establish a simple design formula with which to discriminate between acceptable and unacceptable floors. Although dynamic response parameters were considered, the selected relationship is based on static deflection under a concentrated load:

$$y = \frac{7.217}{L^{1.274}} \quad | y \leq y_{limit} \qquad (14.15)$$

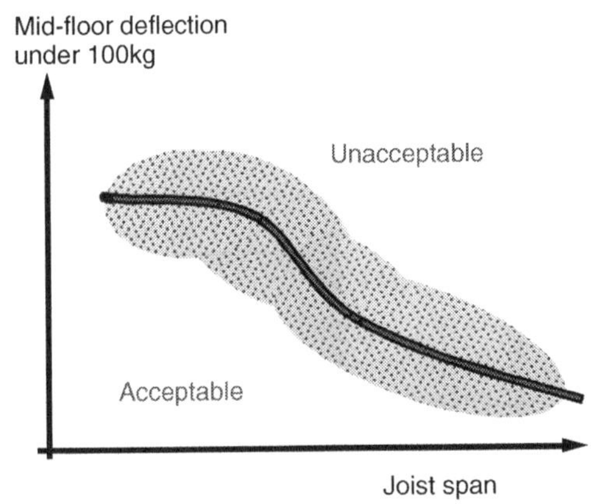

Figure 14.13 Principle of correlation studies (acceptability-based studies)

where y = maximum allowable deflection for the floor system (mm) under a concentrated mass load of 100 kg, L = clear joist span (m), and y_{limit} = upper limit on y. If actual deflection for a floor exceeds y, it is predicted to have unacceptable performance. The formula in Equation (14.15) has subsequently been revised, but its essence is unchanged (Onysko, 1988, Onysko *et al.*, 2000). Russell (1954) and Hansen (1960) report similar, but less extensive, survey type acceptability studies. Perception-based studies have been carried out by a number of researchers to determine user perception and/or rating of floor vibration with parameters such as the type of loading being varied (e.g. Lenzen, 1966, Polensek, 1970). Such studies should be conducted in a manner that makes observers blind to the application, but that was often not the case. Strictly speaking, there should be no such thing as perception-based assessments of floors, i.e. focus needs to be purely on perception of motion, however it is produced (see Section 14.3). Failure to observe this requirement leads to contamination of data as perceptions can include the effects of audio or visual cues, as well as motion.

As with any other survey, it is important that the sample be adequate and representative of the broader population, and that survey questions be properly structured to avoid bias. The primary weakness of correlations is that any resulting relationship is specific to the survey conditions. This can be illustrate in the context of the effect of floor aspect ratio on acceptability ratings, where aspect ratio is the ratio of floor span to floor width. Onysko's design rule for lumber joisted floors (Equation (14.15)) implies that for rectangular floors supported on all edges vibration performance improves the smaller the width of the floor. This is because reducing the width reduces the peak displacement under a centrally located concentrated static load. However, studies by various researchers demonstrate that increasing the width of a floor has beneficial influence on peak acceleration or peak velocity under impulsive loads. Floors of large width are more tolerable than floors of small width, other factors being equal (e.g. Ohlsson, 1982, Smith and Chui, 1988a). The implication is that the 'Onysko rule' and others are only applicable to aspect ratios encompassed by the survey data on which they are based.

14.7.2 Analytical Predictions (Some Useful Formulae)

Modal Frequencies

The well-known Rayleigh method can be used to estimate the fundamental natural frequency, f_o, of an unloaded rectangular joisted floor covered with semi-rigidly attached wood-based flooring. Assuming ends of equally spaced joists are simply supported and a mass of M_o is located at the centre of the floor (Chui, 1987, Smith and Chui, 1988a):

$$f_0 = f_1 = \frac{\pi}{2L^2}\sqrt{\frac{EI(n-1)}{\rho_S hB + \rho_J A(n_J - 1) + \dfrac{4M_0}{L}}} \tag{14.16}$$

where A = cross-sectional area per joist, B = floor width, EI = flexural rigidity per joist, h = flooring (sheathing) thickness, L = floor span, n_J = number of joists, ρ_J = joist density, and ρ_S = flooring density. The purpose of the mass M_o is to simulate an observer standing on the floor. The above equation neglects strain energy in the flooring, and any composite action between the joists and flooring mobilised by connecting them together. In light-frame construction, flooring is commonly attached to joists using nails at spacings too large to lead to significant composite action. The formula is quite reliable for floors with plywood or Oriented Strand Board flooring and lumber joists placed at up to 600 mm on centre. This is true irrespective of whether or not edges parallel to the joists are supported (Smith and Chui, 1988a). If there is full composite action, as when components are glued together with rigid adhesive, the effective *EI* of the section can be calculated by normal methods (transformed section approach). The stiffening effect of partial composite action in the along joist direction can be predicted using built-up beam theory (Smith, 1980, Smith and Chui, 1988b). Sensitivity of floor response to rigidity of flooring to joist connections diminishes as the size and complexity of systems is increased. It is estimated that modelling the stiffness of mechanical connections in floors and other timber systems to within ±20% is adequate for serviceability analysis (Smith and Chui, 1988b). The f_o for a floor with a uniform live load can be calculated from Equation (14.16) by adding the corresponding mass to the product $\rho_S hB$.

If a floor has flexible supports (e.g. supported on flexible girders), the fundamental natural frequency of the system, f_{system}, can be determined approximately from the relationship:

$$\frac{1}{f_{system}^2} = \frac{1}{f_{floor}^2} + \frac{1}{f_{girder}^2} \tag{14.17}$$

where f_{floor} = fundamental frequency for the isolated floor, and f_{girder} = fundamental frequency for the isolated girder supports calculated by including the mass (but not stiffness) imparted by the floor. An alternative approximate formula for predicting f_o of floor panels simply-supported on girders that themselves are supported by columns is (NRC, 1995):

$$f_0 = \frac{1}{2\pi}\sqrt{\frac{g}{\Delta}} \tag{14.18}$$

where $\Delta = 0.77(\Delta_B + \Delta_G) + \Delta_S$, with Δ_B = elastic deflection of floor panel due to bending and shear, Δ_G = elastic deflection of the supporting girders due to bending and shear, Δ_S = elastic shortening of the supporting columns, and g = gravitational constant = 9.81 m/s^2. For cantilevers and two-way floors, the factor 0.77 is replaced by 0.65. In the case of floor panels and girders that are continuous over supports, Δ_B and Δ_G should be determined assuming adjacent spans deflect in opposite directions, and that the weight supported by each span acts in the direction of deflection.

For a rectangular orthotropic plate simply supported along all edges, the nth modal frequency is given by (Ohlsson 1988):

$$f_n = \frac{\pi}{2}\sqrt{\frac{D_x}{mL^2}}\sqrt{1 + \left[2n^2\left(\frac{L}{B}\right)^2 + n^4\left(\frac{L}{B}\right)^4\right]\frac{D_y}{D_x}} \tag{14.19}$$

where D_x, D_y = plate flexural rigidity for bending in the x and y directions ($D_x > D_y$), and m = mass per unit floor area. For joisted floors, assignment of a value to D_x is relatively straightforward, and in the simplest cases it can be taken equal to the flexural rigidity of an isolated joist divided by the joist spacing. However, it is often appropriate to also take account of the plate layer (flooring) and any composite action between the joist and the plate. Also, assigning a value to D_y is not a simple matter, and allowance should be made for any stiffening in the across the joist direction due to influences such as bridging or cross-bracing, or torsional restraint of joists (Chui, 1987, Hu, 1992, Hu and Tardif, 2000). A simple approach is to use the flexural rigidity of the plate layer alone, but this can be highly inaccurate (Smith and Hu, 1993). Predictions of f_n (n $\geqslant$ 2) are highly sensitive to the ratios D_x/D_y and L/B.

The above-mentioned frequency equations are not accurate if shear deformation or rotary inertia in joists is a significant influence. This can be the case when relatively deep wood I-joists or open web joists are employed (Hu, 1992). Comparison with test data and predictions from more exact models indicates that Equation (14.19) can be quite inaccurate for either lightweight or heavyweight floors with I-joists (Smith and Hu, 1993).

Acceleration

Peak acceleration A_o of a simply supported beam with a partial uniformly distributed harmonic load (Figure 14.14) is approximately (Clough and Penzien, 1975):

$$A_0 = \frac{2P_0\beta_1^2\rho}{\pi m}\sin\left(\frac{\pi x}{L}\right)\left[\cos\frac{\pi c}{L} - \cos\frac{\pi d}{L}\right] \tag{14.20}$$

where P_o = amplitude of the harmonic load, ρ = dynamic magnification factor for the fundamental mode (Equation (14.10)), and m = mass per unit length ($\beta_1 = f/f_o$ as before). The expression has been used in an 'equivalent T-beam analysis' to evaluate the vibrational behaviour of heavy one-way rib stiffened plate floors subjected to harmonic dance loads (Matthews and Montgomery, 1988).

An explicit prediction of the damped frequency weighted rms-acceleration, A_r (m/s^2), due to a square impulsive load (heel drop impact) can be

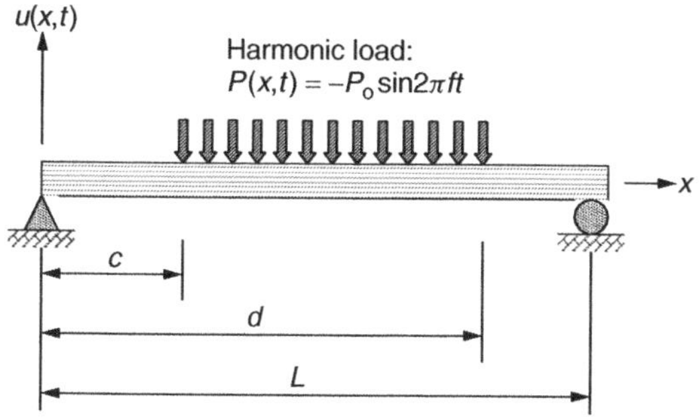

Figure 14.14 Simply supported beam with a partial uniformly distributed harmonic load

made based on the well-known Duhamel integral if certain assumptions are made (Chui and Smith, 1988). Assuming the response is dominated by the fundamental mode (SDF system), the impact is applied at the centre of a rectangular joisted floor, and that the total duration of a heel-drop impact response is 1 second:

$$A_r = \frac{2000K}{m\pi(f_0')^2} \tag{14.21}$$

where

$$m = \frac{L}{2}\left(\rho_S hB + \rho_J A(n-1) + \frac{4M_0}{L}\right)$$

$$K = \sqrt{\frac{\omega_n\left[\frac{1-e^{-2\zeta\omega_n t_1}+(1-e^{-2\zeta\omega_n(1-t_1)})}{(1-2e^{-\zeta\omega_n t_1}\cos\omega_n t_1+e^{-2\zeta\omega_n t_1})}\right]}{4\zeta}}$$

and m = modal mass, f_o' = fundamental natural frequency of the floor with an observer of mass M_o (kg) located at the centre (equation (14.16)), $\omega_n = 2\pi f_o'$, and t_1 = duration of the impulsive force. Other parameters have the same definitions as before. Experiments indicate that normally $0.05\ \text{s} \le t_1 \le 0.07\ \text{s}$ and $\zeta \approx 0.03$.

The above formulae are obviously only a selection of those available, but will give an idea of the type of predictions that can be made with minimal computational effort. For problems that violate assumptions of analytical solutions, the best option is recourse to numerical methods.

14.7.3 Numerical Predictions

Many commercial software packages are available for dynamic analysis of structural systems. These almost all employ the Finite Element Method (FEM), and are based on formulation of appropriate stiffness, mass and damping matrices in the case of free damped vibration analysis, plus a force vector in the case of forced vibration analysis. The governing equations are multiple degree of freedom forms of Equations (14.3) and (14.8), and are solved numerically. FEM models can be used for any problem, being particularly suitable for non-standard problems, such as those where there is an unusual arrangement (e.g. non-rectangular plan, large openings, complex supports) or three-dimensional analysis is necessary (e.g. floors or bridges with sloping surfaces). As is well known, the problem with FEM models is that they can be expensive in terms of time required to discretise the system and interpret output. Although, dynamic analysis tended to be less sensitive than static analysis to fineness of the discretisation, the number of necessary equations can be quite large. An alternative to FEM models is modal synthesis, that works in terms of modal coordinates rather than displacements. As a consequence, it can have significant computational advantages. It has been used to predict the vibration behaviour of rectangular timber floors with simply supported edges, with or without an interior line support (Hu, 1992, Hu *et al.*, 1994, Smith and Hu, 1993, Smith *et al.*, 1993). Analysis has included effects of semi-rigid attachment of the plate (flooring) to the ribs (joists), and shear and rotatory inertia in the ribs. At present, there is no commercial implementation of the software. Verification studies confirm accuracy of numerical models, provide that appropriate allowance is made for physical features of a floor system (e.g. Filiatrault *et al.*, 1990, Hu *et al.*, 1994, Morley and Murray, 1993, Smith and Hu, 1993, Smith *et al.*, 1993).

14.8 CONCLUDING REMARKS

Dynamic behaviour of timber floors and similar systems is a complex issue, with many factors contributing to their excitation, response and acceptability of resultant vibrations. Normally vibrations in floors are a serviceability issue rather than a matter of safety. Focus in research has been, and will likely remain, on behaviour of floors under normal use conditions. Avoiding unacceptable motion is quite straightforward provided some relatively simple 'rules' are observed. Coincidence of forcing and response frequencies causes resonant vibration and avoidance of that, or near resonant vibration, usually precludes serious problems. The choice of appropriate structural arrangements and detailing is also important in order to achieve good results. Various analytical tools and guidance on appropriate detailing are available in the literature. The main reason that problem floors exist is that the knowledge has not been effectively transferred to practice.

REFERENCES

Alisjahbana S.W. (2000) Response of large space building floors to dynamic load which suddenly moves to its new position. *Proceedings 7th International Symposium on Structural Failure and Plasticity*, Pergamon Press, Oxford, UK, pp. 865–870.

Allen D.E. and Rainer J.H. (1976) Vibration criteria for long-span floors. *Canadian J. Civil Engineering*, **3**(2), 165–173.

Allen D.E., Rainer J.H. and Pernica G. (1985) Vibration criteria for assembly occupancies. *Canadian J. Civil Engineering*, **12**(3), 617–623.

Allen D.E., Rainer J.H. and Pernica G. (1987) Building vibrations due to human activities. Structures Congress of ASCE, Orlando, FL, USA.

ATC (1999) Minimizing floor vibration. Design Guide 1, Applied Technology Council, Redwood, CA, USA.

Bachmann H. (1988) Practical cases of structures with man-induced vibrations. *Proceedings Symposium/Workshop on Serviceability of Buildings*, National Research Council, Ottawa, ON, Canada, Vol. I, pp. 419–434.

Bachmann H. and Ammann W. (1987) Vibrations in structures induced by man and machines. Structural Engineering Document No. 3e, International Association for Bridge and Structural Engineering, Zurich, Switzerland.

Brown C.B. and Yao J.T.P. (1983) Fuzzy sets and structural engineering. *ASCE J. Structural Engineering*, **109**(5), pp. 1211–1225.

BSI (1984) Evaluation of human exposure to vibration in buildings (1 Hz to 80 Hz). British Standard BS6472, British Standards Institute, London, UK.

Chen P.E. and Robertson L.E. (1972) Human perception thresholds of horizontal acceleration. *ASCE J. Structural Engineering*, **98**(ST8), 1681–1695.

Chui Y.H. (1987) Vibration performance of wooden floors in domestic dwellings. PhD thesis, University of Brighton, Brighton, UK.

Chui Y.H. and Smith I. (1988) A serviceability criterion to avoid human discomfort for light-weight wooden floors. *Proceedings Symposium/Workshop on Serviceability of Buildings*, National Research Council, Ottawa, ON, Canada, Vol. I, pp. 512–525.

Chui Y.H. and Smith I. (1989) Quantifying damping in structural timber components. *Proceedings Pacific Timber Engineering Conference*, University of Auckland, New Zealand, Vol. 1, pp. 57–60.

Chui Y.H. and Smith I. (1990) Influence of rotatory inertia, shear deformation and support condition on natural frequencies of wooden beams. *Wood Science and Technology*, **24**, 233–245.

Clough R.W. and Penzien J. (1975) *Dynamics of Structures*. McGraw-Hill, New York, NY.

Ebrahimpour A. and Sack R.L. (1988) Group-induced dynamic loads. *Proceedings Symposium/Workshop on Serviceability of Buildings*, National Research Council, Ottawa, ON, Canada, Vol. I, pp. 451–464.

Eriksson P.-E. and Ohlsson S.V. (1988) Dynamic footfall loading from groups of walking people. *Proceedings Symposium/Workshop on Serviceability of Buildings*, National Research Council, Ottawa, ON, Canada, Vol. I, pp. 497–511.

Farah A. (1977) Human response: A criterion for assessment of structural serviceability. PhD thesis, University of Warterloo, Waterloo, ON, Canada.

Filiatrault A., Folz B. and Foschi R.O. (1990) Finite-strip free vibration analysis of wood floors. *ASCE J. Structural Engineering*, **116**(8), 2127–2142.

Foschi R.O. and Gupta A. (1987) Reliability of floors under impact vibration. *Canadian J. Civil Engineering*, **14**(5), 683–689.

Galbraith F.W. and Barton M.V. (1970) Ground loading from footsteps. *J. Acoustical Society of America*, **48**(2), 1268–1292.

Greimann L.F. and Klaiber F.W. (1978) Dynamic forces induced by spectators. *ASCE J. Structural Engineering*, **104**(ST2), 348–351.

Hansen H. (1960) Deflection characteristics of wood-joist floors. Translation NRC-TT883, National Research Council, Ottawa, ON, Canada.

Hansen B.J., Reed J.W. and Vanmarcke E.H. (1973) Human response to wind-induced motion of buildings. *ASCE J. Structural Engineering*, **99**(ST6), 1259–1272.

Holicky M. (1988) Fuzzy concept of serviceability limit states. *Proceedings Symposium/Workshop on Serviceability of Buildings*, National Research Council, Ottawa, ON, Canada, Vol. I, pp. 19–31.

Holicky M. and Ostlund L. (1993) Probabilistic design concept. *Proceedings International Colloquium on Structural Serviceability of Buildings*, International Association for Bridge and Structural Engineering, Zurich, Switzerland, Reports Vol. 69, pp. 91–98.

Hu L.-J. (1992) Prediction of vibration responses of ribbed plates by modal synthesis. PhD thesis, University of New Brunswick, Fredericton, NB, Canada.

Hu L.-J. and Tardif Y. (2000) Effectiveness of strong-back/wood I-blocking for improving vibrational performance of engineered wood floors. Paper 4.6.4, *Proceedings World Conference on Timber Engineering*, University of British Columbia, Vancouver, BC, Canada.

Hu L.-J., Chui Y.H. and Smith I. (1998) Serviceability of wood floor systems with plain concrete topping. *Proceedings World Conference on Timber Engineering*, Swiss Federal Institute of Technology, Lausanne, Switzerland, pp. 2.750–2.751.

Hu L.-J., Chui Y.H. and Onysko D. (2001) Vibration serviceability of timber floors in residential construction. *Progress in Structural Engineering and Materials*, **3**(3), 228–237.

Hu L.-J., Smith I. and Chui Y.H. (1994) Vibration analysis of ribbed plates with a rigid intermediate line support. *J. Sound and Vibration*, **178**(2), 163–175.

Irwin P.A., Ferraro V. and Stone G.K. (1988) Wind induced motion of buildings. *Proceedings Symposium/Workshop on Serviceability of Buildings*, National Research Council, Ottawa, ON, Canada, Vol. I, pp. 200–213.

IS0 (1989) *Evaluation of human exposure to whole-body vibration – Part 2: Human exposure to continuous and shock-induced vibrations in buildings (1 to 80 Hz)*. ISO Standard 2631-2, International Organization for Standardization, Geneva, Switzerland.

Kanda J., Tamura Y. and Fujii K. (1988) Probabilistic criteria for human perception of low-frequency horizontal motions. *Proceedings Symposium/Workshop on Serviceability of Buildings*, National Research Council, Ottawa, ON, Canada, Vol. I, pp. 260–269.

Lenzen K.H. (1966) Vibration of steel-joist-concrete slab floors. *AISC J. Engineering*, **3**(3), 133–136.

Matthews C.M. and Montgomery C.J. (1988) A dance floor with unsatisfactory dynamic response. *Proceedings Symposium/Workshop on Serviceability of Buildings*, National Research Council, Ottawa, ON, Canada, Vol. I, pp. 533–546.

Morley L.J. and Murray T.M. (1993) Predicting floor response due to human activity. *Proceedings International Colloquium on Structural Serviceability of Buildings*, International Association for Bridge and Structural Engineering, Zurich, Switzerland, Reports Vol. 69, pp. 297–302.

Murray T.M. (1979) Acceptability criteria for occupant-induced floor vibrations. *J. Sound and Vibration*, November, 24–30.

Murray T.M. (1988) Practical aspects of floor serviceability design. *Proceedings Symposium/Workshop on Serviceability of Buildings*, National Research Council, Ottawa, ON, Canada, Vol. I, pp. 495–496.

Nawar G. (1988) Performance criteria for small buildings. *Proceedings Symposium/Workshop on Serviceability of Buildings*, National Research Council, Ottawa, ON, Canada, Vol. I, pp. 11–18.

NRC (1995) Commentary A: Serviceability criteria for deflections and vibrations. In: *User's guide – National Building Code of Canada 1995 Structural Commentaries (part 4)*, National Research Council, Ottawa, ON, Canada, pp. 1–7.

Ohlsson S. (1982) Floor vibrations and human discomfort. PhD thesis, Chalmers University of Technology, Gothenburg, Sweden.

Ohlsson S. (1984) Svikt, søngningar & styvhet hos bjølkag – dimensionerings-metoder (Springiness, vibrations and stiffness properties of floors – design methods). Publication T20:1984, The Swedish Council for Building Research, Stockholm, Sweden (*English translation 1987*).

Ohlsson S. (1988) Ten years of floor vibration research: A review of aspects and some results. *Proceedings Symposium/Workshop on Serviceability of Buildings*, National Research Council, Ottawa, ON, Canada, Vol. I, pp. 435–450.

Onysko D.M. (1982) Serviceability criteria for residential floors based on field study of consumer response. Project Report 03-50-10-008, Forintek Canada Corp., Ottawa, ON, Canada.

Onysko D.M. (1988) Performance and acceptability of wood floors – Forintek studies. *Proceedings Symposium/Workshop on Serviceability of Buildings*, National Research Council, Ottawa, ON, Canada, Vol. I, pp. 477–494.

Onysko D.M. and Bellosillo S.B. (1978) Performance criteria for residential floors. Final Report to Central Mortgage and Housing Corporation on Grant No. 120-74, Eastern Forest Products Laboratory, Ottawa, ON, Canada.

Onysko D.M., Hu L.-J., Jones E.D. and Di Lenardo B. (2000) Serviceability design of residential wood frame floors in Canada. Paper 4.6.1, *Proceedings World Conference on Timber Engineering*, University of British Columbia, Vancouver, BC, Canada.

Pernica G. (1983) Dynamic live loads at a rock concert. *Canadian J. Civil Engineering*, **10**(2), 185–191.

Polensek A. (1970) Human response to vibration of wood-joist floor systems. *Wood Science*, **3**(2), 111–119.

Rainer J.H., Allen D.E. and Pernica G. (1988) Dynamic response of floors due to people in motion. *Proceedings Symposium/Workshop on Serviceability of Buildings*, National Research Council, Ottawa, ON, Canada, Vol. I, pp. 566–572.

Russell W.A. (1954) Deflection characteristics of residential wood-joist floor systems. Housing Research Paper 30, US Housing and Home Finance Agency, Washington DC.

Smith I. (1980) Series type solutions for built-up beams with semi-rigid connections. *Institution of Civil Engineers Proceedings: Part 2*, September, 707–719.

Smith I. and Chui Y.H. (1988a) Design of light-weight wooden floors to avoid human discomfort. *Canadian J. Civil Engineering*, **15**, 254–262.

Smith I. and Chui Y.H. (1988b) Serviceability calculations for timber building components with mechanical connections. *Proceedings Symposium/Workshop on Serviceability of Buildings*, National Research Council, Ottawa, ON, Canada, Vol. I, pp. 547–557.

Smith I. and Chui Y.H. (1992) Construction methods of minimising vibration levels in floors with lumber joists. *Canadian J. Civil Engineering*, **19**(5), 833–841.

Smith I. and Hu L.-J. (1993) Prediction of vibration serviceability of ribbed plates by modal synthesis. *Proceedings International Colloquium on Structural Serviceability of Buildings*, International Association for Bridge and Structural Engineering, Zurich, Switzerland, Reports Vol. 69, pp. 243–250.

Smith I., Hu L.-J. and Schriver A.B. (1993) Free flexural vibration analysis of one-way stiffened plates by the free interface modal synthesis method. *Canadian J. Civil Engineering*, **20**(6), 885–894.

Taylor S.B. and Hua G. (2000) Dynamic performance of wood framed floor systems with poured toppings. Paper 4.6.3, *Proceedings World Conference on Timber Engineering*, University of British Columbia, Vancouver, BC, Canada.

Wiss F.J. and Parmelee A. (1974) Human perception of transient vibrations. *ASCE J. Structural Engineering*, **100**(ST4), 773–787.

15

Design for Earthquake Resistance

Erol Karacabeyli and Marjan Popovski

15.1 INTRODUCTION

Among the natural hazards faced by humans on earth, an earthquake is one of the most disastrous and unpredictable actions. The occurrence of a major earthquake often results not only in enormous material damage, but also in substantial physical and psychic trauma and loss of life. On average, about 10 000 people die every year because of earthquakes around the world, with property damage of approximately \$1 billion (Naeim, 1989). Although some areas have a much higher seismic risk than others, earthquakes occur all over the world. Countries where wood is extensively used in residential construction such as the United States, Japan, Canada and New Zealand have regions of high seismicity. Even Australia, with a much lower seismic risk than these countries, has experienced some large earthquakes in the last 50 years. In North America, many people wrongly believe that earthquakes are primarily confined to California when, in fact, more than 70 million American citizens in 44 states are at risk from earthquakes (Foliente, 1997). Although significant earthquakes are infrequent east of the Rocky Mountains, several severe quakes have occurred on this side of North America.

In Europe, earthquake activity is widespread. To the south in Turkey, Greece, the countries of Former Yugoslavia, Italy, Spain and Portugal, a large number of strong earthquakes has occurred over the last few centuries. An earthquake off south-west Iberia on November 1st 1755 produced a great ocean wave, which alone caused many of the 50 000–70 000 deaths occurring in Lisbon, Portugal. A devastating earthquake hit Messina, Italy, on December 28th 1908, causing 120 000 deaths and widespread damage. On December 27th 1939, in Erzincan, Turkey, 23 000 lives were lost from a major earthquake. Similar devastating earthquakes have occurred in Turkey in recent years. North of the Mediterranean margin, Europe is much more stable. Destructive earthquakes, however, do occur occasionally in countries such as Romania, Austria, Germany, Switzerland and the United Kingdom (Bolt, 1988).

Almost half of the world's population lives in densely populated urban areas, many of which are in high seismic active regions. This fact, more than ever, makes people exposed to potentially damaging and life-threatening earthquakes. The Northridge and Kobe earthquakes that occurred during mid 1990s provide grim reminders of how strong earthquakes can adversely affect communities and individuals in heavily populated urban areas. Because earthquakes as natural phenomena are unpreventable and unforeseen, the only course of action for engineers to take is to build seismically resistant structures. These structures should be designed to behave in a desirable and predictable manner when subjected to earthquake ground motion. Over the last few decades, engineers and scientists have been able to learn by observing and studying seismic events, and have identified the dominant factors that affect the seismic behaviour of structures. As a consequence, measures have been devised for successfully resisting the destructive effects of earthquakes. Reconnaissance studies of past earthquakes have contributed greatly to this advance by identifying the causes of distress of structures and initiating the steps to rectify and avoid these shortcomings in the future. For example, experiences from earthquake-prone areas indicate that timber structures generally exhibit good seismic performance. Particularly, the positive performance of platform-frame wood buildings makes them suitable in seismically active areas.

This chapter provides an overview on important issues related to the seismic design of timber structures. Since seismic design of structures is a code-related issue that differs from one country to another, the intention of the authors is to stress the general design rules and present the information code-free, as much as possible. In the beginning, a brief introduction on the earthquake phenomenon is given, followed by the basic characteristics of seismic behaviour of timber structures. Furthermore, past seismic performance of wood structures is presented, as well as the current knowledge on the seismic design of timber buildings, including connections. The text is intended to serve as state-of-the-art information on seismic behaviour and seismic design issues in timber structures, as well as a reminder to owners, designers, builders, code writers and regulators that seismic design of timber structures cannot be taken for granted. To achieve good seismic performance, coordinated anti-seismic measures are needed by all who contribute to the design, construction and maintenance of the structure.

15.2 THE EARTHQUAKE PHENOMENON

15.2.1 Plate Tectonics

An explanation of the uneven distribution of significant earthquakes is given by the theory of plate tectonics. The basic idea is that the Earth's outmost part (called crust or *lithosphere*) was once concentrated in a single continent mass that broke into a number of large and fairly stable slabs of relatively rigid rock called *tectonic plates*. These plates, that usually extend to a depth of about 80 km, and have a horizontal movement relative to neighbouring plates at rates varying from 5–130 mm per year, provide the mechanisms for a great deal of the seismic activity of the world. At certain parts of the world, the contact or collision between adjoining plates results in one plate descending or *subducting* beneath the other. At these places, called *subduction zones*, large deforming (tectonic) forces are created which, when released, cause very strong earthquakes. In other parts of the Earth, tectonic plates just slide past each other along what are called *transform faults*, causing severe earthquakes. All these earthquakes that occur in the marginal zones along the edges of tectonic plates are referred to as *interplate earthquakes*.

Close to 90% of all earthquakes on Earth are interplate earthquakes, most occurring along the Circum-Pacific seismic belt, popularly known as the Pacific Ring of Fire (Chile, Peru, the Eastern Caribbean, Central America, Southern Mexico, California, Southern Alaska, the Aleutian islands, the Kuriles, Japan, Taiwan, Indonesia, the Philippines, and New Zealand), and along the Alpine-Caucasia-Himalayan belt stretching from the island of Java, through the Himalayas and all the way to the Mediterranean. Figure 15.1 shows the representation of the subduction process along the coast of British Columbia, Canada and Washington State.

While the plate tectonics theory is important for the understanding of earthquakes, it does not explain all seismicity in general. Smaller numbers of earthquakes, including a few with a large magnitude, occur within the plates producing the so-called *intraplate earthquakes*. Such major internal seismic activity indicates that tectonic plates are not rigid or free of internal rupture. Many of the earthquakes within plates occur along offsets of geological structures called *faults*. The source (*focus*, or *hypocentre*) of intraplate earthquake typically takes place in the Earth's crust, and may be associated with trust faulting that can be up to 50 km deep. Although the nature of the earthquakes along the plate boundaries is well explained by the theory of plate tectonics and kinematics, the nature of intraplate earthquakes, including major damaging ones, is not well understood.

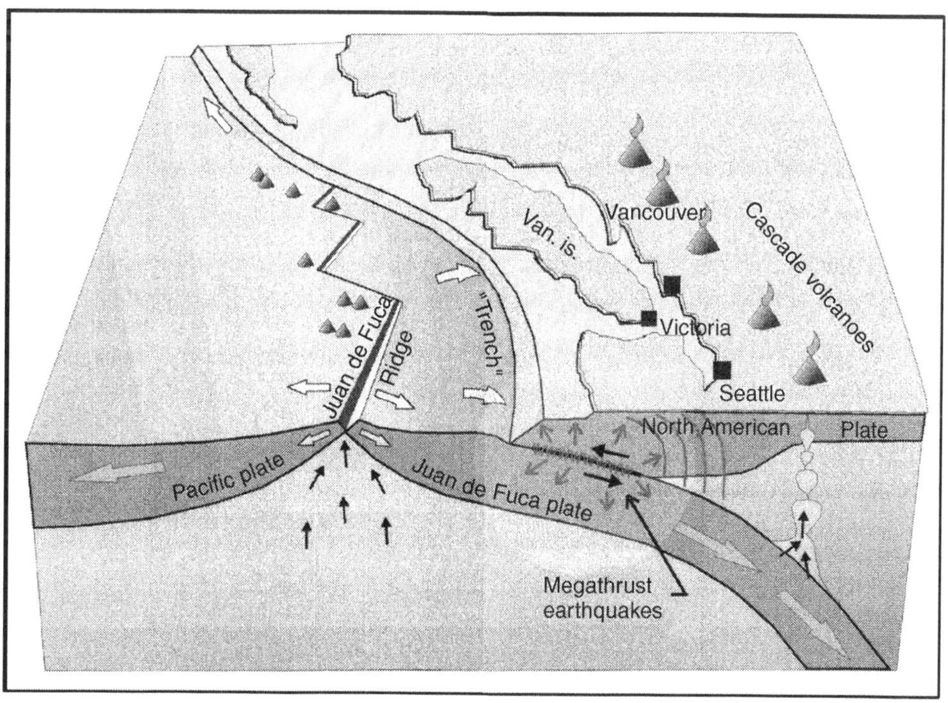

Figure 15.1 Representation of the earthquake generation mechanism along the west coast of Canada and Washington State (Reproduced by permission of Geological Survey of Canada)

15.2.2 Earthquake Magnitude and Intensity

The earthquake *magnitude* is a measure of the size of an earthquake, more specifically a measure of the amount of the strain energy that is released during the sudden rupture of the Earth's crust. The traditional Richter Magnitude M (or *local magnitude* M_L) is defined as the logarithm to base ten of the maximum seismic wave amplitude recorded on a Wood–Anderson seismograph at a distance of 100 km from the earthquake *epicentre*. The epicentre is the point on the Earth's surface directly above the focus. Using the logarithmic scale, every time the magnitude goes up by one unit, the amplitude of the earthquake waves increases ten times. From the definition of the magnitude M_L, it has no theoretical lower or upper limit, although the largest size of an earthquake is certainly limited by the strength of the rocks of the Earth's crust.

It should be mentioned, though, that the fundamental period of the Wood–Anderson seismograph is 0.8 seconds, so it selectively amplifies only those seismic waves with a period ranging approximately from 0.5 to 1.5 seconds. For this reason, the magnitude method has been extended to apply for a number of different types of seismographs around the world. Consequently, a variety of magnitude scales have emerged, such as Surface Wave Magnitude (M_s), Body Magnitude (m_b), and Moment Magnitude (M_w). The first three types of magnitudes, M_L, M_s and m_b, progressively underestimate the strength of earthquakes with magnitudes greater than 7, which are usually produced by long fault ruptures. The moment magnitude (M_w) is the only magnitude not to be affected by this phenomenon, referred to as *magnitude saturation*, and as such is the only one recommended for earthquake size comparison. Because the natural periods of many structures are within the range of the Wood–Anderson seismograph, the Richter magnitude M_L still remains of value to engineers. In this century, some of the largest earthquake events include the 1960 Chile earthquake ($M_w = 9.5$), the 1964 Alaska earthquake ($M_w = 9.2$), the 1957 Aleutian earthquake ($M_w \doteq 9.1$), and the 1952 Kamchatka earthquake ($M_w = 9.0$). Earthquakes usually have to attain magnitudes of more than 5.5 before significant damage occurs near the source of the earthquake.

The earthquake intensity, on the other hand, is a subjective measure of the effect of the earthquake at a specific location. An earthquake can have one M_w magnitude, but ranges of intensities, based on reports of ground and building damage and on interviews with people in different locations. The most commonly used method of assigning intensity is the *Modified Mercalli Intensity Scale* (MMI scale). This scale codes earthquakes by Roman numerals ranging from I (not felt by anyone), to XII (total destruction).

15.2.3 Earthquake Effects

The main effects associated with earthquakes are: surface faulting, ground shaking, ground instability (landslides and liquefaction), and tsunamis. Surface deformations from earthquakes can be produced by ground rupture on the primary fault or displacements from the nearby faults. Surface ruptures are local manifestations of a major earthquake, and would definitely damage buildings that are located directly on the fault line. Most damage in earthquakes, however, is not caused by surface faulting, since only a small percentage of buildings is located directly on a fault line. Ground shaking produces both primary and secondary geological responses, both causing damage to structures. It is also considered to be the most hazardous aspect of an earthquake because of its great area of influence. The amplitude of an earthquake ground motion as a primary response is measured by strong motion accelerographs, which record one vertical and two orthogonal horizontal acceleration components at a particular location. The instruments may be located on free field or mounted in structures. The velocity and displacement time histories are then obtained by integration of the digitised acceleration histories.

Ground shaking also includes secondary responses in unstable sites such as liquefaction and landslides. Soil liquefaction occurs when the pore water pressure of saturated sand, subjected to ground vibrations, builds to the point at which it is equal to the overburden pressure. At this stage the sand develops a liquefied state and completely loses its strength to carry any load. Structures built on such soil conditions virtually sink into the ground, in the case of liquefaction. Failures of natural slopes, or landslides, during earthquakes are also recognised as significant hazards. Tsunamis are huge sea waves caused by earthquakes that originate under the ocean floor. This phenomenon damages the buildings of waterfront communities by the impact of waves and the debris of trees, cars and boats that is carried by the water.

15.2.4 Characteristics of the Ground Motion

The characteristics of ground motion at the building site is governed by a number of factors. Some of the most important factors include:

- Size (length), nature, type and orientation of the fault line in the Earth's crust.
- Amount and type of movement along the fault line.
- Depth of the focus (earthquake source) below the ground surface.
- Strength and deformational characteristics of the ground at the source.
- Attenuation characteristics of the geological deposits between the source and the site.
- Depth and nature of local soil deposits at the building site.
- Distance between building site and the earthquake epicentre (epicentral distance).

The outcome of this complex process is the actual earthquake ground motion experienced at the site. The characteristics of earthquake ground motion that are important in earthquake engineering applications include: peaks of the ground motion (peak ground acceleration, peak ground velocity, and peak ground displacement); duration of strong motion; and the frequency content of the motion. Each of these parameters influences the response of the structure. Peak ground motion primarily influences the vibration amplitudes. The peak ground acceleration at the ground surface is affected not only by the amount of energy released at the source, but also by the distance from the source and by the local soil conditions at the building site. In general, the deeper the earthquake and the further away it is, the weaker the ground motion at the building site. Peak Ground Acceleration (PGA) has been most widely used in engineering design practice because it directly relates to the inertial forces on the structure generated by the earthquake motion.

The duration of strong motion has a pronounced effect on the severity of shaking. A ground motion with a moderate peak acceleration and long duration may cause more damage than a ground motion with a larger acceleration amplitude but a shorter duration. In addition, when the fundamental period of vibration of a building is very close to the predominant period of vibration of the soil at the site, the building will be in a state of *resonance* or partial resonance. In this case, the structural motion is amplified the most, and the structure vibrates strongly. When these periods are far apart, even a major earthquake will induce only a minor structural response. Usually, locations on a rock and those closer to the earthquake epicentre tend to experience high frequency motions, while those further away and on deep soil deposits experience low frequency ones. In addition, soil deposits, especially the soft ones, tend to amplify the earthquake motion induced on the structure, as in the case of the Mexico, 1985 earthquake. This is why some buildings suffer more earthquake damage when located on soft soil than on bedrock.

Some of these effects, such as the amplitude of the motion, frequency content and local soil conditions, are best represented through a *response spectrum*. A response spectrum describes the maximum response (acceleration, velocity or displacement) of a damped single-degree-of-freedom oscillator with various frequencies, subjected to a specific earthquake motion. The response spectra (usually acceleration ones) from a number of records are often averaged and smoothed to obtain the *design spectrum*, which is used in most seismic design codes to determine the seismic design forces on a structure with a given fundamental frequency. It should be noted that in some cases, determining the shape of the design spectra for a particular location is complicated, and caution should be used when

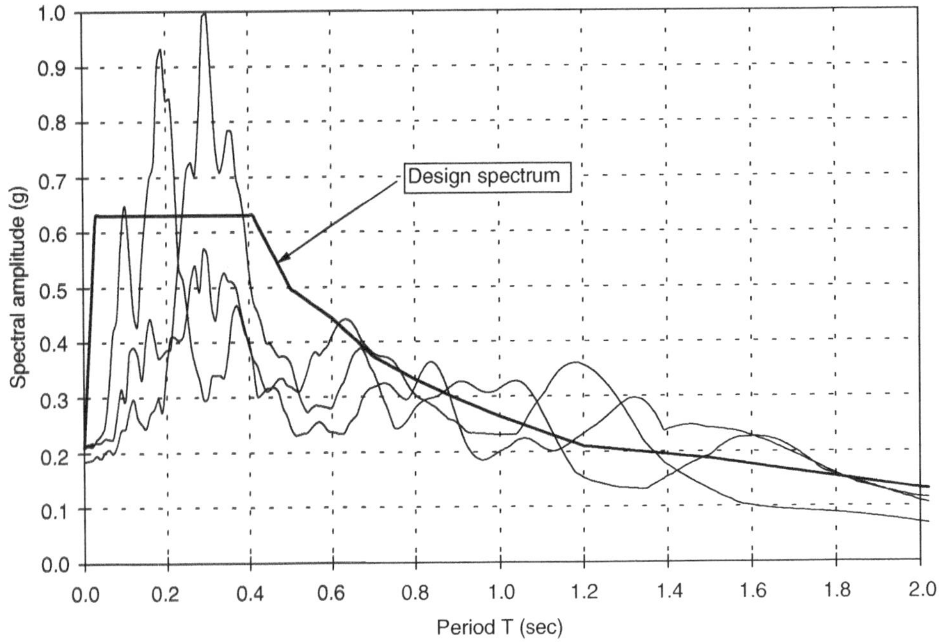

Figure 15.2 Example of three earthquake spectra and a design spectrum for a site

arriving with a representative set of records. It is not unusual practice that a different set of records is used for different frequency regions of the spectrum. An example of acceleration spectra for three different earthquakes and a smoothed design spectrum is shown in Figure 15.2.

15.2.5 Evaluating the Seismic Hazard at a Site

Evaluating the seismic hazard at a particular site is generally based from three main sources: (i) the recorded ground motion; (ii) the history of seismic events in the vicinity of the site; and (iii) the geological data and fault activities of the region. For most regions of the world this information, particularly from the first source, is very limited, and may not be sufficient to predict the size and recurrence intervals of future earthquakes. Nevertheless, earthquake engineers have relied on this limited information to establish some acceptable levels of seismic hazard. The seismic hazard analysis usually begins by developing mathematical models that are used to estimate recurrence intervals of future earthquakes with a certain magnitude and/or intensity. These models, along with appropriate soil attenuation relationships, are commonly utilised to estimate the ground motion parameters at a certain site. If this process is applied over wide areas, it produces seismic hazard maps for important ground motion characteristics such as the peak ground acceleration and velocity corresponding to a specified probability and return period. The seismic hazard maps that are currently used in most of seismic codes in the world are drawn for an earthquake with a return period of 475 years. In other words, the maps present earthquake characteristics that have a probability of exceedance of 10% in 50 years.

Although the structural response and damage potential are closely related to peak ground acceleration according to fundamental laws of dynamics, certain response characteristics may relate more appropriately to ground velocity or displacement. For that reason, for example, the seismic forces on a structure at a certain site according to the National Building Code of Canada (NBCC,

1995) are calculated based on the seismic hazard map showing the peak horizontal velocities in m/sec, with a probability of exceedance of 10% in 50 years. For a number of reasons that are beyond the scope of this text, the current trend in seismic design codes is to use seismic hazard maps developed on a probability of exceedance of 2% in 50 years.

15.3 SEISMIC BEHAVIOUR OF TIMBER STRUCTURES

The seismic response of a timber structure in general is a complex issue, involving many different interacting factors, which need to be understood and quantified. The ground acceleration, velocity and displacements (referred to as ground motion), when transmitted through a structure, cause inertial forces and displacements which may exceed those that the structure can sustain. Observations from past earthquakes supported by theoretical and experimental considerations have shown that, besides the characteristics of the ground motion that were discussed previously, many other parameters influence the seismic response of timber buildings. The following are thought to be of utmost importance:

- Properties of wood as a structural material.
- Building configuration and structural irregularities.
- Dynamic characteristics of the building.
- Deformational characteristics of the building.
- Damping and energy dissipating mechanisms.
- Redundancy.

An understanding of how these factors influence the response of the structure is essential for a safe and economic seismic design.

15.3.1 Properties of Wood as a Structural Material

The physical properties of wood depend upon its complex internal structure. In addition to natural variability and anisotropy, wood is also porous and inhomogeneous. The strength and stiffness properties of wood vary with the relative orientation of the applied load to the wood fibres, which cause the mechanisms of failure to vary accordingly. Tension failures either parallel or perpendicular to grain tend to be abrupt, so these must be avoided under seismic loading (Figure 15.3).

Failures in shear and some of the failures in bending are also brittle, and should be avoided.

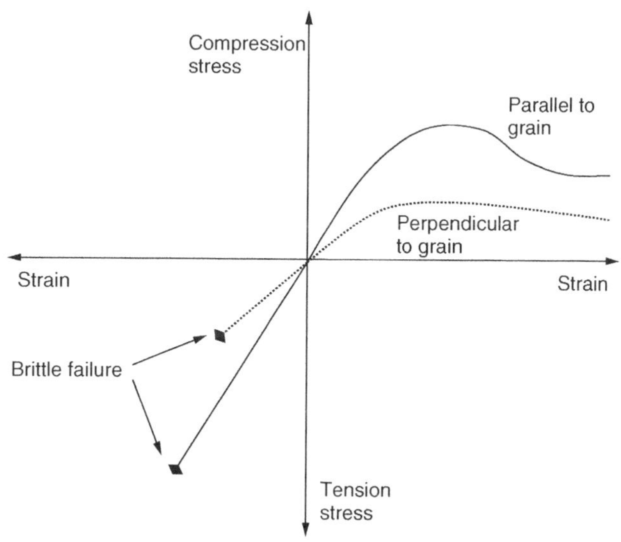

Figure 15.3 Stress-strain relationship for wood (Buchanan, 1988)

The type of bending failure depends upon the ratio of compression to tension strength, while column behaviour can be brittle or ductile depending on the ratio of compression to bending stress. Only the compression perpendicular to grain failures show a ductile behaviour, and are preferable for seismic design purposes. In addition, the moisture content can affect the properties and strength of wood members. An increased moisture content in wood reduces the modulus of elasticity, modulus of rupture and compression strength parallel to grain. Only the toughness of wood increases with an increase in moisture because it involves ductility and strength, and dry wood is less ductile than green wood (Keenan, 1986). The duration of loading has an important effect on the ultimate strength of wood. The ultimate strength of wood decreases under long duration load, or stated otherwise, the wood has enhanced strength under short-term loading. This fact, along with the high strength-to-weight ratio of timber as a material, contributed the most to the satisfactory performance of timber buildings during the past earthquakes.

15.3.2 Building Configuration and Structural Irregularities

It has long been recognised that buildings that have regular configuration and framing in plan and elevation lead to better seismic performance than buildings with irregularities (Paulay and Priestley, 1992). The irregularities encountered in plan or elevation may be due to geometry (structural configuration), or to uneven distribution of mass, strength and stiffness of particular storeys or parts of the building. For instance, when the building plan and the distribution of the structural components make the centre of the rigidity and the centre of mass coincide, the building will not twist, regardless of the direction from which the earthquake motion comes. If the centres do not coincide, usually when one side of a building is significantly stiffer than the opposite side, or when one side of a building contains large openings and the other does not, twisting (torsion) of the building occurs. The building with a rectangular plan will experience torsional stresses that vary along the sides. Buildings with a circular, octagonal or square plan experience less torsional stresses than buildings with irregular plans (e.g. 'L' shape). The best solution to avoid torsion is to design buildings with a symmetrical plan, strength and stiffness properties. If this is not possible, the effects of torsion can be counteracted, but this requires special attention in the design phase.

Structural continuity in elevation is also very important, since vertical irregularities in strength and stiffness properties of the building can result in creation of a 'danger zone'. The dissipation of the input energy during a seismic event will be overly concentrated in this zone, which, for example, may result in creation of a *soft* and *weak storey* collapse mechanism. A soft storey effect occurs when the stiffness of one of the storeys in a building (usually the bottom one) is much lower than the stiffness of other adjacent storeys (Figure 15.4). This implies that the first storey exhibits most of the structural deflection and energy absorption induced by the earthquake motion. Usually, the demand on this storey is much larger than its deformation capacity, resulting in partial or total storey failure. A good rule of the thumb is that a building can potentially be considered as prone to the soft storey mechanism if the lateral stiffness of the load resistant system in a storey is less than 70% of the stiffness of any adjacent storey. Similarly, a weak storey is one in which the storey shear strength is less than in the storey above. The storey shear strength is the total strength of all seismic resisting elements of the lateral load resisting system for any particular direction of loading.

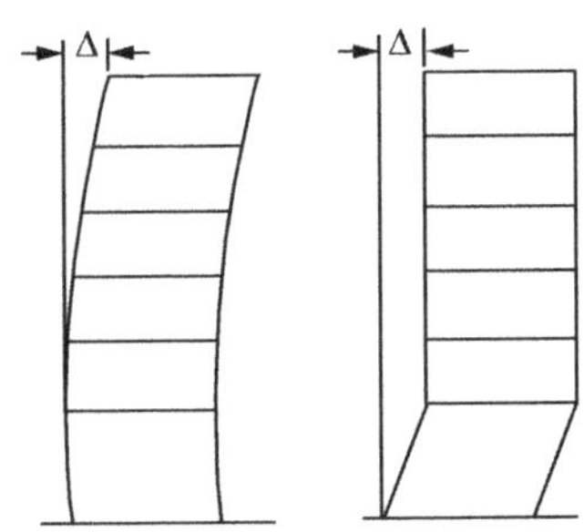

Figure 15.4 Soft storey mechanism in a structure

15.3.3 Dynamic Characteristics of the Building

Dynamic characteristics of the building include parameters such as natural modes of vibration and natural frequencies, or their inverse, the natural periods. These properties, which basically depend upon the structural material properties and connection strength and stiffness properties, will determine the way in which the building reacts when subjected to a certain ground motion. A structural mode shape is the shape in which the structure vibrates if subjected to one of its natural frequencies. Tall buildings tend to have low natural frequencies, or long periods, whereas short buildings tend to have higher natural frequencies, or short periods. If a low building with a high natural frequency is shaken by a high-frequency earthquake, it would respond more severely than if a low-frequency ground motion were involved. If the predominant frequency of the earthquake is almost identical to the natural frequency of the building, resonance might occur, highly intensifying the effects of the earthquake (Chopra, 1995).

The fundamental period of a building is a very important parameter that determines the response of the structure to the input ground motion. The fundamental (initial) period of the structure is also included in all building codes, and is a very important parameter in determining the seismic design forces on the structure. Because in most cases it is difficult or impractical to determine the initial period of the structure that is yet to be built, building codes provide empirical formulas that depend upon the overall dimensions of the building, the selected structural system and material type. A number of experimental studies involving forced vibration tests, shake table tests, ambient vibration tests and free vibration tests were conducted to determine the initial period of wood structures. Some of the studies include: Yokel *et al.* (1973), Dowding and Murray (1981), Yasumura *et al.* (1998), Hirashima (1988), Gramatikov and Gavrilovic (1990), Dolan (1990), Ceccotti and Vignoli (1991), Touliatos *et al.* (1991), Phillips and Itani (1993), Frenette *et al.* (1996), Ellis and Bougard (1998), Kohara and Miyazawa (1998), Durham *et al.* (1999) and Filiatrault *et al.* (2000).

The studies have shown that most timber buildings have an initial fundamental period of vibration in the range of 0.1–0.8 s. This would indicate that resonance with the earthquake motion and local soil conditions would be more likely on a medium to firmer rather than on a softer soil. Nevertheless, timber buildings tend to lose some of the stiffness at the joints over time, especially if some previous shaking has taken place, resulting in an increase of their natural periods. With the softer structures, resonance is therefore more probable on a thicker and softer layer of soil, and consequently, for timber structures built on soft soil, extra measures may have to be taken to ensure structural integrity, particularly at the foundation level.

15.3.4 Deformational Characteristics of the Building

Another important set of properties that affects a building's seismic response is its deformational characteristics, specifically stiffness, strength and ductility. Each of these parameters by itself represents an important property of a building, but also the parameters are interrelated and will also affect the dynamic properties of the building, as well as the connection properties. Adequate stiffness is desirable to limit the deflection and vibration amplitudes from wind, snow, occupancy loads and earthquakes. The high structural stiffness, however, raises the natural frequency of the building, which tends to increase the seismic forces on the structure. This then requires greater lateral strength and ductility in order to prevent collapse in a major earthquake.

Ultimate lateral shear strength is the maximum horizontal force that the lateral load resistant system of the structure is able to resist. Any further load increase will lead to excessive deformations and structural collapse. The strength of a building is affected by the type and quality of material, the size and distribution of the structural members, the quality of the workmanship, and the state of maintenance. A building with rotted wooden posts or corroded steel will have lost some of its strength. Poor workmanship in construction and inappropriate design will also lower the seismic resistance of the structure. The main structural materials (i.e. steel, concrete, masonry and wood) each require different approaches for achieving adequate strength by taking account of their inherent advantages and overcoming potential weaknesses. The pertinent design requirements are largely covered in building codes and in the respective design standards.

Closely linked to strength is the concept of ductility, which is defined as the ability of the structure to yield and undergo repetitive inelastic deformations without any substantial reduction in resistance. When a collapse occurs suddenly without much deformation, the failure is said to be *brittle*. On the other hand, if the structure deflects with considerable plastic deformation, the structure is said to exhibit *ductile* behaviour. The latter is much the preferred response to seismic forces. Ductile behaviour adds to the energy absorption capacity, and also limits the forces that act upon the structure. A consequence of relying on ductility, however, is that severe cracking and permanent deformation of the structure can occur (Bertero, 1988). For that reason, the majority of building codes limit the maximum lateral deflection of a single storey (interstorey drift) to a certain percentage of the storey height. This percentage is about 2–3% of the storey height for regular buildings, and can be as low as only 1% of the height for *post-disaster buildings*. These buildings, such as hospitals or fire halls, are expected to remain fully operational after a major earthquake.

15.3.5 Damping and Energy Dissipating Mechanisms

Damping of the building is a measure of how much energy the structure can absorb and dissipate when subjected to an earthquake motion, thus lowering the seismic impact. The amount of damping significantly affects structural seismic response. In general, structures with low damping will respond more strongly to a given earthquake than those with high damping. The level of damping in a building depends mainly upon the material of construction (damping within the micro structure of the wood itself), the type of connections (slip or friction damping at surfaces in contact), connection deformability (hysteretic damping), and the presence of other elements (often called non-structural) such as cladding and interior finishes (Yeh *et al.*, 1971). The damping, expressed in terms of equivalent viscous damping, varies from 3 to 20% of the critical damping, depending on the structural system and the displacement amplitude. Critical damping is the smallest amount of damping that inhibits the structural oscillation completely. For example, as a part of the CUREE-Caltech research project that was undertaken at the University of California, San Diego, shake table tests on a two storey wood-frame structure revealed the natural frequencies to be between 3.0 Hz and 6.5 Hz, with a damping ratio from 5.1 to 10.4 of the critical (Filiatrault *et al.*, 2000). This is in compliance with the findings from the research projects on dynamic behaviour of shear walls at Forintek Canada Corp. (Karacabeyli and Ceccotti, 1996), and at the University of British Columbia (He *et al.*, 1999). These damping values show that a large amount of energy can be absorbed in a timber wood-frame structure during an earthquake. In other structural systems, such as braced frames, moment resistant frames, and timber arches, the amount of damping is much lower, and is within the range of 3 to 5% (Popovski *et al.*, 1998a).

Energy dissipating mechanisms are closely associated to the damping mechanisms. A large amount of energy is imparted to a structure during the earthquake ground motion. The ability of the building to absorb and dissipate this seismic input energy strongly affects its performance. In timber structures ductility and energy dissipation can be attained through hysteretic damping in timber-to-steel mechanical connections, or through friction between various construction parts. A well-distributed system of energy dissipation is preferable to dissipation concentrated in one location only. Localised energy dissipation can result in exhausting the dissipation capacity of that particular structural element or storey, which can lead to partial or total building collapse. Modern devices have been recently installed in buildings to improve their energy dissipating mechanisms.

An example of these would be the sliding friction devices that have been installed in buildings to improve their energy dissipation mechanisms. The friction devices act as a structural damper to control the amplitude, and as a safety valve to limit the force exerted to the structure.

15.3.6 Redundancy

The lateral redundancy plays an important role in seismic performance of timber structures. A redundant design will almost certainly offer more parallel load paths that can transmit the applied lateral loading on the building down to the foundation. By having alternate load paths, the internal forces can be redistributed within the structure whenever a member or connection yields. The redistribution of internal forces allows the structure to resist higher lateral loads than its actual design values. Internal force redistribution also increases the over-strength of the structure by spreading the energy dissipating regions throughout the structure, instead of being localised in few critical areas. The detailing of connections is very important because the more integrated and interconnected the structure is, the more load distribution possibilities there are. We should, however, never forget that the building's structural integrity is only as good as the weakest link in the load transmission path.

15.4 PAST SEISMIC PERFORMANCE OF TIMBER BUILDINGS

The first step required in the design of an earthquake-resistant structure is acquiring a thorough understanding of the behaviour of structures during previous earthquakes. Fortunately, many of the earthquakes in the last 30–40 years have yielded a significant amount of instrumental data on earthquake ground motions and how the buildings responded, along with surveys of damage or non-damage. For example, Rainer and Karacabeyli (1999) conducted a survey on the performance of the platform-frame wood construction during recent earthquakes and, wherever possible, related the location of damaged or undamaged buildings to the ground motion data in the form of contour maps, and compared the performance of similar types of buildings.

Many timber buildings that were constructed without the benefit of proper engineering design or before engineering calculations were established have generally performed well during past earthquakes. Old residential timber buildings in the Mediterranean region and pagodas and temples in Japan and China have survived 1000 or more years of use and environmental loading, including strong earthquakes (Foliente, 1997). Most Chinese pagodas and temples performed well because of their symmetrical plan, almost uniform elevation, their light weight (attracting lower inertial forces), structural integrity and redundancy (some pagodas are supported by 32 columns), and use of effective joining systems. The Japanese pagodas have the same features as above, and those built during the 17th century had, in addition, a suspended central column which was independent of the surrounding structural frames (Tanabashi, 1960). Under seismic ground motion, that column acts as a suspended pendulum, which is a passive control device that reduces the structural response. In addition, single-family dwellings using the traditional North American light-frame (platform) construction system have performed relatively well during the strong earthquakes of this century.

Satisfactory performance of timber buildings in general can be partially attributed to the material characteristics of wood itself, and to the lightness and high redundancy of some wood-based structural systems. The idea, however, that all timber structures are safe against earthquakes should be looked upon with reserve. While wood-framed buildings generally performed well in the past, there have always been some timber structures that performed badly. Examples of this can be found in the post World War II built, traditional Japanese post and beam structures, or in larger framed buildings with irregular shapes. By inspecting those buildings after a major earthquake, we can learn from the failures and introduce changes to the design and construction procedure in the future. A summary of observations from seven earthquakes, with special emphasis on platform-frame wood construction, is presented below.

15.4.1 Alaska Earthquake, 1964

The Alaska Earthquake of Richter magnitude $M = 8.4$ was one of the largest earthquakes in North America in the twentieth century, and was characterised by intense ground shaking that caused major landslides in the populated area of Anchorage. Both ground settlement and large horizontal and vertical ground displacements caused significant damage to foundations and to the structures supported thereon, resulting in a large number of collapsed buildings and some deaths of inhabitants (Soltis *et al.*, 1980). Unfortunately, no recordings of ground motions were obtained. The box-like wood-frame structures stayed largely intact and 'survived' the earthquake, although often coming to rest in an unusable orientation for immediate occupancy. Other than these spectacular failures due to large ground displacements, wood construction suffered only minor and repairable damage (Karacabeyli and Rainer 2000).

15.4.2 San Fernando Earthquake, California, 1971

This earthquake of magnitude $M = 6.7$ occurred in a northerly suburban area of Los Angeles, and consequently affected a large number of single-family homes, as well as many non-residential and commercial buildings, resulting in a total of 64 deaths. The unreinforced masonry five-storey Veterans Hospital collapsed, killing 46 occupants. Other important structures built of reinforced concrete also collapsed, or were severely damaged to the point where they had to be demolished, such as the Olive View Hospital. The residential areas of San Fernando experienced peak horizontal ground accelerations of 0.6 of gravity (0.6 g) or more (Pacific Fire Rating Bureau, 1971). Older wooden houses in the San Fernando area suffered damage ranging from minor to partial collapse. Newer two-storey apartment buildings with large ground-level openings were also severely affected (Figure 15.5). These two storey residences with large garage openings on the lower floor suffered extensive damage. The greatest deficiency of the wood frame construction was observed to be its lack of resistance to torsional racking caused by large openings and unsymmetrical building plans. This was typically the case in the bottom storey, leading to the so-called *weak* and *soft storey* effect. The importance of such deficiencies in residential construction was first widely recognised in this earthquake, and many of these weaknesses have since been rectified. Following the San Fernando Earthquake, significant changes were made to the Uniform Building Code (UBC), which is used throughout the western USA.

Soft and weak first storeys, however, seem difficult to eliminate, since many occupants like the feeling of openness provided by large windows and the convenience of ground-level parking. Large openings can be accommodated if the rest of the structure provides an adequate strength. Unless this feature is specifically addressed, and adequate compensatory strength provided, earthquakes will continue to expose this weakness in the seismic resistance of all types of building. Aside from the specific types of failures summarised in Table 15.1, the majority of the then-modern wooden houses performed well, especially when the life-safety criterion is considered. Even where the building slid off the foundation or the cripple wall collapsed, the remainder of the building stayed intact and prevented serious injury or death to occupants.

15.4.3 Edgecumbe Earthquake, New Zealand, 1987

This event on New Zealand's North Island consisted of a fore-shock of a magnitude of $M = 5.2$, followed seven minutes later by the main shock of magnitude $M = 6.3$, which was followed by four aftershocks with magnitudes greater than 5 (Pender *et al.*, 1987). The quakes were centred in a rural area, so the only ground motion record obtained came from the base of the Matahina Dam, over 20 km away from the epicentre. The peak ground horizontal acceleration recorded was that of 0.32 g. The town of Edgecumbe (population 2000) may have experienced greater accelerations since

Figure 15.5 Failure of a two-storey wood frame building during 1971 San Fernando earthquake due to soft and weak first storey (Photo credit: Hans Rainer. Reproduced by permission)

Table 15.1 A summary of failures and deficiencies of wood buildings during the San Fernando earthquake

Damage	Deficiency
Houses sliding off the foundations	Frame not bolted to the foundation
Collapse of 'cripple walls' in crawl space	Inadequate lateral bracing
Collapse of add-ons such as porches	Inadequate lateral strength and ties
Collapse of masonry chimneys	Incompatibility of stiffness
Major damage and distortion of first storey (soft and weak storey effect)	Large openings, inadequate lateral and torsional stiffness and strength

it is located only about 8 km from the epicentre of the earthquake. Considerable damage resulted in all types of engineered structures such as railways, bridges, residential and industrial buildings, storage tanks, electrical equipment and municipal services. Widespread liquefaction of the ground was observed in a 10 km radius around Edgecumbe. No deaths or serious injuries were reported.

The timber structures affected were predominantly residential wood-frame houses, typically built on concrete strip or concrete block foundations, with a shallow crawl space below the ground floor. Many of the walls were lined on the interior with gypsum board and covered on the outside with a brick veneer. No exterior sheathing was used, although the wood frame incorporated K-bracing or diagonal bracing members. Of the approximately 6500 houses in the affected region, fewer than 50 suffered substantial damage, but none collapsed. Damage consisted of houses sliding off the foundation, cracking and collapse of the brick veneer on the building exterior, collapse of chimneys and failure of foundation posts and roof struts. The seismic weaknesses exhibited here follow those observed elsewhere, namely lack of foundation connection or bracing, incompatibility of brittle and heavy materials (i.e. masonry) with wood construction, and lack of lateral bracing.

15.4.4 Saguenay Earthquake, Quebec, Canada, 1988

Although the Saguenay Earthquake in northern Quebec with a magnitude of $M = 5.7$ was not as powerful as some of the others surveyed here, it was the largest earthquake in the last 50 years in eastern North America. Furthermore, a total of 11 pairs of horizontal and vertical ground motion records were obtained. The epicentre was located in a lightly populated area 150 km north of Quebec City. Peak ground accelerations in the population centres of Chicoutimi, 36 km from the epicentre, and on the north shore of the St. Lawrence River, 100 km away, were of the order of 0.12–0.17 g. The ground motions were of a high frequency type, and due to the low seismic attenuation rates in the earth's crust in eastern North America, the ground motions were felt at distances of 500 km and beyond. Most of the damage to wood-frame buildings up to two storeys was limited to cracks in chimneys, foundations and brick veneer walls. No cases of near-collapse or deaths were reported. The damage pattern correlated strongly with the presence of soft soil deposits, and damage in the houses was attributed mainly to foundation soil displacements rather than structural weaknesses (Lefebvre *et al.*, 1993).

15.4.5 Loma Prieta Earthquake, California, 1989

The Loma Prieta Earthquake, with a magnitude of $M = 7.1$, had its epicentre 100 km south of San Francisco, but its effects were felt as far as Oakland on the north shore of San Francisco Bay. Total casualties were 62 deaths and over 3000 injured; 49 of these died in the collapse of the double-deck freeway in Oakland alone. This earthquake caused considerable damage and some collapses of older unreinforced masonry buildings near the epicentre and as far away as Oakland.

A number of older four-storey wooden apartment buildings located on landfill sites in the Marina Bay district of San Francisco experienced failure of the ground floor that consisted entirely of garage openings, and therefore represented a very weak first storey. The upper three storeys of the buildings, however, stayed intact, sitting on the remains of the bottom storey. Wood-frame houses located in and near the epicentral region survived the shaking largely intact, with only some repairable damage. Some of these houses near the epicentre were likely subjected to peak ground accelerations as large as 0.5 g.

15.4.6 Northridge Earthquake, California, 1994

The Northridge Earthquake with a magnitude $M = 6.7$ was notable for its high ground accelerations, both horizontally and vertically, which in places exceed the acceleration of gravity (1.0 g). The earthquake caused extensive damage to residential, institutional, and commercial buildings, and to the highway and freeway system of this area about 20 km north-west of Los Angeles. This earthquake produced the most intense ground shaking that had so far been recorded in a populated area in North America. Over 30 persons were killed, and estimates of property losses range from US$ 30 to 40 billion.

This earthquake again drew attention to a weakness that had already been recognised, practically, during the 1971 San Fernando earthquake – the weak first storey in multi-storey wood-frame apartment buildings. Whereas in the 1971 earthquake these weak buildings merely distorted severely but remained standing, in this larger 1994 earthquake a number of ground floors collapsed, so for example in one apartment complex 16 occupants were killed. Such collapses are perhaps not surprising when one considers that the horizontal ground accelerations, combined with the vertical accelerations of comparable amplitude, exceeded the nominal horizontal design acceleration of 0.4 g by factors of 2 and more. Four deaths also occurred in three single-family houses that slid down a hillside and collapsed. Because more than 50% of the total damage during the Northridge earthquake was incurred in wood frame construction, a major investigation was undertaken into the performance of wood frame construction in high-risk earthquake zones. Despite some tragic failures, however, the majority of wood structures performed exceedingly well. As in other earthquakes, chimneys were severely damaged, while the rest of the building survived without significant problems (NAHB Research Centre Report, 1994).

15.4.7 Hyogo-ken Nanbu Earthquake, Kobe, Japan, 1995

The earthquake that hit the city of Kobe in the Hanshin area of Japan on January 17, 1995 was so far the most damaging earthquake in modern times, with estimated losses of well over US$ 100 billion and loss of life of nearly 6000 inhabitants. Its magnitude was 7.2 on the scale of the Japanese Meteorological Agency, which corresponds to Richter magnitude of $M = 6.8$. The area had not anticipated an earthquake of this magnitude, and therefore building design and construction was ill-prepared. Extensive damage and major collapses occurred in all types of structures – elevated highways, bridges, port facilities, utility services, rail and subway installations, and low and high-rise buildings of concrete, masonry, steel and wood (Building Research Institute, 1996). Among the widespread damage, there were some recently constructed buildings that survived the earthquake with almost no visible damage. This included high-rise buildings, mostly of steel construction, but also modern wood-frame buildings of two and three storeys in height. These cases demonstrate that modern building construction can survive, even without damage, one of the most severe earthquakes in recent times.

While the epicentre of the earthquake was located on the nearby island of Awaji south-west of Kobe, the rupture zone extended through the centre of the city to its neighbouring city Nishinomiya to the east. Peak ground acceleration as high as 0.8 g was recorded in densely populated areas of southern Kobe and the surrounding cities. These horizontal motions were also accompanied by large vertical accelerations, often even larger than the horizontal ones. This phenomenon is similar to that observed in the Northridge earthquake of 1994, and appears to be characteristic of ground shaking in the epicentral regions of an earthquake.

Among wooden buildings in the affected area, the hardest hit were those constructed before and immediately after World War Two (Prion and Filiatrault, 1995). These structures consisted of post-and-beam wood framing, with walls formed by horizontal boards nailed to the uprights, infilled with bamboo webbing and covered with clay. Exterior finishing was typically a stucco. Roofs were built of burnt clay tiles laid on a bed of brown clay and supported on horizontally nailed roofing boards. This provided excellent protection against typhoons prevalent in this area, but because of the heavy weight it proved disastrous during the earthquake. There was also evidence of rot and termite damage to some of the structural members. Heavy roofs, inadequate lateral bracing, lack of shear wall panels, material decay and soft storeys were the primary cause of collapse for many of these wooden buildings.

Among the sea of devastation of older style houses, there were examples of modern wood-frame construction houses that showed no visible signs of distress (Figure 15.6). These houses were located in areas of severe shaking, and demonstrate that modern wood-frame construction can withstand severe earthquakes with peak ground acceleration of 0.6 g or more with little or no damage. The wood-frame demonstration projects popularly known as Seattle-Vancouver village in the Nishi-Ward neighbourhood of Kobe city and the Hankyu Nishinomiya Housing Park, contained modern wood housing, the former only wood-frame construction, the latter both post-and-beam and wood frame types. Both areas were shaken severely, the Seattle-Vancouver village in the neighbourhood of 0.3–0.5 g, the Hankyu Nishinomiya Housing Park even higher than that. The wooden houses in both of these projects showed no visible damage, except for some isolated roof tile breakage.

The Kobe earthquake has again demonstrated the devastating power of an earthquake when it strikes a population centre. Every weakness of

Figure 15.6 Three undamaged modern wood-frame buildings in the background and a collapsed older building in foreground in Nishinomiya, Japan (Photo credit: Hans Rainer. Reproduced by permission)

design, construction and maintenance came to light, and caused massive damage to all types of structures. A notable exception to heavy damage was modern wood construction. A quote from a reconnaissance report of the Kobe earthquake (APA, 1995) reads: "The performance of Western-style wood-frame residences was nearly perfect, with no damage to this type of construction observed or reported".

15.4.8 Summary on Past Earthquake Performance of Wood Structures

The survey on the performance of platform-frame wood construction in recent earthquakes shows that wood buildings can resist very severe ground shaking with peak ground acceleration well over 0.6 g, with very low risk to life and limb. Such buildings thus satisfy the objective of life safety. Many instances of low damage have also demonstrated the potential of wood-frame construction for satisfying the more stringent criterion of damage control. However, the weak first storey phenomenon needs to be addressed to further reduce potential collapses and casualties and damage to wood buildings.

A summary of the total number of casualties caused by the earthquakes presented and a number of deaths in wood-frame (platform) construction is presented in Table 15.2. Evidently, the number of deaths in wood-frame buildings is very low, considering the large number of buildings severely shaken. Good performance expectations are, however, contingent on appropriate design, quality workmanship and proper maintenance. Favourable past performance should also not be taken to imply that further improvements are no longer needed. To ensure satisfactory performance in the future, the design codes and construction practices need to be periodically reviewed and revised to reflect new knowledge from theoretical and applied research, supported by field observations after earthquakes. It is important that future earthquake damage reports also focus on what has not been damaged, and relate the performance of buildings to the intensity of ground motion as much as possible.

Table 15.2 Overview of casualties in some recent earthquakes (Karacabeyli and Rainer, 2000)

Earthquake	Magnitude	No. of fatalities (approx.)		No. of wood-frame buildings shaken (estimated)
		Total	In wood-frame buildings	
Alaska, 1964	8.4	130	<10	
San Fernando, 1971	6.7	63	4	100 000
Edgecumbe, 1987	6.3	0	0	7000
Saguenay, 1988	5.7	0	0	10 000
Loma Prieta, 1989	7.1	66	0	50 000
Northridge, 1994	6.7	60	16 + 4	200 000
Hyogo-ken Nambu (Kobe), 1995	6.8	6300	0	8000

15.5 CONNECTIONS

15.5.1 Important Seismic Considerations

The main challenge in the seismic design of timber structures is the design and construction of appropriate connections. Because of the cyclic nature of the loading, the wood will inevitably be stressed in a direction of weak strength, which presents a particularly challenging problem. The fact that most collapses of timber structures during earthquakes were associated with connection failures emphasises the importance of studying the behaviour of different types of connections under cyclic loading. For example, nails, dowels or bolts in the connections can be designed to remain linear elastic or yield during the seismic action, so the corresponding structures can have a linear or a nonlinear ductile response. To obtain ductile structural behaviour, it is of great importance to use ductile connections that are weaker than the wood members they connect. The hysteretic behaviour of connections is governed by the material properties of the fastener and by the embedding behaviour of wood. Mild steel connectors with a large deformation capacity are generally more suitable for energy dissipation than high strength steel connectors with a brittle failure mode. The properties of wood vary with the orientation of loading, and therefore the failure mechanism is more complex than for isotropic materials. Consequently, connections that stress the wood in tension perpendicular to the grain have to be avoided, because they can result in brittle failures, even at relatively low load levels.

The load-displacement curves of ductile timber connections are smooth and continuous from the onset of loading to the ultimate load. Compared to the curves of other materials (steel, for example), there is no clearly defined yielding point, thus a much more flexible definition of yielding and ductility is used. The hysteretic load-deformation curves of timber joints or components are mildly to severely pinched, depending on the fasteners used (Figure 15.7). The pinching effect results in thinner loops in the middle compared to near the ends. This phenomenon is caused by the loss of stiffness at small joint slips, where a cavity around the fastener is formed by crushing of the wood. In this case, the fastener shank, without wood support within the deformation level of the slip, provides the resistance to the applied load. The stiffness of the connection gradually increases again as the fastener makes contacts with the surrounding wood at higher deformations.

Considering that the area inside the hysteresis loop for each cycle represents the amount of energy dissipated during that cycle, the pinching effect reduces the hysteretic damping of the structure. Some recent test results have shown, however, that beside the shape of the hysteresis curve (pinching), the ability of timber structures to sustain large deformations without significant strength deterioration is also very significant (Buchanan 1988). Deam and King (1994) supported this statement after considering the low damage ratio experienced by timber buildings in past earthquakes. A progressive reduction of stiffness in each loading cycle (stiffness degradation) and reduction of strength when cyclically loaded to the same displacement

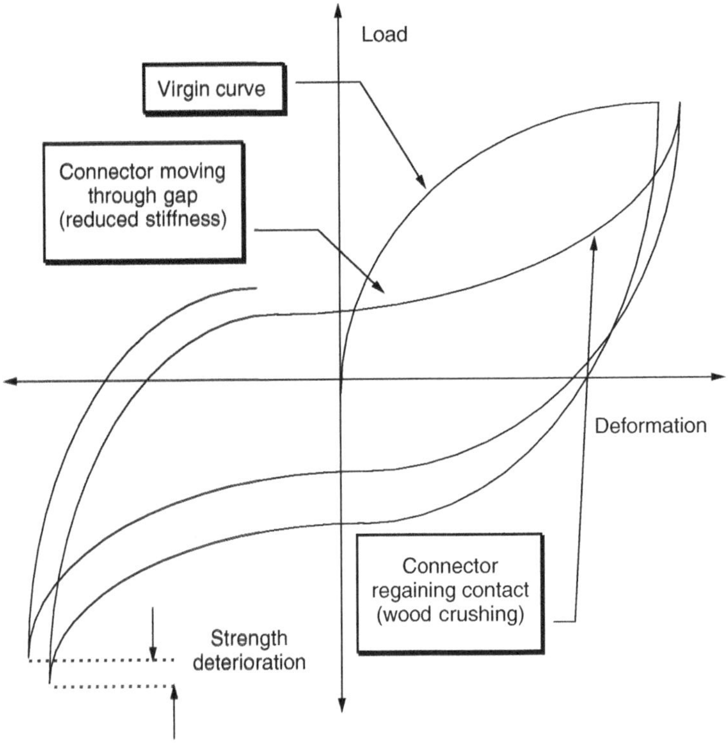

Figure 15.7 Typical hysteretic behaviour of a timber structure component

level (strength degradation), are very common characteristics of the load-deformation (hysteretic) response of timber connections under repetitive cyclic loads. A very important but still controversial topic on the response of timber connections is that the response at a given time depends not only upon instantaneous displacements, but also on past history or the earlier input and response. This phenomenon, also known as *memory*, was observed in nailed and bolted timber joints under irregular short or medium term lateral loading (Foliente, 1997).

In many cases, the mechanical and physical properties of timber connections related to their seismic performance are typically affected by the testing procedures used to evaluate these very parameters. The lack of cyclic testing standards or guidelines for timber joints and assemblies in the past, has made the consistent assessment of seismic performance of connections very difficult. Large sets of data have been obtained in the past where selection of loading histories, test details, and decisions on the presentation of results were done differently. Interpretation of other scientist's results can be a long and frustrating process, limiting the potential usefulness of the data collected. This problem was well recognised during the last decade, and efforts were made, resulting in drafts of proposed standards for cyclic tests of connections at the national level, such as in Australia (AS, 1995; CSIRO, 1996), Europe (CEN, 1995) and the United States (ASTM, 1993). Most notably, Working Group 7 of the ISO Technical Committee on Timber Structures developed an international standard test method for cyclic testing of timber joints with mechanical fasteners (ISO, 2001). It is strongly recommended that this international testing standard be considered in research projects to ensure comparability between results from different investigations.

The influence of the rate of loading on the strength and behaviour of timber connections is of vital importance in seismic design. McLain (1975) tested single-nail joints with a variety of member combinations, and found no statistically significant difference in load-deformation relationships between static and low frequency cyclic loading. Bolted wood-to-wood joints, on the other hand, were found to be considerably stiffer under rapid loading than under static loading. His results also showed connections exhibited higher strength values at the same displacement level under dynamic loading, which indicates an apparent rate of loading effect. Girhammar and Anderson (1988) tested nailed joints monotonically considering two wood species and five deformation rates (from 2 mm/min to 1250 mm/min). The results showed that the rate of loading effect was greater for wood bearing than for nail bending. In another study, however, no evidence was found to support short-duration time-effects for nailed connections subjected to lateral loads (Rosowsky and Reinhold, 1999). Although the rate of loading effect in lumber alone has been identified (Karacabeyli and Barrett, 1993), its direct application on timber connections (as is the case in many timber design codes) still remains a topic for future research.

Fatigue, which is a tendency of a material to break under repeated stress, is another problem in timber connections as it reduces the short-term or residual connection strength due to cumulative damage after a period of cyclic loading. Fatigue was observed to be a potential problem in joints with high-density members or members with steel side plates (Williams, 1984, Dolan and Madsen 1992, Ni and Chui, 1994). The fatigue of a connector is very much dependent on the stress levels induced on the connection and the number of load cycles. Fatigue problems in some joint or assembly configurations under earthquake-type loading need additional research. Critical joint configurations and material combinations that may be susceptible to fatigue still need to be identified.

15.5.2 Connection Types used in Seismic-prone Regions

The most common connectors used in timber construction in seismic-prone areas can be divided into three main categories, namely *dowel type*, *surface type* and *bearing type*. Dowel type connectors are cylindrically shaped connectors that penetrate deep into the wood and transfer the load between members by a combination of wood bearing and connector bending. This type of connector includes nails, spikes, staples, bolts, pins and screws. Surface type connectors combine the dowel-type action with the metal plates so they can collect and transfer the load near the surface of the member. Surface type connectors include punched metal plates, tooth plates, and timber (glulam) rivets. Bearing type fasteners are designed to transfer the forces relying solely on the shear or bearing resistance of the wood, parallel or perpendicular to grain. Examples include split rings, tooth rings and shear plates. Extensive experimental research has been conducted to determine the performance of different connections, although almost exclusively under static monotonic loading. The interest in the cyclic properties of connections or building components is fairly recent.

Nails loaded in shear are typically ductile connectors capable of resisting many reversals of cyclic loading. Under cyclic loading, they usually develop slackness and pinched hysteresis loops as the nails bend and cause local crushing of the adjacent wood fibres (Figure 15.8a). Large displacements, however, can be achieved without connection failure, even though some pullout of the nail is to be expected (Figure 15.8b). Several parameters such as wood density, nail length, nail strength, spacing, etc., influence the connection behaviour (Dolan and Gutshall, 1997). Static tests on nailed connections showed that increased nailing density (smaller spacing) in the connection leads to the formation of a brittle *block-tearing* failure mode with reduced capacity (Kangas, 1999). Cyclic tests also showed that nailed connections containing high density wood members may experience failure due to fatigue breakage of the nails (Ni and Chui, 1996). Due to the large surface area required to accommodate the large number of nails needed, nailed connections are only suitable for deep and slender timber, glulam or Parallel Strand Lumber (PSL) members. Finally, nails are used as the main connectors in wood-frame shear walls, the main

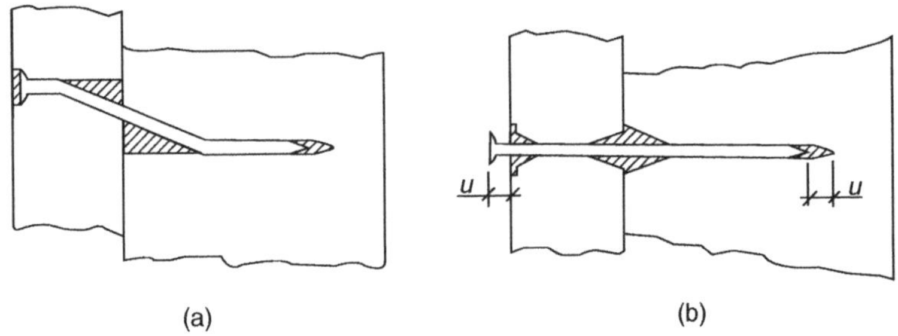

Figure 15.8 Formation of a cavity around the nail during cyclic loading and the pullout effect (courtesy of Dr. Ario Ceccotti)

lateral load resisting system in North American platform type construction. Numerous static, cyclic and shake table tests on nailed shear walls showed a large energy dissipating capacity of these structural elements.

Dowels are well known connectors that can be used effectively for connecting sawn lumber, glulam or PSL elements. The yielding capacity of the dowel in combination with the bearing strength of the surrounding wood can provide high ductility and excellent energy dissipation characteristics under cyclic loads. Test results on the cyclic behaviour of dowel connections showed that best results are obtained with slender dowels and a relatively large spacing between them. Joints with stocky dowels or small spacing tend to fail in a brittle mode before large deformations are reached. It was also found that the shape of the hysteresis loops of dowel connections changes progressively with increased displacement (Figure 15.9). Dowels can be used with a steel shear plate inside the member in the case of an axial or a moment resisting connections (Prion and Foschi, 1994). These aesthetically pleasing connections showed to be very effective in the construction of moment resisting frames in high risk seismic zones as well (Frenette *et al.*, 1996). Successful applications were noted in Europe in the construction of high capacity connections using large numbers of small diameter dowels (Aasheim, 1994). In addition, a research project was carried out in Switzerland to optimise the design of high efficiency dowel connections for use in sawn lumber, glulam and PSL (Mischler, 1998). Multiple shear dowel connections with several slotted-in steel plates were tested. The results showed that these connections, which are very efficient for large timber members, can exhibit ductile behaviour if the dowel strength and spacing are adapted to the timber properties.

Several testing programs were performed on a variety of improved dowel-type connections. Rodd (1988) explored the possibility of increased friction around the dowels to improve the embedment strength and avoid splitting of the wood. In another study, dowel connections with steel plate and fibreglass reinforcement were examined (Leijten, 1988), however, with limited success. The strengthening effect of fibreglass was found

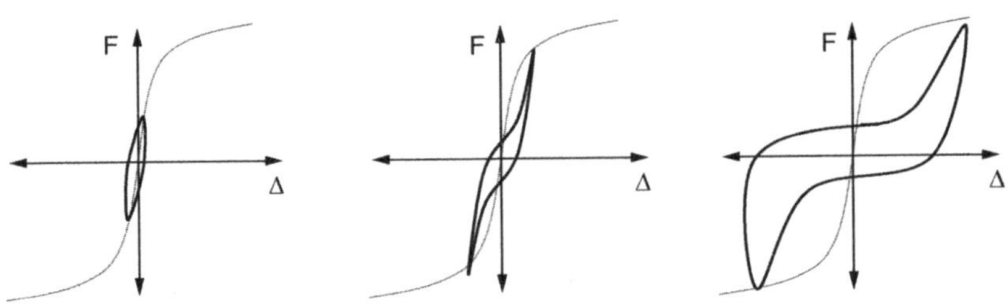

Figure 15.9 Typical hysteresis loops of a dowel connection at different displacement levels

insufficient, while the steel plates, though very effective, were regarded as not being suitable for practical application. Finally, strengthening with a Densified Veneer Wood (DVW) was studied on a dowel-type connector with expanded tubes (Leijten, 1996). Due to the ability of the tube to absorb the deformation energy, these connections showed good ductility performance. Similarly, a hollow dowel fastener injected with resin was tested, and showed sufficient strength and relatively high ductility capacity (Guan and Rodd, 1996).

Bolted connections are frequently used in heavy timber frames, particularly in axially loaded braces and struts. Numerous studies have been carried out to determine the factors that influence the strength and behaviour of multiple-bolted connections under static and seismic loads. Parameters such as wood and bolt properties, end distance, edge distance and fastener spacing, both parallel and perpendicular to the grain, were investigated. Tryer (1932) presented design formulae for bolted joints with steel side plates based on test data (Moss, 1997). He was the first to note that the interaction of the bending of the bolt and the crushing of the wood affected the joint performance.

In 1949, Johansen (1949) introduced a theory to predict the ultimate strength of a bolted (dowel) type joint based on equilibrium equations derived from a free body diagram of a bolt in a wood member. The European Yield Model, based on Johansen's theory and subsequently refined by Larsen (1973), is used in many timber design codes in the world. Larsen's prediction equations assumed that a bolted connection could fail by full bearing (crushing) in one or more wood members (for short bolts), or by yielding of the bolt (for longer bolts), accompanied by local wood crushing near the member interfaces. The model considers several possible failure modes for a given joint configuration, based on the relative strength between the connector and the wood.

Yasumura *et al.* (1987) investigated the influence of end distance, fastener spacing and number of bolts on the ultimate static and dynamic properties of bolted joints in glued-laminated timber. He found an increase in the ultimate load per bolt proportionally to the increase of the bolt slenderness. Another trend observed was that, as the number of bolts in the connection increased, the ultimate load per bolt decreased. The test results also showed that longer bolts showed greater load reduction with a reduction of bolt spacing.

The behaviour of bolted connections was also studied in the past few years at the Royal Military College of Canada (RMC), in Kingston, Ontario, Canada. The research work covered the influence of joint configuration parameters on the strength of perpendicular-to-grain loaded connections (Mohammad and Quenneville, 2000), the influence of cyclic loads on strength and stiffness properties (Mohammad and Quenneville, 1998), and predicting the failure modes and strength of brittle bolted connections (Quenneville, 1998). Research findings at RMC were introduced into a design procedure for predicting the strength of bolted connections in glulam.

A study at the University of British Columbia (UBC) has shown that the ductility and seismic resistance of a single bolted connection in Parallel Strand Lumber (PSL) can be enhanced significantly by adding reinforcement in the form of a truss plate, fibreglass, or a layer of glued-on plywood (Hockey *et al.*, 1999). Another simple and inexpensive way to reinforce a bolted connection is to insert reinforcing rods perpendicular to grain and to the bolts. The reinforcement could be done with either glued rods with epoxy or with threaded rods driven into tight fitting holes. Another study at UBC dealt with various reinforcement patterns in bolted connections in glulam, including the use of reinforcing bars. The results showed that when reinforcement bars were added halfway between the bolts, an improvement of over 25% was achieved in connection strength, and almost 60% in connection ductility (Hockey *et al.*, 1999).

Finally, a joint study between UBC and Forintek Canada Corp. (Popovski *et al.*, 1998b) confirmed that the seismic behaviour of the bolted connection is dependent on the bolt slenderness ratio (width of the wood member versus the bolt diameter). It confirmed that small diameter (slender) bolts tend to behave more like nails or dowels as their shaft can bend without inducing wood fracture,

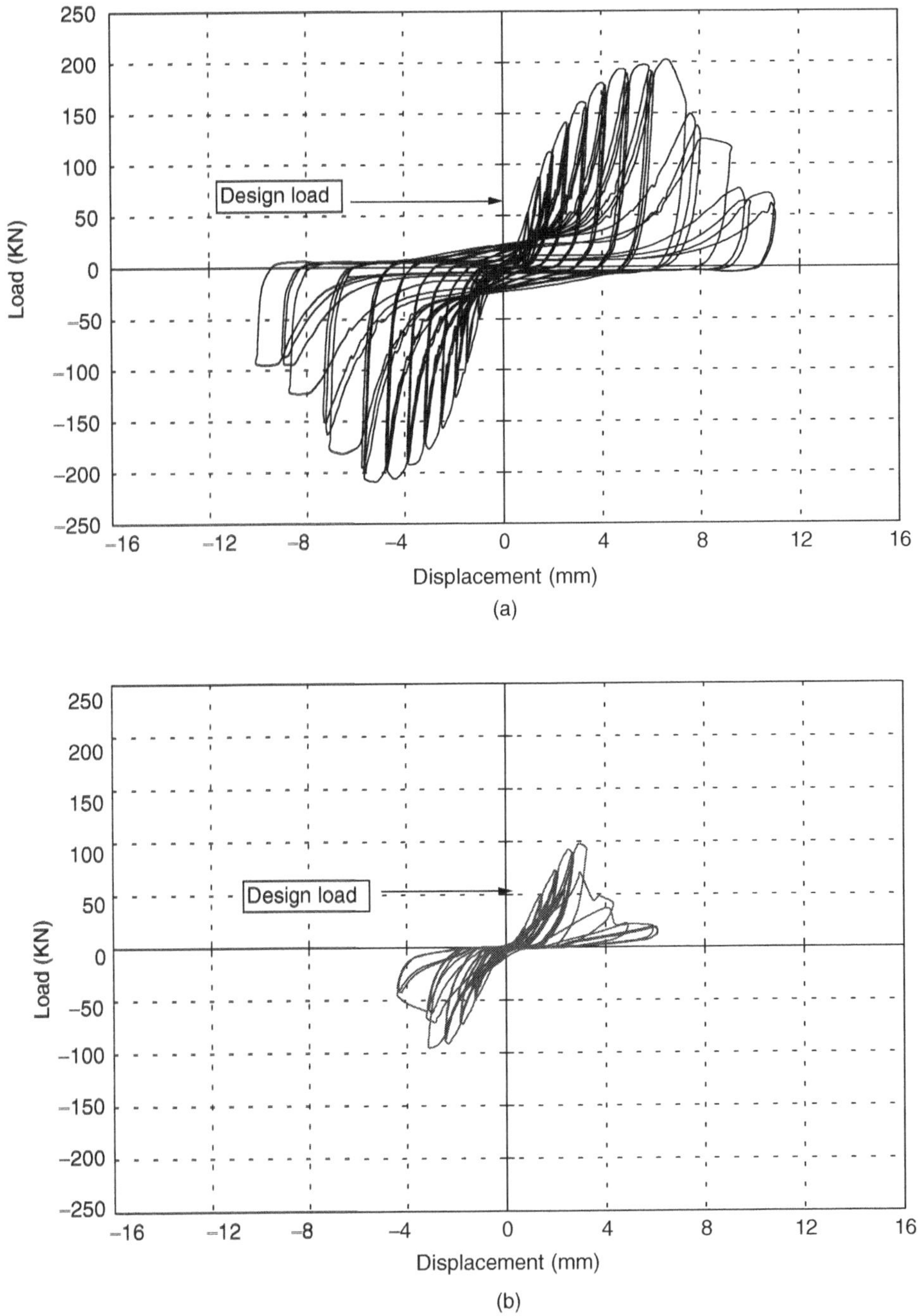

Figure 15.10 Typical hysteretic behaviour of connections with (a) slender and (b) stocky bolts

thus exhibiting ductile behaviour and relatively large energy dissipation (Figure 15.10a). When the slender bolts yield, they also allow for better redistribution of the load among all the fasteners, resulting in improved load sharing. However, when larger diameter bolts are used, the inelastic behaviour and energy dissipation of the connection depends upon the embedding behaviour of the wood alone (Figure 15.10b), which often leads to brittle failures. In addition, oversized holes and large fabrication tolerances can cause non-uniform load distribution and precipitate fracture of the wood. In this case, the lack of ductility or overloading of one individual bolt can lead to the initiation of a fracture that precipitates through the group.

In bolted connections with a lower slenderness ratio (7 and lower), the high flexural rigidity of the bolts resulted in more rigid connections with high level of stress in the wood. This precipitated abrupt wood splitting and sudden loss of bearing capacity. Connections with bolts of higher slenderness ratios (10 and higher) exhibited a more desirable behaviour in that more wood crushing could occur before fracture, although eventually wood splitting was consistently the failure mechanism for all bolted connections tested. In addition, the energy dissipation in bolted connections with small diameter bolts was higher than in connections with larger diameter bolts. Based on these results, it was recommended that slender bolts with slenderness ratio of 10 and more are adequate for seismic design of timber frames.

Timber rivets, also known as glulam rivets or griplam nails, are high strength nail-type fasteners developed in Canada, especially for use in glued-laminated timber construction. Over the years, for a number of designers, they have become the favoured fasteners for glulam because of numerous advantages over other connectors. They are stiffer and provide greater load transfer per unit contact area than most wood fasteners. The wood members need not be drilled or grooved, which simplifies the fabrication and field assembly. Although they are made of high strength steel with a high yield point, they are soft enough to provide very ductile connections. Figure 15.11 shows a typical column to beam glulam riveted connection.

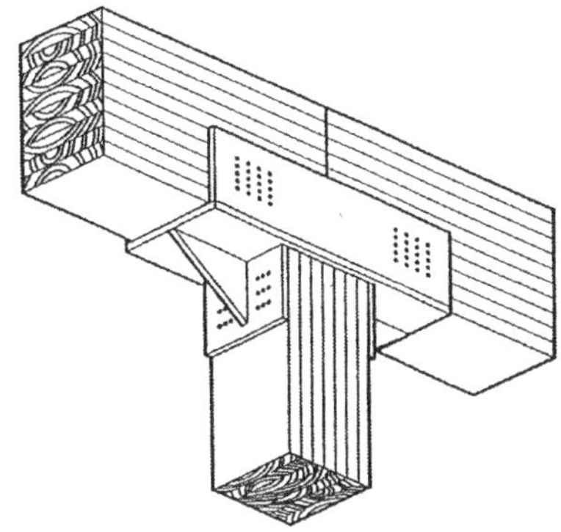

Figure 15.11 Typical column to beam glulam riveted connection (Reproduced by permission of Borg Madsen)

The development of timber rivets, invented by Borg Madsen, dates back to the 1960 (Williams and Karacabeyli, 1996). Design of glulam rivet connections in the current edition of the Canadian Code for Engineering Design in Wood (CSA-O86-01, 2001) is based primarily on the method developed by Foschi and Longworth (1975). According to this method, the ultimate capacity of the connection is governed by either a rivet yielding or a wood failure mode. The rivet yielding failure mode, where the rivets bend and yield while the wood under their shanks fails in crushing, is the desirable failure mode for seismic design. This failure mode was first studied by Foschi (1974). Using finite element analysis, he also studied the effect of rivet penetration and the effect of direction of loading with respect to grain orientation. For the wood failure mode, Foschi and Longworth (1975) investigated the stress distribution around the rivet cluster, and derived the formulae for calculating maximum stresses in the member around the rivets. Weibull's weakest link theory was used to determine the ultimate wood strength. They verified their analytical model with test results on riveted connections in Douglas-fir glulam, and offered the following conclusions:

- rivet spacing controls the failure mode;
- larger spacing results in a rivet yielding failure mode, while a smaller spacing produces sudden wood failures usually a block-shear around the group of rivets, and

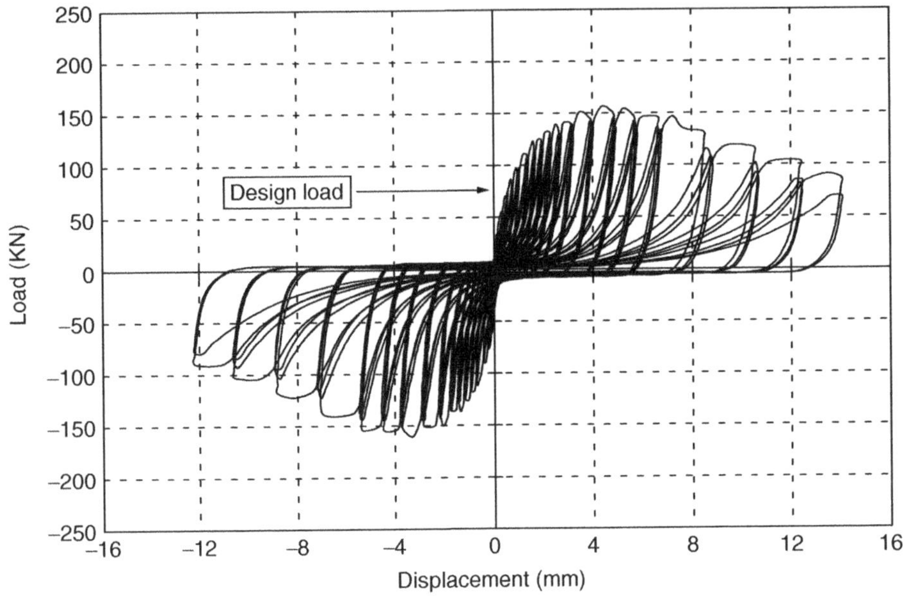

Figure 15.12 Typical hysteretic response of a riveted connection

- for the same rivet spacing, a larger end distance leads to an increase in the ultimate load based on the wood shear failure.

The research results by Foschi were complemented by further experimental studies. Fox and Lincoln (1979) investigated the effect of plate thickness and hole size on the rivet yielding capacity. Karacabeyli and Foschi (1987) performed a theoretical and experimental study on eccentrically loaded glulam riveted connections. They developed a simplified model for predicting the load carrying capacity for the rivet yielding mode, and made recommendations to avoid a wood failure mode in moment connections. Karacabeyli *et al.* (1994) carried out a research study on the use of riveted connections in sawn timber. They also investigated the effect of plate thickness on the joint capacity, and determined the withdrawal strength of the rivets.

Finally, a joint study between UBC and Forintek Canada Corp. on the influence of connections on behaviour of braced timber frames (Popovski *et al.*, 1999), showed that the glulam riveted connections had superior seismic performance when compared to bolted connections for similar design load levels (Figure 15.12). During quasi-static and shake table tests, glulam riveted connections exhibited a capability of resisting many load reversals without significant strength deterioration. Large displacements were possible before failure, which permits some warning before any potential structural failure. In addition, glulam riveted connections were able to dissipate the highest amount of seismic input energy generated by the earthquake motion.

15.6 STRUCTURAL SEISMIC DESIGN CONSIDERATIONS

15.6.1 Fundamentals

One of the most important considerations in seismic design of timber structures is to provide a system that can absorb large amounts of energy and thus lower the earthquake-induced forces, while maintaining adequate stiffness to avoid excessive deformations. In other words, adequate stiffness is needed to control the elastic drift and avoid non-structural damage when a structure is subjected to minor shaking, while adequate strength is needed to avoid excessive inelastic deformation during moderate shaking to limit the structural damage. In addition, adequate ductility is needed, large enough to significantly lower the seismic forces and dissipate the seismic input energy, but small enough to maintain the overall integrity of the structure and its ability to support the gravity loads during and after the earthquake. Therefore, to ensure good structural performance, a structure needs to be designed with careful balance of *stiffness*, *strength* and *ductility*.

The above principles of seismic response of buildings are incorporated in a unified and reasonably simple manner into the codes and standards, that form the basis for the building regulations. Consequently, the content of these codes and standards, and how well they are implemented in the design and construction, will significantly affect the performance of buildings in an earthquake. Most countries develop their own set of codes and standards. In Canada, for example, the National Building Code of Canada (NBCC, 1995) serves as a model code which can then be adopted by each province to become the legal building regulation within its jurisdiction. Other countries such as the USA have several building codes: The Uniform Building Code 'Structural Engineering Design Provisions', referred to as UBC-97 (ICBO-1997); The NEHRP Recommended Provisions for Seismic Regulations for New Buildings and Other Structures, referred to also as FEMA 368 (NEHRP-FEMA, 2001); and The International Building Code, referred to as IBC2000 Code (IBC, 2000). Sometimes foreign provisions or standards are adopted when they are found to be applicable and acceptable. In addition, regional and international codes are being developed, such as The EUROCODE 8 'Design Provisions for Earthquake Resistance of Structures' (ENV, 1998), and those of the International Standards Organization 'ISO/DIS 3100 – Seismic actions on structures' (ISO/DIS, 2000). Some of the codes contain the particular seismic design information on timber structures (such as FEMA 368, Chapter 12), while others just make a reference to the adequate timber design code (such as the NBCC 1995, which makes reference to CSAO86.1).

The fundamental aim of earthquake-resistant design of buildings according to most building codes mentioned above is to provide structures with an acceptable level of safety for public use. This is achieved by specifying design loads and detailing requirements, so that the probability of building collapse or injury to people is acceptably low when the structure is subjected to a certain level of ground motion. Since the occurrence of a major destructive earthquake is usually rare during the lifetime of a structure, the seismic provisions do not usually require all buildings to have the necessary strength and capacity to withstand the largest possible earthquake without suffering any damage, which is often economically not viable. For example, the NBCC and many other national codes use the generally accepted philosophy in designing standard occupancy structures in earthquake prone regions. This philosophy aims to minimise the probability of injury and loss of life by adopting the precept that the structures should be designed:

- to withstand a minor earthquake without any damage;
- to resist a moderate earthquake without significant structural damage, but possibly with some non-structural damage, and
- to resist a severe earthquake without collapse or major failure.

This philosophy aims to achieve a balance between the opposing requirements for safety and economy by prescribing minimum requirements for seismic design to prevent injury to and death of people. This generally means prevention of collapse, but at the same time the structure may be severely damaged during an earthquake, and may even have to be demolished. Thus, under design objectives of *life safety*, if no deaths or serious injuries have occurred, such a situation could be viewed as a 'success'. If *operational readiness* or *immediate occupancy* was the design objective, however, the same severe damage would not represent a satisfactory outcome. Similar types of objectives will form the basis of *performance-based seismic design* that will be included in future building codes (Foliente *et al.*, 1999).

Codes and standards are revised periodically to accommodate new knowledge and insight about a product or structure, changing economic or social conditions, or because new building products require different approaches from previous practice. For example, new knowledge in seismology periodically requires changes in seismic zoning, because the public increasingly demands better damage protection for the buildings and equipment, and not merely life safety. In addition, new materials and devices require new rules for to be effectively employed. Consequently, buildings that were constructed when inadequate or no seismic provisions were in effect may not possess adequate seismic resistance according to today's expectations. Such buildings should be evaluated and, if found inadequate, should be upgraded to acceptable level. Many of the observed failures in past earthquakes can be traced to inadequate old code provisions or to lack of implementation of the ones that did exist.

15.6.2 Methods of Analysis

In seismic design of structures, different approaches can be followed to achieve the desired structural performance. A complex approach involves a complete time history dynamic analysis of the structure under several probable earthquakes. The complete Response History Analysis (RHA), with information on peak response, time of peak occurrence, and number of excursions over a specified response level, can therefore be obtained. In this case, each structural element is modelled by a nonlinear hysteretic model that adequately represents the load deformation properties of that element. A number of hysteresis models for different timber structural components have been developed over the years (Dowrick, 1986, Foliente, 1995). For most structures, a Response Spectrum Analysis (RSA) can also provide the information required for the design. Based on the estimates of mass and elastic stiffness properties of the structure, the RSA procedure allows for computation of the peak response values during an earthquake, represented by its response spectrum. It reduces the dynamic problem to a series of static analyses, taking into account the initial vibrational properties and damping ratios of the structure, as well as characteristics of the ground motion.

The most common design approach is to convert the dynamic loads on buildings to equivalent static lateral loads. This method is often used in seismic design codes around the world. Buildings are designed to resist specified equivalent static forces related to the properties of the structure and seismicity of the region. Based on the estimate of the fundamental period of vibration of the building, formulae are specified for the total base shear and distribution of lateral forces over the height of the building. These forces do not represent the peak dynamic forces that may be exerted on the structure during a typical earthquake for that site. It should only be expected that a structure designed to resist these equivalent static loads would perform satisfactory in accordance with the design objectives. In most seismic codes, the equivalent static loads on a structure with a certain fundamental period of vibration, are usually calculated as the inertia forces generated during its elastic response on a site with certain seismic hazard parameters and soil conditions given by the design spectrum, further reduced by a factor that accounts for the nonlinear behaviour of the structure.

As shown in Equation (15.1), the design base shear force V specified in the NBCC is calculated from the corresponding elastic structural response V_e, reduced by a force modification factor, or so-called *R-factor*. Different R-factors are assigned to different types of structural systems reflecting their design, construction experience and performance during past earthquakes. The factor accounts for the energy absorption capacity of the structural system through damping and inelastic action during load reversals. The types of structural systems that performed well during past earthquakes are assigned higher values of R-factor in NBCC:

$$V = \frac{V_e}{R} \cdot U \tag{15.1}$$

where V_e = equivalent lateral seismic force representing elastic response, R = force modification factor ($R \geq 1.0$), and U = calibration factor ($U = 0.6$).

Equivalent factors to the R-factors in North American codes are the behaviour factor q in EUROCODE 8 (EC8), and a K_D factor in ISO/DIS 3010. In other words, the total lateral design force that acts on a building V, is specified in most building codes as a function of the design weight of the structure W, its fundamental period T, seismic site parameters expressed in terms of Peak Ground Acceleration (PGA) or Peak Ground Velocity (PGV) at the site, with a certain level of exceedance (currently 10% in 50 years, but proposed to be 2% in 50 years in future codes), and the type of the lateral load resistant system in the building expressed by the R-factor. The higher the building weight and shorter the building period, the higher the design base shear. On the other hand, the higher the R-factor, the lower the design shear force.

15.6.3 Force Modification Factors

Force modification factors account for the capability of the structural system to absorb energy within acceptable deformations and without failure, and thereby reduce the structural seismic response. If there are more locations in the structure where energy can be dissipated, the risk of collapse is reduced when some individual member fails or becomes severely damaged. A building designed with a value of R greater than 1.0 is presumed capable of undergoing inelastic cyclic deformations. Connections in such systems must be therefore detailed to accommodate these deformations in a ductile manner.

There is very little theoretical or experimental background for the numerical values of the currently assigned R-factors in the codes. Therefore, the R-factors assigned to different structural systems reflect the design and construction experience, as well as the evaluation of the performance of structures in major and moderate earthquakes. The effects of other parameters in addition to general structural systems are not quantified, so the definition of R-factors requires considerable individual judgement. Consequently, the values of R-factors given in different codes vary significantly. For example, NBCC states that nailed shear walls with plywood, waferboard or OSB panels should be assigned an R-factor of 3.0, while concentrically braced timber frames or moment resistant frames with 'ductile' connections should be assigned a factor of 2.0. Other timber structures with ductile connections are entitled an R-factor of 1.5. Neither NBCC nor its counterpart for timber design in Canada CSAO86.1, however, specify what are the requirements (ductility level, strength and stiffness deterioration) for a connection to be classified as a 'ductile'.

Similarly, four different structural types are specified in EC8: (i) non-dissipative structures ($q = 1.0$); (ii) low-dissipative structures ($q = 1.5$); (iii) medium-dissipative structures ($q = 2.0$); and (iv) well-dissipative structures ($q = 3.0$). The EC8 is, however, more specific on the ductility requirements, stating that the properties of 'ductile' joints under seismic actions are required as a rule to be demonstrated through testing. By such tests, it must be shown that the ductility is sufficient and the joint properties are stable under cyclic loading under at a reasonably high load-deformation level. To ensure sufficient ductility, it is required that the ductility obtained from the cyclic tests should be greater than the assumed behaviour (force modification) factor q multiplied by a factor of 3. The multiplication factor is reduced to 2 for shear wall panel structures, because of their highly positive effect of damping caused by friction and compression perpendicular to grain between parts. These effects are believed to give a damping ratio higher than the usual value of 5% of the critical. In addition, it is stated that the connections must be able to deform plastically for at least three fully reversed cycles at the ductility levels mentioned above without an impairment of their strength of 20% or greater.

Recent studies have shown that the R-factor is a function of many different parameters. Some parameters include the initial period of the structure, its ductility, the shape of the load deformation relationship, the degree of damping, the level of over-strength in the structure, and the characteristics of ground motion and of the $P - \Delta$ effects. $P - \Delta$ effects are second-order effects that are caused by the additional moment due to large displacement and the gravity load. In the case of multi-degree-of-freedom systems, failure mechanism, torsional effects, vertical strength and stiffness irregularities, as well as soil-structure interaction, also affect the force modification factors. From all these parameters, ductility and over-strength appear to have the largest impact on R values.

The dependence of the force modification factors on the initial period of the structure is important for the design of relatively rigid structures, which have fundamental periods of vibration T lower than a specified value of $T_m = 0.5$ sec (Figure 15.13). This range represents most low to medium rise timber structures, especially the shear wall type ones. Period dependence is also important when considering the higher mode effects on more flexible structures. For a reduction of the linear elastic spectral values for intermediate and long period buildings (equal displacement range, $T > T_m$), there is a widely accepted approach to reduce these values by a factor μ (so $R = \mu$), where μ is the displacement ductility factor for a single degree of freedom system.

For short period structures, on the other hand, there is little agreement on recommended R values. The proposition to reduce the elastic spectral values by a constant factor of $R = \sqrt{2 \cdot \mu - 1}$ in the intermediate and short period ranges (equal energy range) has been widely used, in spite of the fact that the relation between the elastic and inelastic spectrum is not constant. This reduction can only be justified if the structure is subjected to relatively short acceleration pulses with respect to the fundamental period of the structure, or when the input energy for the linear elastic structure is the same as that for an equivalent inelastic (perfectly plastic) structure (Bertero, 1986). Unfortunately, these assumptions are not realistic for most cases of seismic response. Paulay and Priestley (1992) suggested that for long period structures ($T > 0.7$ sec), the relation $R = \mu$ is a reasonable assumption. At $T = 0$ and very short periods, they suggested a factor of $R = 1.0$, regardless of ductility. Between $T = 0$ sec and $T = 0.7$ sec, a linear increase in the R factor as a function of T can be assumed according to relationship (15.2):

$$R = 1 + (\mu - 1) \cdot \frac{T}{0.7} \tag{15.2}$$

To assess the appropriateness of force modification factors for wood-frame shear walls, a research project consisting of experimental testing and dynamic analyses of wood-frame shear walls was recently undertaken in Canada (Karacabeyli and Ceccotti, 2000). In the first part of

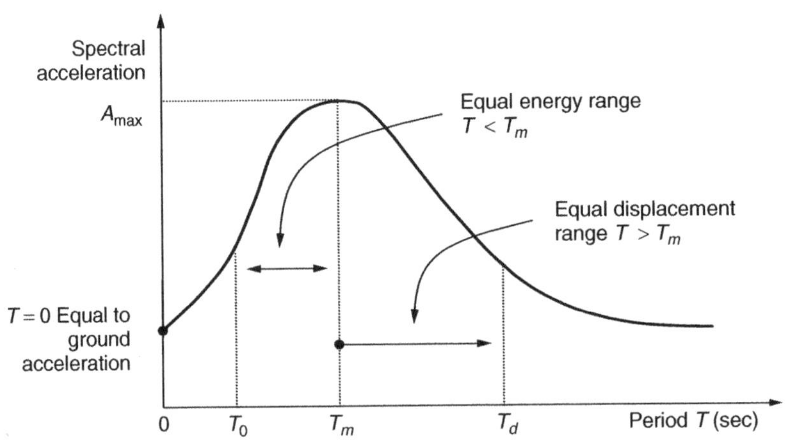

Figure 15.13 Influence of structural period on ductile force reduction

the project, a database was obtained by testing wood-frame nailed shear walls under monotonic and cyclic displacement schedules. The testing program included wood-frame shear walls sheathed with plywood, oriented strandboard and gypsum wallboard. Based on the shear wall cyclic test data, an analytical hysteretic model was calibrated and implemented in a computer program for static and dynamic analysis of structures. A typical wood frame building designed in Vancouver, BC, Canada was selected as a basic structural model for the analyses. A series of two-dimensional time-history dynamic analyses was then performed utilising 28 earthquake records, specially selected to be adequate of the seismic characteristics of a site such as Vancouver.

The results confirmed that the current seismic force modification factor ($R = 3$) in NBCC and the European behaviour factor ($q = 3$) in EC8 were well assigned for lateral load resisting systems comprised of plywood nailed walls. The results also showed that the presence of walls sheathed with gypsum wallboard generally had a positive influence on the response of the structure designed with plywood shear walls only. When the contribution of the gypsum wallboard was accounted for in the seismic resistance of the building, a seismic force modification factor of $R = 2$ was found to be appropriate.

In a similar study, static and cyclic tests were conducted on a variety of connections in Spruce-Pine glulam, utilising different diameter bolts and high strength glulam rivets (Popovski, 2000). These types of connections were chosen because they are typically used in braced timber frames. Shaking table tests were also conducted on single storey and multi-storey models of braced timber frames. The resulting hysteresis curves obtained from the tests were used to calibrate the analytical models that were later used in nonlinear static and dynamic analyses to determine the response of braced frames to the input of different earthquakes. From the analyses, it was possible to determine the influence of different connection details on the seismic response of the frames. In addition, based on the results, an estimate was made on the appropriate force modification factors in NBCC for earthquake resistant design of braced timber frames.

Results showed that the seismic response of braced timber frames is heavily influenced by the behaviour of the connections. It was suggested that braced frames with different connections should be assigned different R-factors. Braced timber frames with mild steel (ASTM A-307) bolted connections with a slenderness ratio of (l/d) of 10 or higher, showed far more adequate seismic performance than those frames that used bolts with lower slenderness ratios. Further research is needed, however, to study the effects of other parameters such as end distance, spacing, number of rows, number of bolts in a row, etc., for the development of general recommendations that will ensure ductile behaviour of braced frames with bolted connections. Until such research is undertaken, an R-factor of 1.5–1.75 appears to be reasonable for braced timber frames with slender bolts, and an R-factor of 1.0 for frames with stocky bolts. On the other hand, glulam riveted connections consistently showed non-brittle deformations in the wood, along with yielding of the connectors from small to large displacement levels. Based on the findings, it was suggested that braced frames with glulam riveted connections designed in rivet yielding mode may be assigned an R factor of 2.0, in recognition of their high and more consistent ductility capacity.

15.6.4 Capacity Design Philosophy

Finally, in seismic design of timber structures, the concept of capacity design is of major importance. This design approach, first introduced in New Zealand, is based on the simple understanding of the way a structure sustains large deformations under severe earthquakes. By choosing certain modes of deformation, we can ensure that the brittle elements have the capacity to remain intact, while inelastic deformations occur in selected ductile elements. These 'fuses' or energy absorbers act as dampers to reduce force level in the structure. It is also very important to have these ductile locations distributed in the structure at strategic locations, so that the vital parts of the lateral resistant system are not destroyed.

In steel structures, where the ductility of the connections is often questionable, the members are typically designed to yield before the connections. Beam failure mechanisms are preferred since they provide sufficient structural ductility without creating a mechanism of collapse. In timber structures, however, the failure of wood members in tension or bending is not favourable because of its limited brittle characteristics. The potential of the wood to absorb energy in compression perpendicular to grain has to be exploited, and ductile connectors that are weaker than the wood members have to be used. Since earthquakes induce numerous reversing load cycles on the structure, it is important to emphasise not only the ductility achieved during the first load cycle, but also the behaviour during the subsequent cycles. According to capacity design principles, seismic design should also assure a number of different load paths in the structure, able to transfer the inertial forces induced on the structure down to the foundation. In addition, connections that are expected to deform significantly have to be identified and designed in a highly ductile manner. Their eventual failure should not result in a structural collapse.

15.7 CONCLUDING REMARKS

The state of the art on seismic performance and design of timber structures is presented. First, causes of earthquakes are discussed along with their effects on the structures. That is followed by a description of the parameters that influence the seismic behaviour of timber structures, and observations on the seismic performance of timber buildings during past earthquakes. A brief summary on the relevant research findings on seismic behaviour of timber connections with different fasteners from various research institutions is also presented. Finally, fundamental seismic design considerations for timber structures are presented, with particular emphasis on capacity design philosophy and force modification factors.

It has been shown that wood buildings can resist very severe ground shaking when properly designed. Currently, there is a renewed interest, particularly in platform-frame wood construction, primarily due to its excellent life-safety performance observed during recent strong earthquakes. Good performance expectations are, however, contingent on appropriate design, workmanship and proper maintenance. Favourable past performance should also not be taken to imply that further improvements are no longer needed. To ensure satisfactory performance in the future, design codes and construction practices need to be periodically reviewed and revised to reflect the new knowledge obtained from theoretical and applied research, and from field observations after earthquakes.

REFERENCES

Aasheim E. (1994) Glulam trusses for the 1994 Winter Olympics. *Proceedings of Pacific Timber Engineering Conference*, Gold Coast, Australia, Vol. 1, pp. 379–386.

APA – The Engineered Wood Association (1995) *Kobe Earthquake – Report of the performance of residential and commercial structures*, Tacoma, WA.

AS (1995) *Australian Standard (draft)*. Timber methods of test for mechanical joint systems. Standards Australia.

ASTM (1993) *Proposed standard tests method for dynamic properties of connections assembled with mechanical fasteners (3rd draft)*, American Society for Testing and Materials, Philadelphia, Pennsylvania, USA.

Bertero V.V. (1988) Ductility based structural design, state-of-the-art report. *Proceedings of 9th World Conference on Earthquake Engineering*, Tokyo-Kyoto, Japan. Vol. 8, pp. 673–687.

Bertero V.V. (1986) Evaluation of response reduction factors by ATC and SEAOC. *Proceedings of the 3rd U.S. National Conference on Earthquake Engineering*, Charleston, SC, Vol. 3, pp. 1663–1673.

Bolt A.B. (1988) *Earthquakes, 2nd edition*. W. H. Freeman and Co.

Building Research Institute (1996) *A survey report for building damages due to the 1995 Hyogo-ken Nanbu Earthquake*. BRI, Ministry of Construction, Japan.

Buchanan A.H. (1988) Practical design of timber structures to resist earthquakes. *Proceedings of the International Timber Engineering Conference*, Seattle, WA, Vol. 1, pp. 813–822.

Ceccotti A. and Vignoli A. (1991) Seismic design of low-dissipative engineered timber structures. *International Meeting on Earthquake Protection of Buildings*, Ancona, Italy.

Ceccotti A. (1995b) Timber connections under seismic actions. *Timber Engineering STEP 1 – Lecture C17*, 1st edition. Centrum Hout, Netherlands.

CEN (1995) *Timber structures test methods – cyclic testing of joints with mechanical fasteners.* EN TC 124.117, Comite' Europeen de Normalisation, European Committee for Standardization. Brussels, Belgium.

Chopra A.K. (1995) *Dynamics of Structures.* Prentice Hall, Englewood Cliffs, NJ.

CSA O86.1-94 (1994) *Engineering Design in Wood (Limit States Design).* Canadian Standards Association, Etobicoke, Ontario, Canada.

CSIRO (1996) *Timber – evaluation of mechanical joint systems. Part 3 – Earthquake Loading.* Commonwealth Scientific and Industrial Research Organization, Division of Building, Construction and Engineering, Highett, Victoria, Australia.

Deam B. and King A. (1994) The seismic behaviour of timber structures, state-of-the-art. *Proceedings of Pacific Timber Engineering Conference*, Gold Coast, Australia, Vol. 1, p. 215.

Dolan J.D. and Gutshall S.T. (1997) Cyclic lateral dowel connection tests for seismic evaluation. *Earthquake Performance and Safety of Timber Structures*, Forest Products Society, Madison, WI.

Dolan J.D. and Madsen B. (1992) Monotonic and cyclic tests of timber shear walls. *Canadian Journal of Civil Engineering*, **19**(3), 415–422.

Dolan J.D. and Filiatrault A. (1990) A mathematical model to predict the steady-state response of timber shear walls. *Proceedings of International Timber Engineering Conference*, Tokyo, Japan.

Dowding C.H., Murray P.D. and Atmatzidis D.K. (1981) Dynamic properties of residential structures subjected to blasting vibrations. *Journal of Structural Engineering*, **107**, 1233–1249.

Dowrick D.J. (1986) Hysteresis loops for timber structures. *New Zealand Journal of Timber Construction*, **2**(3), 14–19.

Durham J., Prion H.G.L., Lam F. and He M. (1999) Earthquake resistance of shearwalls with oversize sheathing panels. *8th Canadian Conference on Earthquake Engineering*, Vancouver BC, Canada, Vol. 1, pp. 161–166.

EC8 (1994) European Committee for Standardization, Eurocode 8. *Design provisions for earthquake resistance of structures, Part 1: General Rules; Part1-1: Seismic actions and general requirements for structures; Part 1-2: General rules for buildings; Part 1-3: Specific rules for various materials and elements.* Central Secretariat, Brussels, Belgium.

Ellis B.R. and Bougard A.J. (1998) *Dynamic testing and stiffness evaluation of the TF2000 building.* BRE Report, UK.

Filiatrault A., Uang C.M. and Seible F. (2000) Ongoing seismic testing and analysis program in the CUREe-Caltech wood-frame project in California. *Proceedings 6th World Conference on Timber Engineering*, Whistler, Canada.

Foliente G.C., Leicester H.L. and Pham L. (1999) Concepts for performance based timber building codes and standards. *Proceedings PTEC'99 International Conference on Timber Engineering*, Rotorua, New Zealand, pp. 137–143.

Foliente G.C., editor (1997) *Earthquake performance and safety of timber structures.* Forest Products Society, Madison, WI.

Foliente G.C. (1995) Hysteresis modelling of wood joints and structural systems. *J. Structural Engineering*, **121**(6), 1013–1022.

Foschi R.O. and Longworth J. (1975) Analysis and design of griplam nailed connections. *J. Structural Division*, **101**(ST12), 2537–2555.

Foschi R.O. (1974) Load-slip characteristics of nails. *Wood Science*, **7**(1), 69–74.

Fox S.P. and Lincoln R.G. (1979) Effect of plate and hole size on griplam nail load capacity. *Canadian J. Civil Engineering*, **6**(3), 390–393.

Frenette C., Foschi R.O. and Prion H.G.L. (1996) Dynamic behaviour of timber frame with dowel connections. *Proceedings 4th International Wood Engineering Conference*, New Orleans, Louisiana, Vol. 4, pp. 89–96.

Girhammar U. A. and Anderson H. (1988) Effect of loading rate on nailed timber joint capacity. *J. Structural Engineering*, **114**(11), 2439–2457.

Gramatikov K. and Gavrilovic P. (1990) Full scale shaking table tests of a truss-frame structure. *Proceedings International Timber Engineering Conference*, Tokyo, Japan.

Guan Z.W. and Rodd P.D. (1996) A simplified FE model for double shear joints made with a hollow dowel fastener. *Proceedings of 4th International Wood Engineering Conference*, New Orleans, Louisiana, Vol. 1, pp. 164–167.

He M., Magnusson H., Lam F. and Prion H.G.L. (1999) Cyclic performance of perforated wood shear walls with oversize OSB panels. *J. Structural Engineering*, **125**(1), 10–18.

Hirashima Y. (1988) Analysis of observed earthquake response of post-and-beam wood structure. *Proceedings International Timber Engineering Conference*, Vol. 2, p. 235. Seattle, WA.

Hockey B., Popovski M., Prion H.G.L. and Lam F. (1999) Ductile connections for heavy timber structures. *Proceedings 8th Canadian Conference on Earthquake Engineering*, Vancouver, Canada, pp. 143–148.

IBC2000 (2000) *International Building Code*, International Code Council™, Inc., Falls Church, VA.

ICBO (1997) *Uniform Building Code UBC-97*, International Conference of Building Officials, Whittier, CA.

ISO/DIS 3010 (2000) *Basis for design of structures – Seismic actions on structures.* Draft International Standard, International Organization for Standardization.

ISO/DIS16670 (2001) *Timber structures – Joints made with mechanical fasteners – Quasi-static reversed-cyclic test method.* ISO Technical committee on timber structures.

Johansen K.W. (1949) *Theory of timber connections.* International Association for Bridge and Structural Engineering, Zurich, Switzerland, Vol. 9, pp. 249–262.

Kangas J. (1999) *Proceedings of Pacific Timber Engineering Conference*, Rotorua, New Zealand, pp. 341–348.

Karacabeyli E. and Ceccotti A. (2000) Dynamic analysis of wood-frame shear walls. *Proceedings 12th World Conference on Earthquake Engineering*, Auckland, New Zealand.

Karacabeyli E. and Rainer J.H. (2000) Performance of North-American platform frame wood construction in earthquakes. *COST Action E5 – Workshop on Seismic Behaviour of Timber* Buildings, Venice, Italy.

Karacabeyli E. and Ceccotti A. (1996) Test results on the lateral resistance of nailed shear walls. *Proceedings 4th International Wood Engineering Conference*, New Orleans, Louisiana, Vol. 2, pp. 179–186.

Karacabeyli E., Fraser H. and Deacon W. (1994) *Rivet connection design with sawn lumber.* Report No. 14, prepared for Canadian Forestry Service, Forintek Canada Corp, Vancouver, BC.

Karacabeyli E. and Barrett J.D. (1993) Rate of loading effects on strength of lumber. *Forest Products Society J.*, **43**(5), 28–36.

Karacabeyli E. and Foschi R.O. (1987) Glulam rivet connections under eccentric loading. *Canadian J. Civil Engineering*, **14**(5), 621–630.

Keenan F.J. (1986) *Limit States Design of Wood Structures.* Morrison Hershfield Limited.

Kohara K. and Miyazawa K. (1998) Full-scale shaking table test of two story wooden houses. *Proceedings 5th World Conference on Timber Engineering*, Montreux, Switzerland, Vol. 2, pp. 548–555.

Lantos G. (1969) Load distribution of a row of fasteners subjected to lateral load. *Wood Science*, **1**(3), 129–136.

Larsen H.J. (1973) The yield of bolted and nailed joints. *Proceedings Division V Conference*, International Union of Forestry Research Organisations, pp. 646–654.

Lefebvre G., Poultre P., Devic P. and Côté G. (1993) Statistical analyses of damages in the 1988 Saguenay earthquake. *Canadian J. Civil Engineering*, **20**(6), 988–998.

Leijten A.J.M. (1996) The concept of the prestressed DVW reinforced joint with expanded tubes. Innovative design with prestressed DVW reinforced joints. *Proceedings 4th International Wood Engineering Conference*, New Orleans, LA, Vol. 2, pp. 295–299.

Leijten A.J.M. (1988) Steel reinforced joints with dowels and bolts. *Proceedings International Timber Engineering Conference*, Seattle, WA, Vol. 2, pp. 474–488.

McLain T.E. (1975) Curvilinear load-slip relations in laterally loaded nailed joints. PhD thesis, Department of Forestry and Wood Science, Colorado State University, Fort Collins, CO.

Mischler A. (1998) Doweled steel-to-timber joints with high efficiency. *Proceedings 5th World Conference on Timber Engineering*, Montreux, Switzerland, Vol. 1, pp. 858–860.

Mohammad M. and Quenneville J.H.P. (2000) Influence of joint configuration parameters on strength of perpendicular-to-grain bolted timber connections. *World Conference on Timber Engineering*, Whistler, Canada.

Mohammad M. and Quenneville J.H.P. (1998) Influence of cyclic loads on strength and stiffness of bolted connections. *Proceedings 5th World Conference on Timber Engineering*, Montreux, Switzerland.

Moss P.J. (1997) *Multiple-bolted joints in wood members.* General Technical Report FPL-GTR-97, US Department of Agriculture, Forest Products Laboratory.

Naeim F. (1989) *Seismic Design Handbook.* Van Nostrand Reinhold.

NAHB (1994) Assessment of damage to residential buildings caused by the Northridge Earthquake. Report prepared by NAHB Research Centre for US Department of Housing and Urban Development, Washington, DC, HUD-1499-PDR.

NBCC (1995) *National Building Code of Canada, 1995 Edition.* Institute for Research in Construction, National Research Council of Canada, Ottawa, Ontario.

NEHRP (1997) Recommended provisions for seismic regulations for new buildings and other structures. Prepared by the Building Seismic Safety Council, Washington, DC, for the Federal Emergency Management Agency-FEMA-302.

NEHRP (2001) Recommended provisions for seismic regulations for new buildings and other structures. Prepared by the Building Seismic Safety Council, Washington, DC, for the Federal Emergency Management Agency-FEMA-368.

Ni C. and Chui Y.H. (1996) Predicting the response of timber joints under reversed cyclic loads. *Proceedings 4th International Wood Engineering Conference*, New Orleans, LA, Vol. 3, pp. 98–105.

Ni C. and Chui Y.H. (1994) Response of nailed wood joints to dynamic loads. *Proceedings Pacific Timber Engineering Conference*, Gold Coast, Australia, Vol. 2, p. 9.

Pacific Fire Rating Bureau (1971) *San Fernando Earthquake February 9, (1971)* San Francisco, CA.

Paulay T. and Priestley M.J.N. (1992) *Seismic Design of Reinforced Concrete and Masonry Buildings*. Wiley, New York.

Pender M.J. and Robertson T.W., editors (1987) *Edgecumbe Earthquake: Reconnaissance Report*. New Zealand Society for Earthquake Engineering, Wellington, Vol. 20 (3), 201–249.

Phillips T., Itani R. and McLean D. (1993) Lateral load sharing by diaphragms in wood-framed buildings. J. Structural Engineering, **5**, 1556–1571.

Popovski M. (2000) Seismic performance of braced timber frames. A thesis submitted in partial fulfilment of the requirements the degree of Doctor of Philosophy, University of British Columbia, Canada.

Popovski M., Prion H.G.L. and Karacabeyli E. (1999) Braced timber frames as lateral load resistant system. *Proceedings PTEC'99 International Conference on Timber Engineering*, Rotorua, New Zealand.

Popovski M., Prion H.G.L. and Karacabeyli E. (1998a) Seismic behaviour of braced timber frames. *Proceedings 6th U.S. National Conference on Earthquake Engineering*, Seattle, WA, p: 140.

Popovski M., Prion H.G.L. and Karacabeyli E. (1998b) Earthquake resistance of braced timber frames. *Recent Advances in Understanding Full – Scale Behaviour of Wood Buildings*. Forest Products Society, Madison, WI.

Prion H.G.L. and Filiatrault A. (1995) Performance of timber structures during the Hyogo-ken Nanbu earthquake of 17 January 1995. *Canadian J. Civil Engineering*, **23**, 652–664.

Prion H.G.L. and Foschi R.O. (1994) Cyclic behaviour of dowel type connections. *Proceedings Pacific Timber Engineering Conference*, Gold Coast, Australia, Vol. 2, p. 19.

Quenneville J.H.P. (1998) Predicting the failure modes and strength of brittle bolted connections. *Proceedings 5th World Conference on Timber Engineering*, Montreux, Switzerland.

Rainer J.H. and Karacabeyli E. (1999) Performance of wood-frame building construction in earthquakes. Special publication SP-40, Forintek Canada Corp.

Rodd P.D. (1988) Timber joint made with improved circular dowel fasteners. *Proceedings International Timber Conference*, Seattle, WA, pp. 26–37.

Rosowsky D.V. and Reinhold A.T. (1999) Rate-of-load and duration-of-load effects for wood fasteners. *J. Structural Engineering*, **125**(7), 719–724.

Soltis L.A., Gromala D.S. and Toumi R.L. (1980) Seismic performance of low-rise wood buildings. *Proceedings Workshop on Seismic Performance of Low-Rise Buildings*, Chicago, IL.

Tanabashi R. (1960) Earthquake resistance of traditional Japanese wooden structures. *Proceedings 2nd World Conference on Earthquake Engineering*, Vol. 1, pp. 151–163.

Touliatos P.G., Tsakanika E.P. and Carydis P.G. (1991) On the Greek experience concerning the structural behaviour of timber constructions in seismic zones. *Workshop on Full-scale Behaviour of Wood-framed Buildings in Earthquakes and High Winds*, Watford, UK.

Tryer G.W. (1932) The bearing strength of wood under bolts. *Technical Bulletin No. 332*, U.S. Department of Agriculture.

Welchert W.T. and Hinkle C.N. (1966) The effects of change of moisture content on the strength of nailed wood joints. *J. Structural Engineering*, **9**(6), 774–781.

Williams G. and Karacabeyli E. (1996) Glulam rivet connections in wood construction. *Proceedings 4th International Wood Engineering Conference*, New Orleans, LA, Vol. 4, pp. 347–354.

Williams R.L. (1984) Design of timber structures to resist high wind and earthquake. *Proceedings Pacific Timber Engineering Conference*, Auckland, New Zealand, pp. 439–445.

Yasumura M., Murota T., Nishiyama I. and Yamaguchi N. (1988) Experiments on a three-storied wooden frame building subjected to horizontal load. *Proceedings International Conference on Timber Engineering*, Seattle, WA, Vol. 2, pp. 262–275.

Yasumura M., Murota T. and Sakai H. (1987) Ultimate properties of bolted joints in glued laminated timber. *Proceedings Meeting Twenty of the International Council for Building Research Studies and Documentation, Working Commission W18 – Timber Structures*, Dublin, Ireland.

Yeh C.T., Hartz B.J. and Brown C.B. (1971) Damping sources in wood structures. *J. Sound and Vibration*, **19**(4), 411–419.

Yokel F.Y., His G. and Somes N.F. (1973) Full-scale test on a two-story house subjected to lateral load. National Bureau of Standards Building Science Series (No. 44). Vol. 72-600301/Coden:BSSNBV, Washington, DC.

PART THREE

Joints and Structural Assemblies

16

Introduction: Fasteners, Joints and Composite Structures

Hans J. Larsen

16.1 GENERAL

Joints are crucial points in many timber structures. As the length of structural timber is generally shorter than the spans, splicing or composite structures (e.g. trusses) must be used. Forces between members are most often transferred through lap joints or gusset joints, either by adhesives (glues) or by laterally loaded dowel-type fasteners (nails, bolts, screws, dowels or nail plates).

The load-carrying capacity of bonded (glued) lap joints is limited due to stress concentrations. Adhesives are therefore used mainly in continuous bond lines. However, bonded steel rods are attracting increasing interest as an effective jointing method.

Timber joints are often less effective than corresponding joints in steel, for instance, because of the relatively low embedding strength and low strength in shear, and particularly in tension, perpendicular to grain. Relatively large spacing and distances to end and edge are therefore needed to avoid splitting. This means that the necessary load transfer areas are quite large, and may often determine the size of members. This shows that great attention must be paid to the design of joints.

16.2 TRADITIONAL TIMBER JOINTS

In old timber structures, the main forces were preferably transferred directly through compression; steel was expensive. Some common joint details are shown in Figure 16.1. In some cases, the joints are quite complicated, especially where there are (small) tension forces. These designs have remained out of use for a period, but due to the introduction of CNC wood-working machines, some may again be of interest because of their

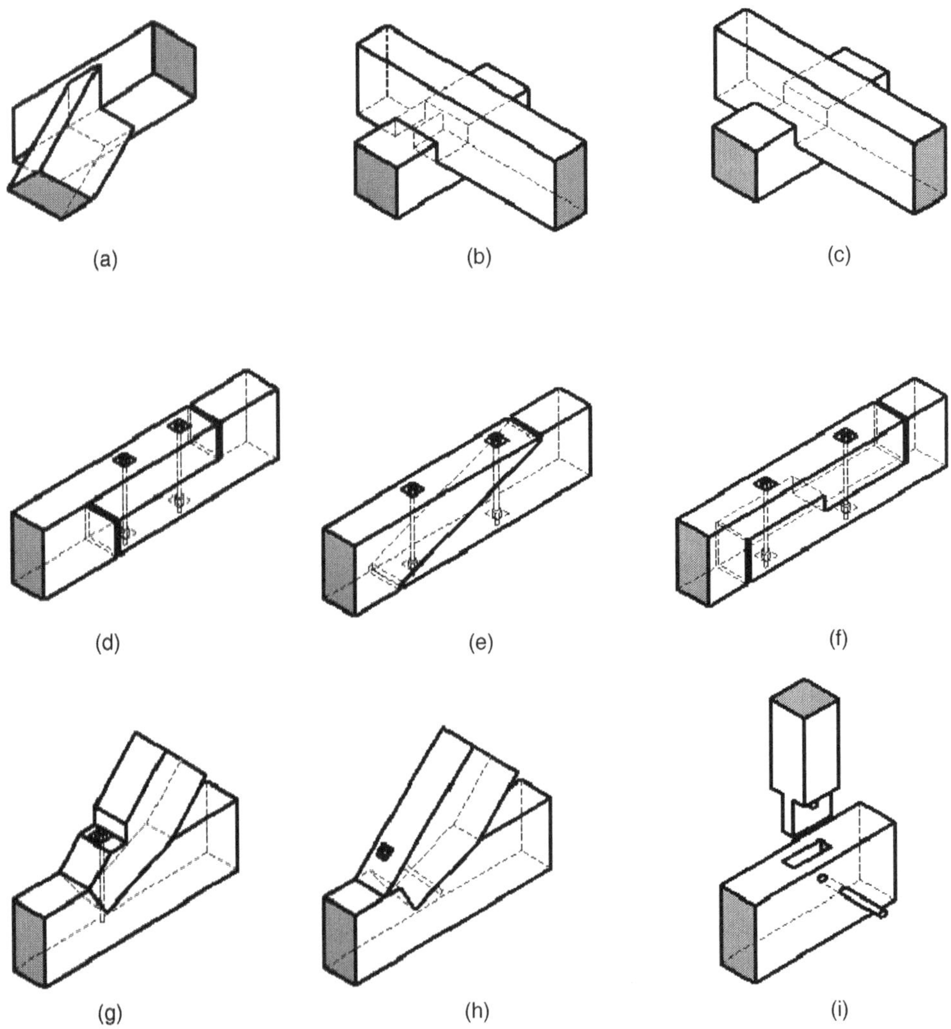

Figure 16.1 Examples of traditional timber joints

fire resistance, or for use in structures with many similar compression members, e.g. lattice shell structures.

16.3 GLUED JOINTS

Correctly made bonded joints on timber surfaces parallel to grain have the same properties as the wood. Likewise, joints between panel materials and timber and panel materials have the same properties as the weakest of the materials. This is the basis for glued laminated timber (glulam), and for built-up members such as box- and I-beams.

The strength of lap or gusset-plate joints is reduced because of stress concentrations that result in considerable size effects. These joints are, therefore, used only in relatively small structures such as trusses with moderate spans. Examples are shown in Figure 16.2.

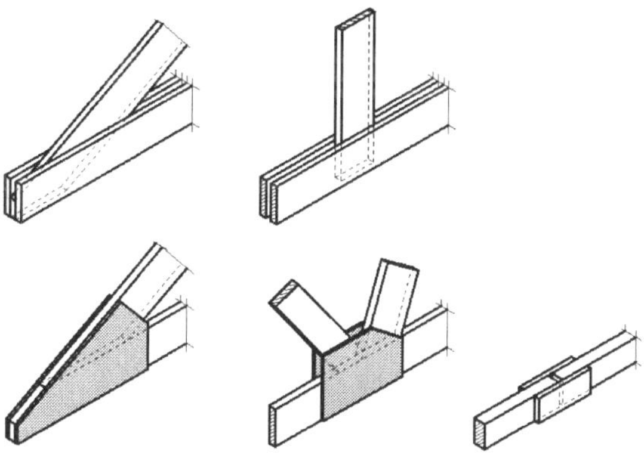

Figure 16.2 Bonded joints with and without gusset plates

Figure 16.4 Finger-jointed frame corner (Reproduced by permission of Lilleheden Advance A/S)

As shown in Figure 16.3, end joints can be made as either scarf joints, which can also be used to join panels, or as finger joints, which can be used for instance in laminations for glulam, but also for larger cross-sections (see Figure 16.4).

A more detailed treatment of bonded joints is given in Chapter 18.

16.4 NAILED JOINTS

Nails are the most common fasteners. They usually have a circular cross-section with a diameter of between 2 mm and 6 mm (in some countries 8 mm), and a length of between 30 mm and 130 mm (in some countries up to 180 mm). Smooth nails are made by cold drawing a steel wire, which results in very high strength values – Eurocode 5 is based on a tensile nail strength of around 600 MPa. Small-diameter nails have the highest strength. The withdrawal resistance is quite

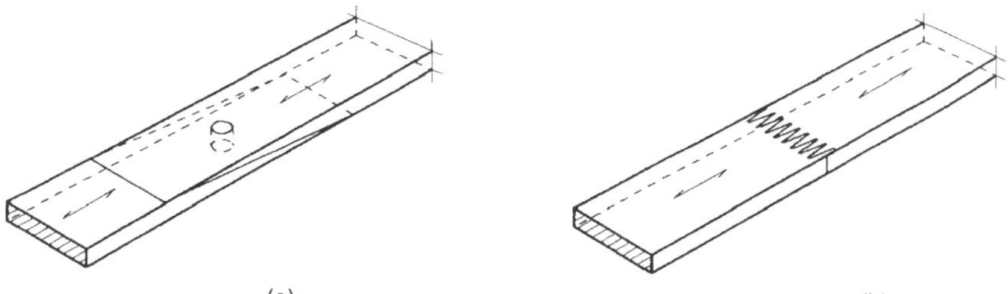

(a) (b)

Figure 16.3 Glued end joints

low. Improved pullout properties can be obtained by corrugating the surface, e.g. by rolling. The most common types are ringed shank nails and screw nails (see Figure 16.5). These nails are normally hardened to obtain a high strength. Staples are used in secondary, light structures.

Characteristic load-carrying capacities in ordinary softwoods are between 0.3 and 3.5 kN. The load-carrying capacity of a common nail (4 mm diameter) is about 1.5 kN. Nails can be placed quite closely together, are very effective and are relatively cheap because they can usually be driven into place directly without pre-drilling. Examples of joints are shown in Figure 16.6. Often, gusset plates of steel, sold as nailing plates with zinc coating and pre-punched holes, or plywood are used

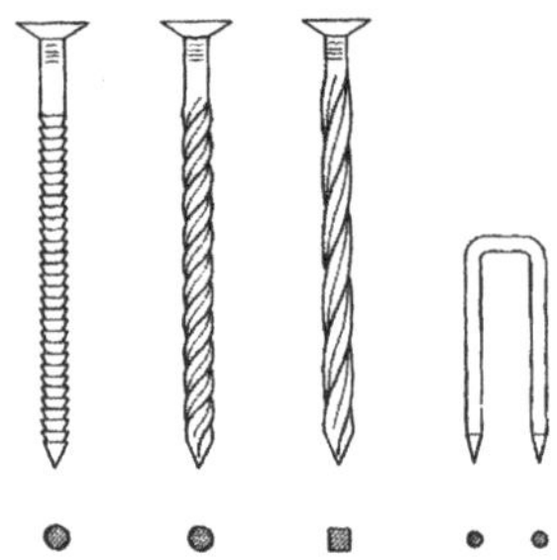

Figure 16.5 Nails and a staple

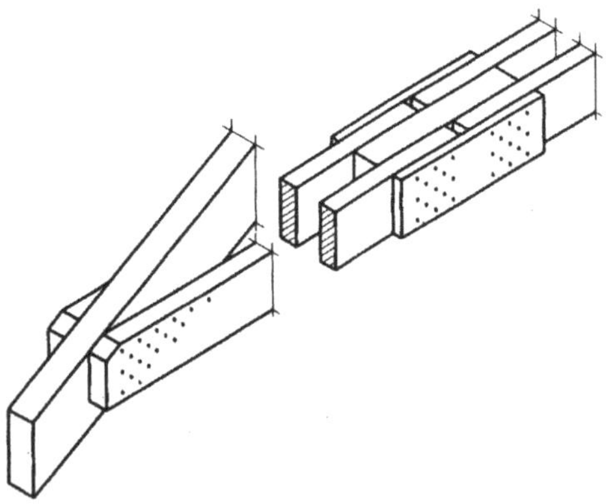

Figure 16.6 Nailed joints in trusses

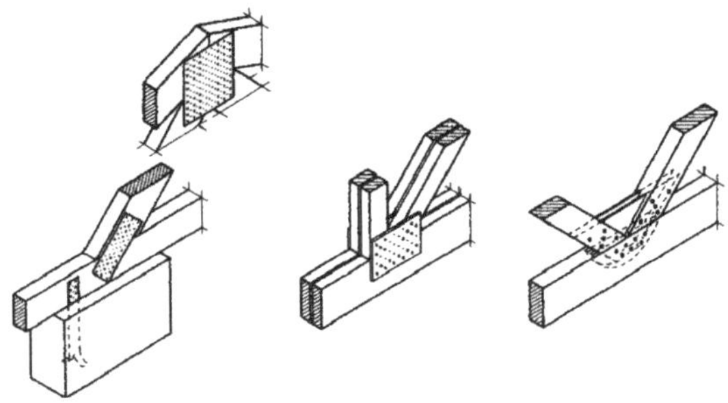

Figure 16.7 Joints with plane nailing plates

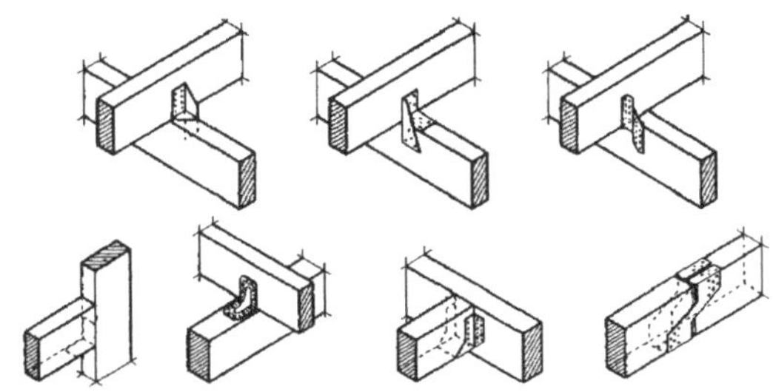

Figure 16.8 Joints with three-dimensional hangers, anchors and angles

Figure 16.9 Thick steel plate with nails. The nails are positioned manually and then driven in pneumatically

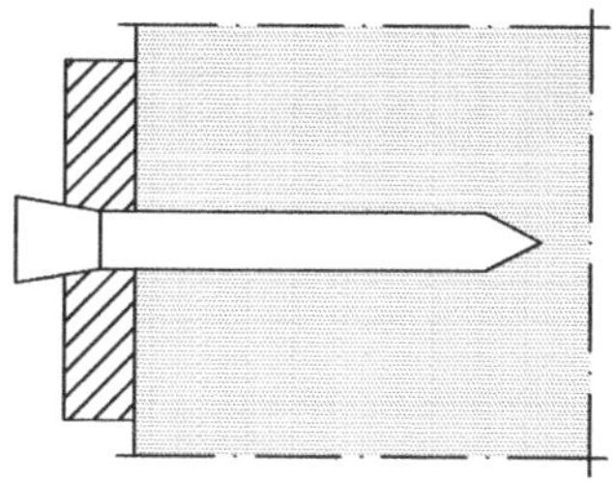

Figure 16.10 Glulam rivets. The rivets are 4 × 6 mm and placed in oblong holes

(see Figure 16.7). By using thick plates it is possible to transfer very big forces, see Figure 16.9. The Canadian Glulam rivet is based on the same principle, see Figure 16.10. Three-dimensional joist hangers and framing anchors and angles are also common (see Figure 16.8).

16.5 JOINTS WITH BOLTS AND DOWELS

Bolts and dowels are used in heavy timber structures, see Figure 16.12. Bolts are usually ordinary machine bolts (M12–M24 with a coarse thread) with washers that have a side length of about $3d$ and thickness of $0.3d$, where d is the bolt diameter. Lag bolts are also used. Holes are usually 1–2 mm oversized, which naturally reduces strength and stiffness (see Chapter 17), and dowels are often preferred to overcome these problems. Dowels are circular (turned) rods of steel, or carbon-reinforced plastics, with precise diameters, and are driven into identical or marginally undersized holes (Figure 16.11). These holes must either be drilled through all members in one operation or made using CNC machines.

Dowels have been used in very impressive structures, e.g. in the buildings that housed the Winter Olympics in Norway in 1994 (see Figures 16.13 and 14).

Bolted joints are often strengthened by connectors in the joint surfaces. Examples are shown in Figures 16.15–16.18. A complete overview of connectors on the European market and their load-carrying capacities is provided in European standards EN 912 and EN 13 271.

16.6 SCREW JOINTS

Screws were originally used where the withdrawal resistance was important, to fasten members to

Figure 16.11 Bolts and dowels

massive parts (see Figure 16.19), or instead of nails to be able to disassembly parts. With the advent of selftapping screws that can easily be driven into place by simple hand-held cordless power tools, screws have become very popular and are replacing nails in many cases. In Chapter 17, Blass describes screws and new types of screwed joints in more detail.

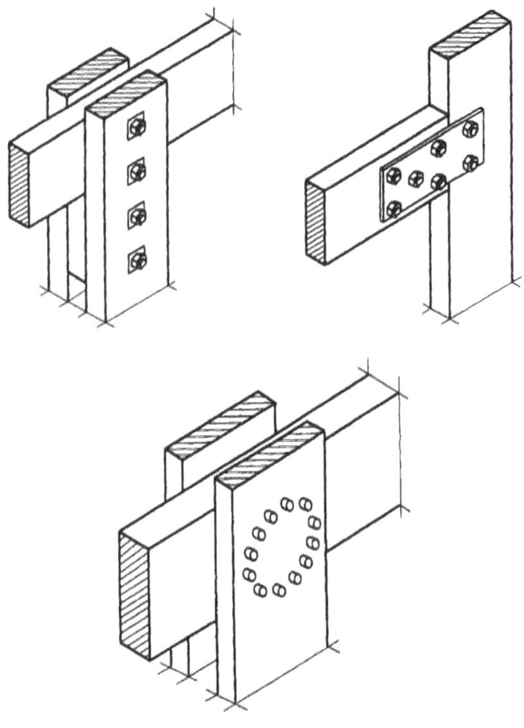

Figure 16.12 Bolted joints

Figure 16.13 Ice skate rink with trusses with doweled joints. Moelven Glulam (Reproduced by permission of Erik Aasheim, NTI)

Figure 16.14 Joints in glulam beams. Up to 11 slotted-in steel plates and 70 dowels were used per joint giving a tensile capacity of about 7500 kN. Moelven Glulam (Reproduced by permission of Erik Aasheim, NTI)

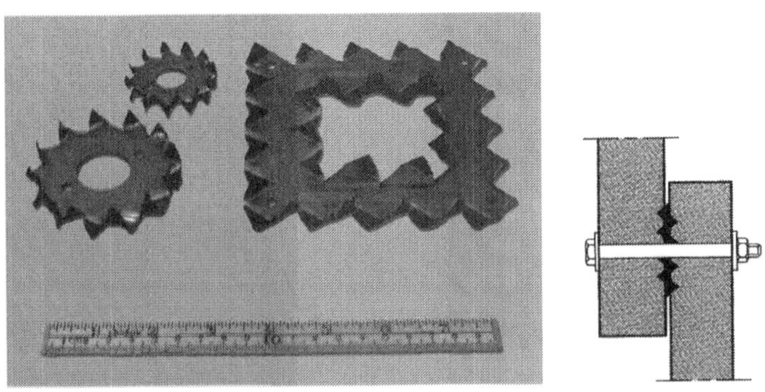

Figure 16.15 Double-sided toothed plates with oversized holes

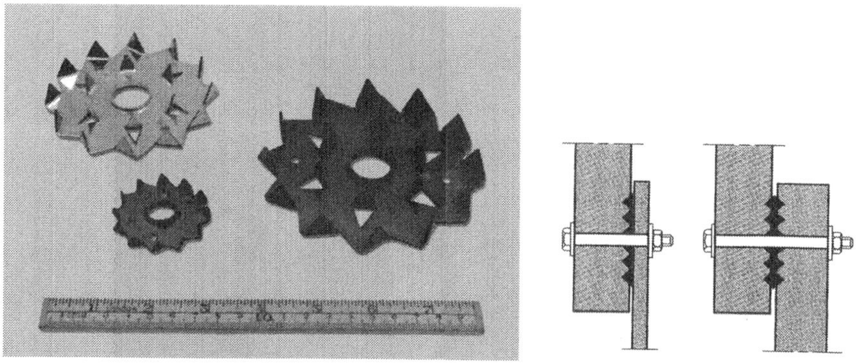

Figure 16.16 Single-sided toothed plates. The holes are of the same diameter as the bolts and the rim is reinforced by a bulb. Two plates are used in timber-to-timber joints whereas a single plate is used in timber-to-steel joints

Figure 16.17 Split ring connectors

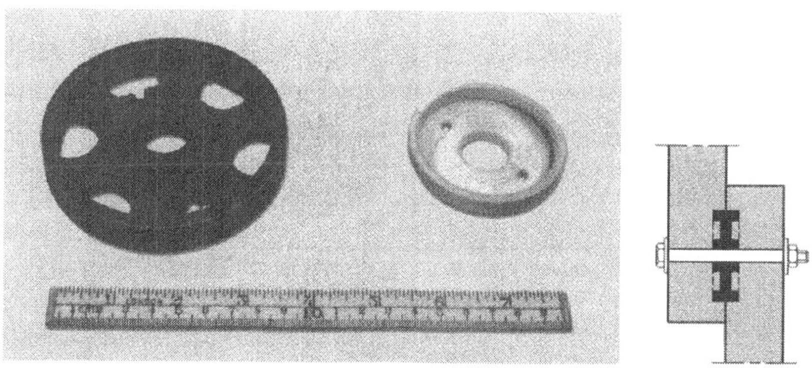

Figure 16.18 Shear plate connectors

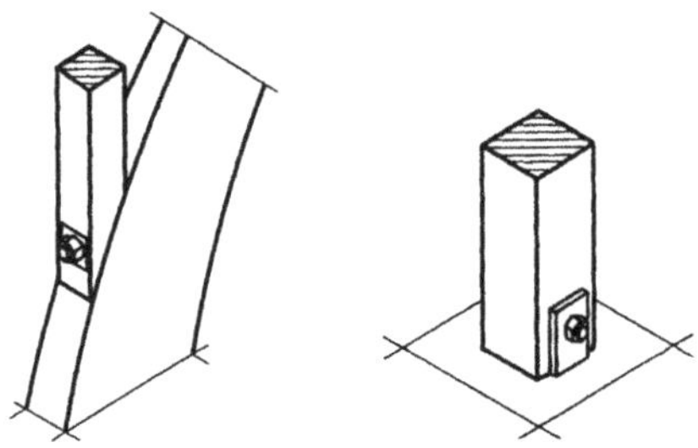

Figure 16.19 Screws used to connect a rafter to a glulam arch, and a column to a steel anchor

16.7 GLUED-IN RODS

Glued-in rods of steel (or in future perhaps of carbon-reinforced plastic) have become very popular. Figure 16.20 illustrates their uses. For a more detailed treatment, see Chapters 17 and 18.

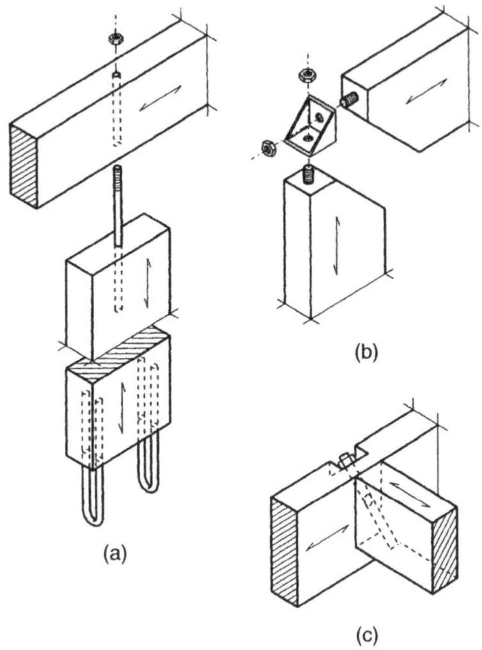

Figure 16.20 Examples of structures with glued rods

16.8 PUNCHED METAL PLATE FASTENERS

Punched metal plate fasteners (nail plates) (see Figure 16.21) dominate in the production of light trusses, as shown in Figure 16.22. A more detailed treatment is provided in Chapter 19.

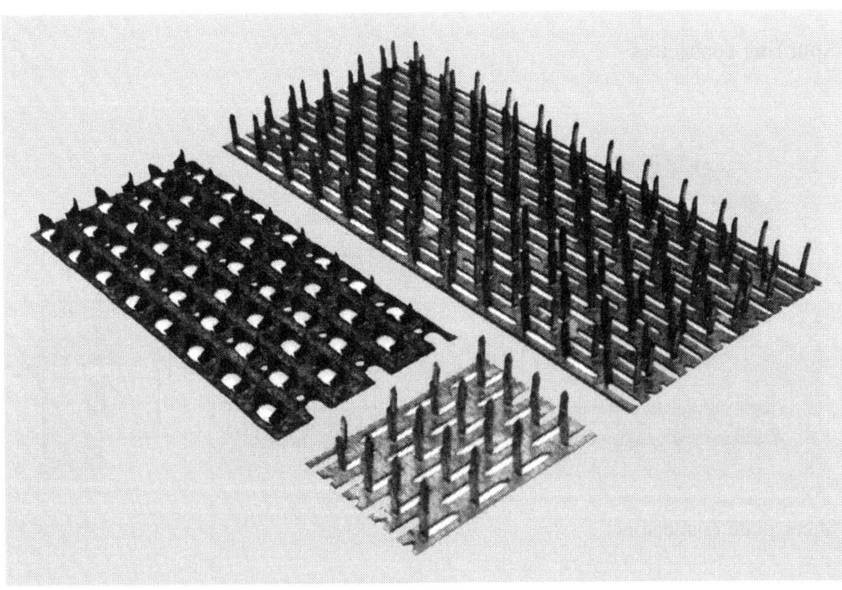

Figure 16.21 Nail plates

Figure 16.22 Trusses made with nail plates

16.9 DUCTILE AND BRITTLE LOAD-CARRYING CAPACITIES

The lateral load-carrying capacity of dowel-type fasteners is determined either by tests or by calculation based on basic material parameters such as the embedding strength and the yield moment of the fastener. For ductile failures, the European Yield Model originally proposed by Johansen (1949) is generally accepted (see Chapter 17).

The Johansen theory assumes full plasticity of both the fastener and the wood under the fastener, and is therefore not valid for brittle failures caused by wood splitting. Brittle failure is most common if the action has a component perpendicular to grain, but may also take place for action parallel to grain, especially where many stout fasteners are in a line (see Chapter 17).

The risk of splitting is reduced by using slender bolts or dowels, and many of the new fastener systems on the market today are based on the use of steel gusset plates and slender self-cutting dowels or screws of high-strength (hardened) steel (see Figure 16.23). However, stout fasteners cannot always be avoided – bolts with a length/diameter ratio of about 3 are the most appropriate for many heavy timber structures.

Figure 16.23 Joint with steel gussets and self-cutting dowels

Minimum spacing and distances are prescribed to ensure ductile failure, but are not always sufficient. The same applies to the many rules of thumb found in codes. Therefore, one of the main research areas is the prediction of load-carrying capacities for actions at an angle (perpendicular) to grain. Based on fracture mechanics (see Chapter 7), Gustafsson has derived the following expression for the failure shear force, $V_{ult,90}$:

$$V_{ult,90} = 0.5\eta b(h_{ef}GG_c)^{0.5} \tag{16.1}$$

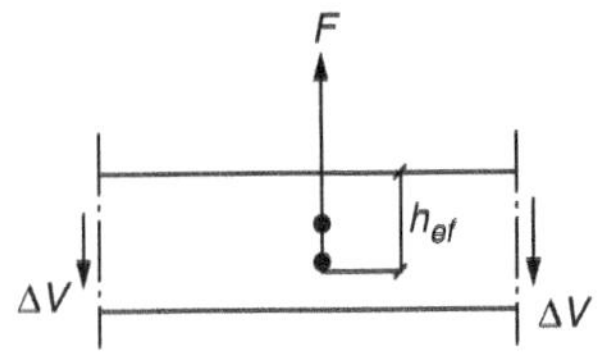

Figure 16.24 Load perpendicular to grain

The symbols mean the following (see also Figure 16.24):

- b width (mm)
- h_{ef} effective depth, i.e. the distance from the loaded edge to the innermost dowel (mm)
- G shear modulus (N/mm^2)
- G_c specific fracture energy (N/mm)
- $V_{ult,90}$ ultimate shear force (N). For a joint with the force $F_{ult,90}$, $V_{ult,90} = 0.5F_{ult,90}$
- η efficiency factor

A slightly modified expression derived by T.C. van der Put (Leijten and Jorissen, 2001) reads:

$$V_{ult,90} = 0.45\eta b(h_{ef}GG_c)^{0.5}/(1 - h_e/h)^{0.5} \quad (16.2)$$

It should be noted that the load-carrying capacity depends not upon a material strength parameter, but on the specific fracture energy. The determining geometric parameter is $h_{ef}^{0.5}$ and not h_{ef}, as seen in all empirical rules of thumb.

16.10 REINFORCEMENT

A reinforcement perpendicular to grain can ensure ductility even for loads that are perpendicular to grain and reduce the required spacing, thus increasing the load-carrying capacity considerably. Nail plates or screws placed perpendicular to grain can be used as reinforcement (see Chapter 19).

Very effective reinforcement can be obtained by gluing glass fibre textiles to the faces of the wood. An example of what can be achieved is demonstrated in Figure 16.25. The reinforcement was either 200 g/m^2 long, parallel (*uni*directional) fibres or 450 g/m^2 *randomly* oriented, chopped fibres of about 50 mm in length (both with tensile strength per unit width of about 125 N/mm). The nails were 4 mm annularly threaded and the spacing and end distance [a] in the fibre direction varied from the minimum code value 15d down to 7.5d. Even for the smallest distances, the load-carrying capacity per nail was more than 75% higher than for the unreinforced joints, and the load-carrying capacity per unit area was increased by a factor of 4 or more. For more information, reference is made to Chen *et al.* (1994) and Haller (1995, 2001).

16.11 COMPARISONS

Figure 16.26 shows typical load-slip curves. The ordinate is F/F_{ult}, where F_{ult} is the load-carrying capacity. Nails, screws and bolts, with or without pressed-in connectors, have similar ductile load-slip curves, and can be used together in a joint. Bolts without connectors should not be assumed to act together with the other mechanical fasteners because of the large initial slip caused by the

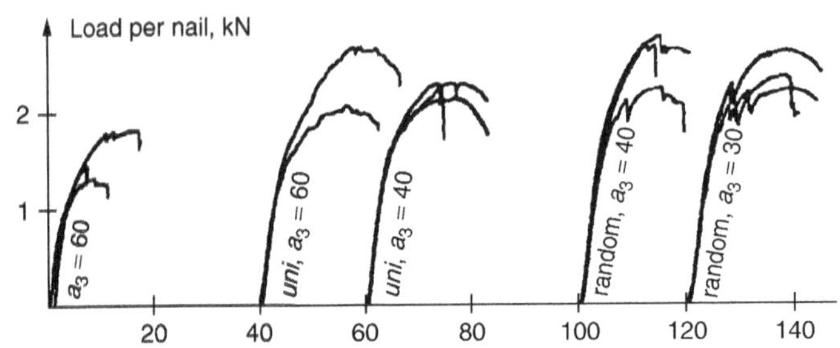

Figure 16.25 Typical load-deformation curves for reinforced nailed joints loaded parallel to grain. The left load-slip curves are for unreinforced joints with end distances of 15 times the diameter ($d = 4$ mm). The other curves are for reinforced joints with end distances of 15d and 10d

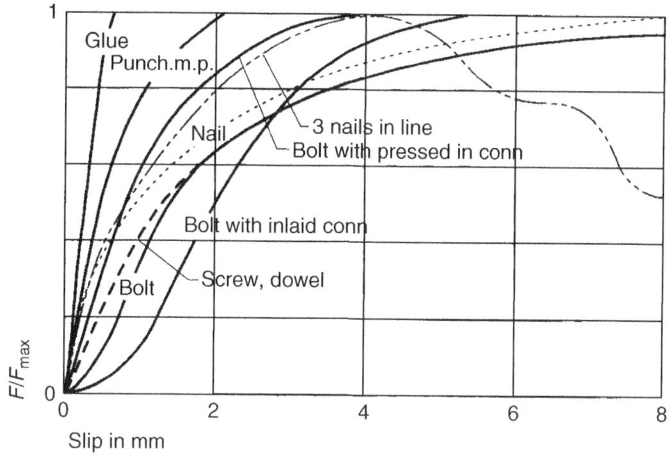

Figure 16.26 Typical load-slip curves

Table 16.1 Approximate efficiencies expressed as the load-carrying capacity (N) per unit area (mm^2)

Bonded gusset plate	2
Bolted joint	0.5–0.7
Joint with pressed-in connectors	0.9–1.3
Nails	2–3
Trusses with punched metal plate fastener	4–5

oversized holes. Bonded joints are too stiff to act together with any mechanical fastener.

Table 16.1 compares efficiencies expressed as the characteristic load-carrying capacity per unit area. The values are approximate and, because spacings are not identical to distances to end and edge, depend upon aspects such as the number of fasteners.

16.12 ASSEMBLIES

For some structures, not only the load-carrying capacity of the joints is of importance, but the load-deformation behaviour plays an important role. For many statically indeterminate structures, the load distribution is influenced by deformations of the members and slip in the joints. The latter contribution is often the bigger. Examples are trusses treated by Nielsen in Chapter 19 and panel structures (diaphragms) treated by Prion and Lam in Chapter 20.

The influence of the deformations is often treated in a very simplified way. For composite structures made by joining members by mechanical fasteners it is, however, necessary with a more rigorous treatment. This is the topic of Chapter 21.

REFERENCES

EN 912: Timber Fasteners – Specifications for connectors for timber.

EN 13 271: Timber fasteners – Characteristic load-carrying capacities and slip-moduli for connector joints.

Haller P., Wehsener J. and Birk T. (2001) Embedding characteristics of fibre reinforcement and densified timber joints. Paper CIB/W18/34-7-7, *Proceedings of Meeting 34*, Venice, Italy.

Johansen K.W. (1949) *Theory of timber connections.* International Association of Bridge and Structural Engineering, Publication No. 9, pp. 249–262.

Leijten A.J.M. and Jorissen A.J.M. (2001) Splitting strength of beams loaded by connections perpendicular to grain, model validation. Paper CIB/W18/34-7-1, *Proceedings of Meeting 34*, Venice, Italy.

17

Joints with Dowel-type Fasteners

Hans J. Blass

17.1 INTRODUCTION

Joints with dowel-type fasteners are the most common joints in timber structures. Dowel-type fasteners include nails, staples, screws, drift pins, threaded rods and bolts. Figure 17.1 shows an opened nailed connection loaded in single shear after a test to determine the load-carrying capacity. Different phenomena are observed:

- A relative displacement between the middle member and the side members occurred.
- The nails show plastic deformations due to bending.
- The timber close to the joint between the members is crushed under the action of the nails.
- The nail on the left is partly pulled out.

From these observations, three main parameters influencing the load-carrying behaviour of joints with dowel-type fasteners like nails can be identified:

- *The bending capacity of the dowel.* The bending capacity is mainly influenced by the dowel diameter and the yield strength of the dowel material. Plastic deformation capacity is essential to provide bending capacity also after considerable deformation of the dowel.
- *The embedding capacity of the timber or wood-based material.* The embedding *strength* primarily depends on the timber density, the

Figure 17.1 Opened nailed connection after a test to determine the load-carrying capacity

embedding *capacity* because of the contact area also on the fastener diameter and penetration depth.

- *The withdrawal strength of the dowel.* Threaded fasteners provide higher withdrawal capacities than smooth fasteners.

A general theory to predict the load-carrying capacity of timber connections with dowel-type fasteners loaded perpendicular to the fastener axis was first developed by Johansen (1949). Johansen considered the dowel as a beam embedded in the timber. Apart from the joint's geometry, the bearing capacity of the joint members and the bending capacity of the fastener determine the failure mode and the load-carrying capacity of the joint. Tensile forces in the fasteners were only taken into account regarding their influence on the friction between the members.

While the load-carrying capacity may be determined using Johansen's yield theory, a prediction of the load-slip behaviour is not possible. Foschi and Bonac (1977) used a nonlinear finite-element analysis to derive load-slip characteristics for nailed connections. The complete load-slip curve not only provides the ultimate load of the connection, but gives additional information regarding, for example, stiffness under service loads or ductility, properties required for a sophisticated nonlinear analysis of statically indeterminate structures.

17.2 JOHANSEN'S YIELD THEORY

17.2.1 Material Properties

Johansen's theory is based on the assumption that both the timber- or wood-based material under embedding stresses and the dowel under bending action behave as rigid-plastic materials (see Figure 17.2). While this assumption considerably simplifies the analysis, it makes little difference to the final results (Hilson, 1995).

The embedding strength f_h is defined as the ultimate load F_{max} divided by the projected area of the dowel in the specimen from an embedment test:

$$f_h = \frac{F_{max}}{d \cdot t} \qquad (17.1)$$

where d is the dowel diameter and t the thickness of the test specimen.

The embedding strength is not a material but a system property, and depends upon the properties of both the timber-or wood-based material and the

Figure 17.2 Idealised rigid-plastic load-displacement behaviour

dowel used in the test. Many researchers have published results from embedding tests. Werner (1993) has summarised the available data from literature, and identifies the following most important parameters for the embedding strength:

- *Density*: the embedding strength of softwood and hardwood timber as well as plywood increases linearly with density.
- *Fastener and hole diameter*: with increasing fastener diameter, the embedding strength decreases. The influence of fastener diameter, however, is lower than the influence of density. The diameter of pre-drilled holes up to 2 mm larger than the dowel diameter has no significant influence.
- *Angle between load and grain direction*: the influence of the load grain angle is more pronounced with larger dowel diameters, and is different for softwood and hardwood timber. Hankinson's equation describes the influence quite well:

$$f_{h,\alpha} = \frac{f_{h,0} \cdot f_{h,90}}{f_{h,0} \sin^2 \alpha + f_{h,90} \cos^2 \alpha} \qquad (17.2)$$

- *Friction between dowel and the surrounding timber*: an increasing surface roughness of the dowel or an adhesive between dowel and timber increases the embedding strength, and reduces the tendency to split for timber loaded parallel to the grain. The influence of the dowel surface is described in Rodd (1973).
- *Moisture content*: like most strength and stiffness properties of timber, the embedding strength depends upon the moisture content below the fibre saturation point.
- *Reinforcement of the timber perpendicular to the grain*: a glued-on wood-based panel or a self-tapping screw oriented perpendicular to the grain prevents timber splitting and hence increases embedding strength and ductility.

The embedding strength for timber based on a large number of embedding tests (Whale *et al.*, 1989) can be expressed depending on the fastener diameter and the timber density for loads parallel to the grain direction:

$$f_h = 0.082 \rho \mathrm{d}^{-0.3} \ \mathrm{N/mm^2}$$

$$\text{without pre-drilled holes} \qquad (17.3)$$

$$f_h = 0.082(1 - 0.01\mathrm{d})\rho \ \mathrm{N/mm^2}$$

$$\text{with pre-drilled holes} \qquad (17.4)$$

In Equations (17.3) and (17.4), ρ is the density in $\mathrm{kg/m^3}$ and d the nail diameter in mm.

For wood-based panels, similar expressions were derived by tests. For plywood, the embedding strength values are about one-third larger than for timber with the same density. The bending capacity of a steel dowel is characterised by its yield moment M_y. Johansen considered as a lower limit the elastic bending capacity of a circular cross-section

$$M_{y,\mathrm{el}} = \frac{\pi}{32} \cdot f_y \cdot d^3 \qquad (17.5)$$

where f_y is the yield stress of the fastener material.

An upper limit is the plastic bending capacity of the circular cross-section

$$M_{y,\mathrm{pl}} = \tfrac{1}{6} \cdot f_y \cdot d^3 \qquad (17.6)$$

which requires large strains, and therefore large bending angles, of up to 45° in the fasteners. Because of strain hardening at large bending angles, the yield stress in Equation (17.6) may be replaced by

$$f_y = 0.8 \cdot f_u \qquad (17.7)$$

In many tests to determine the load-carrying capacity of connections with dowel-type fasteners, the bending angles found were considerably lower than 45°. In general, the bending angle at connection failure increases with decreasing fastener diameter. Based on a theoretical derivation of the fastener bending angle at a joint slip of 15 mm, Blass *et al.* (2001) give the following equation for the bending capacity of dowel-type fasteners with a circular cross-section:

$$M_y = 0.30 \cdot f_u \cdot d^{2,6} \quad (17.8)$$

17.2.2 Johansen's Equations for Timber-to-timber and Steel-to-timber Joints

Observations of tested connections show a limited number of failure modes depending on the geometry of the connection, the embedding strength of the timber and the yield moment of the fastener. Figure 17.3 shows the possible failure modes for timber-to-timber single shear and double shear joints. The failure is either characterised by reaching the embedding strength of one or two members (failure modes (a), (b), (c), (g) and (h)) or by simultaneously reaching the bending capacity of the dowel (failure modes (d), (e), (f), (j) and (k)). If the dowel is bent plastically, either one (failure modes (d), (e) and (j)) or two (failure modes (f) and (k)) plastic hinges per shear plane occur.

The derivation of the load-carrying capacity per fastener per shear plane is shown as an example for failure mode (f) or (k), respectively (see Hilson, 1995).

The load-carrying capacity R is derived on the basis of equilibrium conditions (see Figure 17.4). The part of the dowel between the two plastic hinges is considered. Normal forces in the dowel are neglected, the shear forces in the dowel at the location, where the plastic hinges occur are zero, because the moment at these cross-sections reaches a maximum.

The following notation is used:

t_1 and t_2	is the timber thickness or fastener penetration of member 1 and 2,
$f_{h,1}$	is the embedding strength corresponding to t_1,
$f_{h,2}$	is the embedding strength corresponding to t_2,
β	is the ratio of the embedding strengths, $\beta = f_{h,2}/f_{h,1}$,
d	is the fastener diameter,
M_y	is the fastener yield moment and
R	is the resistance or load-carrying capacity per shear plane.

The sum of the forces perpendicular to the fastener axis yields

$$f_{h,1} \cdot d \cdot b_1 = f_{h,2} \cdot d \cdot b_2$$

$$b_2 = \frac{b_1}{\beta}$$

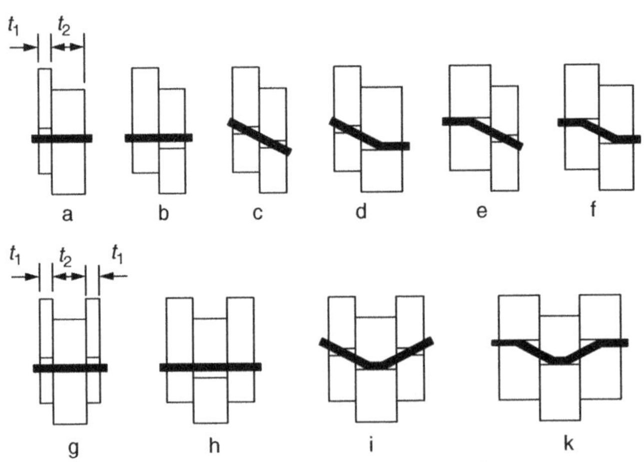

Figure 17.3 Failure modes for timber-to-timber single shear (top) and double shear (bottom) joints

Moment equilibrium gives

$$2 \cdot M_y = f_{h,1} \cdot d \cdot \frac{b_1^2}{2} + f_{h,2} \cdot d \cdot \frac{b_2^2}{2}$$

$$= f_{h,1} \cdot d \cdot \frac{b_1^2}{2} + \beta \cdot f_{h,1} \cdot d \cdot \frac{b_1^2}{\beta^2 \cdot 2}$$

$$= f_{h,1} \cdot d \cdot \frac{b_1^2}{2} \cdot \left(1 + \frac{1}{\beta}\right)$$

$$b_1 = \sqrt{\frac{2 \cdot M_y}{f_{h,1} \cdot d}}\sqrt{\frac{2\beta}{1+\beta}}$$

$$R = f_{h,1} \cdot d \cdot b_1 = \sqrt{\frac{2\beta}{1+\beta}}\sqrt{2 \cdot M_y \cdot f_{h,1} \cdot d} \qquad (17.9)$$

For the other failure modes, the resistance R may be derived similarly (e.g. see Hilson, 1995). Table 17.1 summarises the load-carrying capacity R per fastener per shear plane for the failure modes of single shear joints shown in Figure 17.3. The corresponding expressions for double shear joints are given in Table 17.2. Because of the continuous dowel, failure modes (c) and (e) do not occur in double shear joints.

Tables 17.3–17.7 contain the expressions for R according to Johansen's yield theory for steel-to-timber connections. For these connections, two extreme cases for the support of the dowel by the steel plate are considered: pin-jointed and fixed support. In the case of a simply supported dowel, no clamping is provided by the steel plate and consequently no moment is transferred and the

Table 17.1 Load-carrying capacity R per fastener per shear plane for single shear timber-to-timber joints

$$R = f_{h,1}t_1d$$

(a)

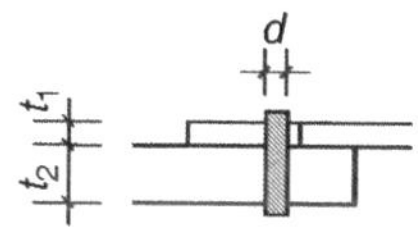

$$R = f_{h,1}t_2d\beta$$

(b)

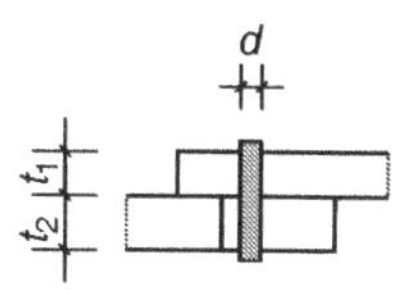

$$R = \frac{f_{h,1}t_1d}{1+\beta}\left\{\sqrt{\beta + 2\beta^2\left[1 + \frac{t_2}{t_1} + \left(\frac{t_2}{t_1}\right)^2\right] + \beta^3\left(\frac{t_2}{t_1}\right)^2} - \beta\left(1 + \frac{t_2}{t_1}\right)\right\}$$

(c)

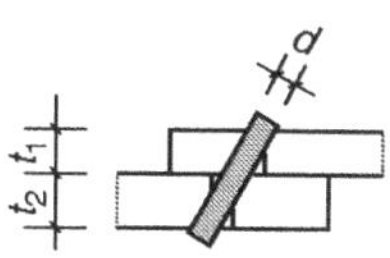

$$R = \frac{f_{h,1}t_1d}{2+\beta}\left[\sqrt{2\beta(1+\beta) + \frac{4\beta(2+\beta)M_y}{f_{h,1}dt_1^2}} - \beta\right]$$

(d)

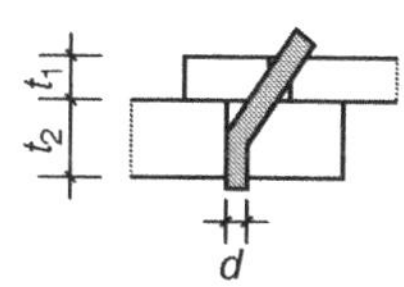

$$R = \frac{f_{h,1}t_2d}{1+2\beta}\left[\sqrt{2\beta^2(1+\beta) + \frac{4\beta(1+2\beta)M_y}{f_{h,1}dt_2^2}} - \beta\right]$$

(e)

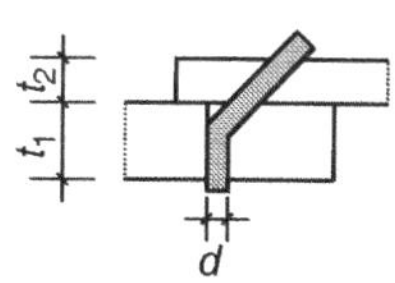

$$R = \sqrt{\frac{2\beta}{1+\beta}}\sqrt{2M_yf_{h,1}d}$$

(f)

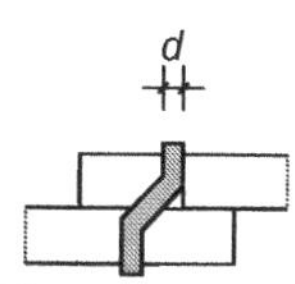

Table 17.2 Load-carrying capacity R per fastener per shear plane for double shear timber-to-timber joints

$R = f_{h,1}t_1d$	(g)	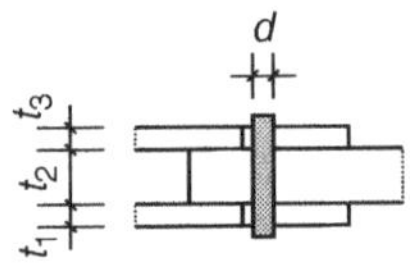
$R = 0,5f_{h,1}t_2d\beta$	(h)	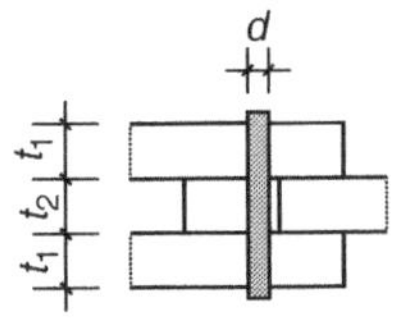
$R = \frac{f_{h,1}t_1d}{2+\beta}\left[\sqrt{2\beta(1+\beta)+\frac{4\beta(2+\beta)M_y}{f_{h,1}dt_1^2}}-\beta\right]$	(j)	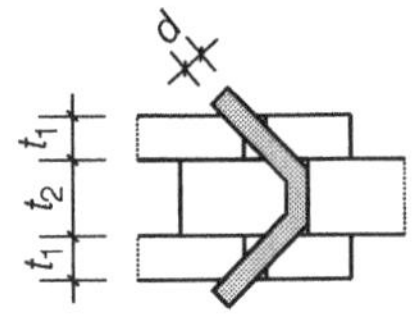
$R = \sqrt{\frac{2\beta}{1+\beta}}\sqrt{2M_yf_{h,1}d}$	(f)	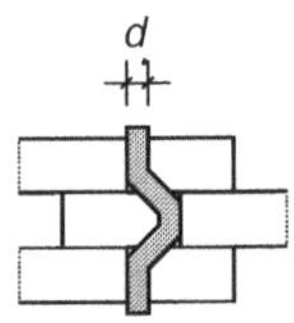

Table 17.3 Load-carrying capacity R per fastener per shear plane for single shear steel-to-timber joints; fixed support of the dowel in the steel plate

$R = f_{h,1}t_1d$	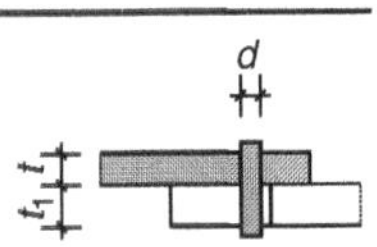
$R = f_{h,1}t_1d\left[\sqrt{2+\frac{4M_y}{f_{h,1}dt_1^2}}-1\right]$	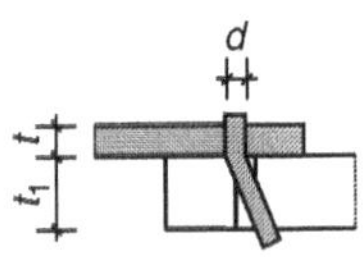
$R = \sqrt{2}\sqrt{2M_yf_{h,1}d}$	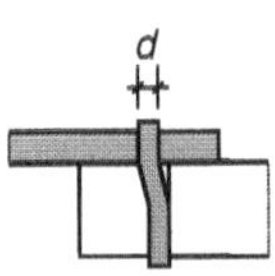

dowel is able to freely rotate in the steel plate. For the case of a clamped support, the steel plate is assumed to provide sufficient rotation resistance to enable a plastic hinge immediately at the surface of the steel plate. In steel-to-timber connections, where only the embedding strength of the timber or wood-based material is reached, the location of plastic hinges for dowels clamped in steel plates is always at the plate surface.

With decreasing bending capacity of the dowel and increasing member thickness and embedding strength, the failure modes change from a pure embedding failure of the timber or wood-based material towards a simultaneous embedding and bending failure. Due to the plastic deformation

Table 17.4 Load-carrying capacity R per fastener per shear plane for single shear steel-to-timber joints; pinned support of the dowel in the steel plate

$R = (\sqrt{2}-1)f_{h,1}t_1d$	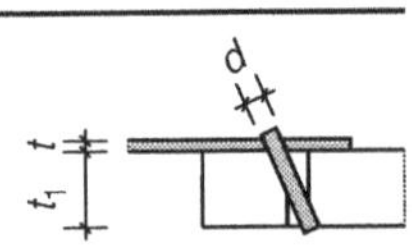
$R = \sqrt{2M_yf_{h,1}d}$	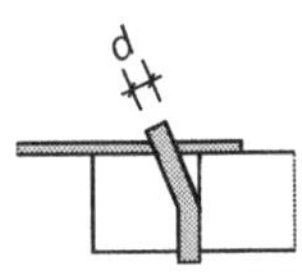

Table 17.5 Load-carrying capacity R per fastener per shear plane for double shear steel-to-timber joints with a steel centre member

$$R = f_{h,1}t_1d$$

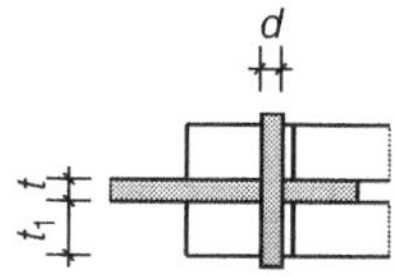

$$R = f_{h,1}t_1d\left[\sqrt{2 + \frac{4M_y}{f_{h,1}dt_1^2}} - 1\right]$$

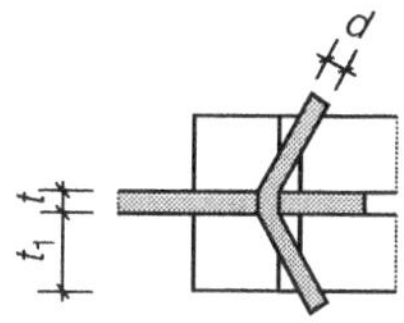

$$R = \sqrt{2}\sqrt{2M_yf_{h,1}d}$$

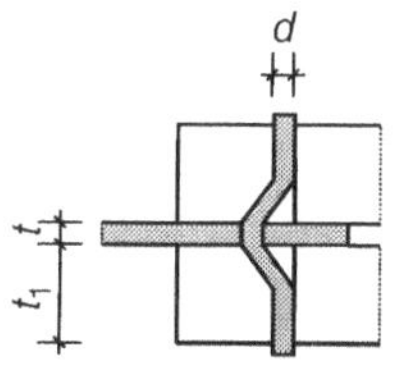

Table 17.6 Load-carrying capacity R per fastener per shear plane for double shear steel-to-timber joints with outer steel members; fixed support of the dowel in the steel plate

$$R = 0.5f_{h,2}t_2d$$

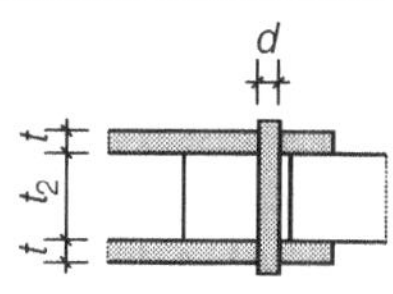

$$R = \sqrt{2}\sqrt{2M_yf_{h,2}d}$$

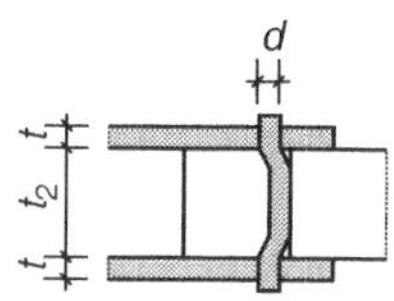

capacity of the steel dowels, failure modes including plastic hinges result in a more ductile failure compared with failure modes where the dowel is not bent. Other benefits of connections with slender dowels are a possible tensile action in the inclined part of the deformed dowel leading to an additional resistance of the connection (see Section 17.2.3), and a decreased splitting tendency of the timber for loads parallel to the grain. Because of these benefits, the preferred failure mode in general shows

Table 17.7 Load-carrying capacity R per fastener per shear plane for double shear steel-to-timber joints with outer steel members; pinned support of the dowel in the steel plate

$$R = 0.5f_{h,2}t_2d$$

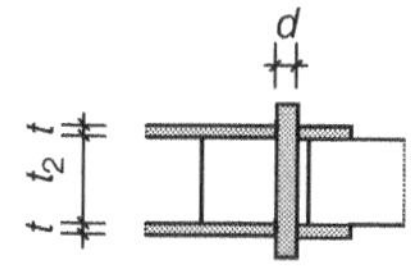

$$R = \sqrt{2M_yf_{h,2}d}$$

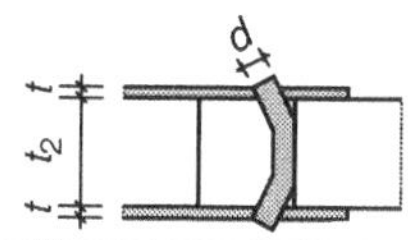

two plastic hinges per shear plane, for steel-to-timber joints with outer steel plates providing no clamping it is one plastic hinge per shear plane.

In an actual joint, the load-carrying capacity corresponds to the lowest value of R from all possible failure modes. The equation giving the lowest capacity also identifies the failure mode.

17.2.3 Additional Resistance Through Tensile Action in the Fastener

When the load-carrying capacity of a connection is reached, in some failure modes the fasteners are either inclined over the whole length (e.g. failure mode (c) in Figure 17.3), or over a part of the fastener length. The latter always occurs in combination with at least one plastic hinge in the fastener. In the inclined part, tensile forces develop depending on the angle of rotation of the fastener axis (see Figure 17.5). These tensile forces either immediately pull the fastener out of the holes in the members or, if there is a sufficient withdrawal resistance, increase the connection capacity by a component parallel to the joint line. The withdrawal resistance depends upon the surface of the fastener shank as well as on an anchor effect of the fastener end, e.g. a nail head or a bolt nut and washer. It is also influenced by the timber density and has to be determined by tests. The capacity is further increased by a friction component R_f caused by the compressive force F_c.

Kuipers and Van der Put (1982) have shown that for long nails with threaded shanks, the theoretical load-carrying capacity can reach up to

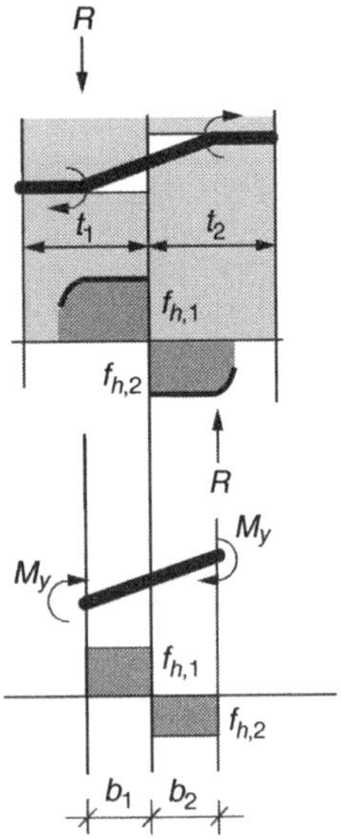

Figure 17.4 Failure mode for timber-to-timber joints with two plastic hinges per shear plane

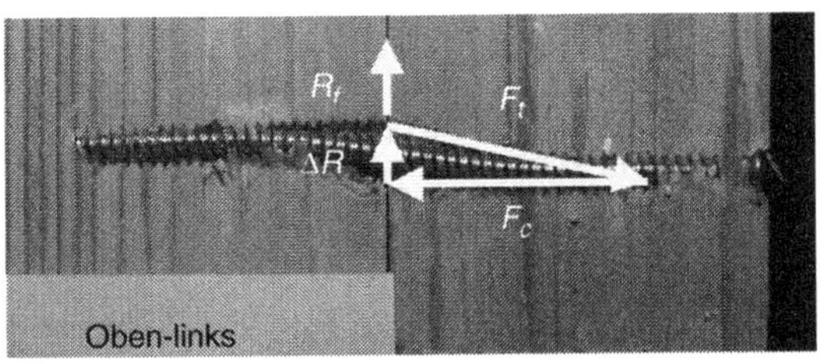

Figure 17.5 Additional resistance ΔR through tensile force F_t and R_f caused by friction in a screw with continuous thread

2.6-times the value based on the equations given above. For single shear steel-to-timber joints with annular ringed shank nails Görlacher (1995) has published test results confirming the increase in load-carrying capacity with increasing nail length and a corresponding increase in withdrawal capacity. In timber-to-timber connections, significant tensile action in fasteners can be expected for bolts, threaded rods and screws. The tensile action of threaded nails depends upon the head pull-through resistance which is mainly influenced by the head diameter and the density of the head-side member.

In nailed steel-to-timber connections, head pull-through does not occur, the tensile capacity is either limited by the withdrawal resistance of the point-side shank or the tensile strength of the nail. Screws with a continuous thread along the shank show the largest withdrawal capacity among the different types of dowel-type fastener.

17.3 GROUP EFFECTS IN JOINTS WITH DOWEL-TYPE FASTENERS

Joints with dowel-type fasteners usually contain more than one fastener. Due to variations in fastener yield moment or timber density and moisture content within the connection area, the individual load-slip behaviour and ultimate fastener loads in a connection are different. The total load-carrying capacity of the connection equals the sum of the single fastener loads at failure. In an optimal connection, the single fasteners all reach their respective ultimate loads and the ultimate load of the connection equals the sum of the ultimate loads of the single fasteners. If the single fastener loads at the failure of the connection show large differences (see Figure 17.6) and some of the fasteners are loaded below their individual load-carrying capacity, the ultimate load of the connection is smaller than the sum of the single fastener ultimate

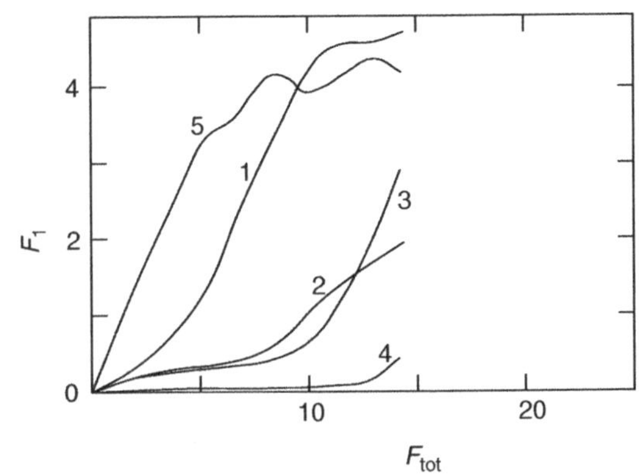

Figure 17.6 Example of a load distribution in a bolted joint according to Wilkinson (1986). F_1 is the single fastener load and F_{tot} is the total joint load

loads. This decrease in load-carrying capacity per fastener is the group effect.

The reason for the different load levels of the single fasteners in Figure 17.6 is the failure mode. The double-shear steel-to-timber connection with outer steel plates and a timber centre member contained a single bolt row with five bolts and was loaded parallel to the grain. It failed before significant plastic embedding deformations took place at every fastener location. The different load take-up at low load levels is caused by the drilling tolerances of oversized bolt holes.

Premature failure before the connection shows a plastic behaviour is often caused by timber splitting along the grain (see Figure 17.7). This failure mode is not considered in the Johansen theory. Tests with multiple fastener timber connections therefore often result in lower ultimate loads than predicted by the yield theory. This is especially true for connections with rows of fasteners arranged and loaded parallel to the grain.

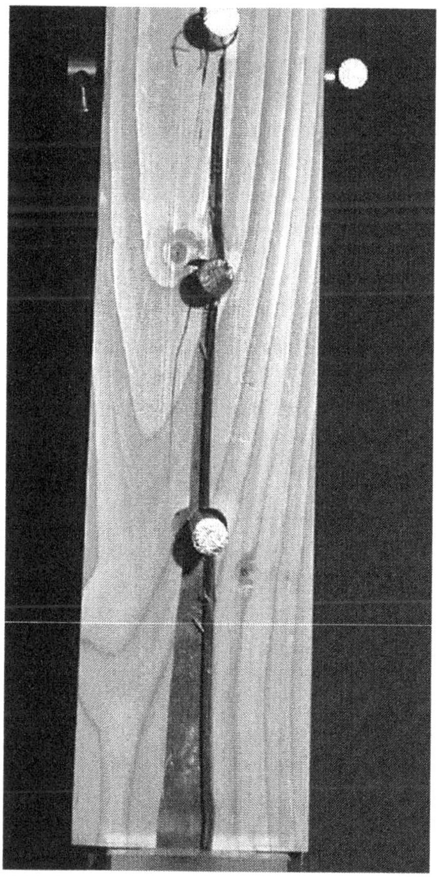

Figure 17.7 Premature splitting in a steel-to-timber connection with dowel-type fasteners

Slender fasteners with a high ratio of penetration depth to fastener diameter cause less splitting than stout fasteners. In particular, stout fasteners act as wedges and cause tensile stresses perpendicular to the grain.

The tensile stresses and the subsequent splitting may be minimised by appropriate spacing and end distances. The larger the spacing, the smaller the tensile stresses perpendicular to the grain caused by the wedge effect of the fasteners. For nailed connections with sufficient spacing an influence of fastener number or arrangement on the characteristic load-carrying capacity per fastener could not be observed (Blass, 1994). Jorissen (1998) has extensively studied the influence of fastener arrangement, number of fasteners and fastener slenderness on the load-slip behaviour of connections with dowel-type fasteners. His comprehensive results lead to the conclusion, that the fastener spacing parallel to the grain is the most important parameter influencing the group effect in multiple fastener timber-to-timber connections loaded parallel to the grain.

Based on Jorissen's research, the effective number of fasteners n_{ef} for one row of n bolts or dowels arranged and loaded parallel to the grain may be calculated as

$$n_{ef} = n^{0.9} \sqrt[4]{\frac{a_1}{10d}} \qquad (17.10)$$

where a_1 is the fastener spacing parallel to the grain and d the dowel diameter.

A possibility to avoid splitting and to guarantee a plastic connection behaviour is to reinforce the timber in the joint area. The tensile stresses are then transferred by a reinforcement perpendicular to the grain. Figure 17.8 shows a connection where the timber side and middle members are reinforced using self-tapping screws with a continuous thread over the whole length of the shank. The screws act like reinforcement bars in concrete and prevent the cracks from further growing. Based on Blass and

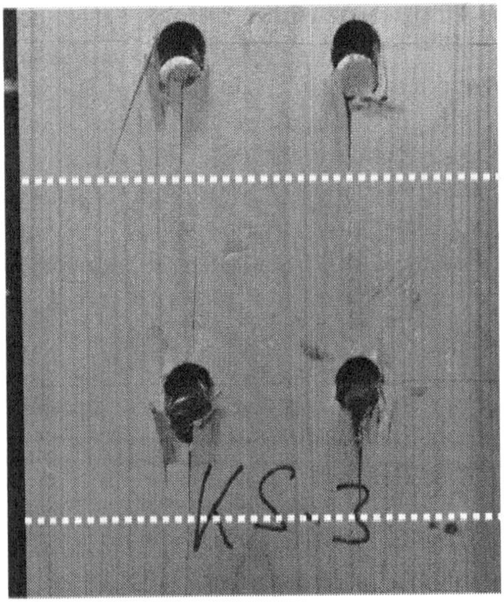

Figure 17.8 Timber-to-timber connection with reinforcement perpendicular to the grain (dotted lines) using self-tapping screws

Schmid (2001), the tensile load in the reinforcing screw may be estimated as 20% of the load transferred by the dowel itself.

Although the dowel slenderness in the side members is extremely low – the side member thickness is only 1.25 times the dowel diameter – the connection behaviour is distinctly plastic. For comparison, the same connection was tested without reinforcement perpendicular to the grain. The resulting failure mode was splitting of the side members. Figure 17.9 shows the load-slip behaviour for both cases. For reinforced connections, the group effect therefore may be disregarded, the total load carrying capacity of the connection equals the accumulated single fastener load-carrying capacities.

17.4 DURATION OF LOAD EFFECTS IN JOINTS WITH DOWEL-TYPE FASTENERS

Like timber, timber connections also experience a loss of load-carrying capacity when loaded at high load levels over a period of time longer than the short-term laboratory test. Apart from the decrease in strength, the slip in loaded connections increases over time. These effects are enlarged by temperature and moisture variations leading to mechano-sorptive effects. Van de Kuilen (1999) has studied the duration of load effects in timber joints based on still ongoing tests started in 1962. Two different duration of load effects for timber joints are distinguished:

- creep as the increase of deformation over time, and
- decrease of load-carrying capacity over time expressed either as time to failure under a constant load or damage accumulation under varying or constant loads.

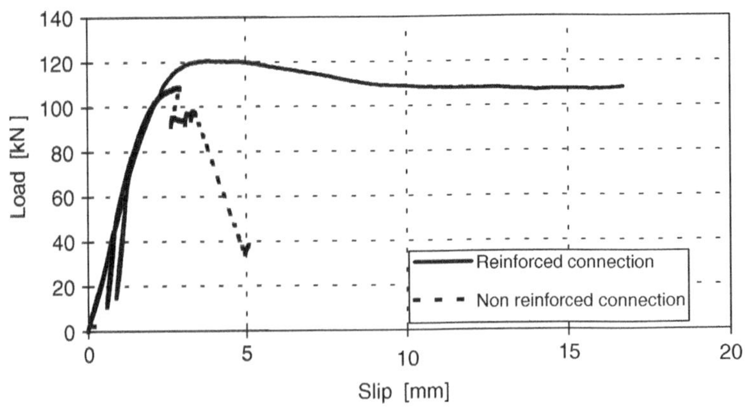

Figure 17.9 Load-slip curves for the connections shown in Figure 17.8

Joint creep as the increase of deformation over time under constant load is mainly influenced by variations in relative humidity. A decreasing timber moisture content leads to increasing deformations and vice versa. Generally, creep rates defined as creep deformation over initial deformation are distinctly larger for joints than for timber members. This can be explained by the fact that the moisture transfer in timber parallel to the grain is much faster than perpendicular to the grain. Consequently, the moisture exchange between timber and the environment through end grain is more pronounced than through side grain. Since connections are often located at member ends and the fastener holes represent additional end grain surfaces, the moisture variations and the resulting local deformations in the immediate vicinity of the fastener are more pronounced than the global deformation of timber members.

Time to failure is defined as the time between loading of a constant load and the subsequent failure of the connection. Although this parameter does not represent a real situation for a member or joint in a timber structure, it was first used to determine the duration of load effects of timber and of timber connections. In 1962 different test series were started at Delft University of Technology with nailed joints, split ring and toothed plate connections. Load levels expressed as ratios between the long-term load in the test and the average short-term load-carrying capacity of identical connections varied between 60% and 95%. After 38 years, the nailed specimens loaded to 60% and 65% have not yet failed. Specimens loaded to levels higher than 65% all failed (Van de Kuilen, 1999).

The slip in the connections was and in 2001 still is measured every two weeks. The average creep deformation in the nailed joints prior to failure was 16.6 mm. Since the original penetration depth of the nails in the timber middle member was only 27 mm, the nails gradually were pulled out of the middle member. The failure in the long-term tests was characterised by a combination of splitting of the side members and nail withdrawal from the middle member. Similar to the connection presented in Figure 17.1, a plastic hinge occurred in the middle member and the nails remained basically straight in the side members (failure mode (d) or (e) in Figure 17.3). A different failure mode was observed in the short-term tests. Here, the failure mode with two plastic hinges per shear plane was governing (failure mode (f) in Figure 17.3). Splitting in the side members was not observed to a large extent in the short-term tests.

The change of failure mode over time in this example is mainly caused by the increasing occurrence of splits in the side members which may mechanically be interpreted as a decrease in embedding strength. Considering the equations in Table 17.1 based on Johansen's yield theory, the change in failure mode is obvious. Since in the expression for R in failure mode (f) the embedding strength appears under the square root, the influence of duration of load is less pronounced as, for instance, in failure mode (a). Consequently, the duration of load effect determining the time to failure in timber joints with dowel-type fasteners is failure mode dependent and decreases with increasing fastener slenderness.

The wedge effect of fasteners in timber connections always causes tensile stresses perpendicular to the grain and shear stresses. Depending on the amount of these stresses the duration of load effect for connections also depends upon the duration of load effect for timber under shear stresses and tensile stresses perpendicular to the grain. The latter is known to be more unfavourable than for timber under, for example, bending or compression. In general, however, the time to failure for timber connections is in the same order as for timber.

In most real timber structures, high loads are not present over long time periods, since the permanent load generally represents only a small portion of the total load. There is no evidence that permanent loads below a threshold value of 30–40% of the short-term strength cause any decrease in timber strength. Damage accumulation accounts for the duration of load effect during those time periods where high load peaks occur over a short time. To study the effects of accumulated loads on the residual strength of timber connections, tests were performed at Delft University of Technology (Van de Kuilen, 1999) to obtain information on the effect of pre-loading on the strength reduction. Test specimens similar to those used in the time to failure tests were loaded to levels between 30% and 50% of the short-term load-carrying capacity for up to eight years, and the residual strength was then determined in short-term tests.

The test results show no reduction in load-carrying capacity and thus no discernible damage accumulation due to the pre-loading periods. This result is supported by the evaluation of the tests using damage accumulation models. The results also show that during most of the lifetime of the time to failure tests the full load carrying capacity is available, and that the damage accumulates over a relatively short time period immediately before failure.

17.5 JOINTS WITH SCREWS

17.5.1 Screws Loaded Perpendicular to the Screw Axis

Two categories of wood screws for connections in timber structures may be distinguished. Traditional wood screws (see Figure 17.10) show a threaded and a smooth part of the shank with distinctly different yield moments. The outer diameter of the thread generally equals the smooth shank diameter which is also the nominal diameter. The diameter of coach screws varies between 8 mm and 20 mm, the diameter of screws with countersunk or round head between 4 mm and 8 mm. The root diameter d_1 usually corresponds to 70% of the outer diameter d (Ehlbeck and Ehrhardt, 1995).

Traditional screws with a diameter of more than 8 mm should be driven into pre-drilled holes to prevent wood splitting and torsion failure of the screw itself caused by the moment necessary for driving the screw. Over the length of the smooth shank, the holes should be pre-drilled with the shank diameter, along the threaded part with the root diameter. Since traditional wood screws are not hardened after cutting or rolling the thread, the bending capacity as well as the torsion capacity are limited. Because of the time-consuming assembly and the comparatively low load-carrying capacity, traditional wood screws are not used to a large extent in timber structures.

Examples of self-tapping wood screws are shown in Figure 17.11. These screws are manufactured with diameters up to 12 mm and lengths up to 600 mm.

After rolling the thread, the screws are hardened. Hardening increases the bending capacity as well as the torsion capacity. Hardened screws are available with threads over a part of the shank length or with continuous threads.

Johansen's equations may be used to determine the load-carrying capacity of screwed connections. The nominal diameter d corresponding to the outer thread diameter may be used except for determining the fastener yield moment. This means that the outer thread diameter determines the embedding area. For traditional wood screws, it is assumed that the thread itself also contributes to the bending capacity, and an effective diameter of 0.9 d should be used to determine the yield moment, provided the root diameter is at least 0.7 d. For hardened wood screws, the yield moment should be determined by bending tests. Generally, the lower value of the yield moment of the smooth shank and threaded shank, respectively, is used to determine the load-carrying capacity of the connection.

Figure 17.10 Coach screw

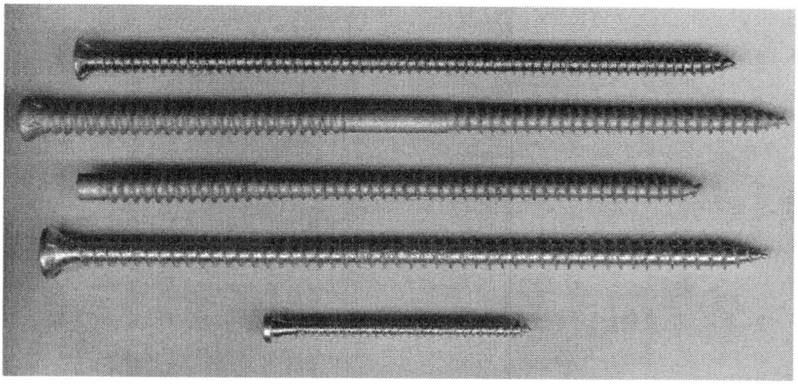

Figure 17.11 Self-tapping wood screws with different threads

17.5.2 Screws Loaded Parallel to the Screw Axis

Screws show the highest withdrawal capacity in timber connections with dowel-type fasteners. Apart from the high load-carrying capacity, the stiffness of connections with axially loaded screws is significantly greater than for joints with screws loaded perpendicular to the fastener axis. Contrary to threaded nails, which are designed to have a low resistance against driving in but a high withdrawal resistance, screws display the same resistance against withdrawal and pushing in.

The withdrawal failure of axially loaded screws is characterised by shear failure of the wood surrounding the thread. Figure 17.12 shows an end grain surface of a glulam cross-section with a wood shear failure along the lower edge. The cut was made after a withdrawal test.

For European softwood timber, the shear strength values from withdrawal tests of screws oriented perpendicular to the grain are in the order of 4–5 MPa. The shear strength mainly depends upon the timber density and the angle between screw axis and grain direction. Wood exhibits distinctly different shear strength values depending on the angle between shear force and grain direction. For screws oriented parallel to the grain, the shear strength parallel to the grain determines the withdrawal capacity. Screws oriented perpendicular to the grain simultaneously activate two different shear mechanisms: rolling shear and cross-grain shear. The rolling shear strength is the lowest shear strength of wood while the cross-grain strength is the highest. Overall, an

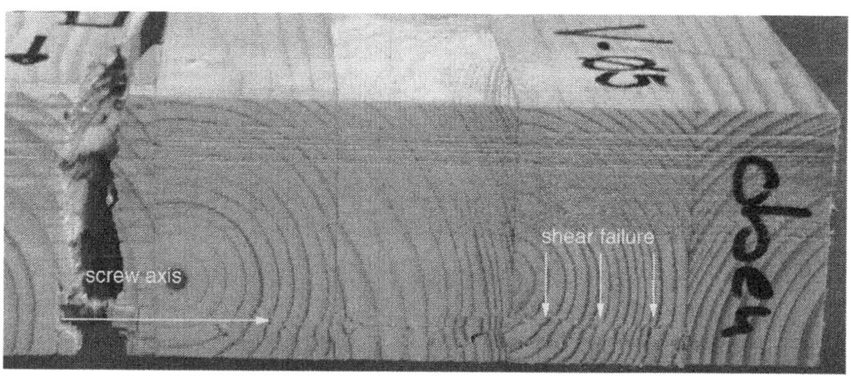

Figure 17.12 Wood shear failure due to withdrawal of a self-tapping screw oriented perpendicular to the grain

orientation perpendicular to the grain leads to greater withdrawal capacities than an orientation parallel to the grain.

For angles between screw axis and grain direction between 0° and 90°, Gehri has proposed to use Hankinson's equation to determine the shear strength in withdrawal:

$$f_{v,\alpha} = \frac{f_{v,0} \cdot f_{v,90}}{f_{v,0}\sin^2\alpha + f_{v,90}\cos^2\alpha} \qquad (17.11)$$

where $f_{v,0} = 0.75\, f_{v,90}$.

Other possible failure modes for axially loaded screws are head pull-through and tensile failure of the screw. The head pull-through resistance depends upon the head shape and outer head diameter and the timber density. The head pull-through resistance in timber or wood-based panels may be significantly increased by using a washer (see Figure 17.13).

17.5.3 Connections Using Inclined Screws

The same principle as shown in Figure 17.14 may be used to connect successive timber members to form continuous purlins. Traditionally, fasteners are placed perpendicular to the joint between adjacent members (see detail in Figure 17.15). Load components acting perpendicular to the roof then cause shear forces in the fasteners, load components parallel to the roof cause axial forces.

If the screws are arranged as shown in Figure 17.16, the force component parallel to the member surface causes tensile and compressive forces in the screws, respectively, while the force

Figure 17.13 Head pull-through failure of a screw with washer

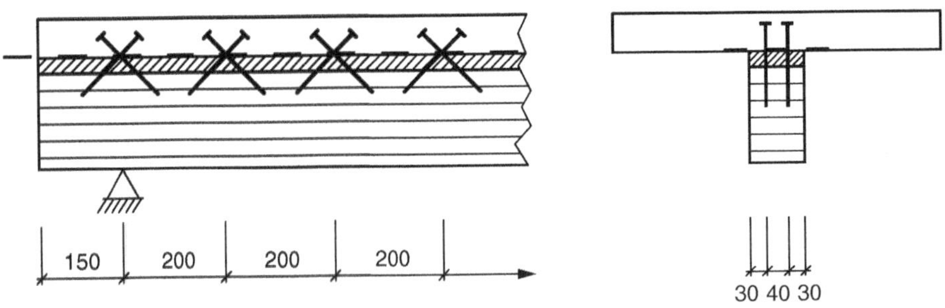

Figure 17.14 Timber-concrete connection with screws oriented under 45° to the grain

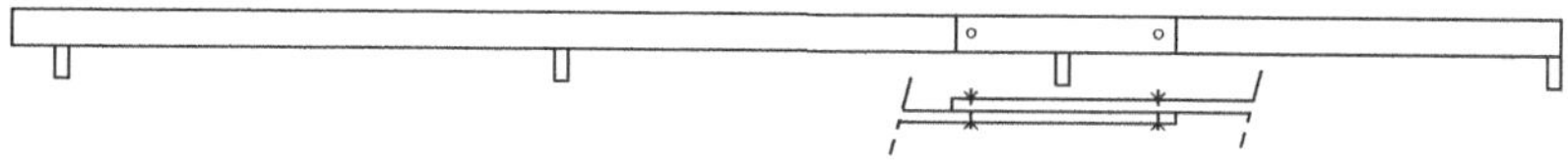

Figure 17.15 Continuous purlin formed by successive timber members

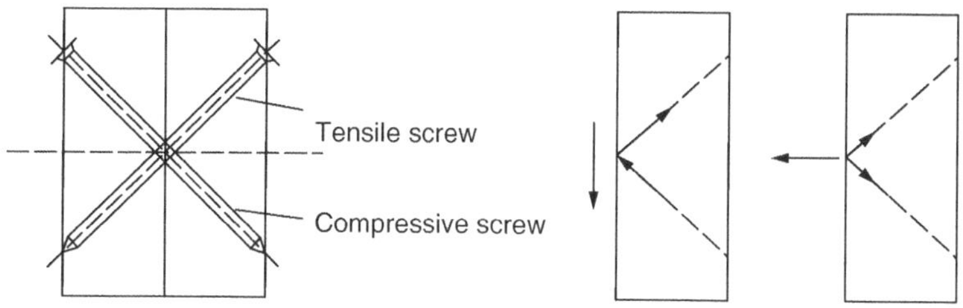

Figure 17.16 Schematic representation of the load transfer in truss-like connections using screws

Figure 17.17 Modern tenon joint as secondary beam-main beam connection

component perpendicular to the member surface causes tensile action in both screws. The tensile screw then governs the design. For the connection shown in Figure 17.16 self-tapping screws with continuous threads are required since the head pull-through resistance is distinctly lower than the withdrawal resistance in the pointside member.

Developments in computer-controlled manufacture of timber members enable a revival of traditional timber connections without steel fasteners. An example of a modern tenon joint automatically machined to connect a secondary beam to a main beam is shown in Figure 17.17.

Because the connection shown in Figure 17.17 experiences large slip under service loads, the load is better transferred using tensile and compressive screws with continuous threads. The screw arrangement is shown in Figure 17.18. Because of the different connection stiffness of the tenon joint and the screwed connection, a load sharing between the two types of connection cannot be expected. The tenon joint only serves as an assembly aid, the joint load is entirely carried by the screws. Figures 17.19 and 17.20 show the different load-slip behaviour of the tenon joint and the connection using inclined screws loaded in tension and compression, respectively.

The use of self-tapping screws with continuous threads arranged under 45° to the grain direction is not limited to the situations shown above. Many connections where dowel-type fasteners commonly are loaded perpendicular to the fastener axis may also be executed using inclined screws loaded in tension and compression. A simple lap joint transferring a tensile force is shown in Figure 17.21. In the right, the screws are arranged in a conventional

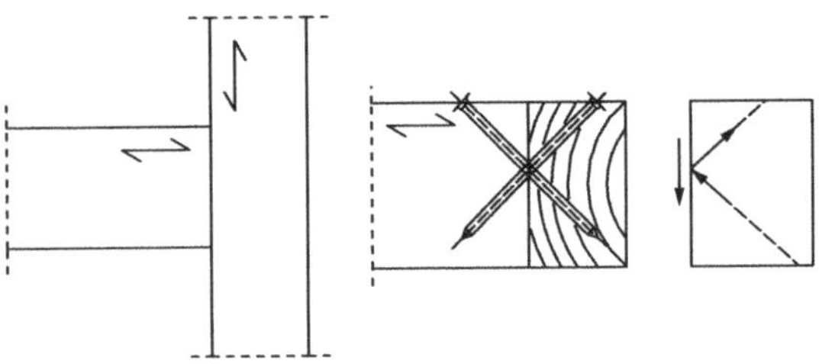

Figure 17.18 Schematic representation of a secondary beam-main beam connection using inclined screws

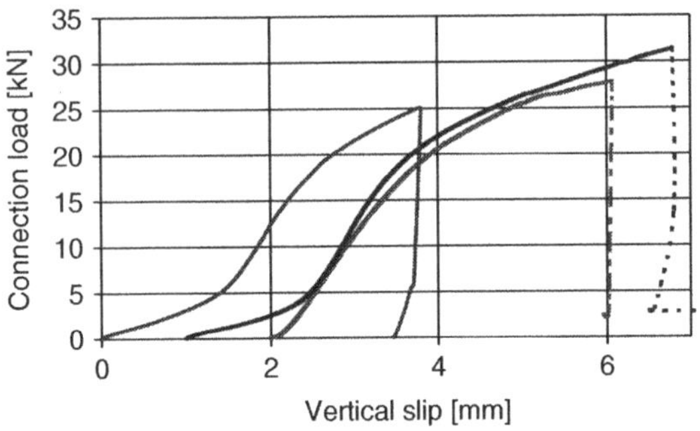

Figure 17.19 Load-slip curves of three secondary beam–main beam connections as shown in Figure 17.17

manner. The load-carrying capacity depends on the screw bending capacity and on the timber embedding strength. In the left, the connection force mainly causes tensile action in the screws and compression between the timber middle member and the side members. Since the withdrawal capacity of the screws exceeds their shear capacity,

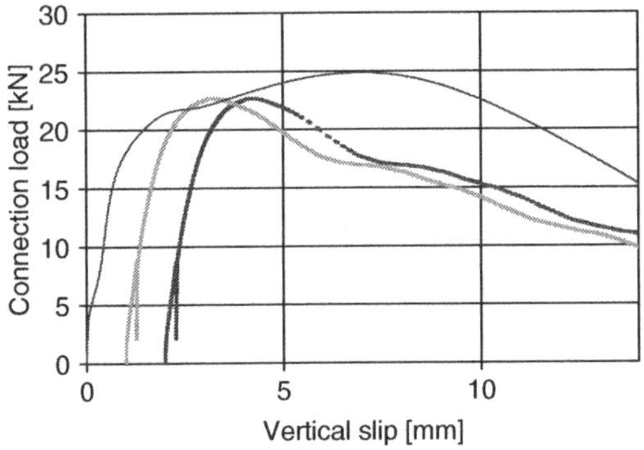

Figure 17.20 Load-slip curves of three secondary beam–main beam connections using inclined screws as shown in Figure 17.18

the version using inclined or crossed screws is generally more economic. Connections with inclined screws show an increase of 50% of the load-carrying capacity and an increase of slip modulus by a factor of up to 12 compared with similar connections with screws loaded perpendicular to their axis.

17.6 SUMMARY

The load-carrying capacity of connections with dowel-type metal fasteners loaded perpendicular to the fastener axis according to the Johansen theory is derived for a basic case. The most important parameters, embedding strength and fastener yield moment, are illustrated. For single shear and double shear timber-to-timber and steel-to-timber

Figure 17.21 Lap joint with self-tapping screws. Left: inclined screw arrangement; right: screws perpendicular to the member surface

connections the load-carrying capacity is given for each possible failure mode.

The reason for the additional resistance caused by tensile forces in the fasteners is explained, and the influence of the withdrawal capacity on the tensile action is outlined. The group effect leading to decreasing values of the load-carrying capacity per fastener with increasing number of fasteners arranged parallel to the load and grain direction is quantified. A method to reinforce the timber in the connection area and to avoid the group effect is given.

The duration of load effect on both the load-carrying capacity and the deformation is described, based on long-term tests with different types of fastener. The creep deformations in mechanical timber connections are distinctly greater than for timber members. The results also show that during most of the lifetime of the time to failure tests the full load carrying capacity is available and that the damage accumulates over a relatively short time period immediately before failure.

New developments regarding self-tapping screws with continuous thread along the shank are presented and possibilities for connections, where the high withdrawal capacity is exploited, are introduced. These connections with the screw axis arranged under about 45° to the interface between the members show higher failure loads and a much greater stiffness than screws loaded perpendicular to their axis.

REFERENCES

Blass H.J. (1994) Characteristic strength of nailed joints. *Forest Products Journal*, **44**(4), 33–39.

Blass H.J., Bienhaus A. and Krämer V. (2001) Effective bending capacity of dowel-type fasteners. *Proceedings PRO 22, International RILEM Symposium on Joints in Timber Structures*, pp. 71–80.

Blass H.J. and Schmid M. (2001) Self-tapping screws as reinforcement perpendicular to the grain in timber connections. *Proceedings PRO 22, International RILEM Symposium on Joints in Timber Structures*, pp. 163–172.

Ehlbeck J. and Ehrhardt W. (1995) Screwed joints. In: *Timber Engineering STEP 1, Basis of design, material properties, structural components and joints*. Centrum Hout, The Netherlands.

Foschi R.O. and Bonac T. (1977) Load-slip characteristics for connections with common nails. *Wood Science*, **9**(3), 118–123.

Görlacher R. (1995) Load-carrying capacity of steel-to-timber joints with annular ringed shank nails. *Proceedings CIB-W18 Timber structures, Meeting 28*, Paper 28-7-3.

Hilson B.O. (1995) Joints with dowel-type fasteners – Theory. In: *Timber Engineering STEP 1, Basis of design, material properties, structural components and joints*. Centrum Hout, The Netherlands.

Johansen K.W. (1949) *Theory of timber connections*. International Association of Bridge and Structural Engineering, Publication No. 9:249-262, Bern, Switzerland.

Jorissen A.J.M. (1998) Double shear timber connections with dowel-type fasteners. PhD Dissertation, Delft University of Technology, The Netherlands.

Kuipers J. and Van der Put T.A.C.M. (1982) Betrachtungen zum Bruchmechanismus von Nagelverbindungen. In: *Ingenieurholzbau in Forschung und Praxis*, J. Ehlbeck and G. Steck, editors, Bruderverlag Karlsruhe.

Meierhofer U. (1993) Tests on timber concrete composite structural elements. *Proceedings CIB-W18 Timber structures, Meeting 26*, Paper 26-7-5.

Rodd P.D. (1973) The analysis of timber joints made with circular dowel connectors. PhD Dissertation, University of Sussex, UK.

Wilkinson T.L. (1986) Load distribution among bolts parallel to load. *J. Structural Engineering*, **112**(4), 835–852.

Van de Kuilen J.W.G. (1999) Duration of Load Effects in Timber Joints. PhD Dissertation, Delft University of Technology, The Netherlands.

Whale L.R.J., Smith I. and Hilson B.O. (1989) Characteristic properties of nailed and bolted joints under short-term lateral load. Part 4 – The influence of testing mode and fastener diameter upon embedment test data. *J. Institute of Wood Science*, **11**(5), 156–161.

18

Structural Adhesive Joints Including Glued-in Bolts

Simon Aicher

18.1 INTRODUCTION

Glued timber joints have been employed for thousands of years, as proven, for example, by sacophargs of Egyptian pharaos, glued with cold-setting adhesives based on gelatine, extracted by hot water boiling from animal bones, skin and tenons (Lucas, 1948). In this contribution, the term *structural adhesive joint* is exclusively related to joints in the more narrow sense comprising the transfer of loads in a confined area between two, and rarely more than four, discrete, generally member-like adherends. Reinforcements, for instance glued-in rods perpendicular to grain in glulam beams, are perceived as a special case of concentrated load transfer. According to the above definition, the adhesive-based assembly (jointing) of monolithic structures such as glulam, multi-layered wooden plate buildups, etc., is not considered to be a discrete structural adhesive joint. Glued connections of continuous webs to the flanges of I-beams (panel web beams), for instance by finger jointing, mark the transition. The reason is that the joint, although spatially highly confined in two dimensions (width and depth of the beam), stretches over the full length, thereby establishing a monolithic element.

The general advantages/disadvantages and some inherent problems of glued joints, as compared to mechanical connections, can be summarised as follows:

- *advantages*: superior rigidity; in general, higher load capacity for equal jointing areas; possibility of highly automated processes; for several types of joints, for instance finger joints in glulam laminations and solid lumber, the sole economic alternative.
- *Disadvantages/problems*: highly sensitive to unskilled manufacture; less obvious manufacturing controls; lower inherent redundancy; on-site manufacture largely restricted; complex mechanical and often very brittle behaviour.

The finger jointing of boards to 'endless' laminations represents the basic technology for the economic manufacturing of glulam. Increasingly, finger jointing technology is used to produce high quality large dimension lumber. Large finger joints are today the most economic and efficient construction method for frame corners. Recent developments show that joints with glued-on gussets of plywood or Laminated Veneer Lumber (LVL) are highly competitive for truss structures. Joints based on glued-in rods belong to the most promising jointing and reinforcement alternatives.

To fully utilise the obviously excellent performance of structural adhesive joints in timber engineering in conjunction with a semi-probabilistic design approach, it is a prerequisite to assess/predict the load capacity of the joints. This strength prediction can be based either on testing or – and far better – on analysis, thus taking into account the different materials, geometries and loading conditions. The prediction of joint strength necessitates the understanding of bond line stresses, strains and essentially of damage mechanisms causing failure. However, this is difficult due to some inherent features of bonded joints. There is, for instance the very small thickness of the glue line, typically in the range of about 0.1–0.5 mm which, especially in the case of wood bonding, shows complex transitions in the interfaces to the wood adherends.

Today, the strength assessment of glued joints in timber engineering applications is based largely on joint tests. In most cases where the (characteristic) joint capacity, as related to a bond line strength, can be computed from a design formula, these equations are simplistic, meaning that the ultimate capacity of the joint is assumed to be the product of the glued area and a size-independent (characteristic) bond line strength value. In rare cases, a size effect is accounted for.

18.2 CLASSIFICATION OF GLUED TIMBER JOINTS

Glued joints in timber engineering comprise a large variety of joint types, differing with respect to manufacture, size, material incorporated, application and failure behaviour. A distinction has to be made between pure wood–wood joints and joints between wood and other materials, especially steel, primarily in the form of rods. The pure wooden joints can be separated into connections where two similar lumber or glulam adherends are jointed directly without extra splices, for instance by finger jointing, and gusset type joints with wood-based materials (e.g. plywood) as connecting material. Table 18.1 gives an overview of the different joint categories.

The sizes of structural adhesive joints cover a wide range, from small finger-jointed lumber cross-sections of about $2 \cdot 10^3$ mm^2 to large finger joints with cross-sections up to about $3 \cdot 10^5$ mm^2. Gusset plate sizes reach from about 10^4 mm^2 to 10^6 mm^2.

The production processes of the joints stretch from fully industrialised production technology for finger joints and special I-beam joints to entire craftsmanship work in the case of large finger joints. The quality control procedures and technologies are very different, varying from continuous proof loading (Eby, 1981) to mere visual inspection.

Table 18.1 Classification of different types of structural adhesive joints in timber engineering

structural adhesive joints

- pure wood-wood joints (lumber, glulam, LVL)
  - end joints, knees, reinforcements
    - scarf joints
    - finger joints
      - glulam laminations
      - (large) lumber cross-sections
    - large finger joints
      - straight beams
      - knee joints
        - with equal materials
        - with corner pieces from plywood, LVL, etc.
    - lap joints
      - end joints
      - beam reinforcements with plywood and LVL straps
    - knee and truss joints with gussets from plywood, LVL, etc.
    - end joints with axially loaded hardwood dowels
  - special (formwork) I-beam joints
    - lattice web joints
    - panel web joints
- hybrid joints
  - joints with glued-in rods
    - threaded steel rods
    - rebars
    - FRP[1] rods

[1]FRP = fibre reinforced plastics

18.3 CONSTITUTIVE LAW OF THE BOND LINE

18.3.1 Elasticity

Due to the very small thickness of the adhesive layer, together with the complex transition interface adhesive/wood see Figure 18.1, and due to the comparable magnitude of the shear modulus of the adhesives and wood, determination of the elastic properties of wood adhesive layers is already a delicate task. A very important work in this field was performed by Clad (1965), who determined the magnitudes of the shear modulus of bond lines with different adhesives in wooden

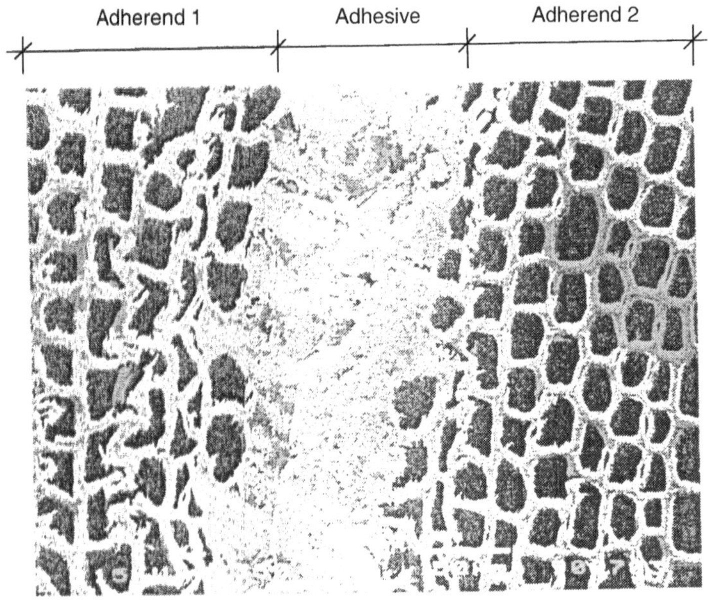

Figure 18.1 A wood adhesive bond line of roughly 0.1 mm in thickness. Scots Pine and phenolic resorcinol adhesive (from Wernersson, 1994)

joints, depending on specific fillers and glue line thicknesses. For urea, melamin and phenolic resorcinol polycondensation adhesives, rather similar values were obtained; for bond line thicknesses of about 0.1, 0.4 and 0.8 mm, the shear moduli were roughly in the ranges of (in N/mm^2) 1300–1800, 700–850 and 600, respectively. It is worth noting that the specified elasticity values were derived by optical measurement of local slip combined with theoretical determination of shear stresses, based on Volkersen theory (see below) for shear stresses.

18.3.2 Non-Elastic and Fracture Behaviour

No real material, not even the most brittle, behaves in a completely linear elastic way until fracture, but either exhibits hardening, ideal plasticity or gradual damage and softening (see Figure 18.2). Epoxies, for instance, can show all types of constitutive behaviour. Some, depending on the type of filler (resp. plasticizer), first behave linear elastically and then show strain hardening and even ideal plastic behaviour. The adhesives predominantly used today for structural timber joints (i.e. phenolic resorcinol and aminoplastic adhesives), on the contrary, are very brittle. The one-component polyurethanes, increasingly used especially for finger jointing, show both a ductile and a softening behaviour.

The constitutive law of the bond line describes the relation between local stresses and local displacements across the adhesive layer. The constitutive relationship depends upon several influences, such as the type of adhesive, thickness of bond line, fibre orientation, type of adherend, surface preparation, curing conditions, etc., and

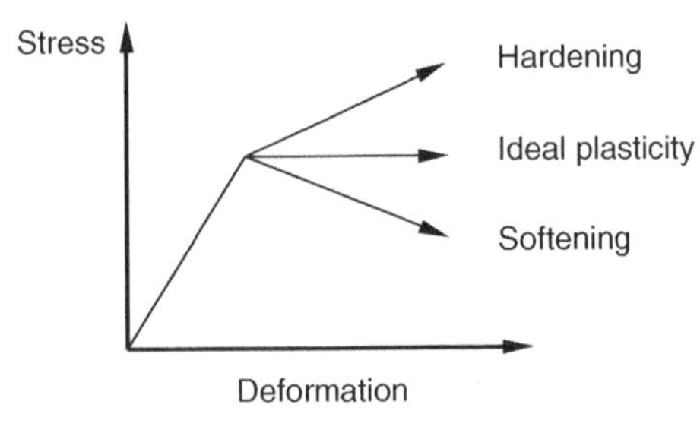

Figure 18.2 Different idealised types of material (resp. bond line) constitutive behaviour

also considerably on the rate of loading. A prerequisite to determining the local behaviour is a stable test and a bond line with almost uniform stress distribution. Further, it is important to include the descending softening branch of the stress-strain relationship in the constitutive law, as a considerable part of the fracture energy, dissipated during progressive damage, may be related to the post-peak area. Fracture energy, G_f, one of the important parameters in Nonlinear Fracture Mechanics (NLFM), is in general defined as the total energy required to fracture a unit area of (bond line) surface. For a general mixed mode situation of a bond line, subjected to inplane shear stress τ and tension stress σ normal to the glue line, the total fracture energy is given by

$$G_f = \int_{\Gamma} (\tau d\delta_s + \sigma d\delta_n) \qquad (18.1)$$

where Γ is the deformation path that results in complete separation of the bond line and δ_s, δ_n are shear and normal deformations. For the decoupled pure shear and pure tension modes, the fracture energies are hence determined by

$$G_{f,\tau} = G_{f,II} = \int_0^{\infty} \tau d\delta_s,$$
$$G_{f,\sigma} = G_{f,I} = \int_0^{\infty} \sigma d\delta_n \qquad (18.2a,b)$$

i.e. the fracture energies $G_{f,\tau}$, $G_{f,\sigma}$ correspond to the areas under the τ-δ_s and σ-δ_n curves, respectively (see Figure 18.3).

The determination of fracture energies in wooden bond lines, and thus of the strength values, τ_f and σ_f, determined from maximum load divided by bond area, is difficult, and was so far been exclusively tackled successfully by Wernersson (1994) and Serrano (1997).

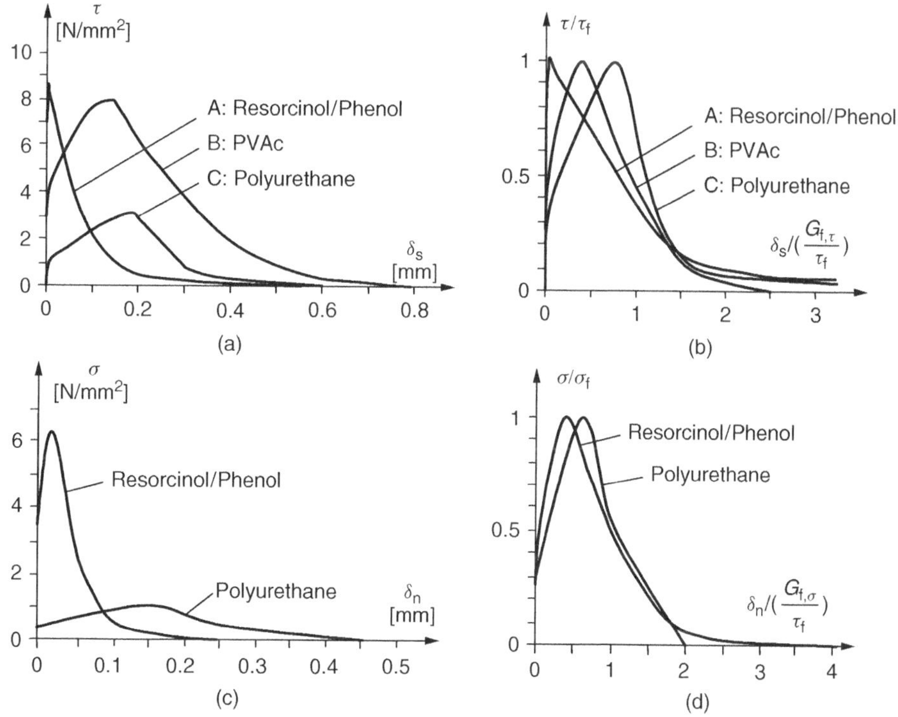

Figure 18.3 Typical stress versus deformation curves (absolute and normalised) for bond lines with different adhesives and loading modes, (after Wernersson, 1994); bond line thickness: 0.1 mm; deformation rate 0.063 mm/min. (a, b) pure shear; 4 days curing time of bond lines, (c, d) pure tension; 14 days curing time of bond lines

Figure 18.3 shows stress vs. deformation curves for pure shear and pure tension stress, respectively, as obtained by Wernersson for bond lines with three (resp. two) different adhesive types; for explanations to the given normalised curves, see below. Similar curves, however, with partly considerably higher strength values, were measured by Serrano (1997, 2000). Tables 18.2 and 18.3 give compilations of the bond line fracture properties, as determined by Wernersson, for different types of adhesives. The pronounced differences in strengths and bond line brittlenesses between the phenolic resorcinol (PR) adhesive and the one-component polyurethane (PUR) are obvious. Bond line shear and tension brittlenesses differ between both adhesives by factors of 8 and 28, respectively.

Table 18.2 Mean values for shear strength, fracture energy and bond line brittleness in pure shear for bond lines of 0.1 mm thickness with three different adhesives and for spruce wood (after Wernersson, 1994); 4 days bond line curing time, 0.063 mm/min deformation rate

Adhesive (or wood)	Shear strength τ_f	Fracture energy in shear $G_{f,\tau}$	Bond line shear brittleness $\tau_f^2/G_{f,\tau}$
–	N/mm^2	Nmm/mm^2	N/mm^3
Resorcinol/Phenol	8.4	0.7	101
PVAC	7.4	2.1	26
Polyurethane[1]	2.8	0.6	13
Wood (picea abies)	8.9	0.95	83

[1]One-component Polyurethane.

Table 18.3 Mean values for tension strength, fracture energy and bond line brittleness in pure tension for bond lines of 0.1 mm thickness with two different adhesives (after Wernersson, 1994); 14 days bond line curing time, 0.063 mm/min deformation rate

Adhesive	Tension strength σ_f	Fracture energy in tension $G_{f,\sigma}$	Bond line tension brittleness $\sigma_f^2/G_{f,\sigma}$
–	N/mm^2	Nmm/mm^2	N/mm^3
Resorcinol/Phenol	6.4	0.36	114
Polyurethane[1]	1.0	0.23	4

[1]One-component Polyurethane.

For the derivation of the NLFM theory of bonded lap joints, subject to Volkersen assumptions of pure shear loading, Gustafsson (1987) introduced the normalisation of the bond line constitutive law by dividing shear stress, τ, by shear strength, τ_f, and further by dividing relative displacement (slip) δ_s by the ratio $G_{f,\tau}/\tau_f$. Then, the constitutive law of the bond line can be defined by τ_f, $G_{f,\tau}$ and a shape function g as

$$\frac{\tau}{\tau_f} = g\left[\frac{\delta_s}{G_{f,\tau}/\tau_f}\right] \qquad (18.3)$$

where peak value τ/τ_f and the area below the nomalised curve are both unity (compare Figures 18.3(b, d) and Figure 18.4). Gustafsson identified four idealised shapes of τ-δ curves ($\delta = \delta_s$), given in Figure 18.4, enabling analytical solutions for load capacities according to generalised Volkersen theory. Some solutions according to this approach for specified curves τ-δ are presented in Section 18.5.

It should be noted that the τ-δ curve No. I in Figure 18.4 corresponds closely to the assumptions then made by Volkersen, and specified below, although the material behaviour in the post-peak regime was not defined there.

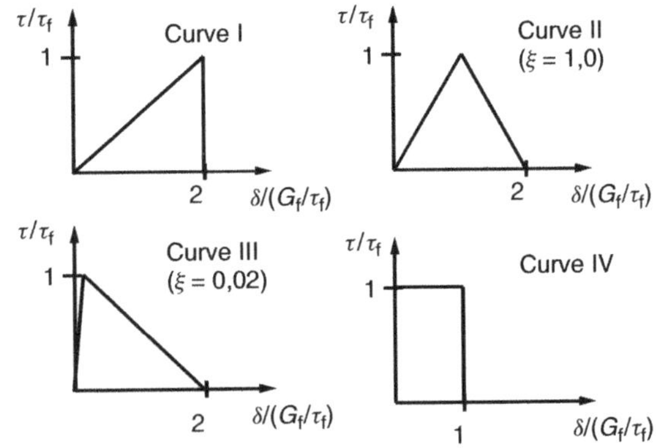

Figure 18.4 Different shapes of normalised τ-δ curves (after Gustafsson, 1987)

18.4 STRESS ANALYSIS

18.4.1 Volkersen Theory

In a pioneering work, Volkersen (1938) derived a uniaxial shear lag theory for single (and double) lap joints (Figure 18.5(a)), whereby the following assumptions were made:

- uniaxial tension of the adherends (no bending),
- pure shear stress and strain state of the adhesive layer,
- linear elastic behaviour of adhesive and adherends,
- isotropic behaviour of the materials.

In the following, the mechanical behaviour of a Volkersen joint is discussed in detail, as this lays the basis for understanding glued joints in general. Geometry and stiffnesses of the joint are given in Figure 18.5(a): E_1, E_2, t_1, t_2 are Young's moduli and thicknesses of the adherends 1 and 2, G and d are shear modulus and thickness of the glue line, and ℓ and b are the bond length and width of the joint.

It can be seen qualitatively from Figure 18.5b that the variation of shear strain γ (or slip $\delta = \gamma d$), and consequently of shear stress τ, is induced by the differences of the adherend normal strains along the bond line, ε_1 and ε_2, decreasing from the maximum to zero at the respective adherend end. Figure 18.5(c) depicts the shear strain and stress distributions along the bond line. In the case of equal stiffnesses of the adherends, $E_1t_1 = E_2t_2$, the shear strain and stress distributions are symmetric to the mid-length of the bond line. In the case of unequal stiffnesses, as depicted in Figures 18.5(a–c), the distributions are unsymmetric, and the respective maximum of γ and τ is located at the end of the stiffer adherend, as there the largest strain difference of both adherends occurs. In the limit case of infinite adherend stiffnesses, we obtain a constant shear strain $\gamma_0 = \tau_m/G$, where τ_m is mean shear stress (Equation (18.8b)).

Quantitatively, the shear strain $\gamma = \delta/d(\delta =$ slip or relative displacement of the two adherends) at a point x along overlap ℓ resulting from the rigid body slip, $\delta_0 = \gamma_0 d$, and from the difference of the strains of the two adherends, ε_1 and ε_2, can be written as

$$\gamma(x) = \gamma_0 + \frac{1}{d}\left(\int_{-\ell/2}^{x} \varepsilon_1(x)dx - \int_{-\ell/2}^{x} \varepsilon_2(x)dx\right) \tag{18.4}$$

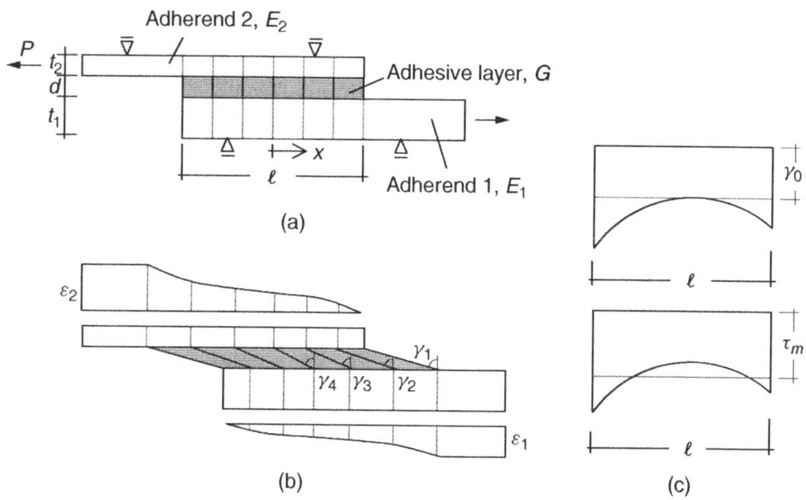

Figure 18.5 Single lap joint according to Volkersen assumptions. (a) Geometry notations and schematic boundary conditions, (b) schematic view of adherend normal strains and of shear strains of the bond line, (c) Shear strain and stress distributions along the bond line

Strains $\varepsilon_1, \varepsilon_2$ can be expressed by shear stress τ, based on force equilibrium of normal and shear stresses of the adherends, giving

$$\varepsilon_1(x) = \frac{1}{E_1 t_1} \int_{-\ell/2}^{x} \tau(x)dx,$$
$$\varepsilon_2(x) = \frac{1}{E_2 t_2 b}\left[P - b\int_{-\ell/2}^{x} \tau(x)dx\right] \quad (18.5a,b)$$

Replacing shear stress τ by (γG), then substituting $\varepsilon_1, \varepsilon_2$ into Equation (18.4) by Equations (18.5a,b), and differentiating twice, gives a second order differential equation

$$\frac{d^2\gamma}{dx^2} - \lambda^2\gamma = 0 \quad (18.6a)$$

where

$$\lambda^2 = \frac{G}{d}\left(\frac{E_1 t_1 + E_2 t_2}{E_1 t_1 E_2 t_2}\right) = \frac{G}{E_1 t_1 d}(1+\alpha) \quad \text{and}$$
$$\alpha = \frac{E_1 t_1 b}{E_2 t_2 b} \quad (18.6b, c)$$

The general solution of Equation (18.6a) is: $\gamma(x) = A \sinh(\lambda x) + B \cosh(\lambda x)$. After determination of constants A and B from boundary conditions, evaluated via Equation (18.4), the shear stress distribution in the glue line is obtained as

$$\tau(x) = \tau_m \frac{\rho}{2}\left[\frac{\cosh(\rho x/\ell)}{\sinh(\rho/2)} - \frac{(1-\alpha)}{(1+\alpha)}\frac{\sinh(\rho x/\ell)}{\cosh(\rho/2)}\right] \quad (18.7)$$

where the substitutions

$$\rho = \lambda\ell = \sqrt{\frac{\ell^2}{E_1 t_1}\frac{G}{d}(1+\alpha)}, \quad \tau_m = \frac{P}{\ell b} \quad (18.8a, b)$$

denote the important so-called Volkersen joint factor and average bond stress, respectively. The maximum peak stress (resp. the peak stress factor at the overlap end) is, in the general case,

$$\frac{\tau_{max}}{\tau_m} = \frac{\rho}{2}\left[\coth\left(\frac{\rho}{2}\right) + \frac{1-\alpha}{1+\alpha}\tanh\left(\frac{\rho}{2}\right)\right] \quad (18.9a)$$

and for the symmetric case, with $\alpha = 1$,

$$\frac{\tau_{max}}{\tau_m} = \frac{\rho}{2}\coth\left(\frac{\rho}{2}\right) \quad (18.9b)$$

Equations (18.7)–(18.9) enable the study of some important basics of glued (lap) joints of which, some are summarised following.

If we investigate, for a constant load, the influence of bond line length ℓ according to Equation (18.7) and keep all other parameters of joint factor ρ fixed, we see that for very small bond length we obtain a rather uniform high shear stress; thus, the peak stress factor $\tau_{max}/\tau_m > 1$ according to Equations (18.9a, b) is not much different from unity. When we increase bond length, an increasingly inhomogeneous stress distribution with pronounced but less high stress peaks at the overlap ends evolves ($\tau_{max}/\tau_m \ggg 1$). The peak stress τ_{max}, as dependent on ρ, tends to a limit value as the hyperbolic functions in Equations (18.9a, b) tend to unity for increasing ρ, so

$$\frac{\tau_{max}}{\tau_m} = \frac{\rho}{1+\alpha} = \frac{\ell}{1+\alpha}\sqrt{\frac{G(1+\alpha)}{E_1 t_1 d}} \quad \text{for } \rho \to \infty \quad (18.10)$$

With sufficient practical accuracy, the limit value of ρ, and hence for bond length ℓ, can be assumed as

$$\rho \geq \rho^* = 5 \quad \text{and}$$
$$\ell \geq \ell^* = \rho^*\sqrt{E_1 t_1 d/[G(1+\alpha)]} \quad (18.11a, b)$$

From Equations (18.10) and (18.11), we see that beyond the limit value ℓ^*, peak shear stress $\tau_{max} = \tau^*_{max}$ does not decrease any more, i.e. has a minimum value; on the contrary, mean shear stress τ_m decreases unlimited with increasing ℓ. The obvious consequences on the joint load capacity are discussed in Section 18.5.

18.4.2 Goland and Reissner Theory, Hart-Smith Analysis

The main weakness of the Volkersen approach, apart from confinement to linear elasticity, is that only the force equilibrium parallel to the joint is accounted for, resulting in pure shear. A first assessment of the peel stresses, necessarily introduced by the bending moment due to the eccentricity of the normal forces, was introduced by Goland and Reissner (1944), accounting for the moment

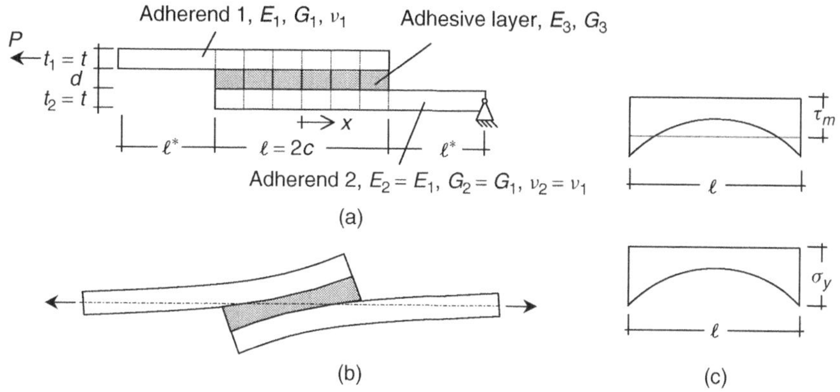

Figure 18.6 Single lap joint according to Goland and Reissner assumptions. (a) Geometry notations and schematic boundary conditions; (b) schematic view of adherend bending due to load eccentricity, and thereby induced normal strains perpendicular to the bond line; (c) shear and normal (peel) stress distributions along bond line

equilibrium, too. Figures 18.6(a), (b) show a single lap joint with geometry notations and boundary conditions according to Goland and Reissner. Similar to as in Volkersen's theory, the materials of the symmetrically built-up joint of width b are considered to be isotropic, and to behave linear elastic. However, now the adherends are treated as cylindrically bent plates. The geometry and stiffnesses of the adherends are denoted by $t = t_1 = t_2$, $E_1 = E_2$, $G_1 = G_2$ and $\nu_1 = \nu_2$; the adhesive layer is defined by thickness d, being small as compared to the thickness of adherends, and, by modulus of elasticity and shear modulus E_3G_3, respectively.

A bending moment factor k was defined which relates the bending moment in the adherends at both edges of the overlap, M_0, to the applied load resp. stress, $P = ptb$, by

$$M_0 = kPt/2 \qquad (18.12)$$

where factor k is dependent on load P, the joint dimension and the physical properties of the adherends, in the manner

$$k = \frac{1}{1 + 2\sqrt{2}\tanh(\mu c)}, \quad \mu = \frac{c}{t}\sqrt{\frac{3(1-\nu_1^2)}{2}\frac{p}{E_1}} \qquad (18.13\text{a, b})$$

As the glue line thickness is neglected in the derivation of the bending moment factor, quantity k is 1 for zero load P, and is reduced during increasing loading to a lower limit dependent on the joint configuration. This means that the joint edge moments are considerably reduced from the initial value $Pt/2$ to about one-third with increasing load. The analysis of Goland and Reissner delivers two independent third and fourth order differential equations for shear and peel stresses τ and σ_y, respectively:

$$\frac{d^3\tau}{dx^3} - \frac{8G_3}{E_1td}\frac{d\tau}{dx} = 0,$$

$$\frac{d^4\sigma_y}{dx^4} + \frac{24(1-\nu_1^2)E_3}{E_1t^3d}\sigma_y = 0 \qquad (18.14\text{a, b})$$

The rather lengthy results for τ and σ_y are specified in Appendix I by Equations (A18.1) and (A18.2). Figure 18.6(c) gives a schematic view of the shear and normal (peel) stress distributions along the bond line. For long overlaps, the Goland and Reissner analysis due to the moment influence gives an elastic shear stress concentration factor, being two times larger than that according to the Volkersen approach.

The most thorough analytical treatment of linear elastic and elastic–plastic stress distributions in lap joints, now also including anisotropy of the adherends and the contribution of the glue line layer on bending, has been presented by Hart-Smith (1973). Comparing the solutions of Hart-Smith for (maximum) shear and peel stresses (not given here) with those by Goland and Reissner, it can be seen that they differ only for a very long bond length, which is due to the fact that the eccentricity factor, k, approaches different limit values, being $k = 0.26$ in the case of Goland and Reissner and $k = 0$ for Hart-Smith. It can be deduced from the more refined stress analysis by Hart-Smith, that the influence of the adherend bending becomes negligible for very long overlaps.

18.5 JOINT STRENGTH ANALYSIS

18.5.1 Linear Elastic Strength of Materials Approach

It is quite obvious that the pronounced shear and peel stress peaks at the overlap ends are problematic for a realistic strength assessment beyond short overlaps and stiff adherends, rather irrelevant for building practise applications. Despite that, to get an intuitive feeling of the behaviour of glued joints, it is worth examining the strength result according to linear Volkersen theory. Conforming to a classic strength of materials approach, the simple failure criterion $\tau_{max} = \tau_f$ is assumed, where τ_f is the 'intrinsic' bond line strength of a very short bond line with assumed uniform shear stress distribution. Then, with $f_v = \tau_{m,max} = P_{max}/(lb)$, the load capacity predicted by Equations (18.9b) and (18.10) is, for the case of $\alpha = 1$,

$$P_{max} = [\tau_f b\ell 2 \tanh(\rho/2)]/\rho \leq P^*_{max} \quad (18.15a)$$

and for the limit value $\rho \geq \rho^* = 5$

$$P^*_{max} = b\sqrt{2\tau_f^2 E_1 t_1 d/G} \quad (18.15b)$$

Global shear strength f_v for the limit case ρ^* is then

$$f_v = \frac{P^*_{max}}{b\,\ell} = \frac{\sqrt{t_1}}{\ell}\left(\sqrt{E_1}\sqrt{\frac{2\tau_f^2 d}{G}}\right) \quad (18.16)$$

Ratio $\sqrt{t_1}/\ell$, termed the 'joint factor' after de Bruyne (1944), showed very good correlations with experimental joint strengths of metal bonds with equal adherend and bond constitutive properties.

The results of Equations (18.15) and (18.16) are depicted in Figure 18.7, showing the dependency of the load capacity P_{max} and of (global) bond shear strength f_v on bond length l and joint factor ρ, respectively.

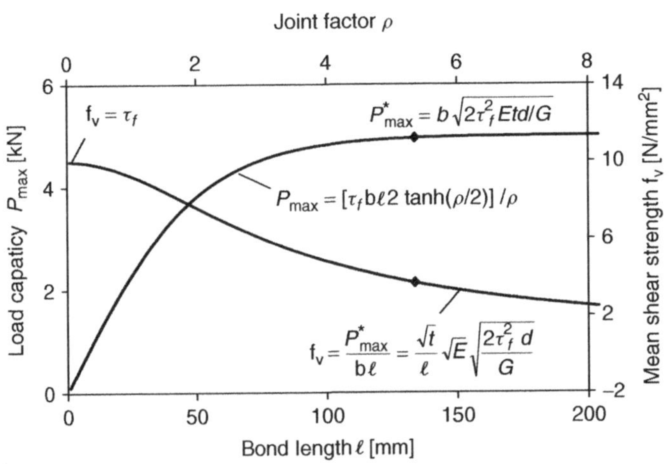

Figure 18.7 Load capacity and bond shear strength of a single lap joint as dependent on bond length ℓ and joint factor ρ, respectively, according to Volkersen theory (here $E = E_1$ and $t = t_1$); the specifically employed parameters are: $E = 15\,000$ N/mm^2, $G = 600$ N/mm^2, $\tau_f = 10$ N/mm^2, $t = 50$ mm, $b = 10$ mm, $d = 1$ mm

18.5.2 Plasticity Theory

An almost ideal elastic–plastic constitutive law is valid for many epoxy adhesives used in mechanical engineering applications. Figure 18.8 reveals qualitatively, for ideal adhesive plasticity, the evolution of global joint stiffness, and especially of shear stress and strain distributions along a bond line of a symmetric double lap joint with increasing load. According to Hart-Smith (1973, 1991), the joint capacity is limited by the critical shear strain. Beyond the joint load P where locally adhesive yielding starts at the overlap ends, stresses and strains develop unequally along the bond line; further, the global load slip behaviour becomes increasingly nonlinear. For numerical investigations on joint strength with strain hardening bond layers, see Harris and Adams (1984).

18.5.3 Linear Fracture Mechanics

The applicability of fracture mechanics approaches for predicting the strength of adhesive joints is widely acknowledged in *mechanical* engineering (Anderson *et al.*, 1988). The general aptness of Linear (Elastic) Fracture Mechanics (LEFM) for assessment of the load capacity of glued timber joints is due to several reasons: the brittleness of many adhesives and also of wood in shear and tension parallel and perpendicular to the grain;

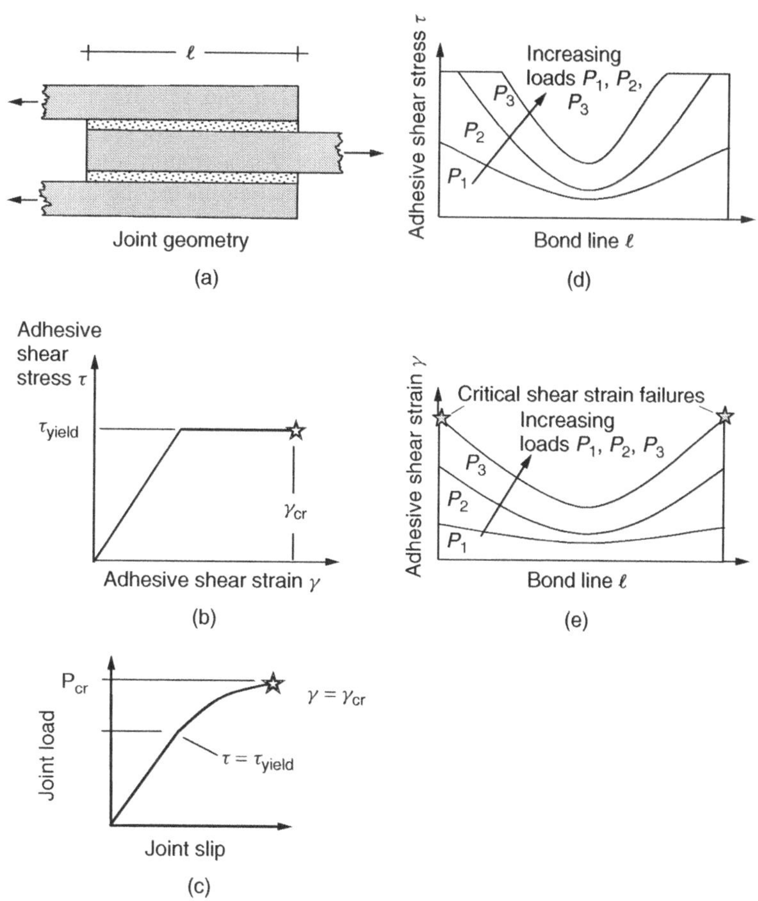

Figure 18.8 Nonlinear load, stress and strain evolution in a double lap joint with increasing load (after Hart-Smith, 1981). (a) Joint geometry, (b) adhesive constitutive law, (c) global load slip behaviour, (d, e) distribution of adhesive shear stress and strain along the bond line

the high stress peaks (singularities) at the overlap ends; and micro-cracks in the end grain faces of the adherend ends, resulting from drying stresses interacting with stresses from loads. The LEFM concept has been applied for load capacity assessment of several types of glued timber joints. Solutions for axially loaded lap joints were given, for instance, by Walsh *et al.* (1973), Leicester (1974) and Komatsu *et al.* (1976). The moment capacity of cross-lapped joints with in-plane torsional shear, constituting a fracture mode III problem, was investigated by Komatsu (1984). Strength assessment and shape optimisation of finger joints was shown by Aicher and Klöck (1991). The specific LEFM approach for double strap lap joints by Walsh *et al.* is presented in Section 18.6. Further, LEFM results as limit cases of nonlinear fracture mechanics are discussed in the following subsection.

18.5.4 Nonlinear Fracture Mechanics: Generalised Volkersen Theory

The assessment of the joint load capacity including the, so far not accounted for, softening behaviour of the bond line was introduced by Gustafsson (1987) and Ottosen and Olsson (1988). According to Gustafsson, the load carrying capacity, P_{max}, of a lap joint subjected to pure shear and with adherends in a state of pure tension is assumed to be influenced by nine geometry and material variables. These are: E_1, E_2, t_1, t_2, b, ℓ being Young's moduli, thicknesses, width and glued length of the adherends and, further, τ_f, $G_f = G_{f,s}$ and g being shear strength, shear fracture energy and the shape function of the normalised constitutive bond line law according to Equation (18.3), respectively. By dimensional analysis, the nine variables can be reduced to four non-dimensional variables, and the normalised load carrying capacity can be described as (α according to Equation (18.6c)).

$$\frac{P_{max}}{\tau_f b\,\ell} = f(\omega, \alpha, g) \tag{18.17}$$

where

$$\omega = \frac{1+\alpha}{2}\,\frac{\ell^2\tau_f^2}{t_1 E_1 G_f} \tag{18.18}$$

For constant variables α and g, the load carrying capacity is governed by ω, termed the 'joint brittleness ratio' of lap joints. High and low brittleness ratios correspond to brittle and ductile failures, respectively. The square root of the joint brittleness ratio, $\varpi = \sqrt{\omega}$, conforms to the Volkersen joint factor ρ according to Equation (18.8a), when employing the important substitution

$$\frac{G}{d} = \frac{1}{2}\frac{\tau_f^2}{G_f} \tag{18.19a}$$

Equation (18.19a) results from the elastically equivalent description of fracture energy by

$$G_f = \frac{1}{2}\tau_f\delta_{s,f} = \frac{1}{2}\tau_f\frac{\tau_f d}{G} \tag{18.19b}$$

where $\delta_{s,f} = \tau_f d/G$ is shear slip at failure. Note that G/d and $\tau_f^2/(2G_f)$ are elastically equivalent descriptions of the apparent bond line slip resistance K_{bond} (units: force/length^3), too. The latter is very helpful for the approximation of joint test results or the prediction of load capacities, as all bond line parameters d, G, G_f and τ_f are properties that are very difficult to measure, and incorporate strong scatter.

It is important to realise that the joint brittleness ratio incorporates the following influences ($\alpha = 1$): influence of absolute size of the joint through ℓ; influence of shape through the ratio ℓ/t_1; influence of adherend compliance by $1/E_1$; and finally, the influence of the bond line fracture properties, τ_f^2/G_f, termed 'bond line (shear) brittleness'. The inverse of the joint brittleness ratio, ($1/\omega$), can be separated into the terms ℓ_m (ℓ_{geo}^{-1}), where

$$\ell_m = \frac{E_1 G_f}{\tau_f^2} \tag{18.20}$$

$$\ell_{geo}^{-1} = \frac{t_1}{\ell^2} \tag{18.21}$$

Here, the quantity ℓ_m is a similar quantity for bond lines as the characteristic length ℓ_{ch} in NLFM for solid materials ($\ell_{ch} = EG_f/f_t^2$), and the ratio $1/\ell_{geo} = t_1/\ell^2$ can be directly compared with the important joint factor $\sqrt{t_1}/\ell$, specified in Equation (18.16). So, the joint brittleness ratio may also be expressed by

$$\omega = \ell_{geo}/\ell_m \tag{18.22}$$

Gustafsson evaluated the load capacity of a single lap joint subjected to the Volkersen boundary conditions shown in Figure 18.5(a) by Equation (18.17) for the four generalised τ-δ curves shown in Figure 18.4. For curve I, being linear up to peak stress and then followed by a sudden drop to zero stress, one obtains

$$\frac{P_{max}}{\tau_f b\ell} = \frac{1+\alpha}{\varpi}\frac{\sinh \varpi}{(\cosh \varpi + \alpha)} \tag{18.23}$$

where

$$\varpi = \sqrt{\omega} = \sqrt{\frac{1+\alpha}{2}}\sqrt{\frac{\ell^2\tau_f^2}{t_1 E_1 G_f}} = \sqrt{\frac{\ell_g}{\ell_m}} \tag{18.24}$$

$$\ell_g = \frac{b\ell^2}{2}\left(\frac{E_1}{E_2 A_2} + \frac{1}{A_1}\right) \tag{18.25}$$

As is obvious from the above statements, the solution (18.23) fully conforms to the Volkersen result specified for $\alpha = 1$ in Equation (18.15a), when employing the substitution $\varpi = \rho$. Note, that ℓ_g according to Equation (18.25) is identical to ℓ_{geo} according to Equation (18.21) for the case $\alpha = 1$.

Figure 18.9 reveals that the three apparently very differently shaped τ-δ curves I, II and III specified in Figure 18.4 obviously give almost similar results for the joint capacity, whereas the pile-shaped curve IV gives higher differences in the ductile/brittle fracture transition range. The similar magnitude of the joint capacities is related to the assumption of equal fracture energies, which in the case of curves II and III is highly bound to the stable post-peak behaviour. If, in the case of these curves, stress dropped to zero at τ_f, joint capacity due to unactivated fracture energy would drop significantly by 30% (resp. 90%); see Ottosen and Olsson (1986) and, Gustafsson (1987). The limit value for ultimate capacity according to Equation (18.17) for ideal plasticity, i.e. for very small joint brittleness ratios ($\omega < 0.1\ (1+\alpha)$), is

$$P_{max} = \tau_f b\ell \tag{18.26}$$

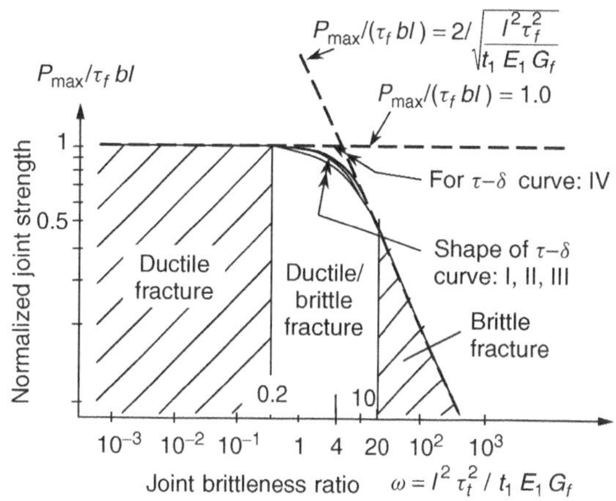

Figure 18.9 Normalised joint strength, $f_v = P_{max}/(\tau_f b\ell)$, versus joint brittleness ratio ω for a symmetric joint buildup ($\alpha = 1$) (after Gustafsson, 1987)

For very high joint brittleness ratios ($\omega > 10(1+\alpha)$), i.e. in the range of linear fracture mechanics where the size of the fracture process zone may be regarded as constant compared to bond line dimensions, one obtains

$$P_{max} = \sqrt{2(1+\alpha)}b\sqrt{t_1}\sqrt{E_1 G_f} \tag{18.27}$$

So, in the case of high joint brittleness ratios (i.e. in the case of LEFM), the load capacity with respect to bond line properties is fully determined by fracture energy G_f, and is not influenced by shear strength or by overlap length ℓ. Formally, quantity $\sqrt{E_1 G_f}$ could be interpreted as the fracture toughness, K_c, of the specific bond line/adherend material combination. Note, that the load capacity (18.27) is equivalent (for $\alpha = 1$) to the Volkersen limit load P^*_{max} according to Equation (18.15b), using the substitution specified in Equation (18.19a).

The brittleness range ω for which validity of NLFM (i.e. the general Equation (18.17)) applies is (compare also Figure 18.9)

$$0.1(1+\alpha) \leqslant \omega \leqslant 10(1+\alpha) \tag{18.28}$$

Concerning numerical solutions of nonlinear strain softening of bond lines, the contribution of Wernersson (1994) has to be mentioned. The author has derived a versatile two-dimensional nonlinear mixed mode fracture mechanics bond line model based on the concept of gradual fracture softening. Applications are shown in Section 18.6.2.

18.6 PURE WOOD JOINTS

18.6.1 Scarf Joints

The scarf joint as shown in Figure 18.10 is, from a purely technical viewpoint, the jointing method delivering the highest strength. The scarf joint is also the forerunner of the finger joint. However, a well designed scarf joint (see below) is uneconomic, due to the high loss of timber (long overlap) and the very costly cramping arrangements, as no self-interlocking of the adherends exists, as in the case of finger joints. Nevertheless, still today, for specific applications, scarf joints can be the best solution for an end joint.

Axial force equilibrium for the mean shear and normal stresses, τ and σ, along the glue line of the axially loaded scarf joint gives

$$\tau = \tau_{\xi\eta} = \sigma_0 \sin\alpha \cos\alpha, \quad \sigma = \sigma_\eta = \sigma_0 sin^2\alpha \qquad (18.29a, b)$$

In the case of axial tension loading, the normal stresses are peel (= tension) stresses, and vice versa in compression loading. Figure 18.11 shows the dependency of the mean stresses $\tau = \tau_{\xi\eta}$, $\sigma = \sigma_\eta$ on the scarf joint angle α, here expressed by the inverse slope ℓ/t. It can be seen that the peel stress σ_η drops rapidly with decreasing slope, and becomes a negligible magnitude beyond about $\ell/t = 10$. Finite element analysis shows that glue line shear stress is distributed very uniformly along the bond line, revealing almost no stress increase at the overlap ends. So, for $\ell/t = 5$ to 10,

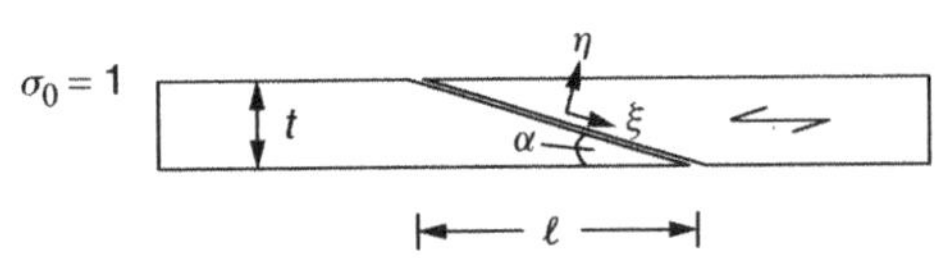

Figure 18.10 Geometry of a scarf joint

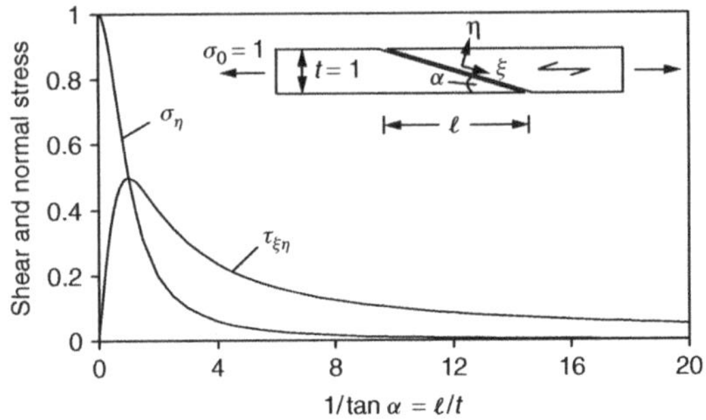

Figure 18.11 Dependency of mean shear and peel stresses, τ and σ, on slope of the scarf joint

the difference between maximum and minimum (at mid-length) stresses is about 10%. Beyond $\ell/t \geqslant 10$ the minimum shear value deviates less than 2% from the average value according to Equation (18.29a). For the peel stresses, the peaks at the overlap ends are more pronounced, but no singularity occurs. To obtain a joint of highest possible tension capacity, a glue line failure should occur at or beyond tension failure parallel to the grain of the timber. So, disregarding the peel stresses for larger ℓ/t ratios, Equation (18.29a) gives (with $\tau = f_v$ and $\sigma = f_{t,0}$)

$$\frac{f_{t,0}}{f_v} = \frac{1}{\sin\alpha \cos\alpha} \qquad (18.30)$$

The (optimum) scarf joint slope is determined by the ratio of tension strength parallel to fibre and shear strength (for small angles $f_v \approx f_{v,\alpha}$). However, it is not sufficient to assume the ratio of (characteristic) strength values of solid wood or glulam in structural sizes, which for most strength classes of softwoods has a value of about 6. This ratio would result in a slope $\ell/t = 5.8$, obviously producing peel stresses that are too high. The clear wood ratio $f_{t,0}/f_v$ for most European softwood species is about 10–12, giving slopes ℓ/t of equal numbers. So, the ratio ℓ/t of an axially tension loaded scarf joint should be at least 10.

18.6.2 Finger Joints in Glulam Laminations and Lumber

A finger joint, as shown in Figure 18.12, is a self-locating end joint formed by machining a number

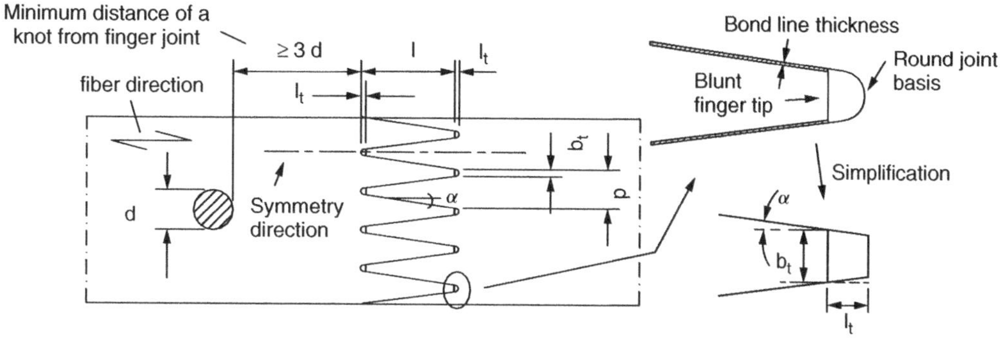

Figure 18.12 Finger joint geometry, notations and minimum distance of a knot from the joint according to EN 385

of similar, tapered symmetrical fingers in the ends of timber members, which are then bonded together. The geometry of a finger joint is defined by three nominally independent parameters, being finger length ℓ, pitch p and tip width b_t. The finger length is the distance between the finger base and the tip of the finger, measured along the centre line of the finger; pitch is the distance between fingers, centre to centre; tip width is the width of the blunt finger tip. The tip gap ℓ_t depends to a large extent upon the production conditions; as a rule of thumb, ratio $\ell_t/\ell \approx 0.03$ can be regarded as an appropriate value. Rigorously taken, only the slope of the finger, α, and pitch, p, are variables intrinsic to the finger cutter. The finger length ℓ, and thus b_t and ℓ_t, are co-determined by the manufacturing process, i.e. by the precision of end cutting and by tool wear.

Finger jointing with various joint geometries and materials (not only solid wood) is certainly the most versatile jointing method widely employed in many types of structural timber products. Finger joints for end jointing glulam laminations are a prerequisite for the economic manufacture of glulam. Further, finger jointing is the basis of quality lumber beyond usual lumber lengths of about 4–5 m. Lumber finger jointing is performed up to considerable cross-sectional dimensions of maximally about 100 × 280 mm. Further, finger joints are employed in the manufacture of special lattice and panel web (formwork) I-beams (see Section 18.6.6). Finger joints used to join full cross-sections of glulam are termed 'large finger joints', whereby the expression large relates to both the size of the cross-section as compared to laminations and lumber, and the size of the finger joint geometry (see Section 18.6.3).

The geometry and the size of the finger joint can vary considerably. The dimensions of typical, industrially used finger joint geometries in Europe are specified in Table 18.4. The tension or bending strength of a finger joint is determined by many factors, of which the most important are the strength and stiffness of the adherends, production quality (determined i.a. by end pressure, cutter wear, end cuts, conveyor transport, climate etc.), finger geometry, adhesive and curing conditions. For assumed invariance in adherend and production quality, different finger geometries and adhesives may forward significantly different strengths. An experimental proof for the effect of geometry on tension strength is depicted in Figure 18.13 for profile types B and C (Aicher and Radovic, 1999). Profile C gave mean and characteristic tension strength values, $f_{t,fj}$, of 42.9 and 29.8 N/mm^2, respectively, exceeding the

Table 18.4 Finger joint geometries (profiles) of the industrially most used finger joints (dimensions in mm)

Profile	Profile geometry		
	ℓ	p	b_t
A	15	3.8	0.42
B	20	6.2	1.0
C	20	5.0	0.5
D	32	6.2	1.0

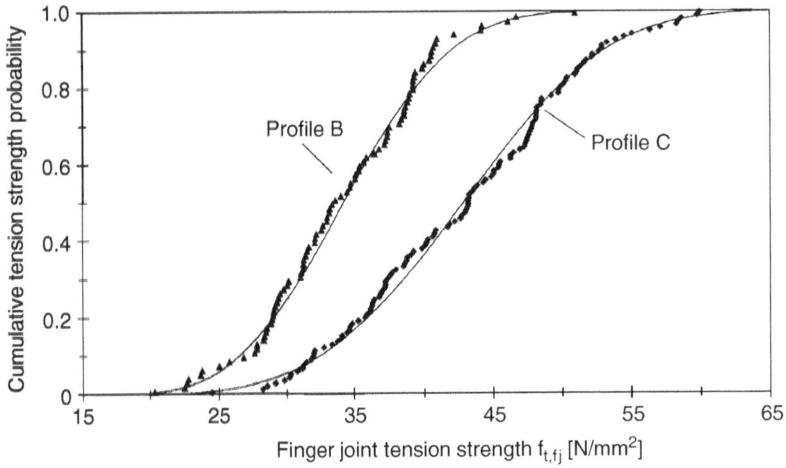

Figure 18.13: Influence of two different finger joint geometries on tension strength of glulam laminations (after Aicher and Radovic, 1999)

corresponding values of profile B by factors of 1.20 and 1.16, respectively. The strength differences obtained are quantitatively fully supported by computational approaches on the effect of finger joint geometry, discussed below.

Apart from finger geometry, the orientation of the finger joint in the cross-section has a significant influence in the case of predominant bending loading (glulam laminations are not considered in the following). Lumber with a rectangular cross-section can be finger jointed in two different ways, whereby the fingers are either visible on the flat or on the narrow side (Figure 18.14). The prevailing jointing feature in Europe is with fingers visible on the flat side. Despite that, the rectangular cross-section is obviously most often bent about the major axis (edgewise bending), resulting in unevenly stressed fingers along cross-sectional depth. Edgewise bending, in general, gives lower bending strength values as compared to flatwise bending, which may be expressed by the ratio

$$k_{f(,k)} = f_{m(,k),\text{flatwise}}/f_{m(,k),\text{edgewise}} \qquad (18.31)$$

where k_f, generally being > 1, is extremely dependent on the specific production conditions. The k_f effect results predominantly from an increased probability of two types of deficiencies/defects at the tension edge (specified below), in combination with small stress redistribution possibilities, as exists in the case of parallel fingers in flatwise bending. One deficiency is related to the fact that, production bound, the outermost finger is often less glued and cramped as are the inner fingers; secondly, a notch often exists when the planed edge hits the finger tip gap. In an extensive experimental study (60 comparative test series, 1200 specimens), factor k_f, determined from the mean (resp. from the characteristic) values of the samples was found to be about 1.25; no significant influence of finger profiles A and B was obtained, as might be supposed (Aicher and Windmüller, 2002).

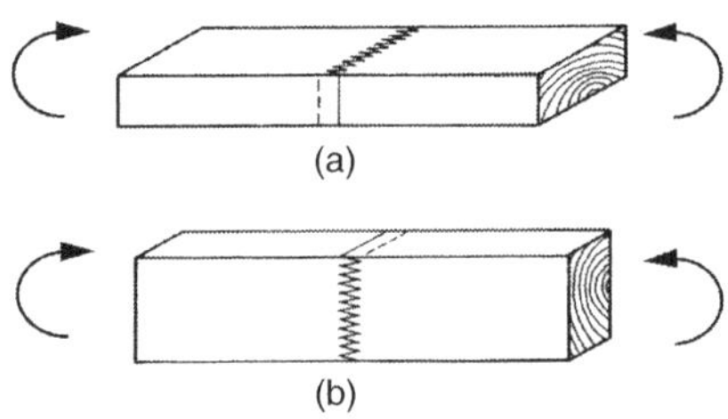

Figure 18.14 Different orientation of fingers in (a) flatwise bending and (b) edgewise bending

Strength verification of finger joints in glulam laminations and lumber is today exclusively performed by testing in the framework of factory production controls. To avoid premature glulam failure, the finger joints in the laminations should show higher strengths than the nominal lamination strength class. This requirement can be assumed to be met when the characteristic finger joint strength in tension or bending, $f_{t,fj,k}$ and $f_{m,fj,k}$, conforms to the following conditions (EN 1194):

$$f_{t,fj,k} \geqslant 5 + f_{t,0,l,k} \tag{18.32a}$$

$$f_{m,fj,k} \geqslant 8 + 1.4 f_{t,0,l,k} \approx 8 + 0.84 f_{m,l,k} \tag{18.32b}$$

where $f_{t,0,l,k}$ and $f_{m,l,k}$ are the characteristic tension strength parallel to grain and the characteristic bending strength of the respective lamination strength class (all strengths in N/mm^2). Note that in Equation (18.32b), validity of the ratio $f_{t,0,l,k}/f_{m,l,k} = 0.6$, as stated in EN 338, is assumed.

Several attempts have been made to optimise the finger joint geometry for load-bearing timber joints since their first use in the late 1930s (Egner, 1942, Selbo, 1963). The optimum design of finger geometry states a minimisation problem of several competing parameters, which are the intrinsic profile cutter dimensions $p, \alpha, \ell + \ell_t$, and then the width of the blunt finger tip b_t. The side ratio of the notch at the finger tip ℓ_t/b_t is also important for the stress concentration and crack initiation at the finger tip (see Figure 18.15(a)). Only three basic geometry parameters are independent and further subject to restrictions related to cutter production and joint manufacture. Roughly, the following conditions can be specified, resulting consistently from different linear continuum and fracture mechanics analysis approaches (Aicher and Klöck, 1990, 1991, Aicher and Radovic, 1999): the cross-sectional reduction due to the blunt finger tip should not exceed a ratio $b_t/p \approx 0.1$. The normalised glue joint area, which can be expressed approximately by $2\ell/p$, should not be less than 8. The slope α of the finger, equally as specified for scarf joints, should at least conform to $\tan\alpha^{-1} = 2(\ell + \ell_t)/(p - 2b_t) > 10$.

Numerical studies on the tension capacity of finger joints based on a nonlinear bond line softening model were first performed by Wernersson (1994). The author investigated the four finger joint geometries given in Table 18.4, employing bond line properties (brittlenesses) for PR adhesive according to Tables 18.2 and 18.3. Figure 18.15(b) shows the calculated global tension stress vs. elongation curves of the different profiles, which are highly linear up to failure in most cases, despite

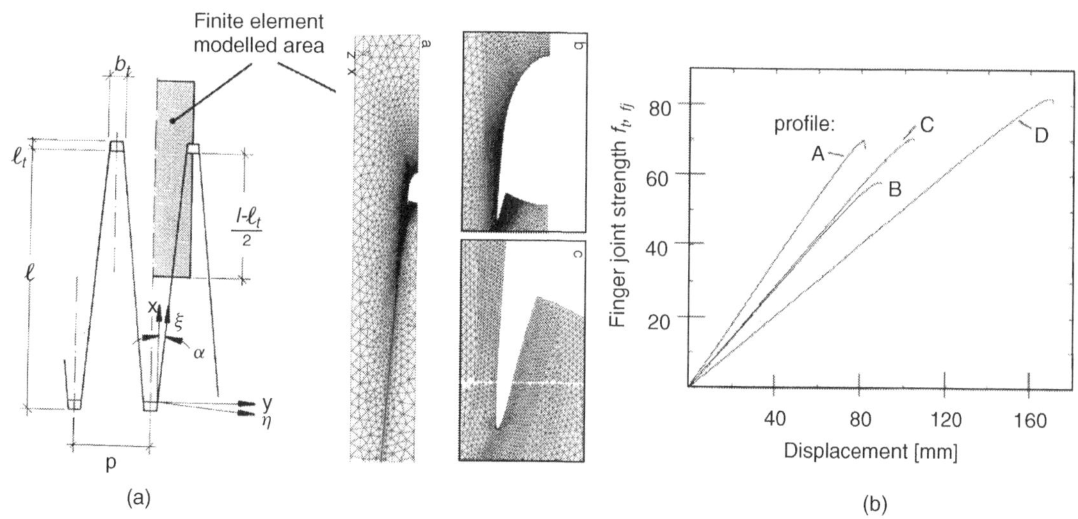

Figure 18.15 Finite Element (FE) modelling of finger joint strength. (a) FE discretised joint area; views of deformed structure with opened initial crack (Aicher and Radovic, 1999); (b) strength results for different finger joint geometries obtained with a nonlinear fracture softening bond line model for phenolic resorcinol adhesive (after Wernersson, 1994)

the expressed nonlinearities of stresses and strains in the bond line (see below). The analysis delivered qualitatively almost the same strength sequence for the different profiles as determined by Aicher and Klöck with linear approaches. However, quantitatively, joint strengths were over-estimated in the simulation. According to the parametric study performed, comprising variations of bond strength τ_f and σ_f, of fracture energies $G_{f,\tau}$ and $G_{f,\sigma}$ and of adherend stiffnesses, load capacity was predominantly influenced by adhesive strengths. Then, consequently, the load capacity limit is partly defined by the wood strength. Increased fracture energy and wood stiffness have a similar but considerably less expressed strength increasing effect as compared to strengths τ_f and σ_f.

Comprehensive finite element studies on the influence of nonlinear bond line properties of different adhesives and the impact of bond line voids on finger joint strength were performed by Serrano (1997, 2000), using the concept of Wernersson's bond line model. For the bond line properties of PR and PUR adhesives employed, the maximum stresses in the bond line, and so the start of gradual bond line damage, were reached at about 25% and 60% of the joint load capacities, respectively. At peak load, about one-third of the bond length at the overlap ends has started to soften in the case of both adhesives. Further, it was shown that finger joint strength is strongly dependent on the presence and location of small defects (voids) in the bond line.

18.6.3 Large Finger Joints

A large finger joint is a joint through the full cross-sectional area at the ends of glulam members bonded together at any angle, and which may include corner pieces of glulam, laminated veneer lumber or plywood (see Figure 18.16). The finger length should be at least 45 mm (EN 387); the dimensions of the most common profile are: $l = 50$ mm, $p = 12$ mm and $b_t = 2$ mm. The predominant use of large finger joints for typical horizontally laminated glulam is for the manufacture of frame corners. However, there are also constructions where large finger joints are employed for joints in straight or slightly curved members, which are primarily loaded in bending and also occasionally in pure tension. The latter application and load case occurs, for instance, in the long span tension band type pedestrian bridge in Essing, Germany (Müller, 2000). The best example for the high reliability of well manufactured large finger joints in straight and kinked connections is probably delivered by a glulam company in the Netherlands, producing vertically laminated glulam. There, due to related production process limitations of glulam

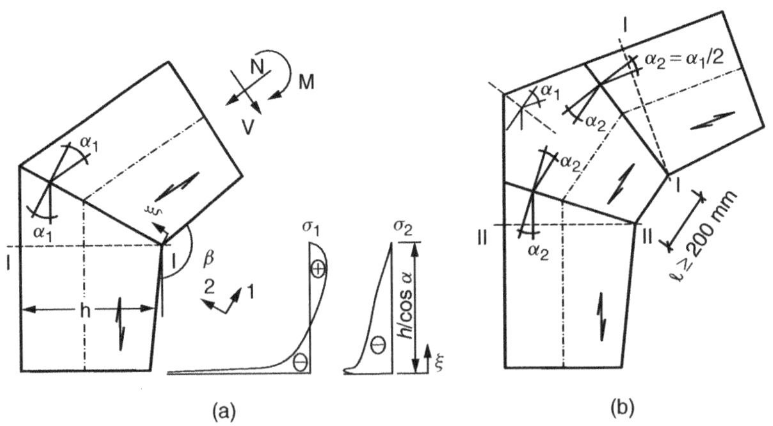

Figure 18.16 Geometries and dimensions of frame corners with one (a) or two (b) large finger joints. Also given are (a) stresses perpendicular and parallel to joint line, σ_1 and σ_2, of the frame corner with one finger joint, subjected to a closing bending moment

member length, several hundred thousand large finger joints were manufactured for the production of large-span members. No failure or other in-service deficiency has been reported from these structures.

The following comments on the load-bearing behaviour and on a simplistic design approach refer to corners subjected to a negative (i.e. closing) bending moment. This loading situation produces high bending compression stresses at the re-entrant corner(s) and compression stresses parallel to the joint line, increasing towards the re-entrant corner(s), as depicted in Figure 18.16(a). The graph shows qualitatively the stress distribution in a frame corner with one large finger joint at about 50% of ultimate capacity. The load capacity of the frame corner with one large finger joint can be increased significantly with a glued-in corner piece (Figure 18.16(b)). To obtain the highest possible strength increase, the end grain edges of the members and of the corner piece have to be cut in such a manner that the angle between the fibre directions of the jointed pieces, being $2\alpha_1$ in the case of a corner with a single joint (Figure 18.16(a)), is reduced maximally; this is obviously the case for $2\alpha_2 = \alpha_1$.

The load-bearing behaviour of a frame corner with a large finger joint, subjected to a negative bending moment, is characterised first by a strictly linear moment vs. rotation relationship. At further loading, a nonlinear behaviour due to increasing compression yielding at the re-entrant corner(s) occurs. Finally, fracture is initiated, in general, by a finger joint tension shear failure at the tensioned salient corner, as shown in Figure 18.17. The strain and stress distributions in the elastic and nonlinear range can be modelled by the Finite Element Method (FEM) in good agreement with full scale tests (Aicher *et al.*, 1997). Obviously, an appropriate design approach should mirror the described failure behaviour, and thus should account for the plasticity-induced effects on the stress distribution, also present at the tension side, and then essentially for the bond line failure; so far, such an engineering design has not been given in the literature.

A rather simplistic and, today, widely-adopted approach for a closing bending moment was proposed by Heimeshoff (1976), then for the deterministic permissible stress design. The design consists of a linear normal force and bending moment interaction at the assumed design relevant location, i.e. the

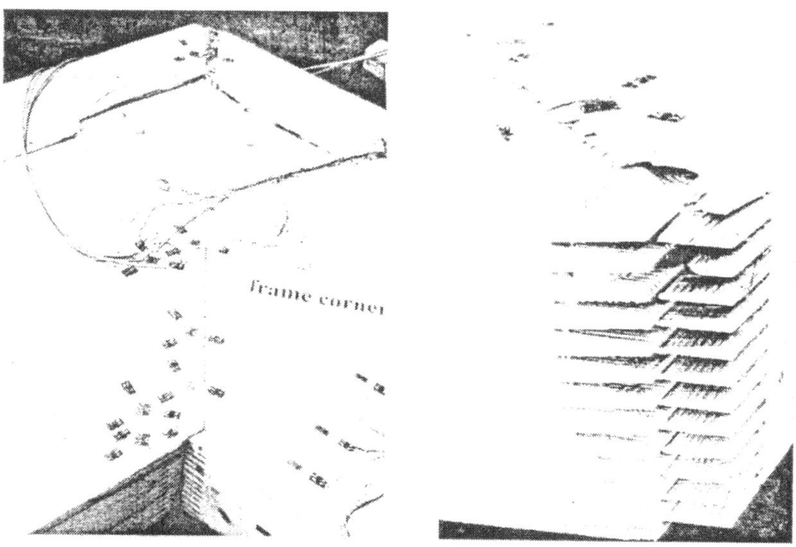

Figure 18.17 Typical failure appearance of a frame corner subjected to a closing bending moment. Tension shear fracture in the joint at the salient corner and nonlinear compression deformations at the re-entrant corner can be seen

re-entrant corner. The compression stresses are calculated according to beam theory. Transposing the design to a semi-probabilistic strength verification, failure is assumed to occur when, at the re-entrant corner, the compression strength at an angle α to grain is reached ($\alpha = \alpha_1$ or α_2 in the case of 1 or 2 finger joints); the limit states equation can then be written as

$$\frac{1}{k_{c,\alpha}}\left(\frac{\sigma_{c,0(,d)}}{k_c f_{c,0(,d)}} + \frac{\sigma_{m(,d)}}{f_{m(,d)}}\right) \leqslant 1 \quad \text{where}$$

$$k_{c,\alpha} = \frac{f_{c,\alpha,k}}{f_{c,0,k}} \qquad (18.33)$$

Factor k_c is a buckling coefficient, being unity when stresses are calculated according to second order theory; $k_{c,\alpha}$ represents the normalised compression strength for an angle α between the load and grain directions.

The (design) stresses (resp. the section forces), $\sigma_{c,0} = N/A_{\text{eff}}$ and $\sigma_m = M/W_{\text{eff}}$, are evaluated for section(s) I–I or I–I and II–II normal to the member axes immediately besides the finger joint. The cross-sectional values are determined with the effective width $b_{\text{eff}} = b(1 - b_t/t)$. To account for defects in the large finger joints, resulting from the finger cutting process, the (design) strength values of the jointed members should be downgraded by one strength class. Defects in the fingers can easily result when the cutter hits knots, which cannot be excluded as in the case of finger joints in laminations, and which may eventually lead to the significant destruction of a larger portion of one or more fingers.

The load capacity predicted by Equation (18.33) depends decisively upon (normalised) compression strength $k_{c,\alpha}$. Several phenomenological equations have been proposed for $k_{c,\alpha}$ in the literature. In most design codes, the Hankinson type formula, based on a linear interaction of normal stresses parallel and perpendicular to grain,

$$k_{c,\alpha} = \frac{1}{\left[\left(\frac{f_{c,0,k}}{f_{c,90,k}}\right)\sin^2\alpha + \cos^2\alpha\right]} \qquad (18.34a)$$

is implemented. Another known relationship, used in the German timber design code, is

$$k_{c,\alpha} = 1 + \sin\alpha\left[\left(\frac{f_{c,0,k}}{f_{c,90,k}}\right)^{-1} - 1\right] \qquad (18.34b)$$

Both equations depend similarly upon the ratio of compression strength values parallel and perpendicular to the grain, $f_{c,0,k}/f_{c,90,k}$, although Equation (18.34a) gives a more pronounced strength decrease for $k_{c,\alpha}$ in the angle range of about $\alpha = 20\text{–}70°$. For glulam, until recently, a strength ratio of $f_{c,0,k}/f_{c,90,k} \approx 5$ was assumed appropriate, and so specified in the European strength class standard (EN 1194:1994); so, for instance, for a glulam beam (GL32) built up homogeneously from C35 laminations, strength values $f_{c,0,k} = 30.5$ N/mm^2, and $f_{c,90,k} = 6$ N/mm^2 are/were[1)] given.

Equation (18.33), despite the deficiency of not incorporating the actual failure mode, enables a fairly good approximation of the load carrying capacity of finger jointed frame corners (Aicher *et al.*, 1997), provided that $k_{c,\alpha}$ roughly conforms to Equation (18.34b) and the ratio $f_{c,0,k}/f_{c,90,k}$ is about 5.

18.6.4 Lap Joints

For double lap joints, an experimentally calibrated mean strength equation was given by Steurer (1981):

$$f_v = \frac{P_{\max}}{2\ell b} = C\sqrt[3]{t_1/\ell^2}[\text{N/mm}^2] \qquad (18.35)$$

where t_1 is the thickness of the respective lateral adherend, and ℓ is the bond length (ℓ and t_1 in mm). The thickness t_2 of the centre stick has to be about $t_2 \approx 2t_1$. The fitted parameter C was specified as 23 and 28 for tension and compression shear, respectively.

An extensive study on glued double lap joints has been performed by Glos *et al.* (1987). The investigation comprised tension and compression shear tests with a solid wood centre stick and lateral adherends of solid wood or beech plywood

[1)] Note that, due to a new European test standard for the determination of compression strength perpendicular to grain, the strength values have been reduced to almost 50% of the former values.

glued at different angles. Based on the experimental study, a phenomenological design rule was suggested, whereby the characteristic joint capacity is made dependent on a basic shear strength value for a reference load case. Further, modification factors are specified taking into account the length of the glued area, the angle between the adherends, the type of adhesive, and the density and width of the joint components (Glos and Horstmann, 1989). The data source was re-analysed by Gustafsson and Serrano (1991), extending the nonlinear fracture mechanics approach given in Section 18.5.4 with an effective modulus of elasticity for the centre stick in order to account for the specific stiffness situation in a truss heel joint. Based on the assumptions of pull-compression boundary conditions (see Figure 18.25(b) and Equation (18.43)), a good approximation of the experimental data was obtained.

A reliability analysis of glued lap joint resistance, evaluated by Volkersen theory, has been shown by Aicher and Klöck (2001). Based on first order reliability analysis, the work gives an assessment of bond strength distributions as being dependent on bond length.

For the double strap lap joint according to Figure 18.18, a design rule based on linear fracture mechanics has been proposed by Walsh *et al.* (1973), as mentioned in Section 18.5.3. The proposal, discussed briefly below, accounts exclusively for the strength-determining effect of the peel stress singularities at the notches of the overlap ends of the straps. For the notch geometry under consideration, any stress component, σ, in the region near the re-entrant notch tip can be described by

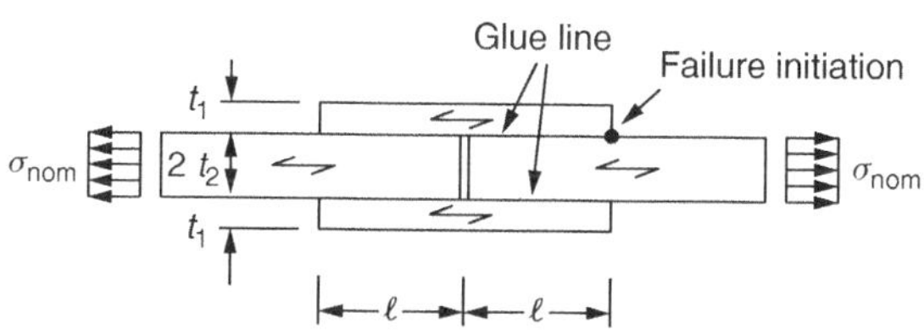

Figure 18.18 Geometry and notations of a double strap lap joint

$$\sigma = K_A(2\pi r)^{-0.45} f_A(\varphi) + K_B(2\pi r)^{-0.103} f_B(\varphi) \tag{18.36}$$

where r and φ are polar coordinates with respect to the re-entrant corner at the strap end, and K_A, K_B are stress intensity factors of the peel normal stress and shear stress fields $\sigma_y(r, \varphi)$ and $\tau_{xy}(r, \varphi)$, respectively. In the case of $\varphi = 0$, the stresses are

$$\sigma_y = K_A(2\pi r)^{-0.45} \quad \text{and} \quad \tau_{xy} = K_B(2\pi r)^{-0.103} \tag{18.37a, b}$$

Equation (18.37a) then gives, for the peal stress intensity factor,

$$K_A = \lim_{r \to 0}(2r\pi)^{0.45}\sigma_y(\varphi = 0) = k_A t_2^{0.45}\sigma_{nom} \tag{18.38}$$

where $k_A = f(t_1/t_2; \ell/t_2)$ is the normalised peal stress intensity factor.

The numerical solutions for K_A, and similarly for K_B, show that K_A is roughly 10 times higher than K_B, and also that stresses σ_y increase, as is obvious from the singularity exponents in Equations (18.37a, b), far more rapidly than stresses τ_{xy}. Consequently, the failure criterion for the strap type lap joint was assumed to be fully governed by the peel stress magnitude (resp. stress intensity). The failure state is then defined by $K_A = K_{AC}$, where K_{AC} is the fracture toughness perpendicular to the bond line and fibre direction. For explicit values of K_{AC} and normalised stress intensity factor $k_A(t_1/t_2; \ell/t_2)$, determined for some Australian softwood and hardwood species, see Walsh *et al.* The significant LEFM size effect specified by Equation (18.38), however, was not verified experimentally in the study.

18.6.5 Gusset Joints

Joints with glued gussets made of plywood or LVL provide excellent possibilities to produce rigid frame corners and truss joints (Figure 18.19). Glued trusses with diagonals jointed by means of plywood gussets were frequently used between about 1950 and 1970, and are now regaining popularity. This is for several reasons: architectural concepts of maximising day-light, wide-span but stiff truss structures, economic and efficient gusset materials (LVL) and, very importantly,

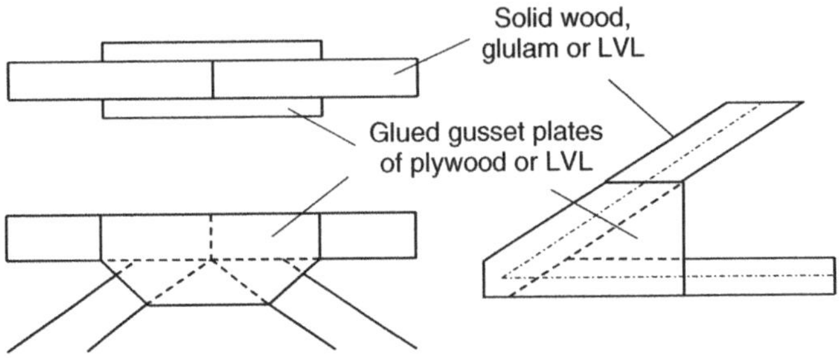

Figure 18.19 Types of glued joints in trusses with gusset plates made of wood-based panels

improved gluing and cramping techniques. The cramping pressure for gluing of the gussets was, until recently, primarily performed by nail-gluing, which affords quite tight nailing (about 1 nail per 65 cm^2 and nail distances $\leqslant$ 100 mm). Today, screw-gluing by means of self-tapping screws is far more efficient and reliable; it enables about a three times larger cramping influence area ($\approx$ 165 cm^2) and enlarged screw distances up to 150 mm.

An extensive investigation on the size effect on global shear strength of plywood gussets glued to lumber diagonals at different angles between grain orientations of the outermost plywood layer and lumber fibre axis was conducted by Kalina (1965). The results of the tests, with about 1000 specimens in compression shear, forwarded a significant size effect and no significant influence of the fibre angle. The global bond shear strength (maximum load divided by glued area) for rectangular glued areas A_{gl} of 20–1000 cm^2 was determined as

$$f_v = 9.1 - 2.2\lg\frac{A_{gl}}{100}[N/mm^2] \qquad (18.39)$$

with A_{gl} in mm^2. Equation (18.39) specifies an equal influence of bond dimensions parallel and perpendicular to the shear direction, and neglects the thickness of the gusset plate (or strap), which is contrary to most other theoretical and experimental findings in the literature (e.g. joint factors given by de Bryne (1944) and Steurer (1981)). Evaluating Equation (18.39) for joint areas of 20 cm^2 and 1000 cm^2, mean bond shear strength values of 6.2 and 2.5 N/mm^2 are obtained. Without discussing the general applicability of Equation (18.39) in more detail, the results support, in a more general judgement, the ideas on a significant size effect.

The bond line design of larger gusset plates can in principle be performed on the basis of a frame analysis, assuming rigid connections in the truss joints. The design is then based on the member end section forces M, N and V, which have to be transferred by the bond line. In an engineering approach, the stresses from normal and shear forces may be added vector-like to the torsional stresses equilibrating the member end moments. Further, a size effect accounting for the dimensions of the glued area has to be considered.

The dynamic behaviour of joints with glued plywood gussets is excellent; extensive investigations with up to 10^6 load cycles on single joints and full scale trusses were performed by Kufner and Spengler (1981).

18.6.6 Special Lattice and Panel Web (Formwork) I-Beam Joints

Figure 18.20 shows the build-up of typical timber formwork I-beams with different types of webs and glued joints for the web to flange connections, used extensively in Central Europe. Strength verification of these joints is today exclusively performed by testing within the framework of technical approvals.

Figure 18.20(a) shows a lattice web beam with diagonal solid wood struds with twin-ended tenons, connected by finger joints and glued into round milled slots of the solid wood flanges; the latter are also finger jointed, in general. Increasingly, the technically/economically very competitive panel

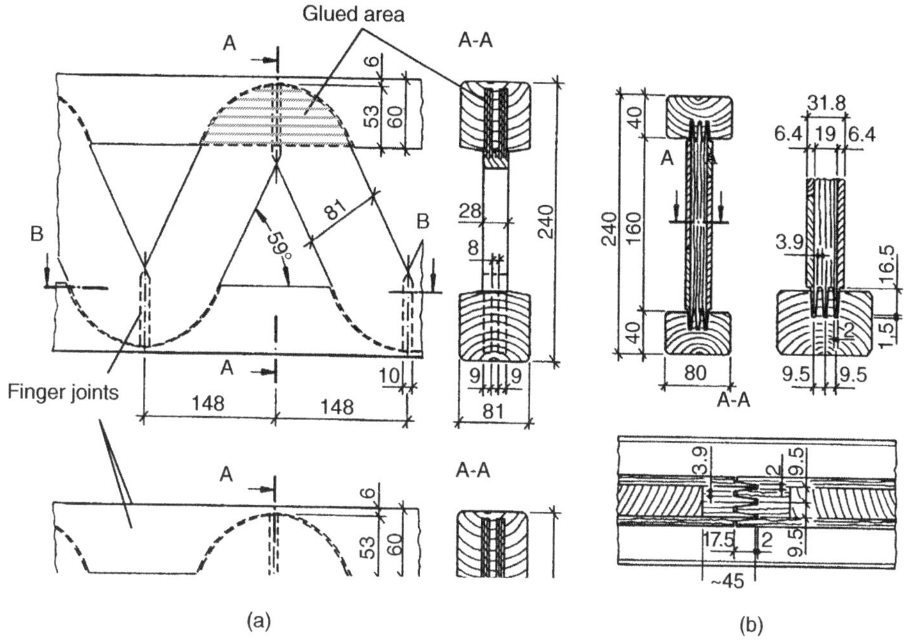

Figure 18.20 Typical formwork I-beams with trussed and continuous webs with different glued connections. (a) Glued lattice web beam (after N.N., 1999), (b) glued panel web beam (after N.N., 2001)

web formwork beams shown in Figure 18.20(b) are used. In these beams, the webs consist of finger-jointed multi-layer wood panels or special particle boards (resin content about 20%); the web is finger-jointed to the flanges. As pointed out in the introduction, the continuous glued flange/web joints mark the transition here from what are regarded as *discrete* structural timber joints to monolithic glued elements, and thus are not discussed further.

The strut/flange joint of the lattice web beam Figure 18.20(a), represents a more refined version of joints in so-called DSB trusses (Handel, 1977, N.N., 1998), shown in Figure 18.21. The graph gives the construction scheme of the truss, construction variants and details of the glued joint build-up. Such glued trusses, standardised to a high extent, were manufactured with millions of running meters in Middle Europe in the 1970s and 1980s, primarily for use in roof structures. Today production is less extensive. The glued tenons of the struts in DSB beams are not finger jointed. Contrary to formwork beams, the (permissible) loading of the glued joints in DSB beams is based on design according to

$$\frac{S}{A_{gl}} \leqslant \tau_{allow.}(\alpha, A_{gl}) \qquad (18.40)$$

where S is the axial force in the diagonal strut, $A_{gl} = 2nA'_{gl}$ is the total glued area of the n strut tonons ($n = 2$ or 3) glued into the circular slots of the flanges, and $\tau_{allow.}(\alpha, A_{gl})$ is the allowable shear stress in the joint depending upon angle α and size of the glued area A_{gl}. The allowable shear stress is 0.6 N/mm^2 for $\alpha < 60°$, reducing to 0.3 N/mm^2 at 90°. The maximum size of the effective glued area is 250 cm^2; in the range of $120 < A_{gl}[\text{cm}^2] < 250$ allowable shear stress is reduced by the factor $320/(A_{gl} + 200)$. It should be noted that the size of the glued joint area, A'_{gl}, is considered small enough to neglect torsional stresses from the strut end moments, which are not taken into account as a virtually ideal truss is assumed.

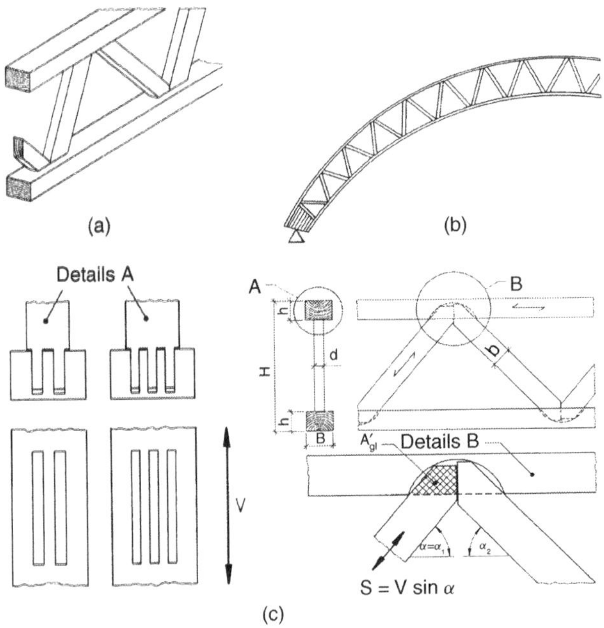

Figure 18.21 Special, so-called DSB truss: (a) perspective view of build-up, (b) curved beam construction variant, (c) details of the glued multiple tenon strut/flange truss joint

18.7 HYBRID WOOD JOINTS

18.7.1 General

Hybrid wood joints are connections or reinforcements of wooden members (solid wood, glulam, LVL) by means of glued-in, non-wooden, high-strength materials such as steel or Fibre Reinforced Plastics (FRP). The connecting materials may be rods (threaded or rebars) or plates; in the latter case, surfaces should be corrugated. The connections of the rods and plates in the case of connecting joints as compared to reinforcements may be performed by screwing and/or welding. Hybrid wood joints offer wide application areas in timber construction. Some examples of different layouts of frame corners are given in Figures 18.22 and 18.23.

Compared to pure wood–wood gluings, hybrid wood joints pose several additional questions/problems arising from the completely different material and processing nature of the adherends. Further, adequate manufacturing technologies, feasible for building construction in terms of *building* practise and economy, have to be considered. One example, therefore, is surface preparations of steel and FRP surfaces. This procedure, being self-evident in aircraft component gluing technology, faces a completely different technical/economical weighting in building construction. So, for glued-in threaded rods in timber, the primary valid assumption, resulting from economy-related considerations, is that the adhesive develops no long-term reliable bond to the basically uncleaned steel surface, but shows a pure mechanical anchorage. Bonding of steel parts based on reliable degreasing and surface preparations can be considered in special applications. Section 18.7.2 deals with glued-in steel rods; for joints with glued-in steel plates see Palm (1991), Aicher (1991), Johansson (1992), Kemmsies (1994) and Aicher *et al.* (1997).

18.7.2 Glued-in Steel Rods

In the following, the case of axially loaded rods is considered exclusively. Laterally loaded rods can be handled similarly to laterally loaded dowels (see Chapter 17).

The global load slip behaviour of axially tension or compression loaded (metrically) threaded

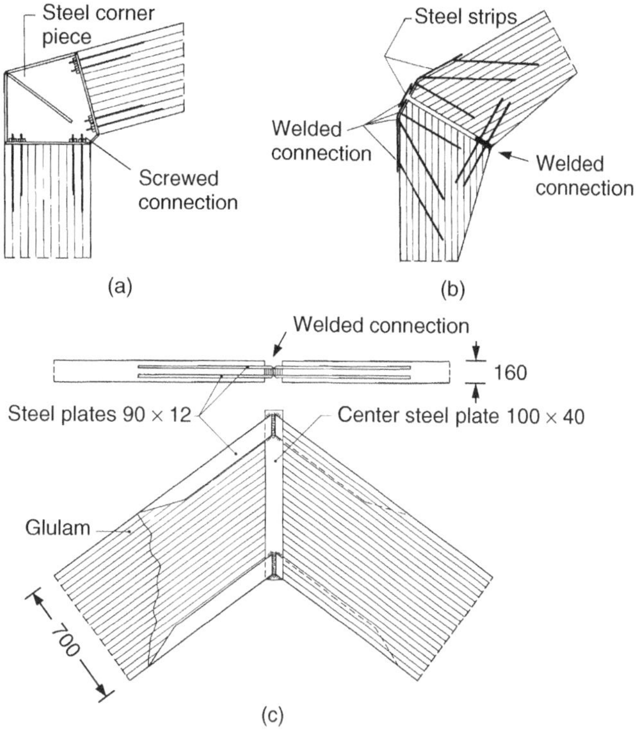

Figure 18.22 Different construction alternatives for frame corners with glued-in rods or steel plates, (a) after Buchanan and Fairweather (1992), (b) after Turkowski (1991), and (c) after Johansson (1992)

rods depends upon the type of adhesive, and also on rod slenderness, as shown in Figure 18.24. The graph gives bond shear stress, τ, vs. slip, δ, curves for a brittle epoxy and polyurethane, and for an extremely ductile/softening special phenolic-resorcinol adhesive (PRFs). Bond shear stress $\tau = \tau_m$ and strength $f_v = \tau_{m,\max}$ are defined as in Sections 18.4.1 and 18.5.1 as global values, derived from axial (ultimate) load $P_{(\max)}$ divided by bond area $A_{\text{bond}} = \ell d\pi$, where ℓ is the anchorage length of the rod and d is the nominal rod diameter, d_{nom}, or hole diameter, d_h, then specified accordingly. Global slip is the slip of the rod vs. the timber measured at the protruding end of the rod. It can be seen that the global compliance of the joint can stretch from extremely brittle with early pre-peak load drops to ideal plasticity and stable softening.

It can be supposed that a glued-in steel rod joint should, in principal, behave as a lap joint. This means that we have a similar inhomogeneous stress distribution along the bond line which is influenced by the cross-sectional and stiffness ratios of the steel rod and the surrounding timber part; boundary conditions have a considerable influence, too. Obviously, the adhesive constitutive law has an enormous influence in the post-elastic range as well.

Concerning bond strength, the inhomogeneous stress distribution results in an adhesive constitutive law-dependent strength reduction with increasing anchorage length ℓ; further, there should be a strength influence of rod shape (= rod slenderness $\lambda = \ell/d$). Contrary to the generally assumed strength increasing effect of higher timber density, widely claimed for glued-in rod connections in the past, this could not be verified in a recent extensive test series in the framework of a European research project on Glued-In RODs (GIROD) in timber structures. Similarly, no consistent effect of the angle between rod and fibre direction, neither in short-term nor in long-term loading was obtained.

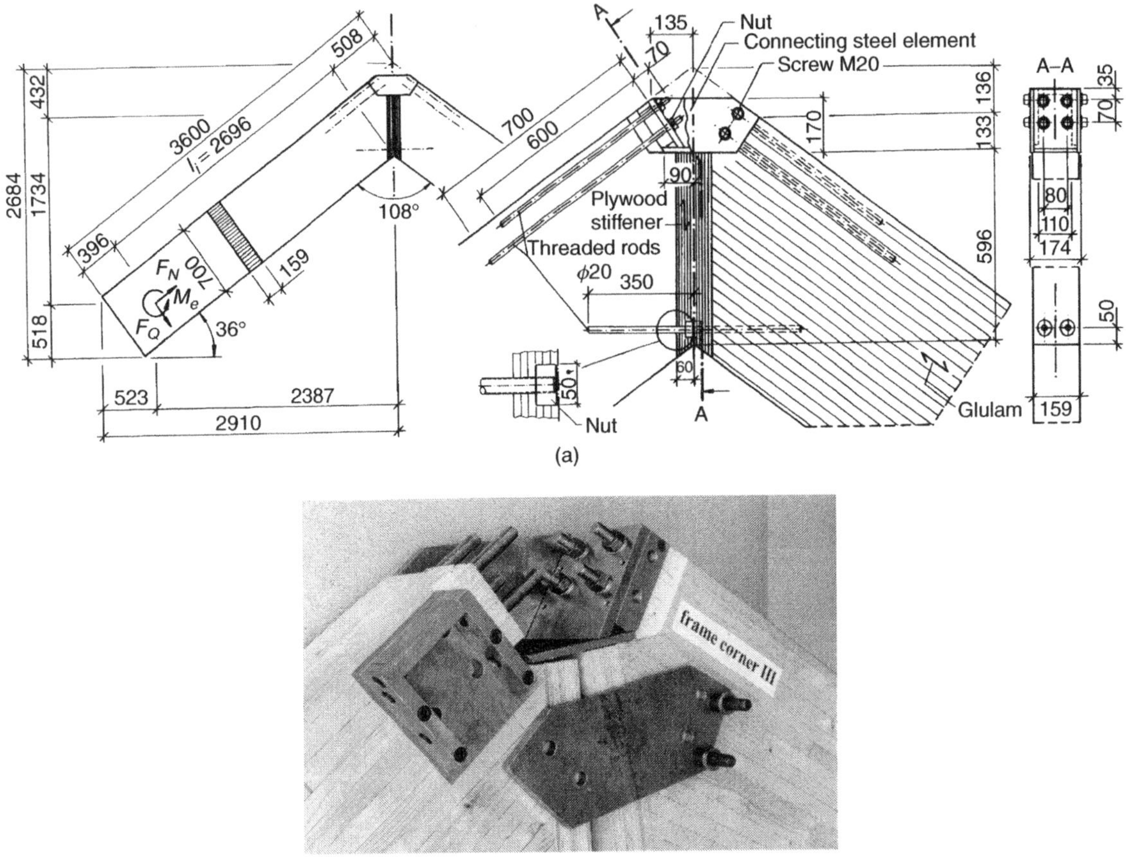

Figure 18.23 Frame corner (demountable) based on glued-in rods designed for a closing bending moment; (a) total view, (b) view of the salient knee with a rather small prototype steel connecting element (after Aicher and Herr, 1998)

The literature states a large number of design equations for mean and few for characteristic load capacities of axially glued-in rods. Most analytical proposals are purely phenomenological (data fits); few incorporate to any extent considerations of the mechanical relevance of the parameters. A recent proposal makes use of Gustafsson's generalised Volkersen approach. In the following, some design equations for the load capacity based on different mechanical assumptions are presented. All stated numerical strength values/coefficients apply to short-term ramp loading of connections that are in a homogeneous state of moisture related to a temperature of about 20 °C and 65% relative humidity. Some comments on the influence of climate and the duration of load follow.

The first profound approach acknowledging the length effect of the bonded rod, validated by tests (Riberholt, 1986, 1988), assumes that the load capacity is proportional to $\sqrt{\ell}$. This corresponds to an assumption of linear fracture mechanics when disregarding the rod shape influence (= one-dimensional rod approximation). The (local) bond fracture resistance is expressed by a quantity which can be regarded as an adhesive dependent fracture toughness $K_{c,\text{adhesive}}$. Further, a linear influence of characteristic density ρ_k was assumed. The characteristic bond shear strength for anchorage length $\ell > 200$ mm was given as

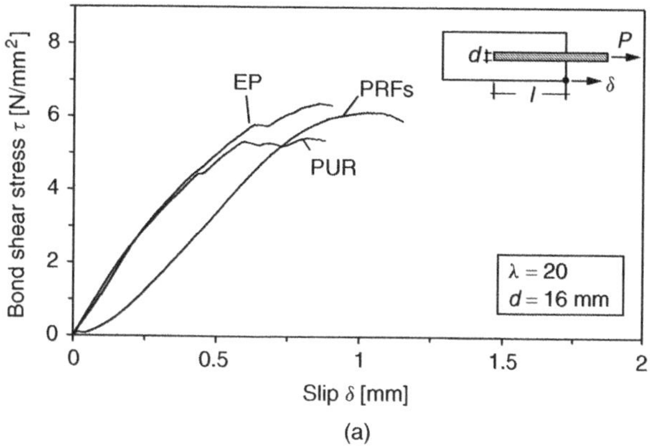

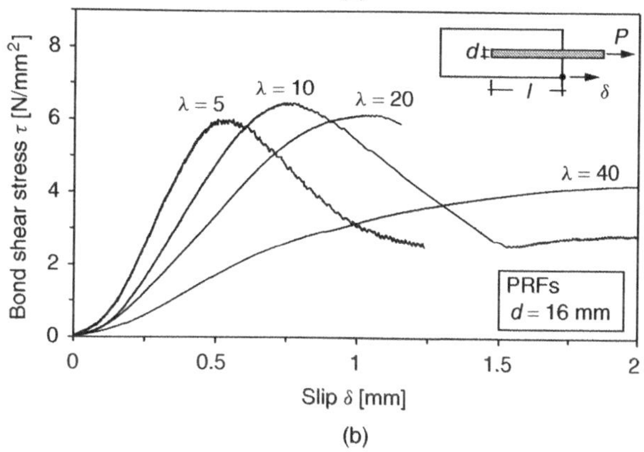

Figure 18.24 Typical bond shear stress vs. slip curves of axially tension loaded threaded rods glued-in glulam parallel to fibre dependent on (a) different adhesive types and (b) rod slenderness (after Aicher *et al.*, 1999)

$$f_{v,k} = \frac{P_{\max,k}}{\pi d\ell} = \frac{1}{\pi} K_{c,\text{adhesive},k}\rho_k \ell^{-0.5} \quad (18.41)$$

where $K_{c,\text{adhesive},k} = 520\ \text{N}/\sqrt{\text{mm}^3}$ and $650\ \text{N}/\sqrt{\text{mm}^3}$ for epoxy and two-component polyurethane adhesives, respectively. For bond length $\ell \leq 200$ mm, a constant shear strength equal to $f_{v,k,\ell=200\ \text{mm}}$ was specified to cut, for a very small bond length, the pronounced exponential LEFM inherent strength increase.

To account for the rod shape influence, the equation

$$f_{v(,k)} = Cd^n\lambda^m = Cd^{(n-m)}\ell^m \quad (18.42)$$

was introduced (Aicher *et al.*, 1999). It can be shown by the generalised Volkersen theory (Gustafsson 1987) that the exponents n and m have mechanically bound limits. So, in the case of an ideal ductile bond, $m = n = 0$, and in the case of an ideal brittle bond $n = -0.5$ and $m = -1.0$. For the general case, the exponents should be in the ranges $-0.5 \leq n \leq 0$ and $-1.0 \leq m \leq 0$. In the GIROD project mentioned, the coefficients/exponents of Equation (18.42) were identified for connections with metrically threaded rods bonded with two different types of adhesives, being an epoxy (300 specimens) and a special softening type phenolic resorcinol (200 specimens). The test results cover rod diameters from 8–30 mm (epoxy: 20 mm) and slenderness ratios λ of 5–40 (Aicher, 2001). The values obtained for coefficients/exponents C, m, n, here specified for characteristic strength $f_{v,k}$ in N/mm², are ($d = d_{\text{nom}}$ and ℓ in mm): 60, −0.46, −0.49 and 12, −0.12, −0.25 for epoxy and phenolic resorcinol, respectively. The strong influence of the adhesive type on parameters ℓ (or λ) and d is obvious. Whereas the brittle epoxy reveals only a minor influence of rod diameter and an almost (one-dimensional) LEFM dependency on anchorage length ($\ell^{-0.46}$), the phenolic resorcinol adhesive shows a quite small and almost similar influence of rod diameter and rod length on bond shear strength.

Besides the above discussed rather/purely phenomenological strength equations, the load capacity of an axially glued-in rod joint (Figure 18.25), which may be perceived as a special case of a lap joint configuration, can be handled by the (generalised) Volkersen approach, sketched in Sections 18.4.1, 18.5.1 and 18.5.4. Here, the substitutions $b = \pi d$, $A_1 = d^2\pi/4$ and $A_2 = a^2$ ($A_2 =$ for square cross-sections with side lengths a; for other geometries see Gustafsson and Serrano (2001)) apply in Equations (18.23)–(18.25). Equation (18.23) relates to the so-called pull-pull configuration (Figure 18.25(a)), being the typical lap joint loading configuration. For the so-called pull-compression load case (Figure 18.25(b)) a more simple and conservative expression is obtained, depending on two parameters τ_f and ϖ, which can be determined from two independent tests with different rod configurations:

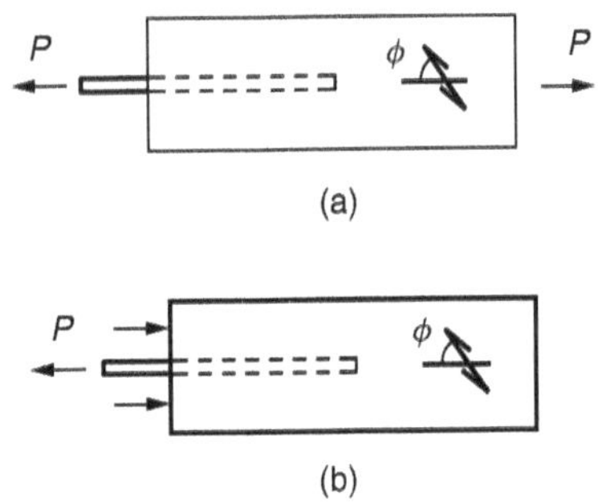

Figure 18.25 Different loading modes for axially loaded glued-in rods: (a) pull-pull, (b) pull-compression

$$\frac{P_{\max}}{\tau_f \pi d \ell} = \frac{\tanh \varpi}{\varpi} \tag{18.43}$$

As calibrated from two sets of pull-compression tests, Gustaffson and Serrano specified for epoxy the fracture parameters $\tau_f = 10.5$ N/mm^2, $l_m = 3600$ mm and $G_f = 1.9$ N/mm. The results according to Equations (18.41)–(18.43) are compared in Figure 18.26 for d = 16 mm and the respective equation parameters given above for epoxy adhesive (for further details see also caption of Figure 18.26).

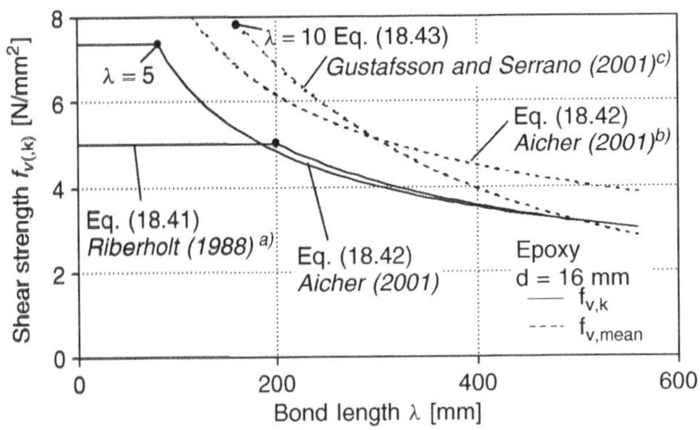

Figure 18.26 Mean and characteristic short-term shear strength of glued-in rods with metrical thread (d = const = 16 mm, epoxy adhesive) dependent on bond length according to different strength equation approaches

[a)] $\rho_k = 0.43$ [g/cm^3] as related to European glulam strength class GL 32 h (C35 laminations) as used in the calibration experiments for the others strength equations specified
[b)] for mean shear strength C = 77 applies (see reference)
[c)] $a = 120$ mm, $E_1 = 210000$ N/mm^2, $E_2 = 11700$ N/mm^2

Concerning the effects of Duration Of Load (DOL), elevated constant and inhomogeneously distributed moisture in the timber and high temperatures (due to ambient climate or fire), the following is briefly stated. Epoxy bonded rods are unaffected by humid climate conditions with moisture content up to 20%; this is valid for both short- and long-term loading. The DOL effect is similar to the long-term strength loss of wood, as described by the so-called Madison curve (see Chapter 8). Elevated temperatures or very low loads result in post-curing and increased short-term bond shear strength. However, when elevated temperatures of about 40 to 60 °C (range strongly adhesive dependent) above about 50 °C act on for the first time connections under load, the bond strength is fully controlled by the glass transition temperature T_g of the specific adhesive. For general purpose adhesives, T_g should be beyond about 65 °C. However, this is not sufficient for fire scenarios (Buchanan and Barber, 1996); for fire design, the rods should be appropriately protected.

Other adhesives, such as specific two-component polyurethanes and softening type phenolic resorcinols, can exhibit significant short-term strength decreases at elevated moistures and temperatures; the DOL effect can be accelerated strongly by both environmental impacts, too. To use conventional timber strength reduction factors for DOL and climate (service) classes for connections with glued-in steel rods, evaluation of the characteristic bond strength values has to be performed at unfavourable conditions of the respective climate classes.

APPENDIX I

Shear and normal stress distributions along bond line $\ell = 2c$ of a single lap joint with equal and isotropic adherends according to Goland and Reissner (1944) are given here. For notations not specified below, see Section 18.4.2.

$$\tau = \frac{pt}{8c}\left\{\frac{\beta c}{t}(1+3k)\left[\cosh\left(\frac{\beta c}{t}\frac{x}{c}\right)\Big/\sinh\left(\frac{\beta c}{t}\right)\right] + 3(1-k)\right\} \quad \text{(A18.1)}$$

$$\sigma_y = \frac{pt^2}{\Delta c^2}\left[\left(R_2\lambda^2\frac{k}{2} + \lambda k'\cosh\lambda\cos\lambda\right)\cosh\frac{\lambda x}{c} \times \cos\frac{\lambda x}{c} + \left(R_1\lambda^2\frac{k}{2} + \lambda k'\sinh\lambda\sin\lambda\right) \times \sinh\frac{\lambda x}{c}\sin\frac{\lambda x}{c}\right] \quad \text{(A18.2)}$$

where

$$\beta = \sqrt{8\frac{G_3}{E_1}\frac{t}{d}}, \quad \lambda = \frac{c}{t}\sqrt[4]{6\frac{E_3}{E_1}(1-\upsilon_1^2)\frac{t}{d}},$$

$$k' = k\frac{c}{t}\sqrt{3(1-\upsilon_1^2)\frac{p}{E_1}}$$

$$R_1 = \cosh\lambda\sin\lambda + \sinh\lambda\cos\lambda,$$

$$R_2 = \sinh\lambda\cos\lambda - \cosh\lambda\sin\lambda,$$

$$\Delta = \tfrac{1}{2}(\sinh 2\lambda + \sin 2\lambda).$$

REFERENCES

Adams R.D., Comyn J., Wake W.C. (1997) Structural adhesive joints in engineering, second edition. Chapman & Hall, London.

Aicher S., Klöck W. (1990) Spannungsberechnungen zur Optimierung von Keilzinkenprofilen für Brettschichtholz-Lamellen. *Bauen mit Holz*, **92**(5), 356–362.

Aicher S. (1991) Investigations on glued structural timber-steel plate joints. *Otto-Graf-J.* **2**, 8–36.

Aicher S., Klöck W. (1991) Finger joint analysis and optimization by elastic, non-linear and fracture mechanics finite element computations. *Proceedings Int. Timber Eng. Conf.*, Vol. 3, London, UK, pp. 3.66–3.76.

Aicher S., Bornschlegl V., Herr J. (1997) Numerical and full scale experimental investigations on glulam frame corners with large finger joints. *Proceedings Conf. of IUFRO, S 5.02 Timber Eng.*, Copenhagen, Denmark, pp. 239–258.

Aicher S., Herr J., Bornschlegl V. (1997) Entwicklung leistungsfähiger, geklebter Verbindungen für Rahmenecken und Stützenanschlüsse unter Verwendung hochfester Anschlußelemente. *Research Report*, Otto-Graf-Institute, University of Stuttgart.

Aicher S., Herr J. (1998) Investigations on high strength glulam frame corners with glued-in steel connectors. *Proceedings 5th World Conf. on Timber Eng.*, Vol. 1, Montreux, Switzerland, pp. 273–280.

Aicher S., Wolf M., Gustafsson P.J. (1999) Load displacement and bond strength of glued-in rods in timber influenced by adhesive, wood density, rod slenderness and diameter. *Proceedings first RILEM Symposium on Timber Engineering*, Stockholm, Sweden, pp. 369–378.

Aicher S., Radovic B. (1999) Investigations on the influence of finger joint geometry on tension strength of finger jointed glulam lamellas (in German). *Holz als Roh- und Werkstoff*, **57**, 1–11.

Aicher S., Windmüller T. (2002) Biegefestigkeiten von keilgezinktem Vollholz bei Flach- und Hochkantbiegung. *Research Report*, Otto-Graf-Institute, University of Stuttgart.

Aicher S., Dill-Langer G. (2001) Influence of moisture, temperature and load duration on performance of glued-in rods. *Proceedings Int. RILEM Symp. Joints in Timber Structures*, Pro 22, RILEM Publications S.A.R.L., pp. 383–392.

Aicher S. (2001) Characteristic axial resistance of threaded rods glued in spruce dependent on adhesive type – a complementary database for the GIROD project. Research Report, Otto-Graf-Institute, University of Stuttgart.

Aicher S., Klöck W. (2001) A reliability based analysis of glued lap joint resistance. *Proceedings Int. RILEM Symp. Joints in Timber Structures,* Pro 22, RILEM Publications S.A.R.L., pp. 473–482.

Anderson G.P., Brinten B.H., Ninow K.J., De Vries K.L. (1988) A fracture mechanics approach to predicting bond strength. In: *Advances in Adhesively Bonded Joints, Winter Annual Meeting of ASME*, Chicago, IL, pp. 93–99.

Buchanan A.H., Fairweather R.H (1992) Epoxid moment resisting connections for timber buildings. *Proceedings IPENZ Conf.*, Vol. 1, pp. 245–249.

Buchanan A.H., Barber D.J. (1996) Fire resistance of epoxied steel rods in glulam timber. *New Zealand Timber Design J.*, **5**(2), 12–18.

Clad W. (1965) On the elasticity of cured glue-lines in wood bondings (in German). *Holz als Roh- und Werkstoff*, **23**(2), 58–67.

De Bruyne N.A. (1944) The strength of glued joints. *Aircraft Eng.* **16**, 329–335.

Deng J.X. (1997) Strength of epoxy bonded steel connections in glue laminated timber. *Civil Eng. Research Rep. 97/4*, University of Canterbury, Christchurch, New Zealand.

Egner K. (1942) Schiftzinkung von Bauhölzern. *Mitteilungen des Fachausschusses für Holzfragen beim VDI und DFV*, **32**, 87–100.

Eby R.E. (1981) Proofloading of finger-joints for glulam timber. *Forest Products J.*, **31**(1), 37–41.

Epple A. (1983) Untersuchungen über Einflüsse auf die Spannungsverteilung in aufgeleimten Holzlaschen und hölzernen Knotenplatten. PhD thesis, University of Hamburg, Germany.

Fairweather R.H. (1992) Beam column connections for multi-storey timber buildings. *Civil Eng. Research Rep. 92/5*, University of Canterbury, Christchurch, New Zealand.

Glos P., Henrici D., Horstmann H. (1987) Strength of wide glued joints (in German). *Holz als Roh- und Werkstoff*, **57**, 355–364.

Glos P., Horstmann H. (1989) Strength of glued lap timber joints. *Proceedings CIB-W18A meeting 22*, East Berlin, Germany, paper 22-7-8, pp. 1–17.

Goland M., Reissner E. (1944) The stresses in cemented joints. *J. Applied Mechanics, Trans ASME*, **66**, A 17.

Gustafsson P.J. (1987) Analysis of generalized Volkersen-joints in terms of non-linear fracture mechanics. In *Mechanical Behaviour of Adhesive Joints*, Edition Pluralis, Paris, France, pp. 323–338.

Gustafsson P.J., Serrano E. (1998) Glued truss joints analysed by fracture mechanics. *Proceedings 5th World Conf. on Timber Eng.*, Vol. 1, Lausanne, Switzerland, pp. 257–264.

Gustafsson P.J., Serrano E. (2001) Glued-in rods for timber structures – Development of a calculation model. *Report ISRN LUTVDG/TVSM – 01/3056 – SE(1-96)*, Lund University, Sweden.

Gustafsson P.J., Serrano E., Aicher S., Johansson C.-J. (2001) A strength design equation for glued-in rods. *Proceedings Int. RILEM Symp. Joints in Timber Structures,* Pro 22, RILEM Publications S.A.R.L., pp. 323–332.

Harris J.A., Adams R.D. (1984) Strength prediction of bonded single lap joints by non-linear finite element methods. *Int. J. Adhesion and Adhesives*, **4**(2), 65–78.

Hart-Smith L.J. (1973) Adhesive bonded single lap joints. Technical Report CR-112236, NASA, Langley Research Center, USA.

Hart-Smith L.J. (1981) In: *Developments in Adhesives – 2.* Kinloch A.J., (ed.). Applied Science, London.

Handel P. (1977) *Konstruktionsgrundsätze und Bemessungstabellen für den Dreieck-Streben-Bau.* Verlag von Wilhem Ernst & Sohn, Berlin.

Heimeshoff B. (1976) Berechnung von Rahmenecken mit Keilzinkenverbindungen. *Holzbaustatik aktuell*, **1**, 7–8.

Johansson C.-J. (1992) Limträramar med dörband av inlimmade stålstänger. *SP Report 1992:60*, Swedish National Testing and Research Institute, Boras.

Kalina M. (1965) The shear stress in plywood gussets at glued trusses (in German). *Holz als Roh- und Werkstoff*, **23** (10), 394–396.

Kangas J. (1994) *Joints of glulam structures based on glued-in ribbed steel rods.* VTT Publications 196, Espoo, Finland.

Kemmsies M. (1994) The influence of adhesives and bonding techniques on the properties of glued timber-steelplate joints. *SP Report 1994:39*, Boras, Sweden.

Komatsu K., Sasaki H., Maku T. (1976) Application of fracture mechanics to strength analysis of glued lap joints. *Wood Research*, **61**, 11–24.

Komatsu K. (1984) Application of fracture mechanics to the strength of cross-lapped glued timber joints. *FRI Bulletin 61*, New Zealand Forest Service, New Zealand.

Komatsu K., Koizumi A., Sasaki T., Jensen J.L., Jijima J. (1997) Flexural behaviour of GLT beams end-jointed by glued-in hardwood dowels. *Proceedings CIB-WI8 meeting 30*, Vancouver, Canada, paper 30–7–1, pp. 1–8.

Kufner M., Spengler R. (1981) Structural frameworks with glued nodes (in German). *Holz als Roh- und Werkstoff*, **39**, 51–62.

Lucas A. (1948) *Ancient Egyptian Materials & Industries*. Edward Arnold, London.

Müller C. (2000) *Laminated Timber Construction*. Birkhäuser, Basel, Switzerland.

N.N. (1998) German Technical approval Z-9.1-185, Deutsches Institut für Bautechnik (DIBt), Berlin.

N.N. (1999) Technical approval Z-9.1-248, Deutsches Institut für Bautechnik (DIBt), Berlin.

N.N. (2001) German Technical approval Z-9.1-167, Deutsches Institut für Bautechnik (DIBt), Berlin.

Ottosen N.S., Olsson K.G. (1988) Hardening/softening plastic analysis of an adhesive joint. *J. Engineering Mechanics*, **114**(1), 97–116.

Palm J. (1991) Glued steel-timber joints – Test and finite element analysis. *SP Technical Note 1991:15, Building Technology*, Swedish National Testing and Research Institute, Boras, Sweden.

Riberholt H. (1986) Glued bolts in glulam. Series R, No. 210, Technical University of Denmark, Lyngby.

Riberholt H. (1988) Glued bolts in glulam – proposals for CIB Code. *Proceedings CIB WI8A Meeting 21*, paper 21–7–2, Parksville, MD.

Selbo M.L. (1963) Effect of joint geometry on tensile strength of finger joints. *Forest Products J.*, 390–400.

Serrano E. (1997) Finger joints for laminated beams. Experimental and numerical studies of mechanical behaviour. *Report TVSM – 3021*, Lund University, Sweden.

Serrano E. (2000) Adhesive Joints in Timber Engineering – Modelling and testing of fracture properties. PhD thesis, Department of Mechanics and Materials, Lund University, Sweden.

Steurer A. (1981) Leimverbindungen. In: *Publication No. 81–1*, ETH Zürich, Switzerland.

Turkowski S. (1991) Prefabricated joints of timber structures on inclined glued-in bars. *Proceedings Int. Timber Eng. Conf.*, Vol. 3, London, UK, pp. 3.143–3.148.

Volkersen O. (1938) Die Nietkraftverteilung in zugbeanspruchten Nietverbindungen mit konstanten Laschenquerschnitten. *Luftfahrtforschung*, **15**, 41–47.

Walsh P.F., Leicester R.H., Ryan A. (1973) The strength of glued lap joints in timber. *Forest Products J.*, **23**(5), 30–33.

Wernersson H. (1994) Fracture Characterisation of wood adhesive joints. *Report TVSM – 1006*, Division of Structural Mechanics, Lund University, Sweden.

19

Trusses and Joints with Punched Metal Plate Fasteners

Jacob Nielsen

19.1 INTRODUCTION

Trusses with punched metal plate fasteners are typically used as the main structural component of roof structures with up to a 30–40 m span. In Figure 19.1 some typical roof trusses for spans about 10 m are shown.

Trusses are also often used as structural components in floors, formwork, scaffolding and bridges (see Norges Byggskole, 1999). The trusses are prefabricated at truss plants, where the timber beams are cut and assembled with punched metal plate fasteners, which are embedded into the wood with a hydraulic pressing tool.

The name 'punched metal plate fastener' (sometimes called 'nail-plate' for short) characterises a thin steel plate with a thickness of 1–2 mm in which nails are stamped perpendicular to the plate. Depending on the stamping tool, one or two nails are made per hole. A punched metal plate fastener with two nails per hole is shown in Figure 19.2. The length of the nails is normally between 8 and 15 mm.

The size of the plates varies from 30 cm^2 up to about 1 m^2. Normally, two symmetrically located plates of equal size are used in a joint (see Figure 19.3).

To minimise the gap between timber parts in each joint and to ensure an effective plate embedding, the timber is planed at all four faces. The thickness of the timber varies usually from 35 to 70 mm.

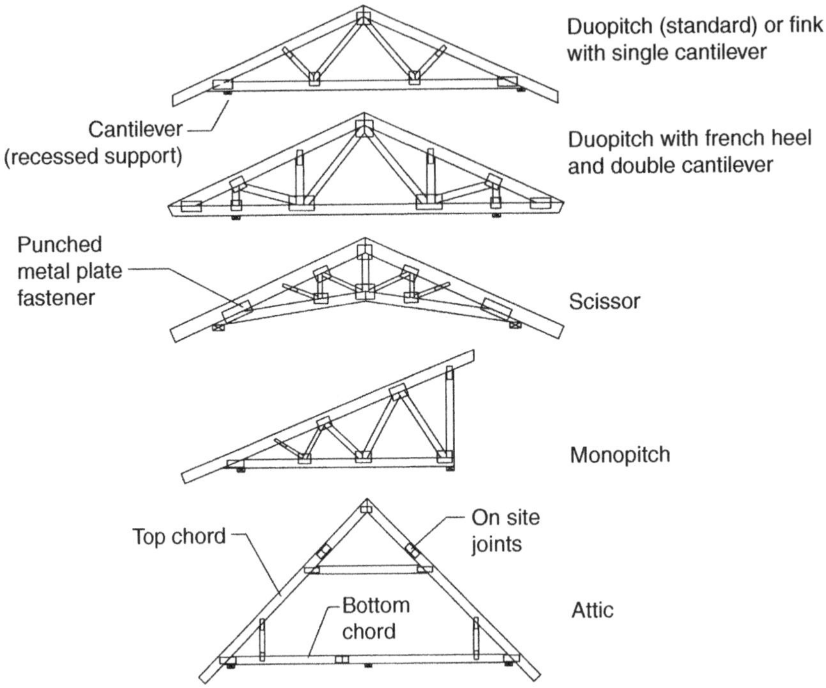

Figure 19.1 A selection of different truss types used for roof construction

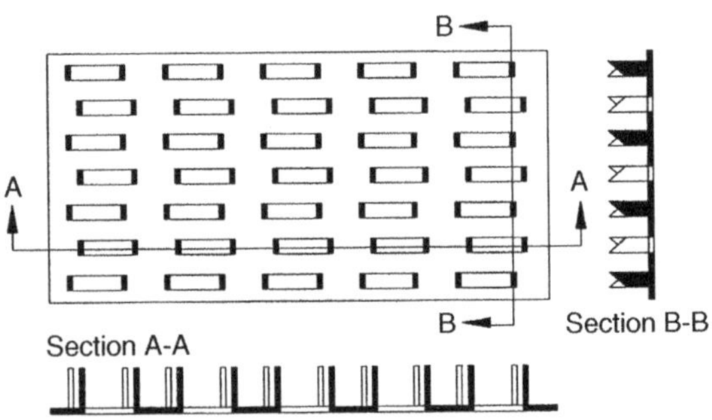

Figure 19.2 A punched metal plate fastener with two nails per hole

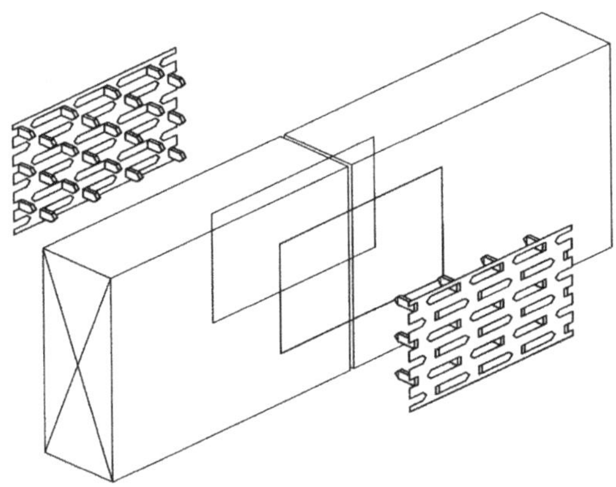

Figure 19.3 Splice joint with two punched metal plate fasteners

In cases with large trusses, production and transportation problems imply that so-called 'on site joints' are required, e.g. see the attic truss in Figure 19.1. The complete truss is then assembled on site. The apex is added to the top of the attic shown in Figure 19.1. An 'on site joint' can be made of perforated steel plates with ringed nails.

19.2 TRUSSES IN ROOF STRUCTURES

In general, a truss can resist in-plane forces only, and therefore special care must be taken to ensure the 3D stability of the roof structure.

19.2.1 Systems to Ensure 3D Stability of Roof Structures

In Figures 19.4, 19.5 and 19.7, three different bracing methods for trusses are shown. The distance between the trusses can vary between 0.3 m and 1.2 m.

The trusses of the roof structure in Figure 19.4 are covered with plywood, which is nailed to the top chord. Out-of-plane forces are transferred to the supports by shear in the plywood.

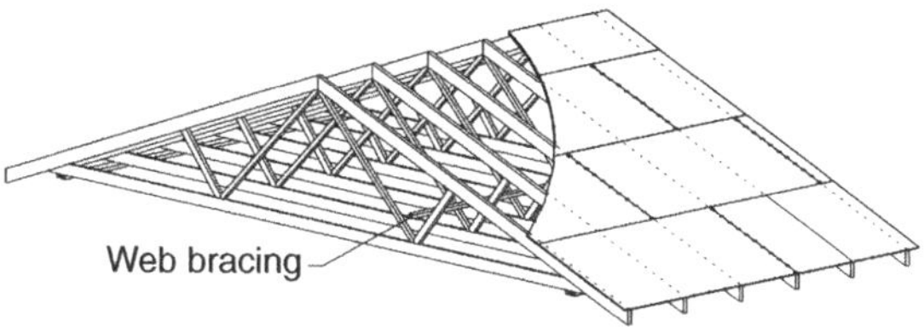

Figure 19.4 Shear effect in structural plywood ensures the 3D stability of the roof structures

In Figure 19.5, a special bracing truss is located in the roof plane and nailed between two vertical standard trusses. The horizontal out-of-plane forces are transferred from the gable through the battens to the stiff bracing truss. At the heel of the standard trusses the bracing truss is fixed to the supports. At the apex the bracing truss is fixed directly or through a compression bar by tension diagonals often made of steel or timber. When the tension diagonals are active, additional compression forces will be added to the top chord of the active trusses within the bracing system (see Figure 19.6).

For the roof structure in Figure 19.7, the out-of-plane stability of the standard trusses is ensured by a hipped end in which the hip trusses are located perpendicular to the standard trusses.

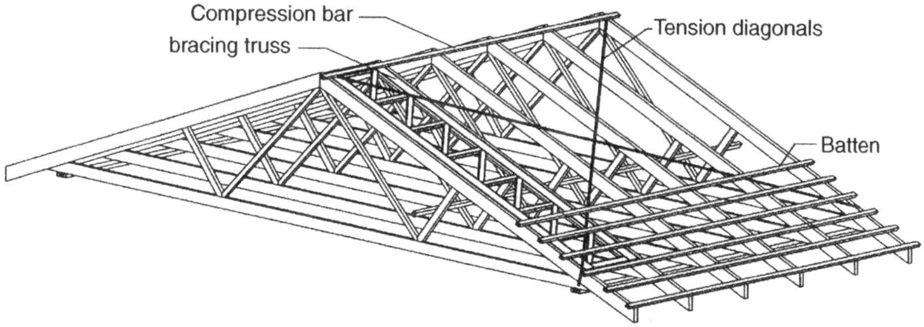

Figure 19.5 A bracing truss and steel tension diagonals ensure the out-of-plane stability

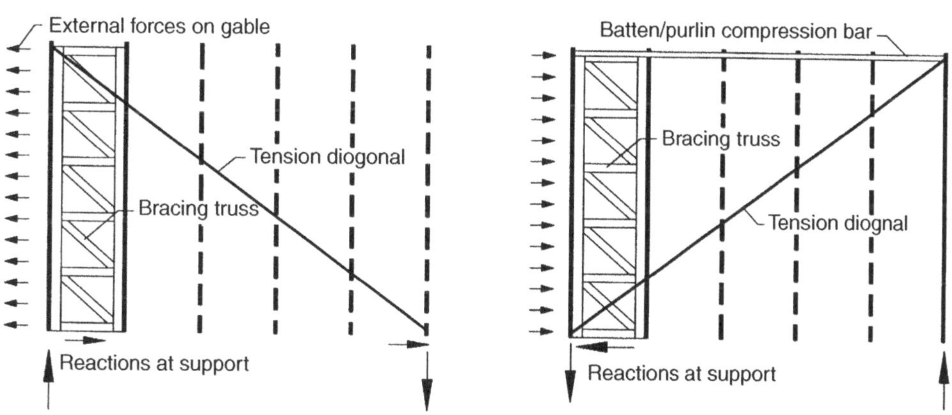

Figure 19.6 External forces and reactions in a system with tension diagonals

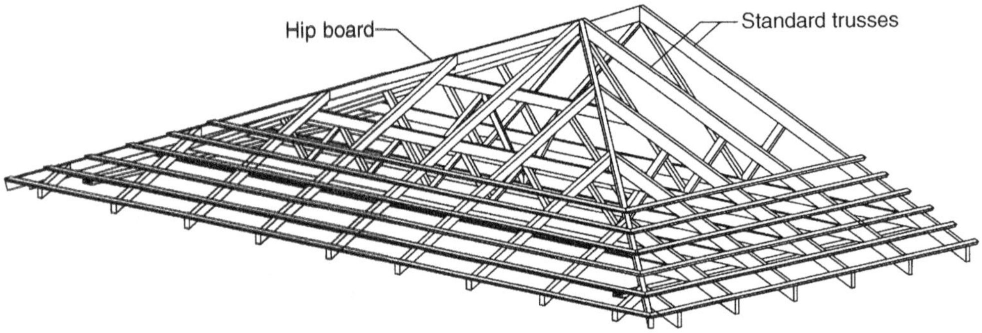

Figure 19.7 A hipped end ensures the 3D stability of the roof structure

The above systems can resist out-of-plane forces at the top chord. Out-of-plane forces of the bottom chord must also be transferred to the supports, and the boarding and ceiling located below the bottom chord may constitute a bracing system. If the ceiling is not able to transfer forces at the bottom chord, it must have another bracing system.

19.2.2 Bracing to Reduce the Buckling Length

Overall bracing systems are also used to increase the critical resistance against buckling by reducing the out-of-plane buckling length of the compression members in the trusses. The size of the out-of-plane forces from the compression members depends upon the stiffness of the bracing system and the prebuckling shape of the trusses. A discussion on bracing forces in roof structures is given in Burdzik (2000). To reduce the buckling length of the webs, a separate bracing system connected to the overall bracing system is used; see the batten on the web in Figures 19.4 and 19.5.

19.2.3 Load Transfer to Beams and Girders

The trusses transfer the external forces to the supports. In cases with large openings in the roof (caused by a window, a chimney or a dormer) or floor (caused by a chimney or a stairwell), the chords may be cut and additional supports must be added (see Figure 19.8).

The additional supports can be made by large timber or steel beams, which transfer the supporting forces to load-bearing walls or girders. In general, a girder is a strengthened truss, which is able to resist additional forces from other trusses, and it can be produced by a single truss with increased dimensions of the timber members and plates, or simply by nailing two or more standard trusses together (see Figure 19.8), where the latter solution is shown. The size of the reaction force at the additional support depends upon the stiffness of the supporting structure, i.e. the size and span of the beam, the number and location of the supports of the beam, and the stiffness of the girder or wall. The additional support should be modelled as an elastic support.

19.2.4 System Effect

The plywood and battens on top chords are mainly used to transfer the forces from the roofing material to the trusses. In simple truss design, the forces are normally assumed to be equally distributed on the trusses. However, as the distances between the trusses are normally below 1.2 m, the system with plywood and battens is a system with considerable bending stiffness. This system is able to distribute the forces to the trusses dependent on the stiffness of each truss, i.e. a stiff truss will be subject to more load than a truss with low stiffness. Due to the strong, positive correlation of stiffness and strength of timber, the above load sharing effect reduces the probability of failure of a truss in the system. The probability of two or more weak

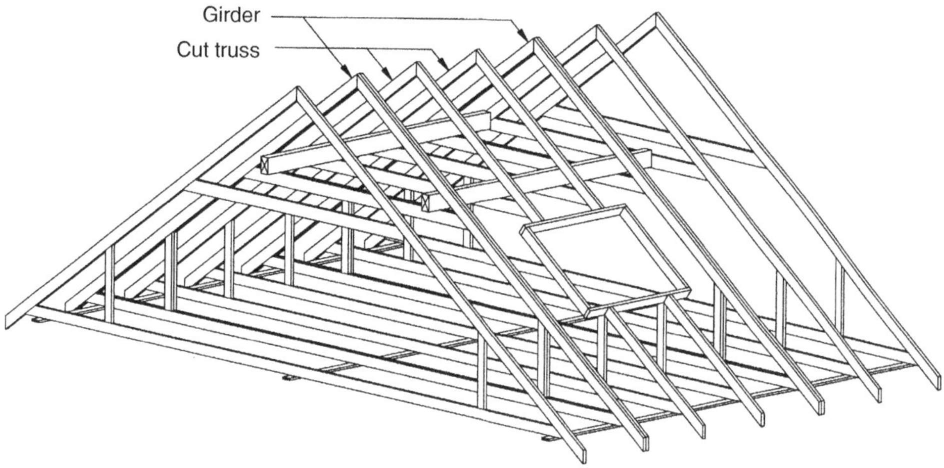

Figure 19.8 Two girders and two beams constitute additional support at the two cut trusses

neighbouring trusses in the system is very low. In Cramer *et al.* (2000), a load sharing factor was found to range from 1.06–1.24, depending on the truss configuration. The trusses were spaced 0.61 m apart and covered with structural plywood sheathing. The load sharing factor is defined as the nominal resistance of an individual sheathed truss divided by the resistance of an individual truss without sheathing.

19.3 TRUSS MODELLING

As described in the previous section, a truss is part of a complex 3D structure and, therefore, a full static model of the system demands a 3D model. The timber beams of the trusses and the battens can be modelled by 3D beam elements[1]. In a 3D model, the stiffness relation at each joint and all eccentricities has to be involved. The trusses may also be subjected to torsion and out-of-plane bending, and the battens are loaded in tension, compression and bending. Structural plywood may be modelled by shell elements. Solution of a full 3D model is a huge and very complex problem, which still needs some research before implementation into commercial programs.

At the moment, commercial programs do not model a roof structure by a real 3D model. In general, the strategy is to split the 3D system into 'simple' 2D static models in which forces are added directly to the truss model and the reactions can be transferred from one 2D model to another. The timber members in the trusses are modelled as linear elastic Timoshenko or Bernoulli beams, although the vast majority use the latter type.

In modelling timber trusses, the beams cause minor problems – the real problem is how to model the joints, which will be the main subject in this section. Only plane trusses and joints loaded in the plane will be discussed.

19.3.1 Modelling of a Truss

To give a brief illustration of the development of truss models during the last three decades, some different truss models for a fink are shown in Figure 19.9. In this Figure, the truss has a minor cantilever on the left-hand side and a major cantilever on the right-hand side.

Model 1 is the most simple, where the beams are modelled by bar elements and the joints are modelled by hinges. The distributed loads are applied as point loads at the hinges. This model

[1] The problem can also be modelled by 2D or 3D solid elements, but this solution is still very time consuming both to model and calculate. A splice and a heel joint are modelled with 2D solid elements in Nielsen (1996).

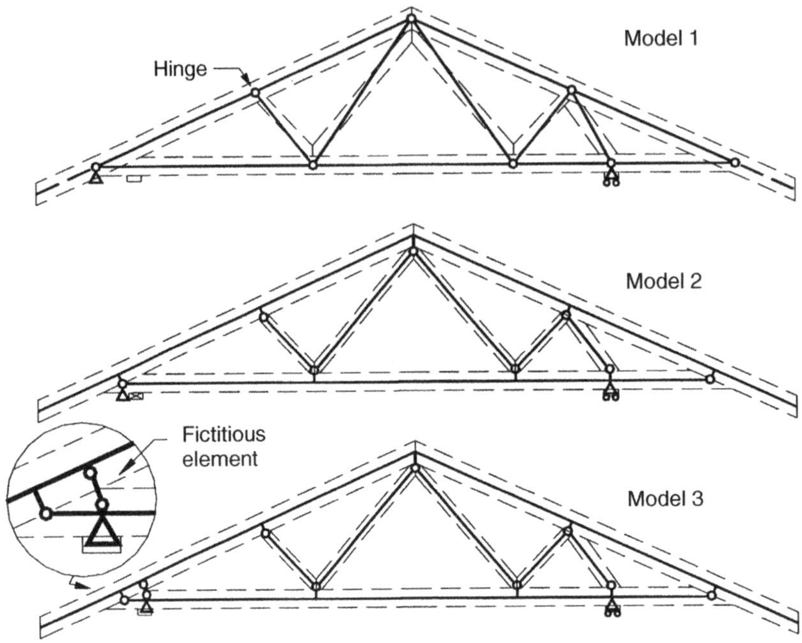

Figure 19.9 Three different truss models for a fink truss

makes it possible to estimate the axial forces by hand calculations or a slow and very expensive computer, which were the conditions in the 1970s. Simple factors are used to estimate the moment distribution.

Model 2 is slightly improved, where the chords are modelled as continuous beams and small auxiliary elements, including the eccentricities between the system lines at each joint. The system lines of the model become closer to the centroid of the timber members. The moment distributions in the chords are given directly. In model 2 the joints are also made with hinges, but they are located at the joint line.

The heel joint is able to transfer a considerable moment, and therefore a fictitious element (a bar element) has been introduced in model 3. The stiffness of the joint, and thereby the moment distribution at the joint, depends upon the location and stiffness of the fictitious element. The fictitious element runs from a node on the top chord close to the contact zone, and down to a node on the bottom chord close to the support.

In general, the same stiffness is chosen for the fictitious element as for the incoming chords. By introducing the fictitious element, an improved model of the support conditions (cantilever) has been obtained.

The heel modelling has great influence on the moment distribution at the dimension controlling areas of the timber, and it has been subjected to detailed studies for many years e.g. see Lau (1986) and Riberholt (1982, 1990). More sophisticated models of the heel have also been adopted (see Figure 19.10).

In the right model, only forces perpendicular to the contact zone will be transferred by the fictitious elements. The axial force in the top chord is adopted at the intersection between the top and bottom chords only.

Today, most of the commercial truss programmes are based on truss models using fictitious elements. The users of the programs have, in co-operation with the program distributors, developed a lot of different joint models based on experience and know-how. The strategy to determine

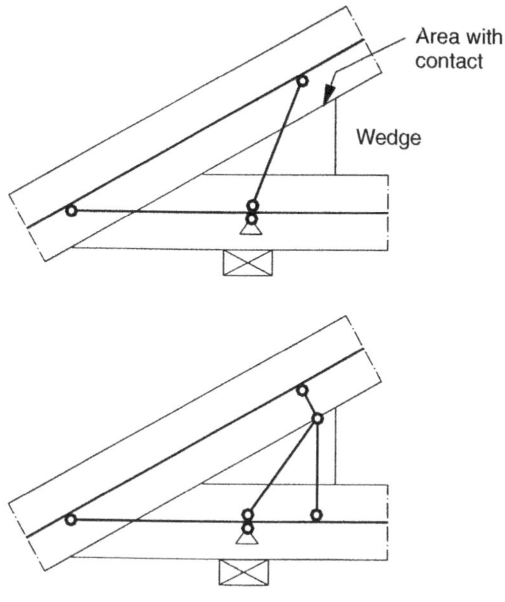

Figure 19.10 Two heel joint models with fictitious elements

the stiffness of each fictitious element and finding the forces in each joint type is a rather 'dark and impenetrable subject'.

All the above models have some mutual problems:

- The forces in each plate, nail group and contact zones are not given by the model and they still have to be determined.
- The deformation of each joint is not included in the overall deflections, as all the above joints are modelled by hinges.

To include joint slip, truss models with springs, modelling the joint stiffness, were introduced. Three mutually-independent springs representing two translations and one rotational stiffness are used at every joint. The springs may be linear or nonlinear, and the stiffness depends upon the size and location of the punched metal plate fasteners. However, the spring models do not overcome the first of the above problems, and in reality the springs are not mutually independent.

Foschi (1977) introduced a model that was able to estimate the stiffness and sectional forces in each nail group and plate. Nail, plate and contact elements are used to join the beam elements. The nail and plate elements are developed with nonlinear load-slip relations. The model has been used in a few programmes used for research (Triche *et al.*, 1988, Nielsen, 1996), but it is still not implemented in commercial truss programmes.

Today, the Foschi model is the most advanced beam model, and it will be described in detail in the following.

19.3.3 Forces in Joints with Punched Metal Plate Fasteners

Before beginning our description of the Foschi model, it is important to explain how the forces are transferred in a joint. Look at Figure 19.11, where the forces at a heel joint with a wedge are shown. Together with the reaction forces and other external forces the sectional forces at the top and bottom chord must form a system in equilibrium. The forces between the top chord and the bottom chord are transferred by:

- *The punched metal plate fastener*: the forces are absorbed in the plate by the nails at one nail group and the plate transfers the forces from one nail group to another (e.g. N_{pl}, V_{pl} and M_{pl} in Figure 19.11).
- *Contact between the timber members*: compression and friction can be transferred at the contact zone. The compression and friction force, F_{com} and F_{fric} are perpendicular and parallel to the contact zone, respectively.

A countless number of different tests have shown that joints with punched metal plate fasteners are able to transfer moments, and therefore, a static model of the above heel joint will be twice statically indeterminate. The forces are then distributed according to the stiffness of the punched metal plate fastener and the timber at the contact zone.

Other types of joints are several times statically indeterminate, e.g. a K-joint with plates covering three incoming timber members (see Figure 19.12).

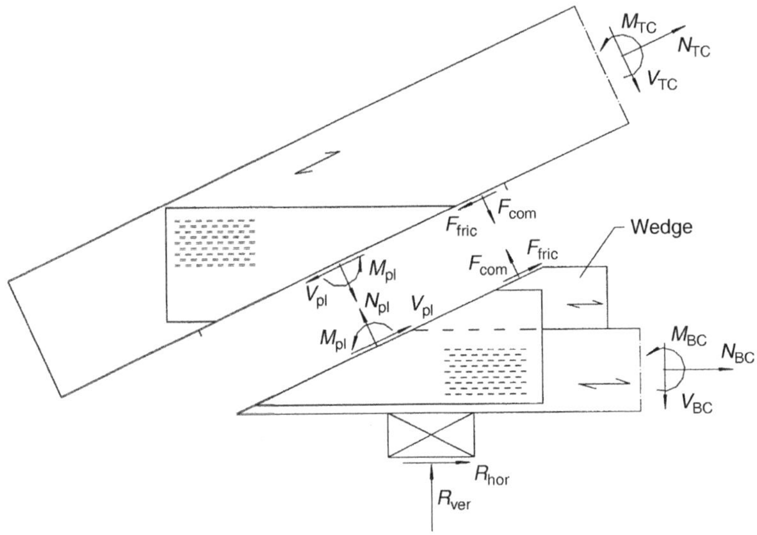

Figure 19.11 Forces at a heel joint with a wedge

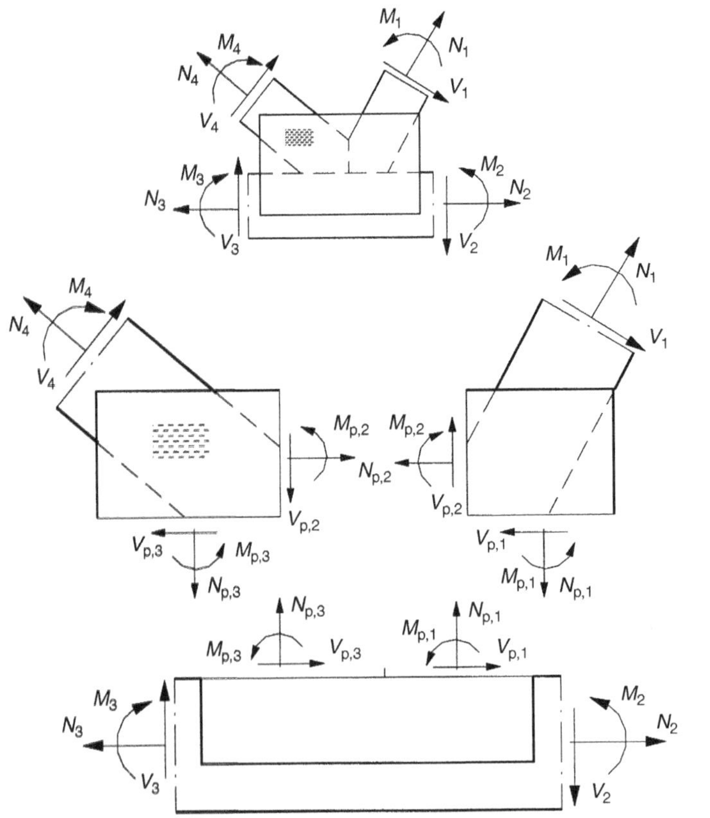

Figure 19.12 Section forces in plate and timber at a K-joint. Contact forces are not shown

19.3.4 The Foschi Model for Joints with Punched Metal Plate Fasteners

In Figure 19.13, the Foschi model is shown for a heel joint. The Foschi model consists of:

- *Beam elements*. This main element is used to model all straight timber beams in the truss. It is based on linear elastic Bernoulli or Timoshenko beam theory, and has a system line that coincides with the centre of gravity of the incoming timber members. The cross-sectional area, A, and the moment of inertia, I, are calculated from the actual dimensions.
- *Auxiliary elements*. These elements model the eccentricities, and they transfer forces from a nail, a contact or a support to the main timber elements. Also, these elements are based on linear elastic beam theory, and are located perpendicular to the main elements. The cross-sectional area, A, and the moment of inertia, I, are assumed to be very large compared to those of the main elements.
- *Contact element*. These elements identify the location of two nodes (points) on the timber and determine if there is contact or not. In the case of contact, the element transfers compression and friction forces. A contact element should be located at all potential contact zones. Some contact elements may only be loaded in compression and should have an axial stiffness EA that reflect the stiffness of the compressed wood at the contact zone.
- *Nail element*. The nail element transfers forces from a node on the timber to a node on the plate. The nail element is located at the gravity centre of the nail group considered. The stiffness of the element depends upon the nonlinear load-slip relation for the nails and the shape of the nail group. Timber and plate are assumed to perform stiff-body displacements.
- *Plate element*. A plate element transfers forces between two nail groups. The stiffness depends upon the plate material located at the joint zone between two nail groups. The stiffness is modelled by a number of small Bernoulli beam elements with linear or bilinear properties.

For modelling two symmetrically located punched metal plate fasteners, two nail elements and one plate element have to be used as a minimum.

In the following, the nail, plate and contact element are described. Some parts of the original model given in Foschi (1977) are developed further (see also Nielsen, 1996).

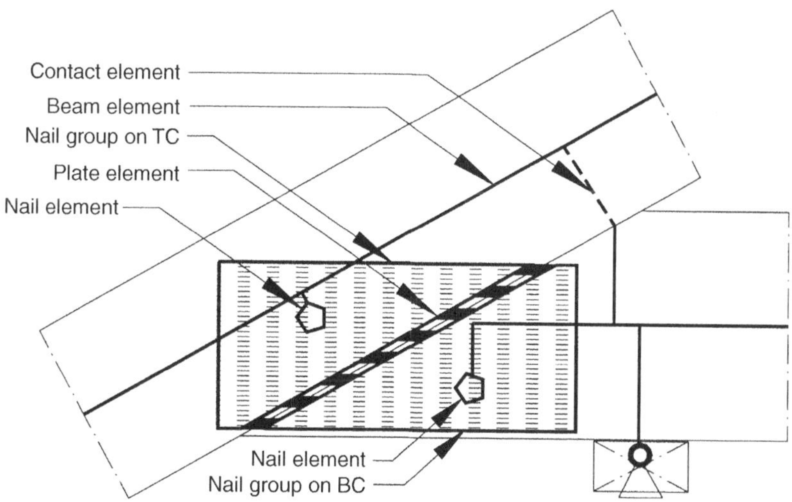

Figure 19.13 Location of nail, plate and contact elements in a heel joint

The Nail Element

Consider the nail group area (A-B-C-D) for a punched metal plate fastener on a beam in Figure 19.14.

The nail element is connected to a node on the beam at point b and to a node at point p on the punched metal plate fastener. It is assumed that the beam and plate perform stiff-body translations and rotations in the joint region. Further, the rotations α_b, α_p are assumed to be small. The displacements of the beam are given by u_b, v_b and α_b at node $b(x_b, y_b)$, and the displacements of the plate are given by u_p, v_p and α_p at node (x_p, y_p). The location of nodes b and p can be chosen arbitrarily.

Consider a nail i at point (x_i, y_i). It can be shown that the relative displacements of point i in the x- and y-directions are given by

$$\Delta_x = \mathbf{q}_x^T \cdot \mathbf{u} \qquad (19.1)$$

$$\Delta_y = \mathbf{q}_y^T \cdot \mathbf{u} \qquad (19.2)$$

where

$$\mathbf{q}_x = [\,1 \quad 0 \quad -(y_i - y_p) \quad -1 \quad 0 \quad (y_i - y_b)\,]^T \qquad (19.3)$$

$$\mathbf{q}_y = [\,0 \quad 1 \quad -(x_i - x_p) \quad 0 \quad -1 \quad (x_i - x_b)\,]^T \qquad (19.4)$$

$$\mathbf{u} = [\,u_p \quad v_p \quad \alpha_p \quad u_b \quad v_b \quad \alpha_b\,]^T \qquad (19.5)$$

The force on a nail, $p(\Delta)$, is given by the expression

$$p(\Delta) = (p_0 + k_1 \cdot \Delta) \cdot \left(1 - \exp\left(\frac{-k_0 \cdot \Delta}{p_0}\right)\right) \qquad (19.6)$$

where Δ is the absolute displacement of the observed nail given by Equation (19.7):

$$\Delta = \sqrt{\Delta_x^2 + \Delta_y^2} \qquad (19.7)$$

The stiffness parameters p_0, k_1, k_0 are defined in Figure 19.15.

The above stiffness parameters can depend upon the angle between the nail force vector $\boldsymbol{p}$ and the grain direction, ϑ, and the angle between the major directions of the plate and the grain direction. Further details on the angle dependence are described in Foschi (1977) and Nielsen (1996).

A small variation in one of the displacement components of the displacement vector $\mathbf{u}$ causes a virtual displacement $\delta\Delta$ in nail i (see

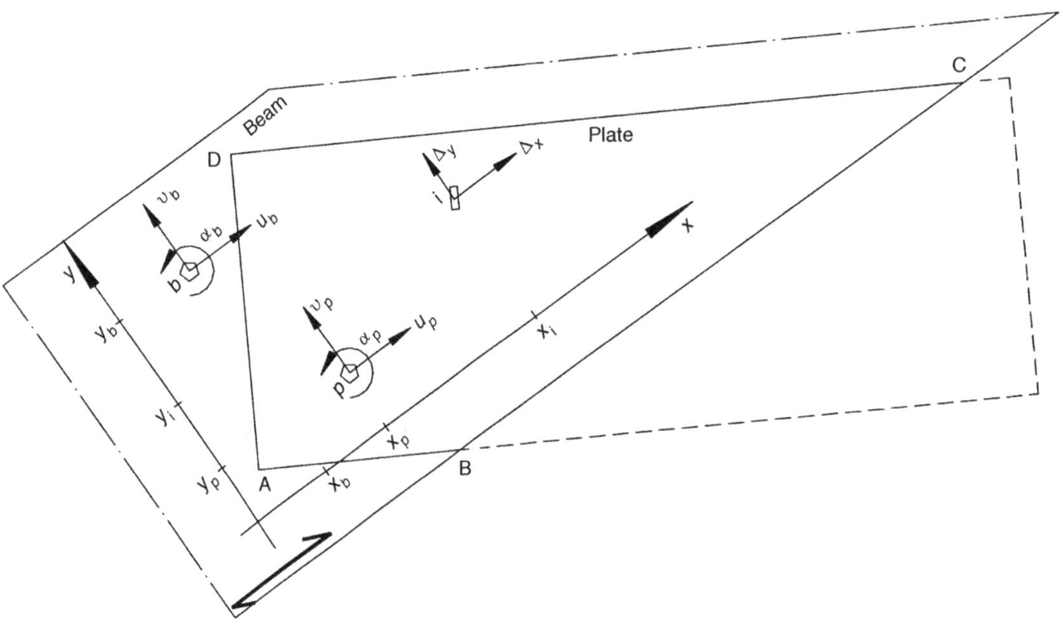

Figure 19.14 Global and local systems for a nail element

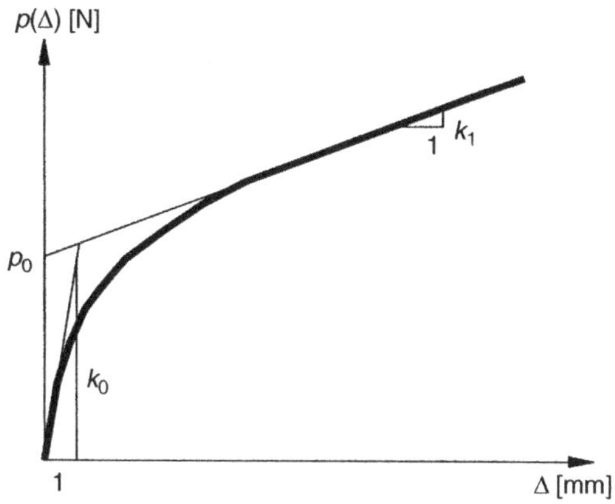

Figure 19.15 Definition of the stiffness parameters for a nail

Figure 19.16). The virtual internal work performed by nail i is then given by Equation (19.8):

$$^{i}\delta W = \mathbf{p}\cdot\delta\Delta = p(\Delta)\cdot\delta\Delta\cdot\cos(\gamma-\vartheta) \quad (19.8)$$

It is assumed that the nail force vector **p** is parallel to the displacement vector Δ. This is not true for nails in an orthotropic material, but the validity of the assumption is good, since the angle between **p** and Δ is small.

The internal work performed by the nails in the plate area A (A-B-C-D in Figure 19.14) is given by Equation (19.9):

$$\begin{aligned} ^{i}\delta W &= \int_A \Im\cdot p(\Delta)\cdot\delta\Delta\cdot\cos(\gamma-\vartheta)\,dA \\ &= \int_A \Im\cdot p(\Delta)\cdot\left(\frac{\Delta_x}{\Delta}\delta\Delta_x+\frac{\Delta_y}{\Delta}\delta\Delta_y\right)dA \end{aligned} \quad (19.9)$$

where $\Im$ is the nail density. By inserting Equation (19.1) and (19.2) into Equations (19.9), and assuming small variations for the internal and external work in all the displacement components in **u**, the local finite element equation is obtained:

$$\mathbf{K}\cdot\mathbf{u}=\mathbf{f} \quad (19.10)$$

where **K** is the local element stiffness matrix, given by

$$\mathbf{K}=\int_A \Im\cdot\frac{p(\Delta)}{\Delta}(\mathbf{q}_x\cdot\mathbf{q}_x^T+\mathbf{q}_y\cdot\mathbf{q}_y^T)\,dA \quad (19.11)$$

and **f** is the element force vector. The above integration is performed by using Gauss quadrature. In Figure 19.14, the nail group is defined by a quadrilateral, however in general, a nail group is shaped like a polygon with $3, 4, 5, 6 \ldots n$ edges. A general procedure to integrate a polygon with arbitrary shape is made by splitting the polygon into triangles and applying Gauss quadrature to each triangle (see Figure 19.17). Three integration points in each triangle are sufficient.

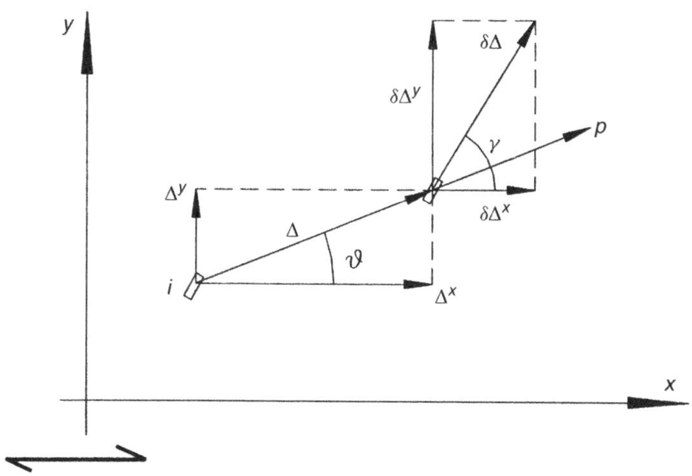

Figure 19.16 Force and virtual displacement of a nail

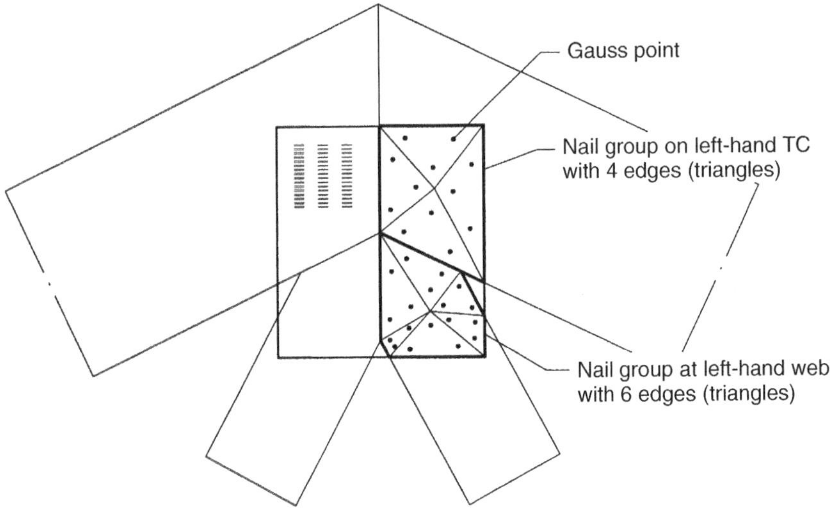

Figure 19.17 Four nail groups at an apex joint. The integration of each nail group is made by splitting the polygon shape of the nail group into triangles

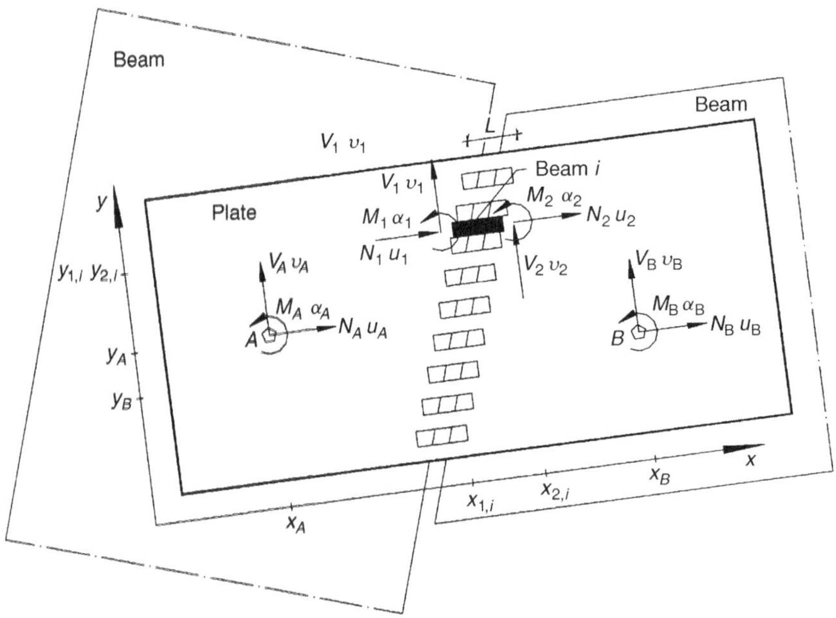

Figure 19.18 Small Bernoulli beams constitute a plate element between two nail groups

Plate Element

Consider the punched metal plate fastener in Figure 19.18.

The plate regions on the timber beams are assumed to perform like stiff bodies. A number of small beams with the length L connect the stiff regions at the joint line. The small beams shall model the stiffness of the plate region above the joint line. The displacements of the left-hand plate region are given by u_A, v_A and α_A at node A. The displacements of the right-hand plate region are given by u_B, v_B and α_B at node B. The location of nodes A and B are chosen to be the same as the nodes used for the nail elements, but they can be chosen arbitrarily. Consider beam i between the two plate regions in Figure 19.18. The element forces and the displacements of beam i, respectively, are given by

$$\mathbf{f}_i = [\, N_1 \quad V_1 \quad M_1 \quad N_2 \quad V_2 \quad M_2 \,]^T \qquad (19.12)$$

$$\mathbf{u}_i = [\, u_1 \quad v_1 \quad \alpha_1 \quad u_2 \quad v_2 \quad \alpha_2 \,]^T \qquad (19.13)$$

The element forces and displacements at nodes A and B, respectively, are given by

$$\mathbf{f} = [\, N_A \quad V_A \quad M_A \quad N_B \quad V_B \quad M_B \,]^T \qquad (19.14)$$

$$\mathbf{u} = [\, u_A \quad v_A \quad \alpha_A \quad u_B \quad v_B \quad \alpha_B \,]^T \qquad (19.15)$$

If small rotations are assumed, the displacements of beam i are given by

$$\mathbf{u}_i = \mathbf{D}_i \cdot \mathbf{u} \qquad (19.16)$$

where

$$\mathbf{D}_i = \begin{bmatrix} 1 & 0 & y_A - y_{1,i} & & & \\ 0 & 1 & x_{1,i} - x_A & & 0 & \\ 0 & 0 & 1 & & & \\ & & & 1 & 0 & y_B - y_{2,i} \\ & 0 & & 0 & 1 & x_{2,i} - x_B \\ & & & 0 & 0 & 1 \end{bmatrix} \qquad (19.17)$$

The forces at A and B are given as a summation of the forces in n small beams between the two plate regions:

$$\mathbf{f} = \sum_{i=1}^{n} (\mathbf{D}_i^T \cdot \mathbf{f}_i) \qquad (19.18)$$

The relation between the element forces and displacements of beam i is given by

$$\mathbf{f}_i = \mathbf{K}_e \cdot \mathbf{u}_i \qquad (19.19)$$

where $\mathbf{K}_e$ is a typical 6×6 element stiffness matrix of Bernoulli beams. The local element stiffness matrix for the plate element is then given by

$$\mathbf{K} = \sum_{i=1}^{n} (\mathbf{D}_i^T \cdot \mathbf{K}_e \cdot \mathbf{D}_i) \qquad (19.20)$$

The properties of the small Bernoulli beams may be linear or bilinear if plastic situations are to be modelled. In a real plate, the situation is not like shown in Figure 19.18. The joint line is not crossed by a well-defined number of beams, and the dimension of a beam is not well defined[2]. In Nielsen (1996), the area and the moment of inertia are based on the geometry of the small plate strips (beams) in the direction of the main axis. The length of the small beams, L, and E-modulus of the beams are parameters to be fitted to test results. For further details of the plate element, see Foschi (1977) or Nielsen (1996). In general, it is very complicated to determine the plate properties of a punched metal plate fastener, because the stiffness properties have been 'destroyed' at the stamping process.

Contact Element

Consider the two parallel beam edges in Figure 19.19.

A contact element is located at a potential contact surface and connected between nodes 1 and 2 with an initial gap size, g, measured perpendicular to the contact surface. During deformation the distance between nodes 1 and 2 is given by

$$g + u_2 - u_1 \qquad (19.21)$$

where u_1 and u_2 are the displacements parallel to the gap g. It is assumed that contact between the timber members is obtained when

$$u_2 - u_1 < -g \qquad (19.22)$$

If in contact, the relation between the element forces and displacements is given by

$$\mathbf{f} = \mathbf{K} \cdot \left(\mathbf{u} - \tfrac{1}{2}\mathbf{g}\right) \qquad (19.23)$$

where

$$\mathbf{f} = [\, N_1 \quad V_1 \quad M_1 \quad N_2 \quad V_2 \quad M_2]^T \qquad (19.24)$$

$$\mathbf{u} = [\, u_1 \quad v_1 \quad \alpha_1 \quad u_2 \quad v_2 \quad \alpha_2 \,]^T \qquad (19.25)$$

$$\mathbf{g} = [\, g \quad 0 \quad 0 \quad -g \quad 0 \quad 0]^T \qquad (19.26)$$

[2] The plate element is to be developed further to describe the elastic and plastic stiffness dependent on the angle between the joint line and the major axes of the plate.

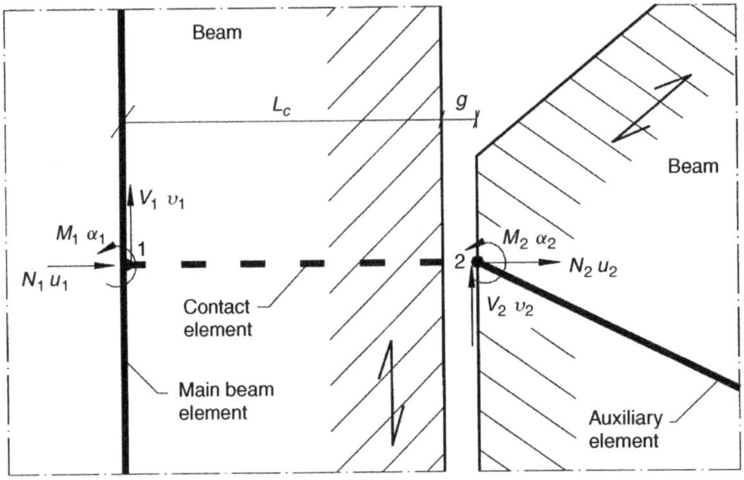

Figure 19.19 A (non-active) contact element between nodes 1 and 2

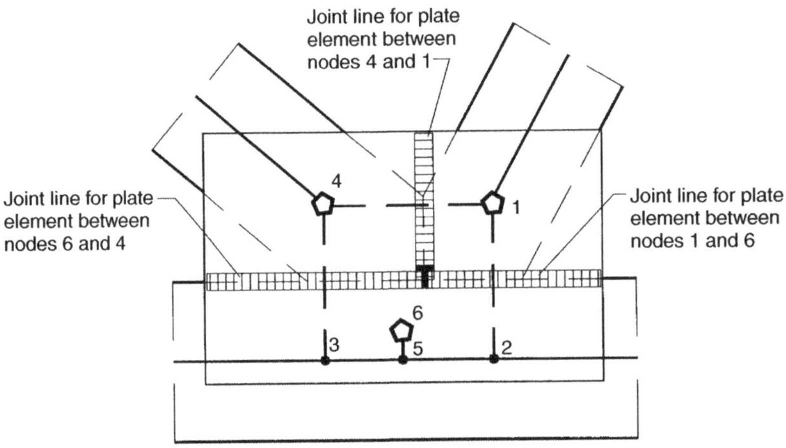

Figure 19.20 Three nail, three plate and three contact elements to model a K-joint

K is the local stiffness matrix given by Equation (19.27):

$$\mathbf{K} = \begin{bmatrix} \dfrac{E \cdot A_c}{L_c} & & & & & \\ 0 & 0 & & \text{sym.} & & \\ 0 & 0 & 0 & & & \\ -\dfrac{E \cdot A_c}{L_c} & 0 & 0 & \dfrac{E \cdot A_c}{L_c} & & \\ 0 & 0 & 0 & 0 & 0 & \\ 0 & 0 & 0 & 0 & 0 & 0 \end{bmatrix} \quad (19.27)$$

where E is the modulus of elasticity of the timber, which depends upon the grain direction. L_c is the length of the contact element, and A_c is the cross-sectional area of the contact zone. L_c and A_c are parameters to be estimated dependent on deformations in the region of the contact zone. The above contact element does not include friction, though this can be implemented using more sophisticated contact elements, where plastic conditions may also be considered.

To illustrate the use of the Foschi model, a K-joint with one pair of punched metal plate fasteners is shown in Figure 19.20.

Two webs are connected to a continuous bottom chord, and each of the timber members has its own nail element. The three nail elements are each connected to a timber node and a plate node located at the points 1, 6 and 4. The nodes are placed in the gravity centre of each nail group. A small auxiliary beam is used to connect the nail node on the bottom chord at point 6 to the timber element at point 5. As the eccentricities between the system lines of the webs and the centre of the nail groups are small, the beam elements of the webs can be directly connected to the nail nodes at points 1 and 4. Three plate elements are located between the three nail groups. The joint lines for each plate element are shown in Figure 19.20. The three plate elements are connected to the plate nodes at points 1, 6 and 4. Also, three contact elements are used between the nodes at the points 1–2, 3–4 and 4–1.

The advantages of the Foschi model are that this model includes the deformation of the joints in the calculations of the sectional forces and deformation. The sectional forces in each beam, plate, nail group and contact zone are given directly, and can be compared to the load capacity given by the national codes or standards. Each type of element has understandable physical properties. In general, the model can be set up simply by looking at the joint.

Results from tests on many different truss and joint configurations have been successfully compared to results from the Foschi model (see Karacabeyli *et al.*, 1990, King *et al.*, 1988, Lau, 1987, Lum *et al.*, 1988, Nielsen, 1996 and Triche *et al.*, 1988).

The limitations of the Foschi model are:

- The plate deformation and failure are assumed to occur at the joint lines. However, some tests have shown that local buckling zones in the punched metal plate fasteners do not always follow the joint lines.
- Splitting and plastic deformations in the timber members are not included.
- All nails in a nail group are assumed to have the same stiffness parameters – also nails close to the timber edges.
- The gap size must be known.

By further developments and tests of the model, it may be possible to make better estimates of the load-carrying capacity of arbitrary truss configurations, and further, it might be able to predict the failure type.

The properties of the nail and plate elements can be determined by tests, some of which will be described in the next section.

19.4 STIFFNESS AND LOAD CAPACITY OF JOINTS WITH PUNCHED METAL PLATE FASTENERS

The properties of each type of punched metal plate fastener are estimated by tests in the laboratory. In standards and national codes, detailed procedures are described to estimate the anchorage properties of the nails and the tension, compression and shear properties of the plate, e.g. see prEN 1075, ANSI/TPI 1-1995.

19.4.1 Anchorage of the Nails

The anchorage properties of the nails depend upon the angles between the main directions of the plate, the grain direction of the timber and the force direction. Detailed analysis of the above properties demands a lot of different test series, and further, as the results should be statistically significant, a number of equal tests within each test series should be performed. In total, a detailed test programme will be very time-consuming and expensive. In practice, there are only made a few basic test series, as shown in Figure 19.21 to get some basic anchorage values to be used in empirical formulas dependent on the above angles. Discussions on empirical formulas are given in Kevarinmäki (2000a).

The plate sizes for the tests are chosen so that anchorage failure is expected. However, in some cases with tension perpendicular to grain, it may be difficult to determine if the failure is caused by withdrawal of the nails or splitting in the wood.

As the calculated anchorage capacity per nail or unit area depends upon the number of nails in

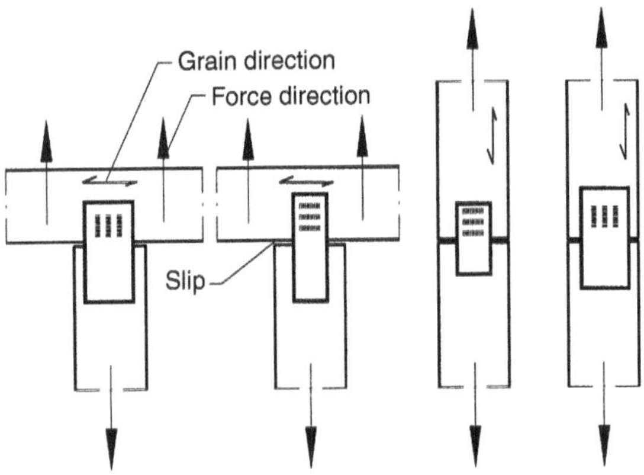

Figure 19.21 Basic test used for determination of nail properties

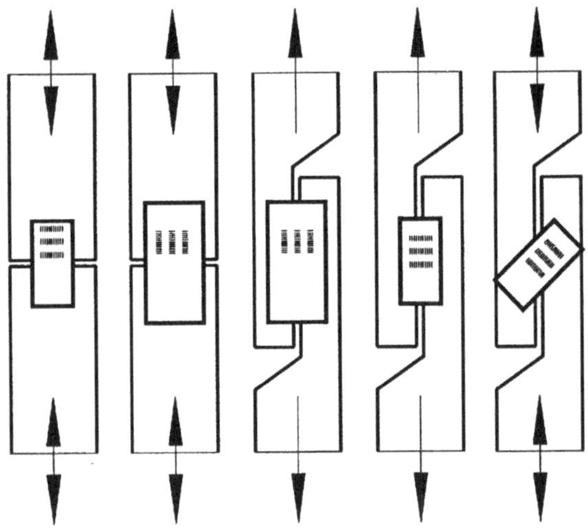

Figure 19.22 Different tension and compression tests for determination of plate properties

the nail group (group effect), the punched plate fasteners used must have a minimum size.

The anchorage properties are highly correlated to the properties of the timber, and therefore the anchorage properties of the nails depend upon the same factors as for the timber, e.g. wood density, moisture content, load duration, etc.

The anchorage values assume that the punched metal plate fasteners are fully embedded in the wood. In practice, some of the plates are only partially embedded, which causes a reduced anchorage capacity. Discussions and references on this subject are given in Nielsen (1999).

During loading the slip is measured (see Figure 19.21). The slip measuring includes deformation of the nail group and the plate. For the Foschi model it is necessary to separate the total displacement into the deformation of the nail group and deformation of the plate at the joint line. Different test arrangements have been used to obtain properties for the plate and nail element (see Ellegaard, 1999).

19.4.2 Plate Capacity

The plate capacity depends upon the angle between the joint line and the main direction of the plate, and the angle between the main direction of the plate and the force direction. The plate size (nail group) is chosen so that failure appears in the plate at the joint line. Both tension and compression tests are made (see Figure 19.22). As a small gap is made at the joint line, buckling failure is obtained in the plate during compression loading. Only a few tests are made to obtain some basic strength values, which are to be used in empirical formulas. Some tests used for determination of the plate properties are shown in Figure 19.22. Discussion and comparison between test results and empirical formulas are given in Bovim *et al.* (1985) and Kevarinmäki (2000a).

19.5 REINFORCEMENT OF TIMBER BY THE USE OF PUNCHED METAL PLATE FASTENERS

A punched metal plate fastener can be used as a local reinforcement in areas with intensively loaded timber. By using a reinforcement, the timber dimensions and thereby the cost may be reduced. The method is still very new, and it has been tested in the following situations (see also Figure 19.23):

(a) Reinforcement to increase the compression capacity perpendicular to the grain direction. At support areas of timber structures, the compression stresses perpendicular to the grain

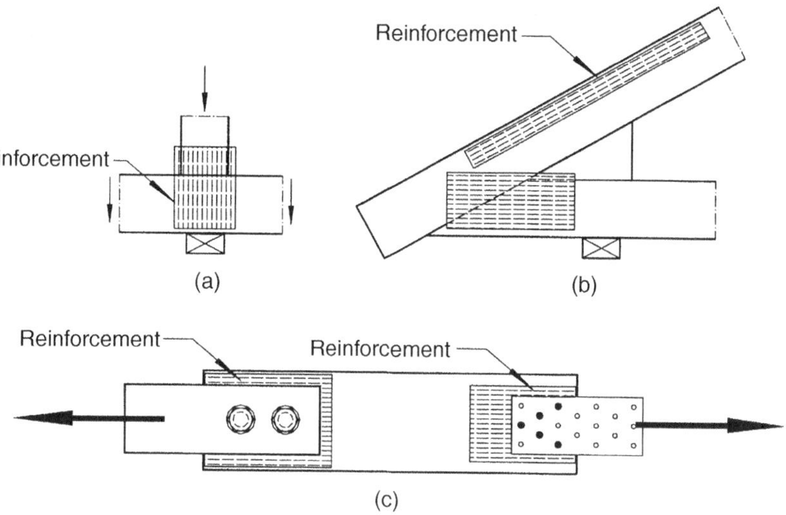

Figure 19.23 Using punched metal plate fasteners as reinforcement of wood

may be high. To reduce the deformation and increase the capacity of the timber, a punched metal plate fastener can be used (see Kevarinmäki, 1991). The compression capacity can be increased by 30–200%.

(b) Reinforcement to increase the bending capacity. Local reinforcements can be added in dimension controlling areas with moment peaks, e.g. at heel joints. However, tests show that a reinforcement in general only has a minor influence on the bending capacity, but the stiffness is increased by up to 20% (see Nielsen, 2000).

(c) Reinforcement to increase the capacity of doweled joints. By using punched metal plate fasteners as interlayers in dowel type joints, the embedding strength increases, the risk of wood splitting is reduced, and the tension capacity of a doweled connection increases by up to 250% (see Blass *et al.*, 2000). Punched metal plate fasteners have also been used at nailed metal plate-to-timber joints (see Kevarinmäki, 2000b).

REFERENCES

ANSI/TPI 1-1995 (1995) National Design Standard for Metal Connected Wood Truss Construction, American National Standard, Truss Plate Institute, USA.

Blass H.J., Schmid M., Litze H. and Wagner B. (2000) Nail plate reinforced joints with dowel-type fasteners. *Proceedings of World Conference on Timber Engineering*, Whistler, British Columbia, Canada.

Bovim N.I. and Aasheim E. (1985) *The Strength of Nailplates*, CIB W18/18-7-6. International Council for Building Research Studies and Documentation.

Burdzik W.M.G. (2000) Bracing of timber roofs – the new South African bracing rules and their impact. *Proceedings of World Conference on Timber Engineering 2000*, Whistler, British Columbia, Canada.

Cramer S.M. and Drozdek J.M. (2000) Load-sharing in metal-plate connected wood truss assemblies. *Proceedings of World Conference on Timber Engineering*, Whistler, British Columbia, Canada.

Ellegaard P. (2000) *Stiffness Analysis of Nail-Plates*, CIB W18/33-7-9. International Council for Research and Innovation in Building and Construction.

Foschi R.O. (1977) Analysis of wood diaphragms and trusses. Part II: Truss-plate connections. *Can. J. Eng.*, **4**, 353–362.

Karacabeyli E., Varoglu E., Lum C. and Olson L. (1990) Structure performance of metal plated glulam trusses. *International Timber Engineering Conference*, Tokyo, pp. 693–700.

Kevarinmäki A. (1991) *Capacity of Support Areas Reinforced with Nail Plates in Trussed Rafters*, CIB W18/24-14-1. International Council for Building Research Studies and Documentation.

Kevarinmäki A. (2000a) Semi-Rigid Behaviour of Nail Plate Joints. PhD. Thesis, Helsinki University of Technology, Laboratory of Structural Engineering and Building Physics Publications, Espoo, Finland.

Kevarinmäki A. (2000b) Nail plate reinforced metal plate-to-timber joints made with nail, screw or pop rivet fasteners. *Proceedings of World Conference on Timber Engineering*, Whistler, British Columbia, Canada.

King C.G. and Wheat D.L. (1988) Deflection and member behaviour of metal-plate-connected parallel-chord wood trusses. *International Conference on Timber Engineering*, 1, pp. 482–497.

Lau P.W.C. (1987) Factors affecting the behaviour and modelling of toothed metal-plate joints. *Can. J. Civ. Eng.*, **14**, 183–195.

Lum C. and Varoglu E. (1988) Testing and analysis of parallel chord trusses. *International Conference on Timber Engineering*, 1, pp. 460–466.

prEN 1075 (1999) *Timber Structures – Test Methods – Joints made with Punched Metal Plate Fasteners*, CEN/TC 124.

Norges Byggskole (1999) *Trekonstruksjoner med spikerplater*. (Timber construction with punched metal plate fasteners) (in Norwegian). Norske Takstolprodusenters Forening.

Nielsen J. (1996) Stiffness Analysis of Nail-Plate Joints subjected to Short-Term Loads. PhD. Thesis, Structural Paper No. 2, Aalborg University.

Nielsen J. (1999) An Unravelling of the Effect of Partially Embedded Punched Metal Plate Fastener on the Lateral Anchorage Strength. Structural Paper No. 3, Aalborg University.

Nielsen J. (2000) *Moment Capacity of Timber Beams Loaded in Four-Point Bending and Reinforced with Punched Metal Plate Fasteners*, CIB W18/33-14-1. International Council for Research and Innovation in Building and Construction.

Riberholt R. (1982) *Guidelines for Static Models of Trusses Rafters*, CIB W18/15-14-1. International Council for Building Research Studies and Documentation.

Riberholt R. (1990) *Analyses of Timber Rafters of the W-type*, CIB W18/23-14-1. International Council for Building Research Studies and Documentation.

Triche M.H. and Suddarth S.K. (1988) Advanced design of metal plate connector joints. *Forest Products J.*, **38**(a), 7–12.

20

Shear Walls and Diaphragms

Helmut G.L. Prion

Frank Lam

20.1 INTRODUCTION

A shear wall or diaphragm is an in-plane or plate-type structural element designed to transmit force in its own plane. Generally, the term 'diaphragm' is applied to horizontal elements such as floors and roofs. The term 'shear wall' is applied to vertical elements such as walls (including partition walls) when appropriately designed and constructed.

Horizontal roof and floor diaphragms are designed to distribute lateral loads to the shear wall elements, which in turn carry the loads to the foundations. Diaphragms do not have to be horizontal. Sloped roofs, hipped roofs and curved roof diaphragms may also be used, but their design requires more complex procedures than those shown here. Simplified arrangements of typical roof shear walls and diaphragms are illustrated in Figure 20.1.

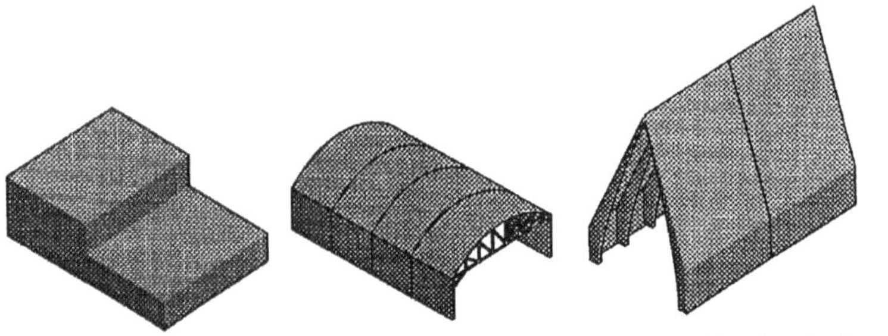

Figure 20.1 Typical shear walls and diaphragms (CWC, 1996)

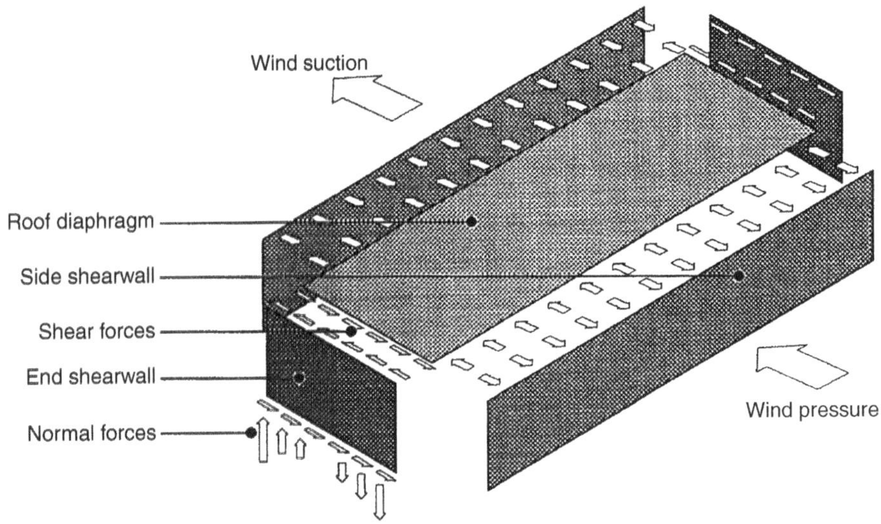

Figure 20.2 Shear wall and diaphragm action (CWC, 1996)

The action of wood shear walls and diaphragms in bracing a simple box-like building subjected to wind load is illustrated in Figure 20.2. The side walls, considered to be simply supported at roof and foundation, transfer one half the total wind load to the roof level. The roof diaphragm, acting as a deep horizontal beam, transmits the load to the end shear walls, which in turn transfer the load to the foundations.

The distinction between shear wall and diaphragm elements arises because of different load and support conditions at their boundaries. The roof diaphragm is subjected to normal forces from wind pressure on the side walls, and is supported by shear forces from the end shear walls.

The end shear walls are subjected to shear forces at roof level from the roof diaphragm, and are supported by shear and normal reactions at the foundations. Shear walls must be anchored to the foundation to resist uplift forces.

In panelised building systems (Figure 20.3), shear walls and diaphragms typically fulfill multiple roles. In addition to forming part of the building envelope or acting as partition walls or floors, they are important load carrying components. In addition to carrying in-plane shear loads, they are often also relied upon to carry loads perpendicular to their plane (occupancy loads on floors and roofs, snow loads on roofs, wind loads on walls and roofs) or axial loads in their plane (vertical

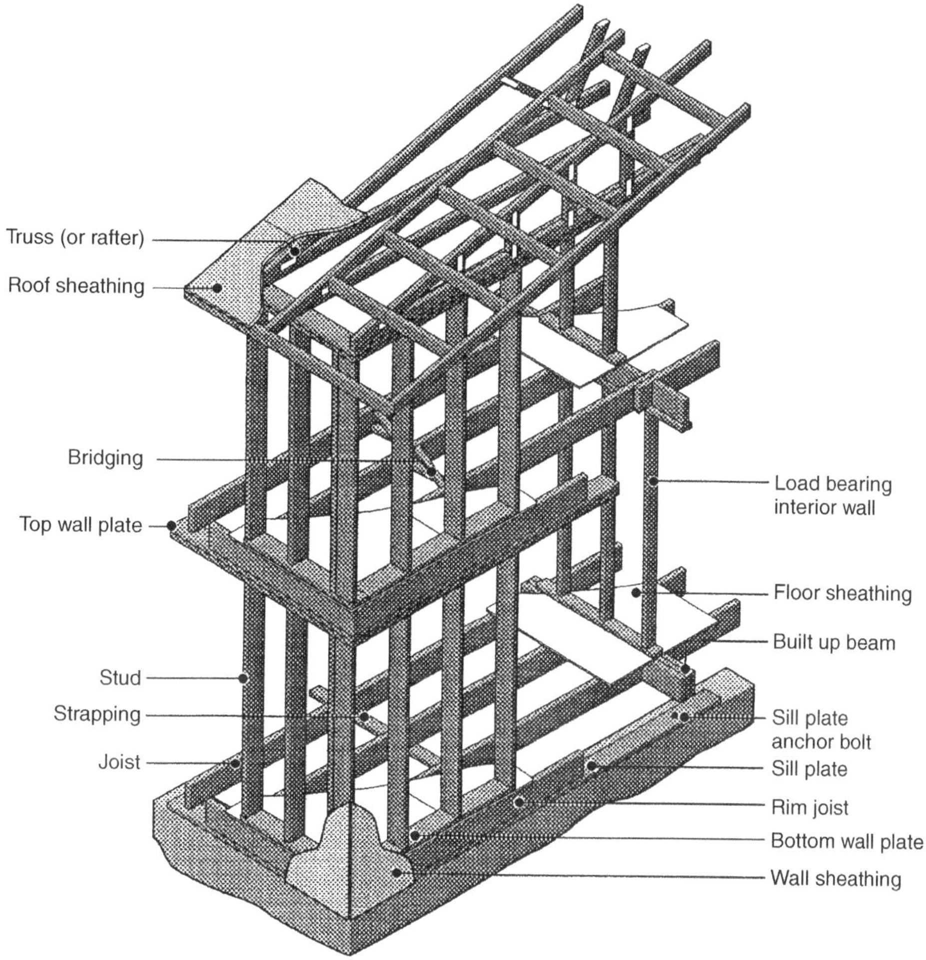

Figure 20.3 Wood frame construction (CWC, 1996)

loads on walls). Not all walls technically qualify as shear walls from an engineering design point of view, although even those that are not specifically designed as shear walls, will carry shear loads to some lesser degree. The same applies to diaphragms. To clarify load paths and ensure proper performance of shear walls and diaphragms, certain detailing requirements have to be adhered to, and engineers will specifically select certain panel elements to fulfil such duties. Detailing requirements will be discussed later in the chapter.

Whereas shear walls and diaphragms are used in many applications in steel, reinforced concrete and timber construction, the application of such elements will here be discussed in the context of light woodframe construction, as it is commonly used in residential construction.

The panelised woodframe construction system, as simple as it seems, is one of the most efficient lateral load resisting systems available, for the following reasons:

- Roof, wall and floor panels fulfil multiple purposes as load carrying elements, while also often acting as the envelope to the structure.
- By virtue of continuous connections between adjacent panels, a three-dimensional box system is created, which is well suited to distribute unsymmetrical loads and overcome discontinuities in the building system.

- Nailed wood shear walls are highly redundant, and are therefore not necessarily governed by the weakest members.
- A relatively low grade of lumber can be used to provide a very reliable system.
- The simplicity of the building system requires neither special equipment nor high carpentry skills and workmanship for construction and erection.
- Wood shear walls are very ductile and forgiving because of the deformability of the nails and the surrounding wood.
- The system is readily adaptable for openings and services.
- Non-structural elements, such as wall cladding, often provide for significant additional resistance.

At this point, the distinction needs to be made between a shear wall and an ordinary stud wall. A *shear wall* in timber construction is a load-bearing wall that is designed to carry racking loads in the plane of the wall (shear loads) in addition to the vertical loads and lateral pressures on its surface. A *stud wall* will only carry vertical and perpendicular loads, and is designed similar to a beam-column. A *partition* wall does not carry any load, other than its own weight, and is used to partition parts of a dwelling according to the needs of the occupants. A *floor* or *roof diaphragm* is oriented in a horizontal or inclined direction, and carries loads perpendicular to its surface, while also providing racking resistance through in-plane shear. From a functional point of view, a diaphragm seems to be similar to a shear wall. The construction detailing is quite different, however, and therefore diaphragms are treated separately. Figure 20.3 shows the major elements in light woodframe construction, while the most important components of a shear wall are shown in Figure 20.4.

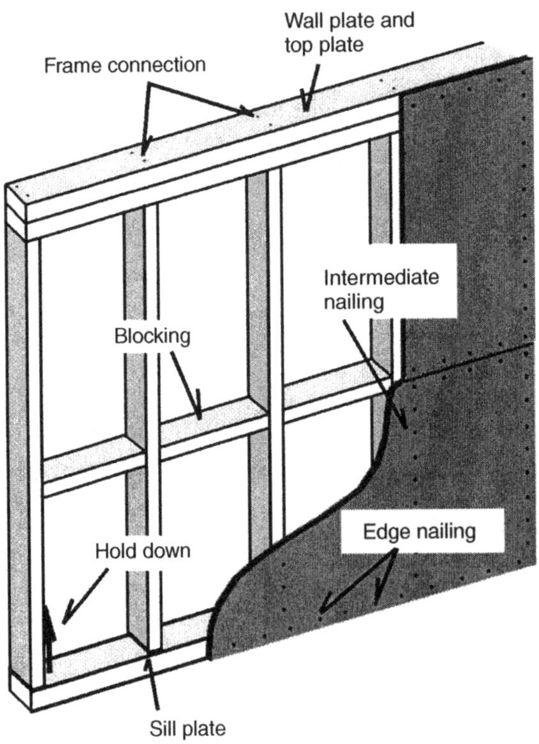

Figure 20.4 Wood frame shear wall with plywood sheathing

20.1.1 The Mechanism of a Wall Plate

A plate consisting of a frame with closely spaced studs or joist, sheathed with panels or lumber sheathing, on one or both sides, normally carries three types of loads:

- loads perpendicular to its surface, such as wind loads for *shear walls* and occupancy, wind or snow loads for *diaphragms*;
- in-plane loads parallel to the studs resulting from gravity loads in a *shear wall* and wind or earthquake loads in a *diaphragm*;
- in-plane shear loads resulting from horizontal loads transferred from other plates and originating from wind or earthquake loads.

The out-of-plane loads are carried through bending in the sheathing to the studs or joists, which are thus loaded in bending.

Vertical loads are carried by the studs, which act as columns and are laterally supported by the nailed-on sheathing. When lateral loads (perpendicular to the surface) are also present, the studs have to be designed as beam-columns.

Racking loads are applied to the *framing* through connections with other plate elements (shear walls or diaphragms), but are carried mainly by the *sheathing* in shear (Figure 20.5). The connections are thus of primary importance to transfer the racking loads to the sheathing. Since the connections between framing members are nominal at best, the

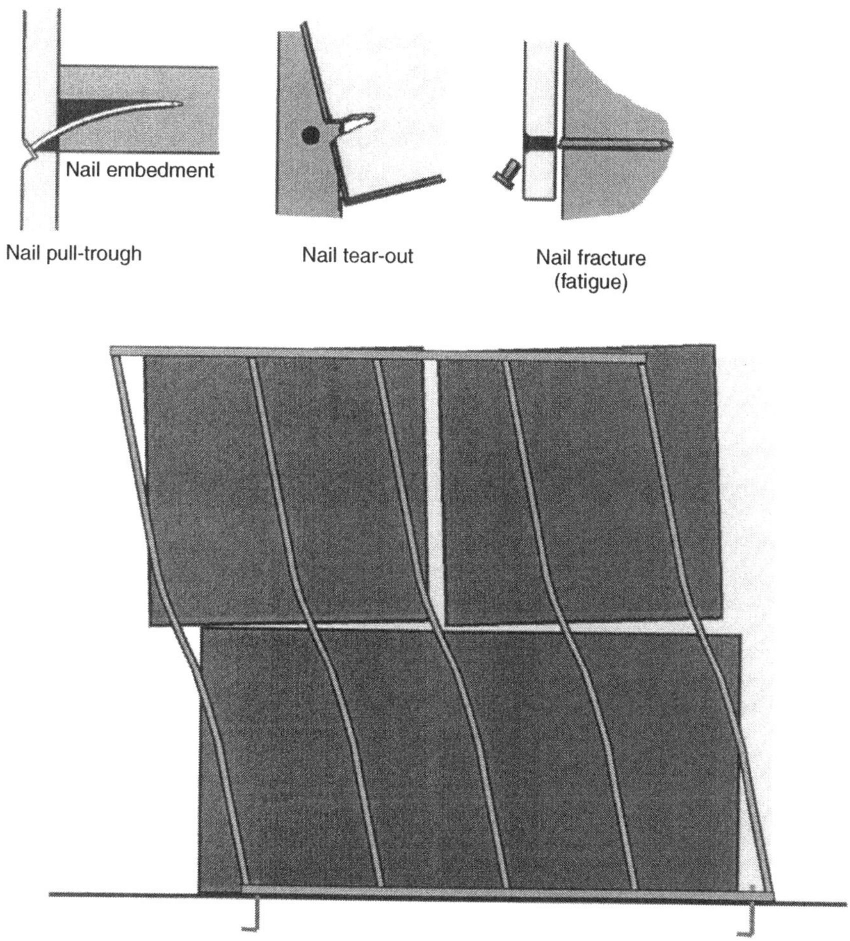

Figure 20.5 Shear load transfer through a shear wall and failure mechanisms of nail connectors

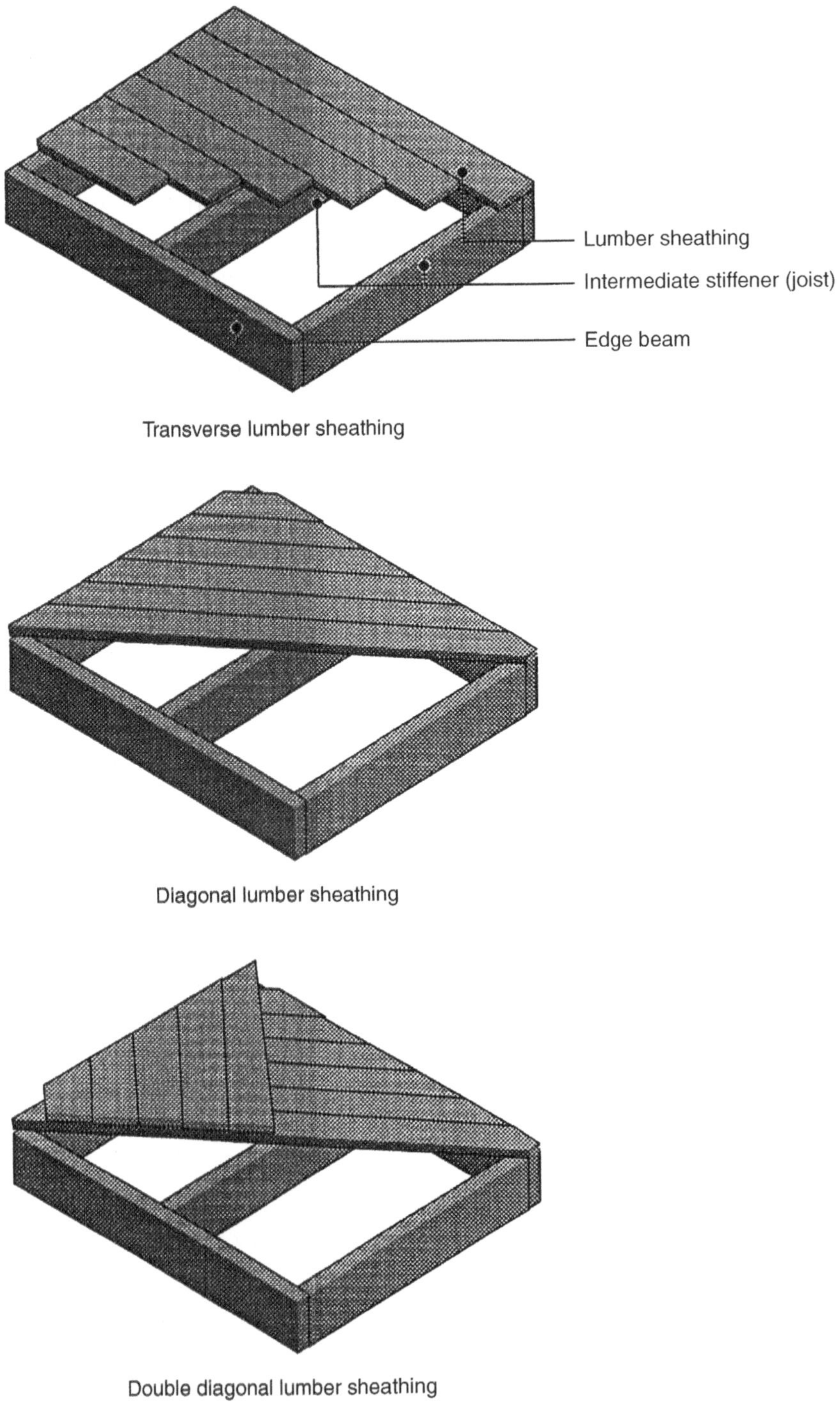

Figure 20.6 Sawn lumber board sheathing (CWC, 1996)

sheathing and its connectors play a crucial role in transmitting loads between framing members as well. A shear wall or diaphragm is thus a highly redundant system when it comes to resisting in-plane racking loads, with framing members, sheathing and nail connectors interacting closely.

Sheathing of diaphragms and shear walls typically consists of plywood or oriented strandboard panels (Figure 20.5), but could also be diagonal or transverse lumber sheathing (Figure 20.6). Whereas *panels* act as almost rigid plate entities, where the edge nails are the important connector elements, *lumber sheathing* has to be designed with individual nails and tongue and groove joints to provide a continuous plate. The latter is less common in modern construction, but may be found in older buildings or in special applications where the sheathing may be exposed and forms part of the visual appearance of a building.

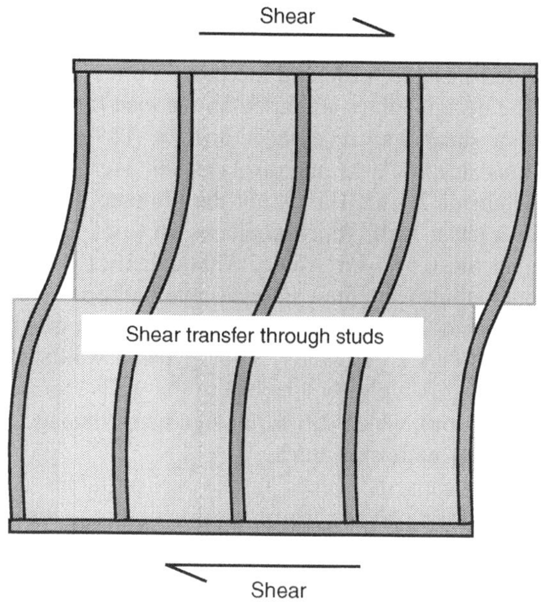

Figure 20.7 Unblocked shear wall

20.1.2 The Role of Blocking

Since sheathing panels (typically 1.2 × 2.4 m or 4 × 8 ft) are usually much smaller than the wall itself, the panel edges at the seams have to be connected to transmit the shear forces from one panel to the next. In shear walls, vertical seams usually coincide with a stud and adjoining panels are nailed to the same framing member to make this connection. In the transverse direction, however, special blocking needs to be provided between studs to achieve the same continuous connection between the panels (Figure 20.4). In diaphragms, the joists provide a means for nailing adjacent panels and thus form a shear joint. Transverse joints between panels are often left free or unblocked, thus forming a weak link in the system. For the transfer of transverse loads it is of importance to provide a shear connection to avoid differential displacement of panel edges. For this purpose, H-clips are often provided, which can also transmit in plane shear forces. When the in-plane shear forces are very high, it is best to provide timber blocking.

The provision of blocking poses some practical problems. Aside from the additional cost, builders prefer to leave a ventilation and expansion gap between panels, which would be closed off by the blocking. Since some form of damp proofing is typically provided anyhow, the ventilation requirement does not seem to be as much of an issue. If blocking is not provided, the racking strength of the wall depends upon the framing members bridging the gap, and the nails in the immediate vicinity of the seam (Figure 20.7). Although a few additional nails in this area would greatly facilitate shear transfer across the opening, it was found that stud failures often limit the amount of shear that can be transferred. Some building codes permit unblocked horizontal *diaphragms*, and provide reduced resistance values for different framing configurations. *Shear walls*, however, are required to have framing members along all the panel edges, i.e. blocking is mandatory.

Recent research by Ni *et al.* (2000) has provided guidelines for reduced design values for unblocked shear walls, and strength adjustment factors have now been included in the Canadian design code.

20.2 SHEAR WALL DESIGN

Due to the complexity of the stresses in a shear wall when subjected to lateral loads, it is not practical for engineers to design the individual components. Design values are thus typically provided in table form, and are based on test results. The design values are based on the load resistance of a shear wall at a shear drift of 0.5%, which translates into a 12 mm displacement over a storey height of 2.4 m. This is not the ultimate capacity of a shear wall, which could be up to 50% more than the 0.5% drift value. Although it is considered an ultimate limit state design condition, shear walls are designed with a serviceability criterion in mind. This is necessary, since gypsum wallboard paneling is typically attached to the walls (for fire protection), which will suffer significant damage at drifts of more than 0.5%.

The capacity values in the design tables are given for a certain type of construction (e.g. members must be 38 mm wide at panel seams), and are dependent on the thickness of the sheathing, the size and spacing of nails, and the species of the framing. Since oriented strandboard and plywood paneling behaves very similar in a shear wall, no distinction is made for the different panel products. The usual modification factors apply for wet service conditions and load duration effects.

A simplified calculation method can also be used, based on the shear capacity of a nail connector. Two different approaches have been suggested (Källsner and Lam, 1995), a linear elastic method and a plastic or limit state approach. Both models assume that the frame members and panels behave as rigid bodies, and that all deformations occur in the panel-to-framing connectors.

Linear Elastic Model

According to this model (Figure 20.8), all the connectors remain linearly elastic and the superposition principle can be used to calculate the maximum force in a connector.

If the origin of the x- and y-axes is chosen at the centroid of the panel and connector group, then the forces in the x- and y-directions are

$$F_{xi} = -Hh\frac{y_i}{\sum_{i=1}^{N} y_i^2}; \quad F_{yi} = -Hh\frac{x_i}{\sum_{i=1}^{N} x_i^2}$$

where x_i and y_i are the coordinates of connector i. Using vector addition, the maximum connector force is

$$F_{\max} = \sqrt{F_{xi}^2 + F_{yi}^2} \leq F_{design}$$

The capacity of the wall is thus reached when one of the connectors (usually the one furthest from the centroid) reaches its limit.

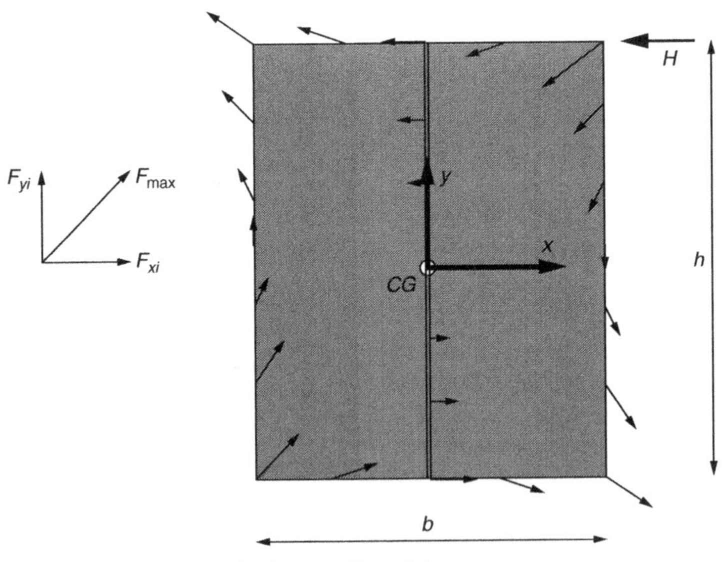

Figure 20.8 Connector forces in linear elastic shear wall model

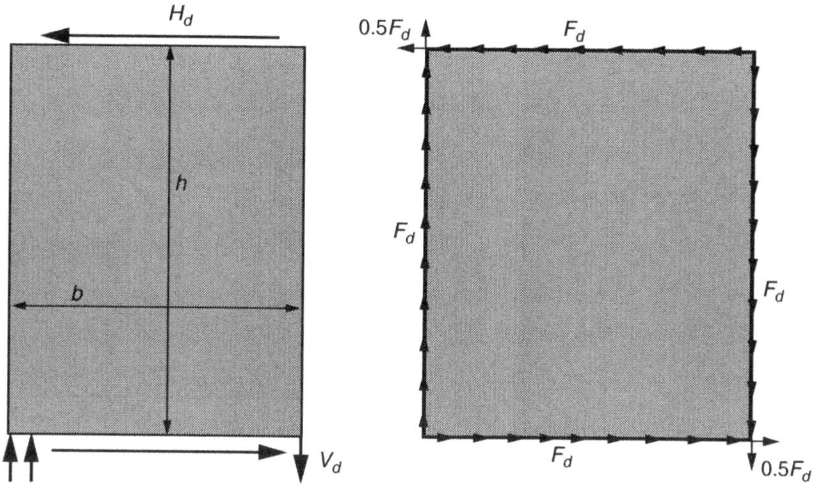

Figure 20.9 Connector forces in plastic shear wall model

Plastic Model

When all the connectors can be assumed to reach their maximum capacity, a simple plastic model can be used, which gives a lower bound solution. A uniform connector shear load is assumed along the edges of a panel, in compliance with equilibrium conditions of the wall (Figure 20.9). All connectors, except for the corner ones, carry the full design load. Therefore the total capacity of the wall is given by

$$H_d = n_H F_d$$

where F_d is the load in each connector and n_H is the number of fastener spacings along the upper edge in the direction of the horizontal force. The hold down force of the panel can also be calculated as

$$V_d = n_V F_d$$

where n_V is the number of fastener spacings along the vertical edge.

Experimental investigations have shown that both methods give comparable results and are somewhat conservative, compared to the test results.

Nailed Board Sheathing Walls

Board sheathing is often used in single family dwellings and low-rise industrial buildings, most of the time because the boards are left over from the concrete formwork that is used for the foundations. When board sheathing, consisting of individual planks, is used to provide shear resistance, these can be arranged either horizontally or diagonally. Diagonal sheathing can be very effective (Figure 20.10), especially when used in two directions (Figure 20.11), and design values are often given in the code for a certain board thickness, nail size and spacing. For non-standard configurations or when values are not provided, the capacity can be calculated from basic nail connection properties using the model as shown in Figures 20.10 and 20.11.

The shear resistance of horizontal or transverse lumber sheathing is relatively limited in comparison to that of panel sheathing, diagonal and double diagonal lumber sheathing. Where transverse sheathing without gluing is used, the design shear must be resisted entirely by nail couples (Figure 20.12). The moment capacity per connection is

$$M_r = N_r s$$

where N_r is the strength resistance of individual nails, and s is the nail spacing (mm).

The lateral resistance of common round wire nails can be determined from information on connection resistance. The factored shear resistance of

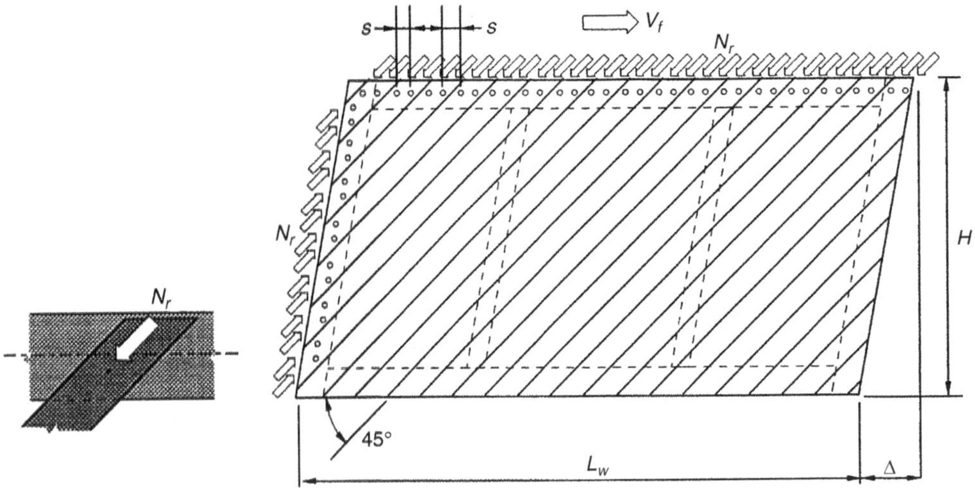

Figure 20.10 Single diagonal board sheathing design (CWC, 1996)

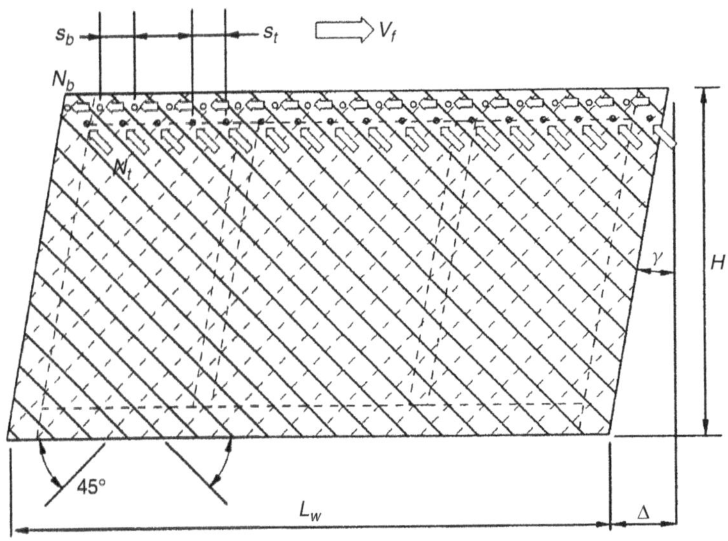

Figure 20.11 Double diagonal board sheathing design (CWC, 1996)

a nailed shear panel with transverse lumber sheathing, V_r, is the equal to the moment capacity per nail couple times the number of nail couples, divided by the height of the wall, H:

$$V_r = N_r s[n][H/b][1/H]$$
$$= (N_r s[n])/b$$

where n is the number of vertical members (end posts plus intermediate stiffeners), and H/b is the number of boards in the shear wall.

Numerically,

$$n = L_w/a + 1$$

Rearranging the above equation for V_r to give the required design nail spacing(s):

$$s = (bV_r)/(N_r n)$$
$$= (bv_r L_w)/(N_r n)$$

where $v_r = V_r/L_w$.

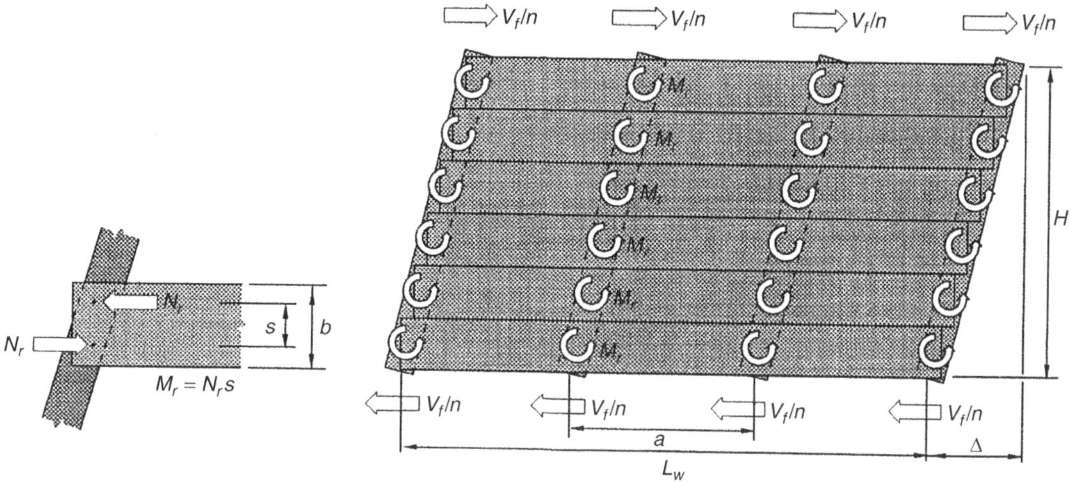

Figure 20.12 Horizontal board sheathing design (CWC, 1996)

20.3 DIAPHRAGM DESIGN

The design of a diaphragm is very similar to that of a shear wall. Whereas a shear wall is treated similar to a cantilever beam, a diaphragm is typically designed as a simply supported beam (Figure 20.13). The procedure outlined below is suitable for a box-type of structure, where the diaphragm is supported by a shear wall at either side. Furthermore, the width-to-depth ratio needs to be sufficient (two or higher) to warrant the beam analogy. The diaphragm is then designed similar to a steel plate girder. The chord members are assumed to resist the bending action, while the sheathing acts as the web to carry the shear. The joists in the diaphragm act as stiffeners of the shear web.

Consider the horizontal diaphragm subjected to a uniformly distributed lateral load w_f along one edge, supported by shear walls at each end.

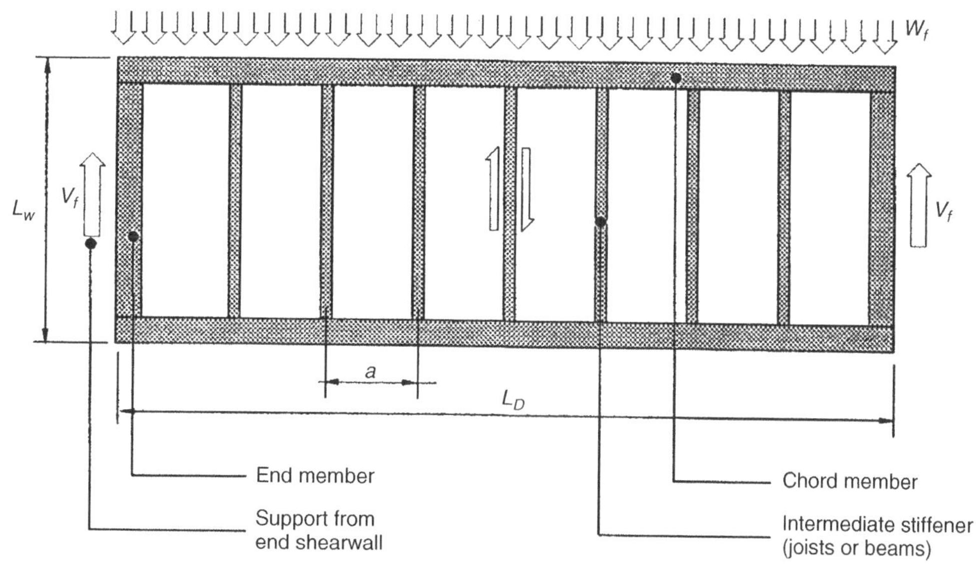

Figure 20.13 Design of a diaphragm (CWC, 1996)

Assuming that the diaphragm acts as a simple beam, and that only the chord members resist the applied moment, the design axial force in the chord member, P_f, is the maximum factored moment divided by the diaphragm width:

$$P_f = (w_f L_D{}^2)/(8L_w)$$

It is important to bear in mind that the chord must be continuous, and properly designed splices are necessary. Such splices typically consist of nailed metal straps.

The above assumptions for calculating chord forces were shown to be reasonable for aspect ratios (L_w/L_D) up to 1.0 (Prion and Lam, 2000). For deeper diaphragms, the sheathing and joists contribute substantially to the bending resistance, and the simplified method of calculating chord forces produces very conservative values. In practice, this is not an important issue, however, since the calculated chord forces are typically very small, due to the large depth-to-span ratio of the diaphragm.

The shear force in the sheathing varies along the length of the diaphragm. The maximum shear force V_f at the supports equals

$$V_f = w_f L_D/2$$

If the shear force is assumed to distribute uniformly along the diaphragm width, the shear force per unit length at the support (shear wall) is

$$v_f = w_f L_D/2L_w$$

This is the maximum design shear force per unit width for the sheathing. Since the applied shear force varies from a maximum value of V_f at the supports to zero at the centreline, the shear strength of a horizontal diaphragm can be reduced towards the middle of the structure.

The design for strength is identical to that for shear walls outlined earlier, and tables for shear resistance are typically given in the design code. Note that blocking is not always used in diaphragms, and values for unblocked diaphragms are typically available.

When a shear wall is not extending over the full depth of a diaphragm, a drag strap or collector member must be provided (Figure 20.14). This

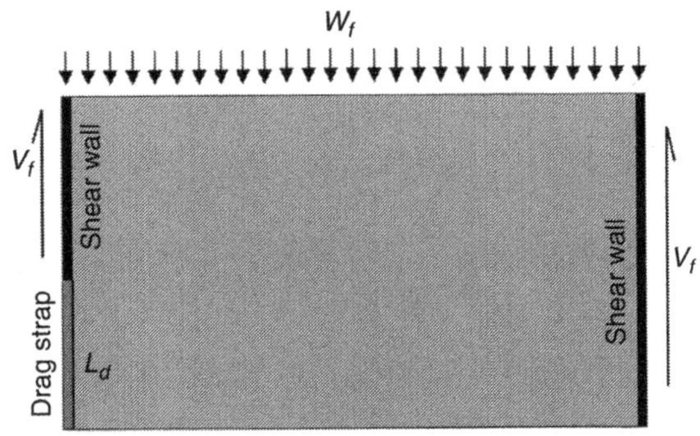

Figure 20.14 The application of a drag strap to transmit diaphragm forces to a discontinuous shear wall

drag strap is attached to the diaphragm and transmits the shear reaction to the shear wall(s). The axial force in the drag strap is calculated as

$$T_f = v_f L_d$$

where L_d is the length of the drag strap.

Of course, a proper connection needs to be designed to transmit this concentrated axial load to the shear wall.

20.4 SHEAR WALL AND DIAPHRAGM ANALYSIS

There are two issues to be dealt with here:

- the load path and force distribution throughout the structure has to be established;
- stiffness and strength values have to be found for shear walls with openings.

20.4.1 Diaphragms

Diaphragms are essential in distributing horizontal forces in the plane of the diaphragm to the shear walls, which in turn transfer these forces to the next storey below or to the foundations. In most cases, a system consists of more than two shear walls, and is therefore redundant. The distribution of forces to the individual shear walls now depends largely upon the stiffness of the diaphragm and the shear walls. If a full finite element analysis cannot be performed, some assumptions have to be made to find a reasonable force distribution. The two extreme cases would be to consider the diaphragm to be either very flexible or infinitely rigid.

- *flexible diaphragms*: in this case, the in-plane diaphragm forces can be assumed to be distributed to the shear walls according to the tributary areas associated with each shear wall (Figure 20.15). This is often the preferred method for engineers because of its simplicity, and since stiffness values for the shear walls are not required.
- *rigid diaphragms*: when the diaphragm is assumed to move as a rigid body, the relative stiffnesses of the shear walls need to be calculated. Furthermore, the diaphragm can translate or rotate and torsional effects are thus taken into account in the analysis. A relatively simple approach is to assume that the stiffness of a wall is proportional to its length. This is a reasonable assumption when all the walls are sheathed in the same way and have the same height.

In comparing the two approaches, it is clear that the force distributions can be significantly different. If in Figure 20.15(b), the middle shear wall is half as long as the outside walls, it represents about 1/5 of the total lateral stiffness, and will thus attract 20% of the load, if no torsion is present. In the case of a flexible diaphragm, however (Figure 20.15(a)), it attracts about 50% of the load. One could, of course, choose a solution in between and assign a somewhat smaller tributary area to the shorter wall. It is clear that the engineer will have to use some judgment in apportioning the loads.

When assuming a rigid diaphragm, the centre of rigidity of the system of shear walls has to be calculated. The offset from the resultant force will then represent the torsional component of the load (Figure 20.16). The translational force is

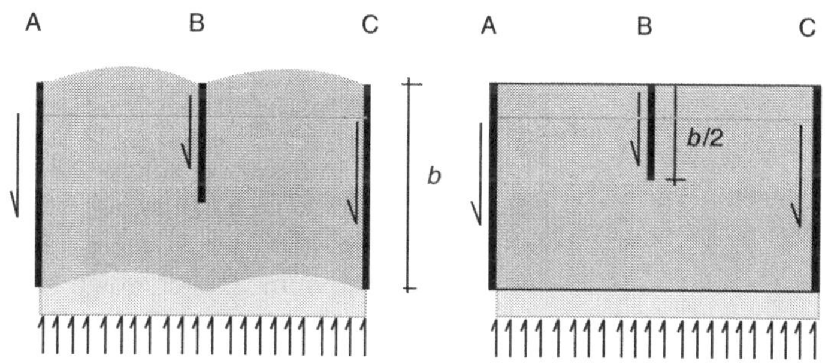

Figure 20.15 Load distribution through a diaphragm. (a) flexible diaphragm, (b) rigid diaphragm

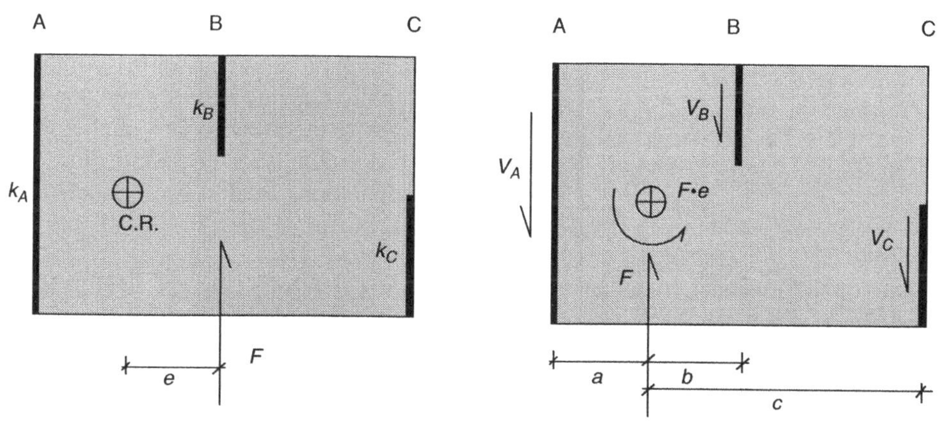

Figure 20.16 Load distribution with a rigid diaphragm

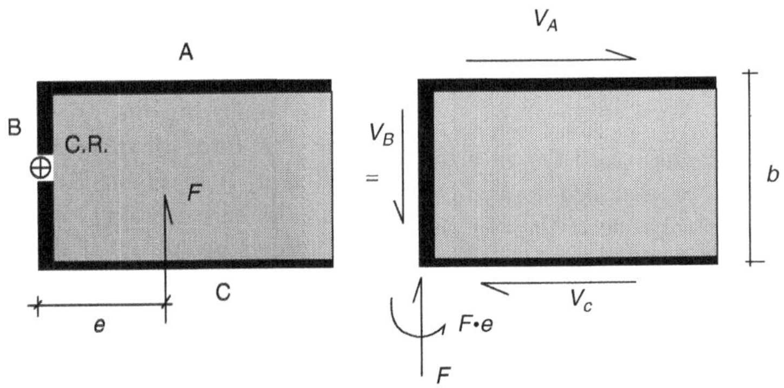

Figure 20.17 Torsion on shear walls in U-shaped building

then apportioned according to each shear wall's stiffness, while the torsional component is dependent on the stiffness and distance from the centre of rigidity.

For the example shown, the direct shear forces would be

$$V_{dA} = (k_A/\Sigma k)F$$
$$V_{dB} = (k_B/\Sigma k)F$$
$$V_{dC} = (k_C/\Sigma k)F$$

and the torsional components

$$V_{tA} = F \cdot e \cdot a \cdot k_A/J$$
$$V_{tB} = F \cdot e \cdot b \cdot k_B/J$$
$$V_{tC} = F \cdot e \cdot c \cdot k_C/J$$

where $J = k_A \cdot a^2 + k_B \cdot b^2 + k_C \cdot c^2$ and $k_i = F_i/\Delta_i$, the stiffness of a shear wall.

The total wall forces are then

$$V_A = V_{dA} + V_{tA}$$
$$V_B = V_{dB} + V_{tB}$$
$$V_C = V_{dC} + V_{tC}$$

Note that some of the torsional components will be subtracted. In practice, the negative components are typically ignored, and each wall will be designed for a force at least its direct component. Because of the uncertainty of eccentricities, beneficial effects should be ignored.

In the case of a U-shaped open core shear wall, loaded as shown in Figure 20.17, the centre of rigidity coincides with the wall parallel to the applied force. The torsional moment is carried by the two walls that are perpendicular to the force, and the direct shear is carried by the wall parallel to the force.

$$V_B = F$$
$$V_A = V_C = F \cdot e/b$$

In a system where walls are positioned parallel and perpendicular to the load direction, a similar approach is taken. The centre of rigidity is determined, after which the load is distributed into direct shear components (according to the wall stiffnesses) and torsional components.

Note that earthquake design codes typically prescribe an accidental eccentricity, which can be in either direction. This is to account for the uncertainty of the centre of mass of the objects on a given floor, and to some degree, the uncertainty of the centre of stiffness. This implies that the worst-case scenario should govern, and several force configurations need to be considered.

20.5 OPENINGS IN SHEAR WALLS AND DIAPHRAGMS

Openings in shear walls and diaphragms can have a marked effect on their performance. Framing should be designed to permit shear transfer around openings. When large openings such as doors divide a shear wall into a number of smaller elements, proportioning of lateral shear among the various elements requires special consideration.

When the difference in length between the wall segments is not too much, the load can be shared in proportion to the wall segment lengths (Figure 20.18). When too large a difference exists, equal displacement of the wall segments is assumed and the shear is distributed according to the stiffness distribution.

Another approach is often used when the openings result mainly from windows (Figure 20.19). In such a case, the above approach would assume a series of narrow tall shear walls, extending over the full height of the building. This, however, does not reflect the behaviour of the wall, which acts more like a moment frame with deep beams, connecting a series of wide columns (walls). It is thus more appropriate to consider the frame action of the entire wall. This approach has been proposed by Sugiyama and Matsumoto (1994), where the strength of a wall is reduced by a factor consisting of the area of the openings versus the area of the wall. It has been shown to be accurate for walls where the ratio of the depth to width in all the

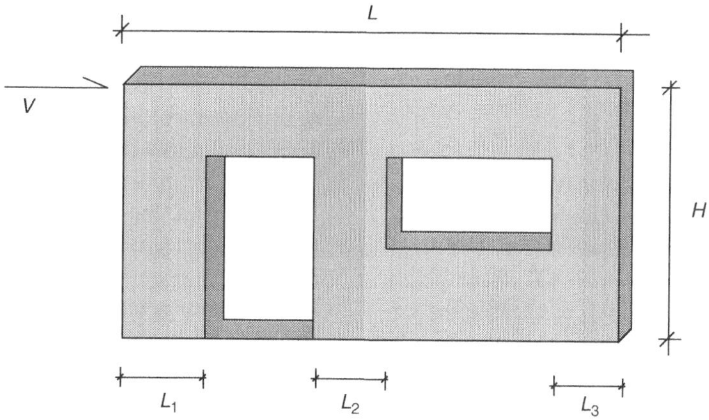

Figure 20.18 Shear wall with large openings

Figure 20.19 Shear wall with large number of small openings

wall spaces above and below openings is not less than 1/8 and the sheathing ratio is not less than about 30%.

The sheathing area ratio r has been defined as

$$r = 1/[1 + (\alpha/\beta)]$$
$$= 1/[1 + A_0/(H\Sigma L_i)]$$

where α is the opening area ratio $A_0/(HL)$, β is the wall length ratio $(L_1 + L_2 + L_3)/L = \Sigma L_i/L$, A_0 is the total area of the openings, and H is the height of the wall.

Three equations have been proposed. Yasumura and Sugiyama (1984) showed that the following empirical equation is applicable for shear deformations of around 1/100 (1% drift), as well as for the ultimate stage:

$$F = r/(3 - 2r)$$

Further research at the University of Tokyo led to a more refined approach (Sugiyama and Matsumoto, 1994) that is less conservative. For small racking displacements of up to 1/300 of the height of the wall (0.33% drift), which is suitable for serviceability limit design, the shear load ratio can be estimated as

$$F = 3r/(8 - 5r)$$

while for larger shear deformations, between 1/150 (1% drift) and 1/60 (1.6% drift),

$$F = r/(2 - r)$$

The shear load ratio F is defined as the ratio of the shear capacity of the wall with openings versus that for the wall without openings.

20.5.1 Diaphragms with Openings

Often it is necessary to have an opening in a diaphragm to allow for services to pass through or to accommodate a staircase. In such a case, the diaphragm will be weakened, and the designer has to use judgment in assigning strength values. Many codes will allow openings of a certain size without the need to adjust the strength of the diaphragm. In such cases, the framing around the opening has to be reinforced to assure the transfer of forces around the opening. One example that illustrates such an arrangement is shown in Figure 20.20. In this case, both the vertical loads (perpendicular to

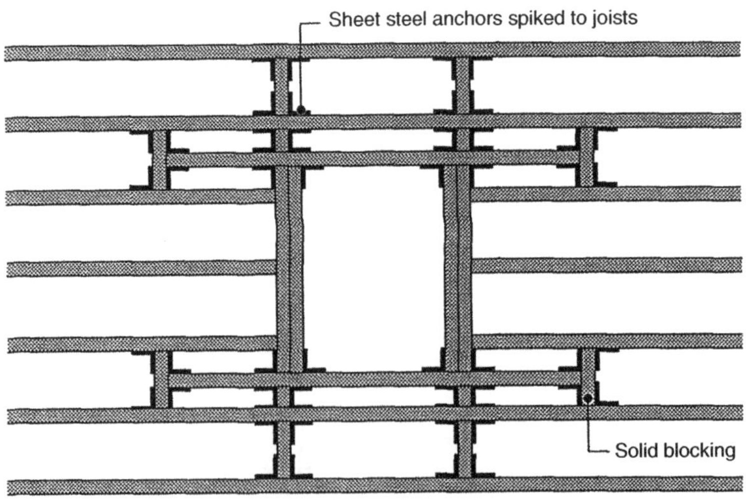

Note
This suggested layout can be used where diaphragms with openings are unavoidable. Other systems of reinforcing the openings can be developed.

Figure 20.20 Diaphragm framing around an opening (CWC, 1997)

the plane of the diaphragm) and the shear loads (in the plane of the diaphragm) need to be considered.

When larger openings are required, a rational approach based on mechanics principles needs to be taken to assure that bending and shear loads are safely carried. An example of such a case is shown in Figure 20.21.

The narrow strip of the diaphragm besides the opening may experience very high shear and bending forces when the load is applied directly to the end of the diaphragm. This would be the case when a wind load on the wall is transmitted directly to the chord of the diaphragm. The edge chord A will carry axial forces from the overall diaphragm bending as well from secondary bending of the narrow strip.

The axial compression force in chord A can be assumed to be

$$P_f = (w_f L_D{}^2)/(8L_W) + (w_f L_O{}^2)/(15a)$$

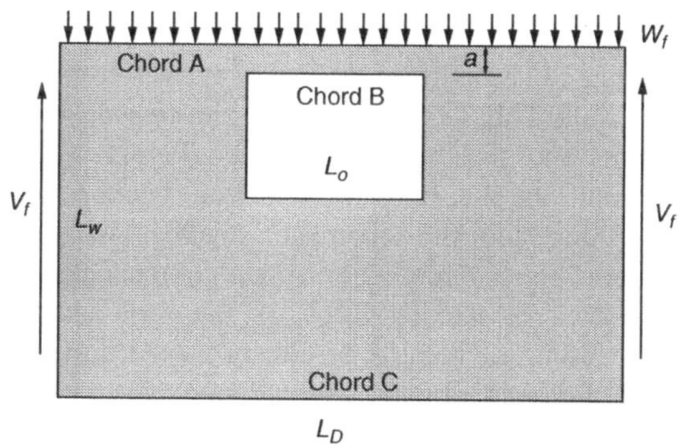

Figure 20.21 Diaphragm with large openings

which is a superposition of the two bending effects and accounts for some fixity of the narrow strip where it attaches to the large diaphragm.

Chord B will experience a tension force of the order

$$T_f = (w_f L_O{}^2)/(15a)$$

which assumes that no axial compression force is received from the overall diaphragm bending. The reason for this assumption is that the compression force would cancel the tension force, which may not be the case, depending on the type of connection of this short chord to the rest of the diaphragm. It is thus a conservative approach to design for the tension load resulting from the localised bending in the narrow diaphragm strip.

20.5.2 Diaphragm with Offsets

Although the end chords are assumed to carry the entire bending load, substantial axial stresses are transmitted through the sheathing and joists. When the edge of the diaphragm has an offset (Figure 20.22) the chord would not be continuous and the question arises whether the simplified method to calculate the chord forces still applies. An extensive finite element study of diaphragms with offsets showed that for offsets of up to 1.2 m no special measures need to be taken to assure chord continuity (Prion and Lam, 2000). When the offset exceeds this distance, the chord must be continued over the offset (Figure 20.22).

It should be borne in mind that a diaphragm is a very complex system and the force distribution depends to a large extent on the framing details

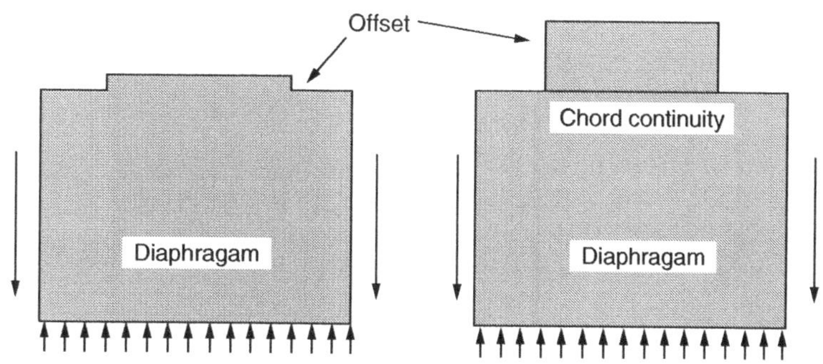

Figure 20.22 Diaphragm offsets

and methods of connecting elements. In general, it is thus advisable to use a conservative and simple method that assures a proper load path.

20.6 DISPLACEMENTS IN SHEAR WALLS AND DIAPHRAGMS

The linear elastic displacements of a wall or diaphragm can be calculated using approximate formulae, proposed by COFI (1979) and Jephcott (1980). The displacements respectively consist of three components, namely a flexural component, a shear component and an expression, representing the contribution from nail deformations (see Figure 20.23).

The following is the expression to calculate the displacement of a simply supported diaphragm:

$$\Delta = 5v_Q L^3/96EAb + v_Q L/4B_v + 0.0006Le_n$$

while the displacement of a shear wall is given as:

$$\Delta = 2v_Q h^3/3EAb + v_Q h/B_v + 0.0025he_n$$

where B_v is the shear through thickness rigidity of the panel, and is typically listed in the property tables in codes, and e_n is the nail deformation in mm. The designer can assume a value for this deformation, or it can be determined from nail property tests. It is also often given in code documents for typical nail connections.

There is disagreement in the literature about the magnitude of the constant associated with the third term. Since e_n represents the deformation of a single nail, and since two rows of nails are used at panel points, twice the nail deformation may be appropriate for the nail calculation of diaphragm deflections. A factor of two has been incorporated in the constant 0.0006. A similar adjustment has been made for shear walls. This therefore implies that the above formulae apply only to fully blocked shear walls and diaphragms.

Whereas the above formulations are valid for linear elastic deformations only, a more detailed analysis of the deformations of shear walls in the inelastic range can be made using the approximate method outlined by Easley *et al.* (1982) or by a finite element analysis as described by Foschi (1977).

20.7 ANCHORAGE

The capacity of a shear wall is greatly influenced by the type of anchorage that is provided to transfer the lateral shear forces and prevent overturning of the wall. This can be achieved through different means, such as a hold down device (Figure 20.26), as well as the horizontal nailing between the sheathing and the sill plate, together with the anchor bolts. The major force components are the shear or racking forces in the wall and the overturning or flexural forces. In a well designed wall, a clear load path is provided to assure lateral stiffness and strength of the wall.

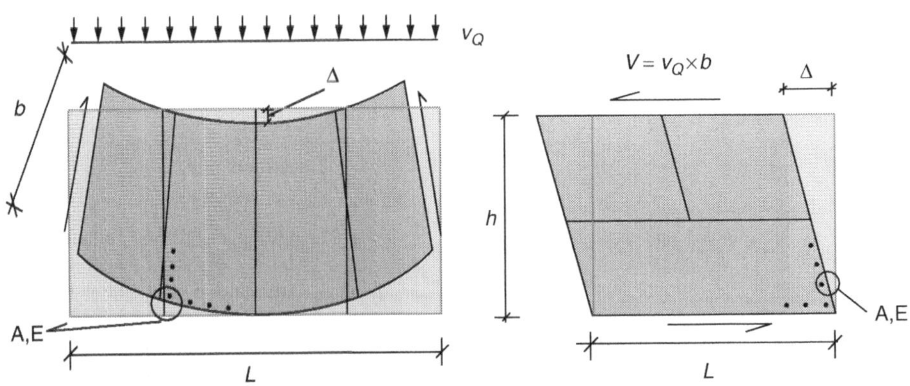

Figure 20.23 Diaphragm and shear wall displacements

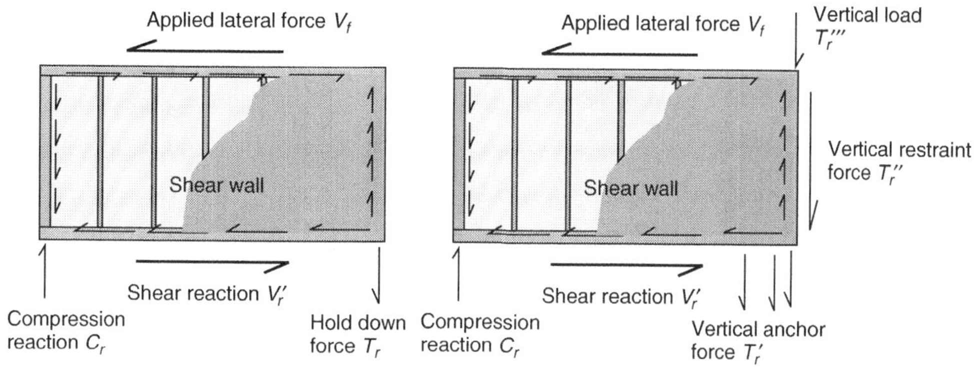

Figure 20.24 Force equilibrium of a shear wall with full and partial overturning restraint (solid arrows represent external forces on framing, dashed arrows represent forces exerted by sheathing on framing)

The diagrams in Figure 20.24 show a simplified set of forces acting on the framing of a shear wall. The sheathing essentially fulfils the purpose of preventing the framing to deform into a parallelogram, in other words, it provides the shear stiffness and strength of the wall. To maintain equilibrium, hold down forces and a shear reaction are needed. The hold down forces can be provided through vertical loads from upper storeys or the roof, or they can come from a hold down device that is attached to the vertical end stud (Figure 20.26). Otherwise, in a less efficient manner, the shear anchor bolts can also provide some vertical reaction force.

When designing shear walls, it is important to understand the difference between hold downs and anchor bolts. *Anchor bolts* provide horizontal shear continuity between the sill plate and the foundation (Figure 20.25). They are not designed to transmit vertical forces to the foundation, although some capacity can be achieved, if necessary. In this case, the bottom row of nails transmits the vertical forces in the sheathing to the sill plate (instead of the vertical end stud) where the anchor bolts will further transmit the forces into the foundation. Because of the eccentric load transfer, transverse bending is created in the sill plate and splitting often occurs. To prevent such a brittle failure mode, large washers (preferably square or rectangular) need to be provided to affect the eccentric load transfer from the sheathing through the nails, into the sill plate to the anchor and foundation.

Hold downs directly connect the vertical end stud (typically a double member) to the foundation. Some typical hold down details are shown in Figure 20.26. Because of the large concentrated forces, these are substantially larger than anchor bolts. As failure of the hold-downs often occurs in a brittle manner, capacity design principles need to be employed to ensure that the wall fails in shear along its nail connectors before any of the hold down connections fail.

Equilibrium of the shear wall free body diagram is employed to calculate the forces in the hold-downs (T_r) and anchor bolts (V_r) (Figure 20.24). When no hold downs are provided, the vertical restraint is provided by the anchor bolts (T_r'), by a perpendicular wall that is attached to the end stud (T_r''), or by vertical loads on the wall (T_r''') from an upper storey or roof. To simplify the derivation, the vertical load from an upper storey or roof is assumed to act on the end stud, and a reasonable contributing width should be assumed by the designer. In the case of anchor bolts providing uplift resistance, the vertical anchor forces are carried by a number of nails along the sill plate, which are thus deemed ineffective for the transfer of horizontal shear forces. Such a shear wall will thus have a reduced shear capacity V_r'. Ni and Karacabeyli (2000) have done extensive testing on shear walls with and without hold downs, and have proposed a reduction formula that is based on a relatively simple mechanics model.

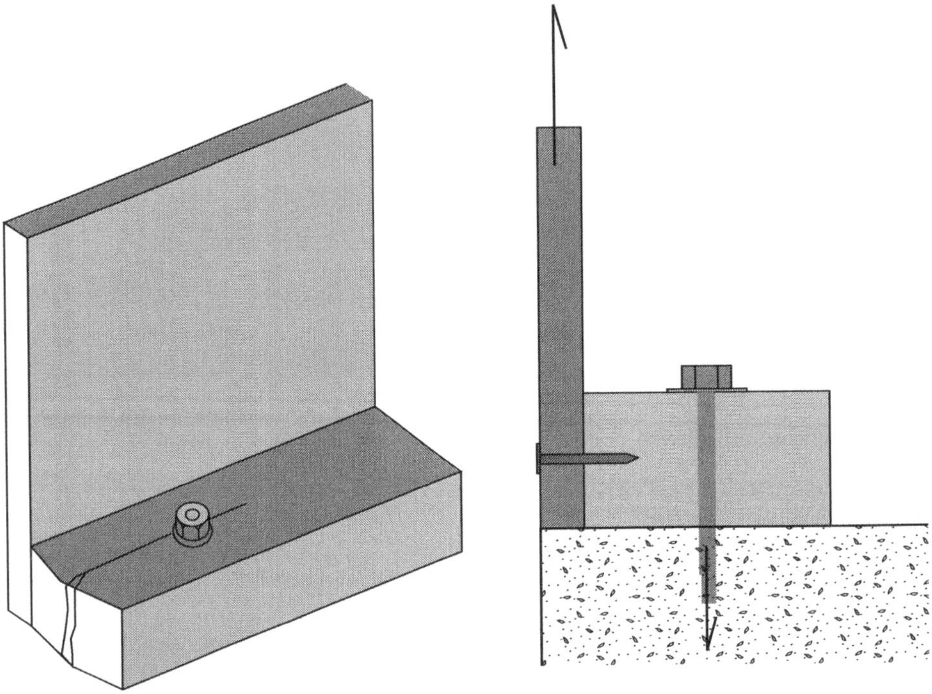

Figure 20.25 Sill plate failure at anchor bolt locations

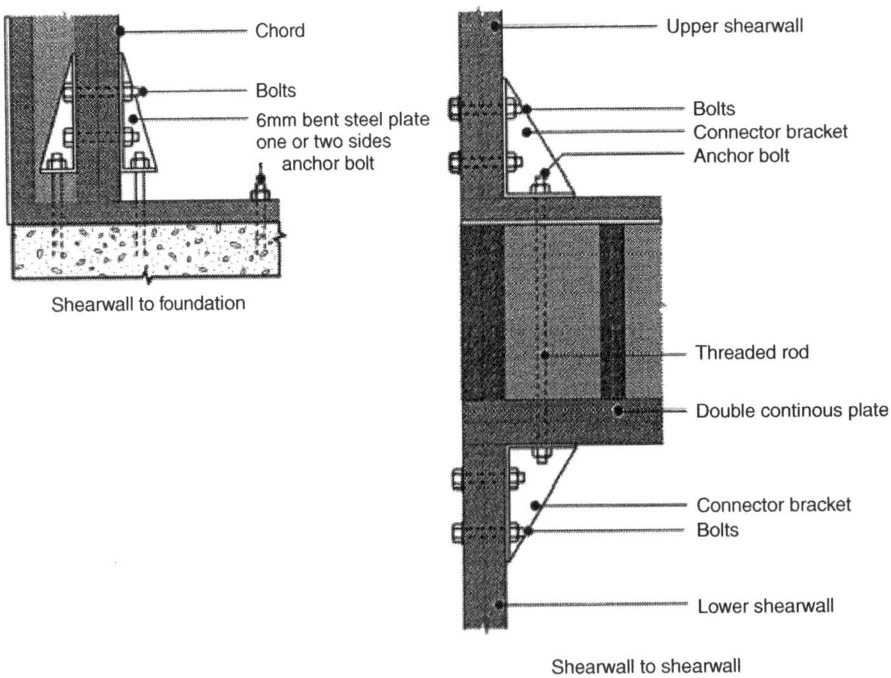

Figure 20.26 Shear wall hold down details (CWC, 1996)

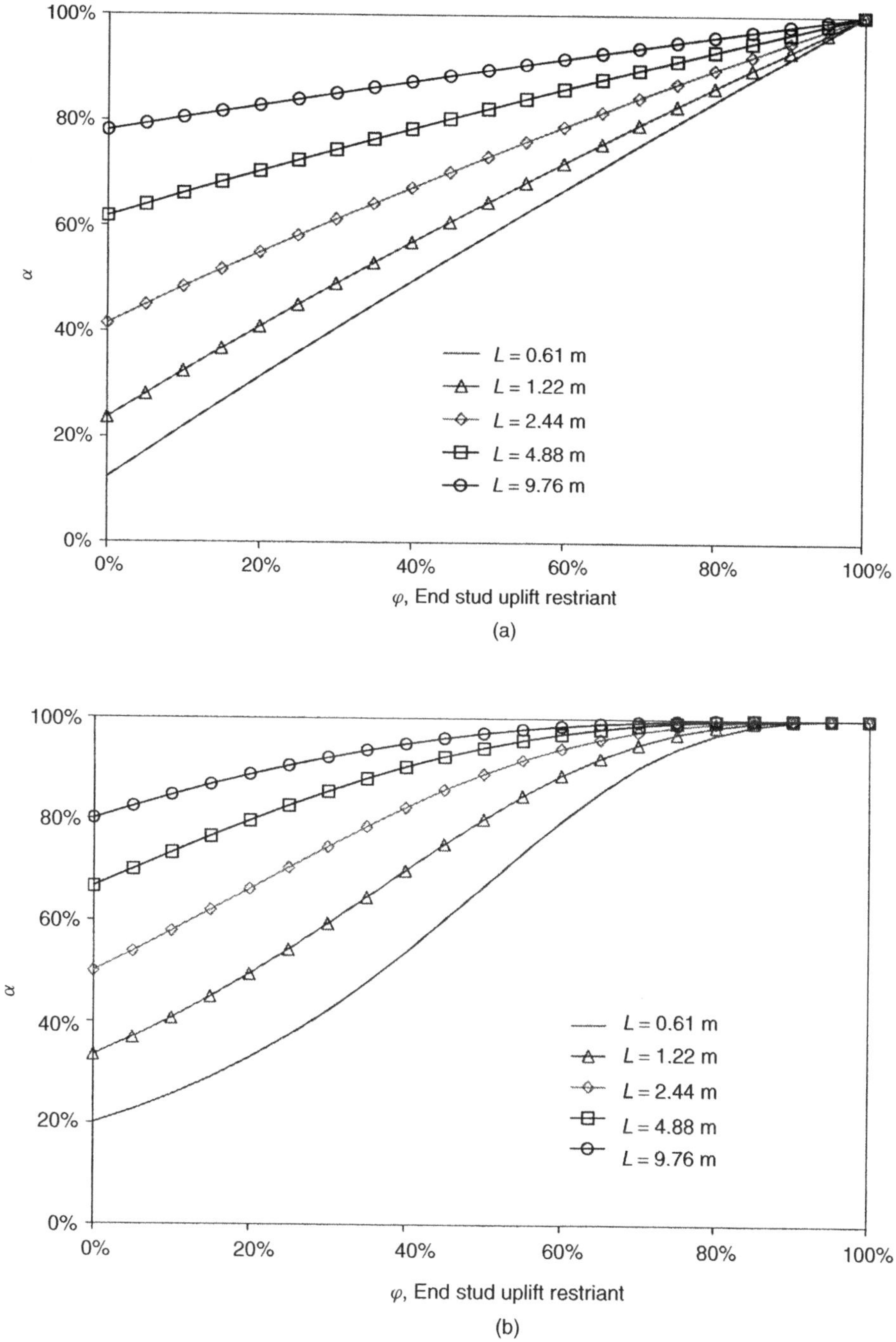

Figure 20.27 Effective shear ratio α of a 2.44 m high shear wall versus the restraint ratio φ for (a) simplified mechanics based approach, and (b) adjusted empirical formulation, calibrated to test results (Ni and Karacabeyli, 2000)

Figure 20.27(a) shows the effect of partial restraint on a 2.44 m high wall for different wall lengths, L, where $\alpha = V_r'/V_r = (1 + 2\varphi\gamma + \gamma^2)^{1/2} - \gamma$, is the ratio of the shear capacities of the partially restrained versus the fully restrained wall, where the wall ratio $\gamma = H/L$ (height/length) and the restraint ratio φ is defined as the ratio of the restraint force on the end stud (i.e. from a hold down, a perpendicular wall or vertical top load) versus the full nail capacity along the height of the end stud. In other words, if φ is 0, all the vertical restraint has to come from the nails in the sill plate and when φ is 1.0, full restraint is provided to the end stud and no anchor force is transmitted through the sill plate nails.

Alternatively, the following empirical formula, which was found to best represent the effective shear ratio of test results, can be used:

$$\alpha = 1/[1 + H/L(1 - \varphi)^3]$$

This relationship is shown in Figure 20.27(b).

When no hold down restraint is provided ($\varphi = 0$), the empirical formula becomes $\alpha = 1/(1 + H/L)$, which represents the points along the left axis of the graph in Figure 20.27(b). Since hold-downs are not always provided in smaller buildings, this method provides a reasonably simple way to find an equivalent shear wall capacity. For multi-storey buildings, it is advised to always provide proper anchorage and keep the load path clear and simple.

20.8 RECENT DEVELOPMENTS

With the expansion of wood frame construction into large-scale applications such as multi-storey residential or commercial buildings, the designer is often faced by a relatively small proportion of shear walls that can be relied upon to carry very high lateral forces from earthquake and wind actions. When the available shear walls are not adequate to carry the lateral loads, auxiliary systems often need to be employed, such as steel frames or reinforced concrete shear walls. While such hybrid systems seem to satisfy the high shear demands, certain disadvantages exist, for example differential shrinkage properties of the different materials, the added cost of engaging different trades (and possibly unions) on a job site, and differences in thermal properties, which can lead to envelope leakage or condensation problems. It is thus of advantage if wood can be used throughout the building.

For this reason several developments have occurred in the recent past, aimed at increasing the lateral resistance of shear wall systems.

Karalic (1997) has developed a modular brace system that can be introduced into a regular wood frame shear wall without disrupting the construction sequence (Figure 20.28). A substantial increase in capacity is achieved; however, the load transfer mechanism also changes by demanding higher capacities from the frame-to-frame nailing, which otherwise would only carry nominal loads.

By utilising the added shear capacity of nails when loaded in multiple planes, Karacabeyli *et al.* (2001) developed the so-called Midply® wall system (Figure 20.29). Very high shear capacities can be achieved as the centre panel nail connections are loaded in double shear. The basic load transfer mechanism of a shear wall is the same with very ductile behaviour. Because of the high shear capacities, however, the failure mechanism of such walls often changes and the hold-down devices become the critical elements. Since such failures are brittle and thus less desirable, special hold-down devices need to be developed that can resist the high overturning forces. The added sheathing in the middle of the wall cavity may also pose some problems with the introduction of insulation and the installation of plumbing and electrical services.

Other systems have been developed to enhance the shear capacity of shear walls. Reinforcement of the sheathing edges to increase the nail holding strength has proven to be very effective (Pryor, 2000). This can be achieved through metal strip reinforcement along the edges of the sheathing that will prevent pull-through of the nails and increase the tear-out resistance. Again, the added capacity places increased demand on the hold-down devices, which need to be designed to these increased overturning forces.

Figure 20.28 Modular brace system in shear wall construction (Karalic, 1997)

Lam *et al.* (1997) have found that oversized sheathing panels can result in substantial shear capacity increases. Oriented strandboard mills nowadays produce panels in sizes of up to 3 m wide and 8 m long, which are then cut into standard 1.2 × 2.4 m panels. When the full sized panels are used, the seams between panels are largely eliminated and significant increases in stiffness and strength can be achieved (Figure 20.30). In walls with openings, such continuity increases the frame action, thus resulting in a much more efficient load resisting system (He *et al.*, 1999).

Durham *et al.* (2001) have shown that large panels (2.4 × 2.4 m) provide a significant improvement for walls subjected to earthquake loading. The inconvenience of handling large panels suggest that the most efficient way to take advantage of such a system is in prefabricated construction where entire wall panels are made in a factory and shipped to the site where a crane is available to help with the erection.

Shrinkage of the wood after construction poses a different problem because hold-down devices become loose, which negates their effectivity when resisting overturning forces. Especially during an earthquake, such slack can cause so-called

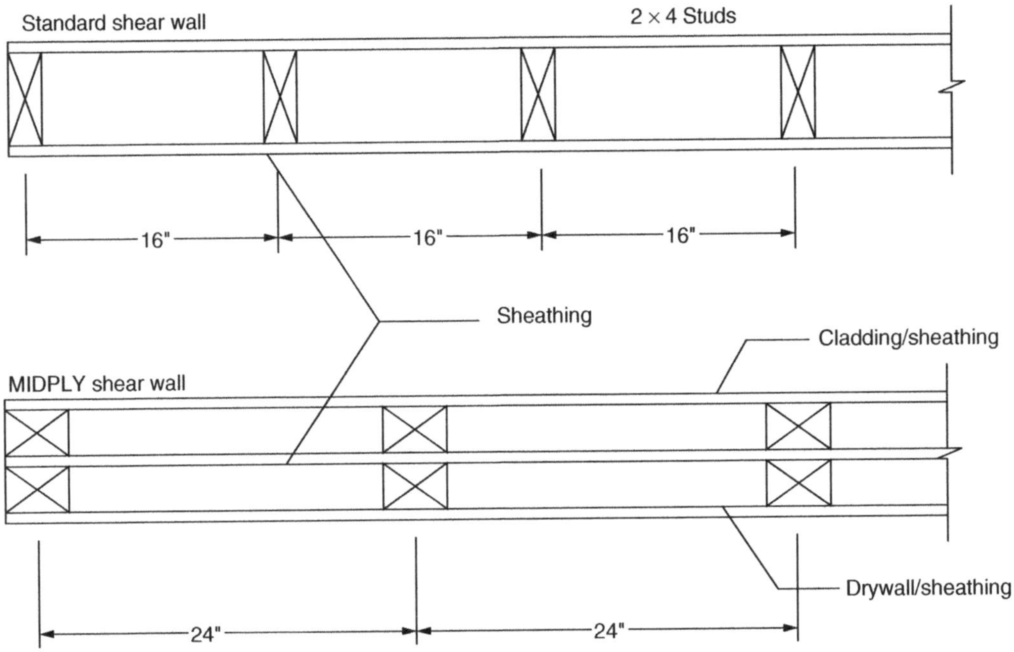

Figure 20.29 Midply® wall system (Karacabeyli *et al.*, 2001)

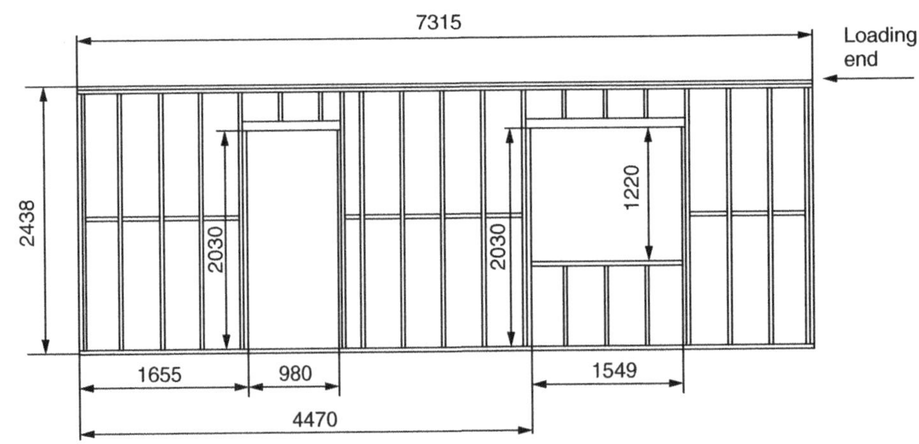

Figure 20.30 Racking resistance of shear walls with large panels (He *et al.*, 1999, Lam *et al.*, 1997)

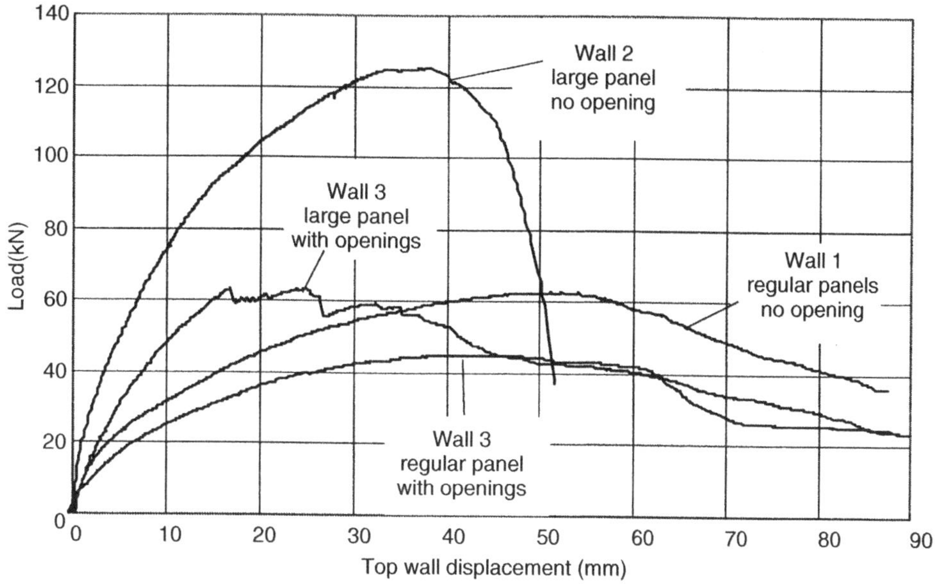

Figure 20.30 (*continued*)

whipping effects, with resulting amplification of ground motions in the building. Since hold-down devices are hidden inside a wall, tightening at a later stage is not a viable option. Some companies have developed self-tightening nuts that are spring-loaded and will maintain a certain hold-down force. It is not the intention to list all the new devices that are available on the market and many may not have been mentioned, as the list invariably is quite long. Designers are encouraged, however, to seek information on the latest developments through the Internet or from trade shows. What is important here is to gain an understanding of the underlying issues and potential problems and have a good idea of what is required to alleviate failures in buildings, be they of a catastrophic nature or a mere inconvenience to occupants.

20.9 SUMMARY

Light wood frame construction typically provides a highly redundant and very forgiving structural system to buildings ranging from small single-family homes to large multi-storey residential or commercial buildings. Because of the box-type panel construction with many contributing elements, such buildings have proven to be very resilient to extreme load conditions like earthquakes, snow and windstorms. Some spectacular failures have reminded us, however, that the integrity of woodframe construction cannot be taken for granted. When large openings weaken the shear wall system or when large eccentricities in building plans introduce torsional effects, large demands may result in building collapses. Most of the failures during earthquakes occurred because of weak ground floor systems (resulting from unbraced parking facilities), inadequate stub-walls (pony walls, cripple walls) or poor anchorage. These could have been avoided if basic engineering principles were applied to identify building weaknesses. Valuable lessons were learned from the widespread destruction of wood frame housing during hurricane Andrew in Florida. Inadequate connections between walls and roofs were blamed for the loss of integrity of the housing units, pinpointing the use of staples, instead of nails, as the culprit. The basic deficiency was rather the lack of overall integrity of the building system and ignorance of architects, engineers, builders and inspectors as to the logical load path in a wood frame system, which could have avoided the weakness of critical elements.

It is hoped that engineers, architects and builders will appreciate the simplicity and forgiving (i.e. redundant) nature of wood frame construction, while at the same time understand the primary and secondary load carrying mechanisms, its advantages and possible weaknesses. In an ever-changing global societal environment, woodframe construction can play an important role in meeting demands for affordable housing. In many parts of the world where highly skilled carpenters are a rare commodity, woodframe construction provides a building system that can provide inexpensive housing using local labour and materials, yet being resilient under catastrophic conditions such as earthquakes, hurricanes and typhoons, and extreme snow loads.

REFERENCES

Ceccotti A. and Karacabeyli E. (2000) Dynamic analysis of nailed wood-frame shear walls. *Proc. 12th World Conference on Earthquake Engineering*, Auckland, NZ.

COFI (1979) Douglas fir plywood diaphragms – description, design and detailing. Council of Forest Industries of British Columbia, Vancouver, BC.

CSA (1994) CAN/CSA-O86.1-94, *Engineering Design in Wood (Limit States Design)*, Canadian Standards Association, Ottawa.

CWC (1995) *Wood Design Manual.* Canadian Wood Council, Ottawa.

CWC (1996) *Introduction to Wood Design.* Canadian Wood Council, Ottawa.

Diekman E.F. (1997) Design and code issues in the design of diaphragms and shear walls. In: *Earthquake Performance and Safety of Timber Structures*, G. Foliente, (editor), Forest Products Society, Madison, WI.

Durham J. Lam F. and Prion H.G.L. (2001) Seismic resistance of wood shear walls with large OSB panels. *J. Struct. Eng*, December.

Easley J.T. Foomani M. and Dodds R.H. (1982) Formulas for wood shear walls. *ASCE J. Structural Division*, **108**(ST11), 2460–2478.

Foschi R. (1977) Analysis of wood diaphragms and trusses. Part 1: Diaphragms. *Canad. J. Civil Eng.*, **4**(3), 345–352.

He M. Magnusson H. Lam F. and Prion H.G.L. (1999) Cyclic performance of perforated wood shear walls with oversize OSB panels. *J. Struct. Engrg.*, **125**(1), 10–18.

Jephcott D.K. and Dewdney H.S. (1980) Analysis methods for horizontal wood diaphragms. *Proc. Workshop on Design of Horizontal Wood Diaphragms*, ATC, Berkeley, CA.

Kaellsner B. and Lam F. (1995) Diaphragms and shear walls. *STEP Lectures: Holzbauwerke nach Eurocode 5–Grundlagen, Entwicklungen, Ergaenzungen*, Fachverlag Holz, Duesseldorf, Germany, pp. 15/1–15/19.

Karacabeyli E. and Ceccotti A. (1996) Test results on the lateral resistance of nailed shear walls. *Proc. Int. Wood Engineering Conference*, New Orleans, LA. Vol. 2, pp. 179–186.

Karacabeyli E. (1997) Lateral resistance of nailed shear walls subjected to static and cyclic displacements. In: *Earthquake Performance and Safety of Timber Structures*, G. Foliente, (editor), Forest Products Society, Madison, WI.

Karacabeyli E. Stiemer S.F. and Ni C. (2001) MIDPLY Shear wall System. *Proc. 2001 ASCE Structures Congress.*

Karalic M. (1997) Analysis and performance of floors and shear walls with new engineered bracing systems. In: *Earthquake Performance and Safety of Timber Structures*, G. Foliente, editor, Forest Products Society, Madison, WI.

Lam F. Prion H.G.L. and He M. (1997) Lateral resistance of wood based shear walls with large sheathing panels. *J. Struct. Eng.*, **123**(12), 1666–1673.

Ni C. and Karacabeyli E. (2000) Effect of overturning restraint on performance of shear walls. *Proc. World Conference on Timber Engineering (WCTE2000)*, Whistler, Canada.

Ni C. Karacabeyli E. and Ceccotti A. (2000) Lateral load capacities of horizontally sheathed unblocked shear walls. *Proc. Meeting 33, Int. Council for Research and Innovation in Building and Construction, Working Commission W18 – Timber Structures (CIB-W18)*, Delft, Netherlands, paper 33-15-1.

Prion H. and Lam F. (2000) Finite Element Analysis of diaphragms with openings and offsets. Report to Canadian Wood Council, Department of Wood Science, University of British Columbia, Vancouver, BC.

Pryor S.E. Taylor G.W. and Ventura C.E. (2000) Seismic testing and analysis program on high aspect ratio wood shear walls. *Proc. World Conf. on Timber Engineering 2000*, Whistler, BC, Canada, paper 1.1.4.

Sugiyama H. and Matsumoto T. (1994) Empirical equations for the estimation of racking strength of a plywood sheathed shear wall with openings. *Annual meeting, Architectural Institute of Japan.*

Tissell J.R. (1990) *Structural panel shear walls.* APA Research Report 154, American Plywood Association, Tacoma, WA.

Yasumura M. and Sugiyama H. (1984) *Transactions of the Architectural Institute of Japan*, 338.

21

Composite Structures

Ario Ceccotti

21.1 INTRODUCTION

For centuries, in order to overcome the limits imposed by the natural dimension of tree-trunk cross-sections, cross-section dimensions were 'enlarged' just by super-imposing elements onto each other (Figure 21.1). That way increased strength and stiffness of beams and columns were obtained, optimising the wood consumption at the same time.

Mechanical joints are deformable, therefore the stiffness (and strength) achieved was never 100% as anticipated, due to the relative slip between connected parts. Only with the advent of structural-use-proof glues has been possible to achieve this goal, e.g. with glulam techniques.

It is a commonly held view that nowadays glued elements have overshadowed the mechanically jointing technique. In fact, the use of mechanical jointing is still very important, both culturally and economically. The composite technique is very much used in case of:

- renovation works, when due to the cultural value of the structure to be strengthened, it is important to preserve original timber elements *in situ*;
- prefabricated elements, where lightness and easy-to-install features are important as much as strength and stiffness;

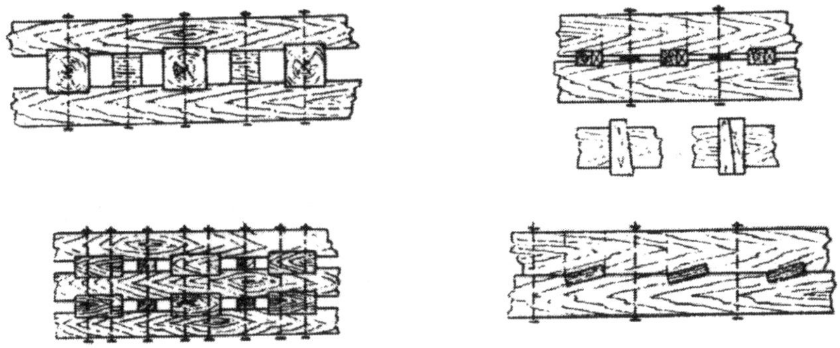

Figure 21.1 Some examples of composite members as used in old timber constructions to increase the strength and stiffness of structural elements, i.e. beams and columns, utilising mechanical fasteners

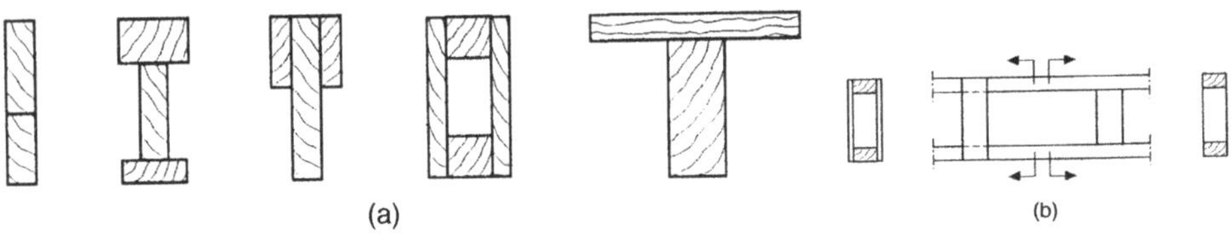

Figure 21.2 (a) Examples of composite sections: beams and columns fully length connected (schematic); (b) schematic example of two flange spaced cross-sections with discontinuous connection: battens (left), packs (right)

- in the case of *in situ* assembled elements when gluing-in-place would be difficult to control, e.g. with big element sections of large-span structures, or even in small construction when the workmanship is not particularly specialised with gluing techniques.

Figure 21.3 Composite action obtained by punched metal plates (Reproduced by permission of NAFI, Australia)

A wide variety of cross-sections can be built (see Figures 21.3–21.9). Parts can be made of solid wood, glued laminated timber, or wood-based materials (e.g. OSB, Plywood, LVL). Members can be fully length connected (Figure 21.2(a)), or just spaced apart with discontinuous connections (Figure 21.2(b)). It is also possible to use concrete plates as flanges or steel elements as web.

The timber-concrete composite technique is usually used as shown in Figure 21.9 (left) where concrete works in compression and timber in tension. A reverse situation may be used in prefabricated houses when local users prefer to walk on a timber floor.

Timber-concrete composite floors are also very effective for renovating building floors of masonry buildings. In-plane rigidity becomes so pronounced that the floor is rigid enough to be able to keep its shape, and consequently the shape of the entire building, if well connected with building walls. This is very important for the survival of the structure under an earthquake (Figure 21.10).

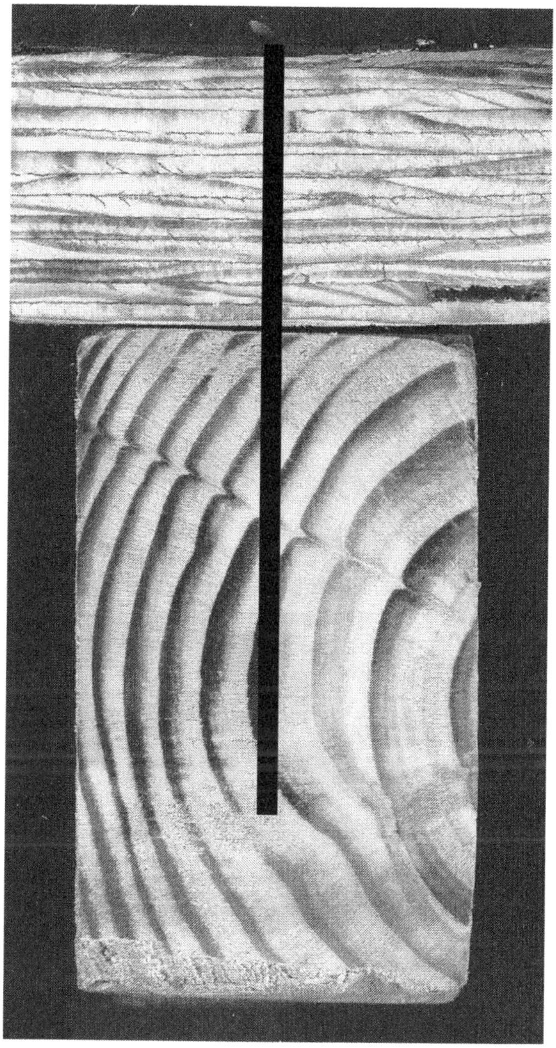

Figure 21.4 LVL flange and solid wood

21.2 MECHANICAL BEHAVIOUR (BASICS)

The connection of members is made by mechanical fasteners such as nails, bolts, dowels, nail plates, etc. In bending, each individual component tends to slip one on the other, and a displacement v (*slip*) between two adjacent parts is achieved. Slip is maximum at the ends and null in the middle of the beam (Figure 21.11). Due to this slip, the classical bending theory of beams is no longer valid for mechanically jointed elements. In fact, the cross-section does not keep its planarity under loading. However, the theory is applicable to individual components. Such a behaviour of mechanical joints is usually called 'semi-rigid' (glued joints are regarded as 'rigid' joints).

To better understand the mechanical behaviour, let us consider a two-pieces member in bending. This is convenient for the sake of simplicity, and does not harm the correct understanding of the phenomenon.

Vertical Behaviour

Deflection w of each member piece is the same at every position along the beam: vertical relative displacements are negligible. Then, referring to the upper component as 1 and lower component as 2 (see Figure 21.11, right, and Figure 21.12), it is possible to write:

$$w_1(x) = w_2(x) \tag{21.1}$$

$$w_1''(x) = w_2''(x) \tag{21.2}$$

$$M_1(x) = -E_1 I_1 w_1''(x) \tag{21.3}$$

$$M_2(x) = -E_2 I_2 w_2''(x) \tag{21.4}$$

and then:

$$M_1(x)/M_2(x) = E_1 I_1 / E_2 I_2 \tag{21.5}$$

Horizontal Behaviour

When two elements are loaded in bending, they tend to slip relative to each other and a system of tangential forces Q develops because the fasteners tend to counteract the slip (Figure 21.12, right).

With no connection at all, the beam behaves like two independent individual components (Figure 21.13(a)), i.e. no-composite action; with perfectly rigid connection the beam behaves like one beam and cross-sections remain plane (Figure 21.13(b)), i.e. fully-composite action. In the case of mechanical fasteners, the relevant tangential action Q ensures the equilibrium of each individual piece along the horizontal direction. This creates, section by section, along the beam length, axial forces N acting in the opposite direction.

Figure 21.5 Impressive results can be obtained with plywood flange nailed to solid wood web (sport arena in Italy, courtesy of A. Dini. Bureau)

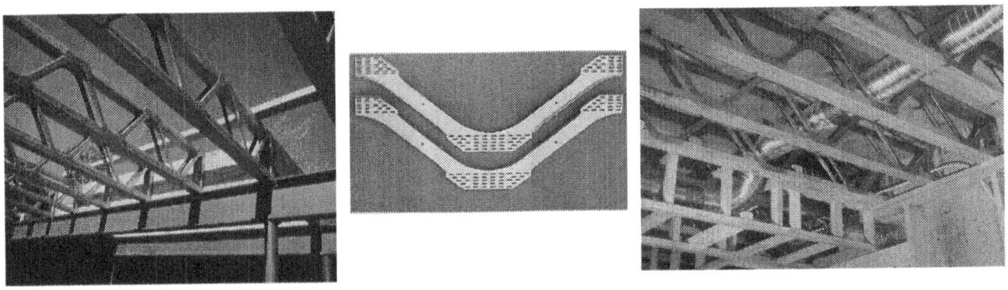

Figure 21.6 V shaped steel elements used as web connecting top and bottom wooden chords allow easy utilities installation (Reproduced by permission of NZTIF, New Zealand)

N takes a maximum value in the middle of the beam. In other words, when loaded in pure bending, the upper part of a composite beam is under compression, whereas the lower part is under tension (Figure 21.12, left). These two forces act as a couple of normal forces whose intensity depends upon the deformability and stiffness of fasteners: zero in the case of no connection at all (Figure 21.13(a)), maximum in the case of very rigid connections (Figure 21.13(b)). The total bending moment acting on the beam's middle cross-section equals the sum of the two bending moments acting on each component, plus the moment provided by the two axial forces. Referring to Figure 21.12, then:

$$N_1 = N_2 = N \quad M = M_1 + M_2 + Nr \tag{21.6}$$

In Figure 21.14 possible resulting strain and stress distributions in the case of semi-rigid connection are shown.

In conclusion, when increasing the connection rigidity, the bending moments on each individual component of the beam decrease, whereas the axial force increases.

The load-slip behaviour of an individual connections is usually expressed as

$$Q = K\nu \tag{21.7}$$

where K is called the 'slip modulus': higher values correspond to rigid connection.

In principle, K is determined by tests. A possible test arrangement for determining the load-slip behaviour of a connection system is shown

Figure 21.7 Undulated steel sheet can be used as web in beams and columns (Reproduced by permission of NZTIF)

Figure 21.8 Parallam composite columns with discontinuous steel packs connections at UBC, Faculty of Forestry, Vancouver, Canada

in Figure 21.15. Load-slip behaviour is usually non-linear (Figure 21.16).

The coefficient:

$$k = \frac{K}{s} \tag{21.8}$$

where s is the fastener spacing, is usually called the 'per-*unit*-length *slip modulus*'.

21.3 COMPUTATION METHODS

21.3.1 Analysis

Equilibrium and deformation solutions have been developed by differential equations for a long time (e.g. Möhler, 1956). The basic assumptions are: for every part the simple bending theory is valid and shear deformation is disregarded. The connection is regarded as continuous, and the cross-sections and joint stiffness are constant in the direction of the beam axis, i.e. uniformly distributed fasteners. For the sake of simplicity, a linear load-slip behaviour of fastener connection is assumed. In Figure 21.17 the basics for the equilibrium and deformation conditions are presented with reference to the case of a two member section, also known as a 'T' section. *Mutatis mutandis* the

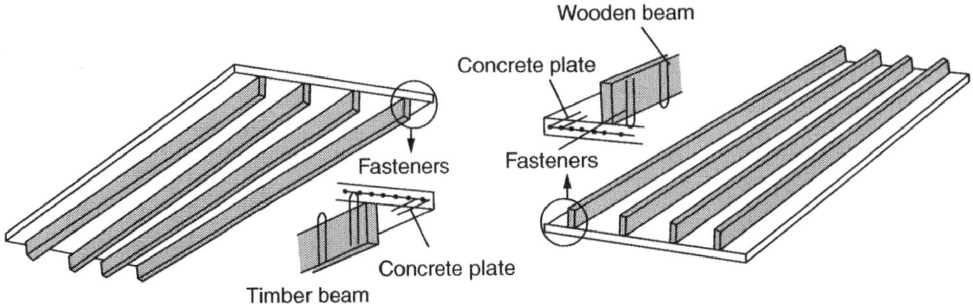

Figure 21.9 Possible timber-concrete solutions for floors

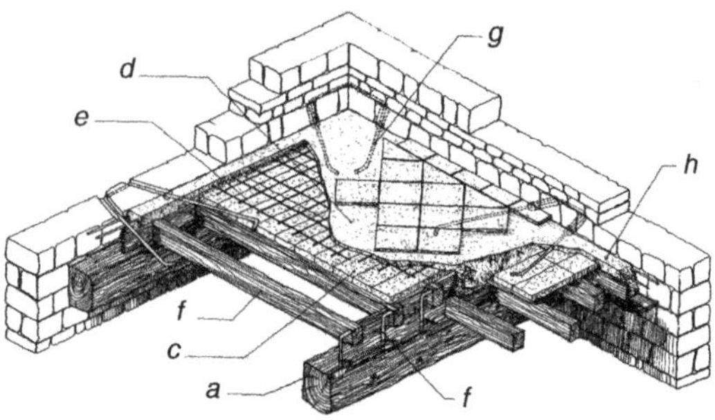

Figure 21.10 Example of earthquake-resistant design with an existing timber floor in a middle-European masonry building. (a) Main beam; (b) secondary beam; (c) brick tiles; (d) concrete slab; (e) steel mesh; (f) steel fasteners epoxy glued in to timber; (g) steel stirrups connecting concrete layer with masonry; (h) all-around reinforced concrete girder

approach for a three member section (e.g. 'I' or 'box' sections) is basically the same.

From the deformation and equilibrium equations, it is possible to find the solution of the problem in terms of differential equations (omitted here; a good reference can be found in Larsen and Riberholt (1983) and Kreutzinger (1994). For instance, for $w(x)$, the deflection along the beam:

$$w'''' - \left(\frac{\delta}{\beta}\right)^2 w'' - \frac{1}{(EI)_0}(q(x) + M(x)\delta^2) = 0$$

$$(EI)_0 = E_1 I_1 + E_2 I_2 \qquad (21.9)$$

where

$$\beta^2 = \frac{(EI)_0}{(EI)_\infty} \qquad (21.11)$$

$$\delta^2 = \frac{k}{(EA)_r} \qquad (21.12)$$

indicating with $(EI)_\infty$ the stiffness in the case of a perfect rigidity (glued section):

$$(EI)_\infty = (EI)_0 + [(EI)_\infty - (EI)_0] = (EI)_0 + (E_1 A_1) {r_1}^2 + (E_2 A_2) {r_2}^2 \qquad (21.13)$$

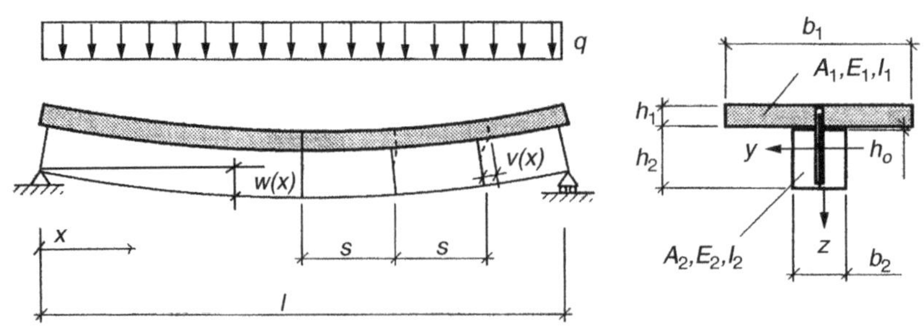

Figure 21.11 Basic behaviour of a composite section with a semi-rigid connection. A slip ν occurs and cross-sections do not keep their planarity

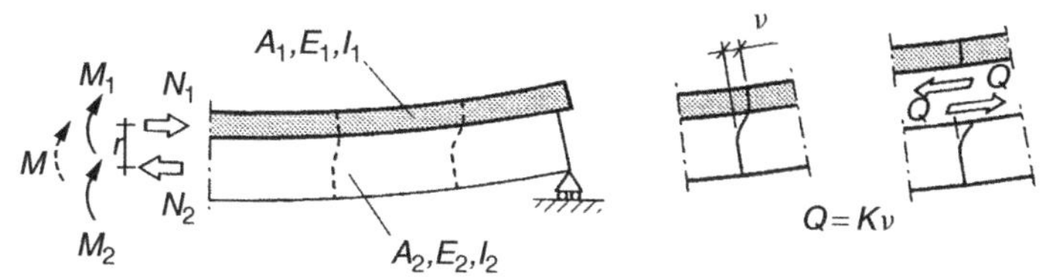

Figure 21.12 Upper layer is under compression and bending, and lower beam is under tension and bending ($N_1 = N_2$); fasteners are under shear

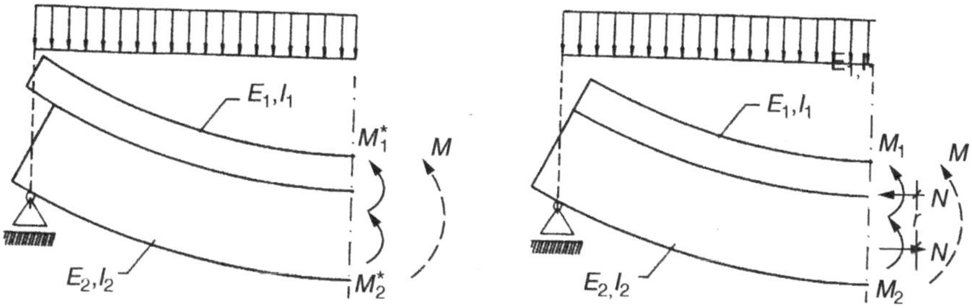

Figure 21.13 (a) no-connection case, (b) totally rigid connection case

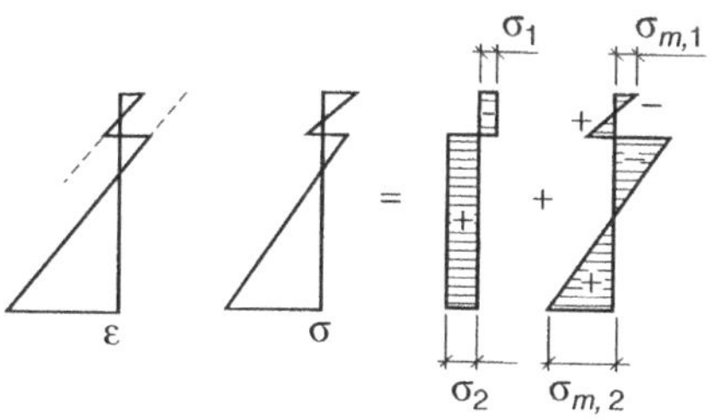

Figure 21.14 The strain distribution has the same slope because the parts have the same curvature; the stress diagram is the result of compression-bending and tension-bending stresses

and

$$(EA)_r = \frac{E_1 A_1 \cdot E_2 A_2}{E_1 A_1 + E_2 A_2} \tag{21.14}$$

For simple load cases, i.e. a simple supported beam with a uniform distributed load q along its length or a concentrated load Q at mid-span, it is possible to find the solution of the differential equation in a closed form.

For a uniform load q:

$$w(x) = w_\infty(x) + \frac{ql^2}{8(EI)_\infty}\frac{1-\beta^2}{\delta^2} \times \left[\frac{8\beta^2}{\delta^2 l^2}\left(\frac{\cos h\dfrac{x\delta}{\beta}}{\cos h\dfrac{l\delta}{2\beta}} - 1\right) + 1 - \frac{4x^2}{l^2}\right] \tag{21.15}$$

where $w_\infty(x)$ is the deflection of the corresponding 'glued' beam. The deflection at mid-span is:

$$w_0 = \frac{5}{384}\frac{ql^4}{(EI)_\infty}\frac{1}{\eta_q} \tag{21.16}$$

$$1/\eta_q = \left[1 + \frac{48(1-\beta^2)}{5\delta^2 l^2}\left(\frac{8\beta^2}{\delta^2 l^2} \times \left(\frac{1}{\cosh\dfrac{l\delta}{2\beta}} - 1\right) + 1\right)\right] \tag{21.17}$$

where $1/\eta_q$ is a quantity bigger than unity.

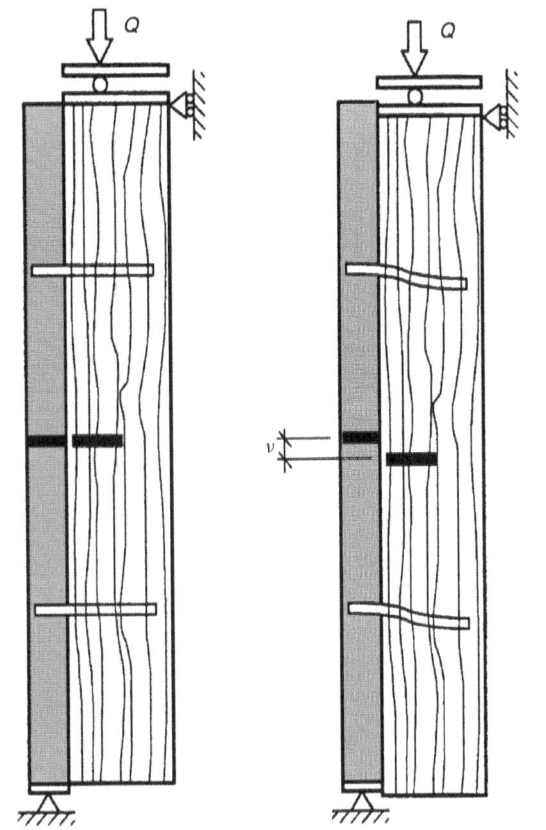

Figure 21.15 A possible test arrangement for determining load-slip behaviour of a connection system

For a concentrated load Q at mid-span:

$$w(x) = w_\infty(x) + \frac{Ql}{4(EI)_\infty}\frac{1-\beta^2}{\delta^2}\left[\frac{2\beta}{l\delta}\left(\sin h\frac{x\delta}{\beta} - \text{tgh}\frac{l\delta}{2\beta}\cos h\frac{x\delta}{\beta}\right) - \frac{2x}{l} + 1\right] \quad (21.18)$$

and the deflection at mid-span is

$$w_0 = \frac{1}{48}\frac{Ql^3}{(EI)_\infty}\frac{1}{\eta_Q} \quad (21.19)$$

where $1/\eta_Q$ is a quantity bigger than unity:

$$1/\eta_Q = \left[1 + \frac{12(1-\beta^2)}{l^2\delta^2}\left(1 - \frac{\text{tgh}\dfrac{l\delta}{2\beta}}{\dfrac{l\delta}{2\beta}}\right)\right] \quad (21.20)$$

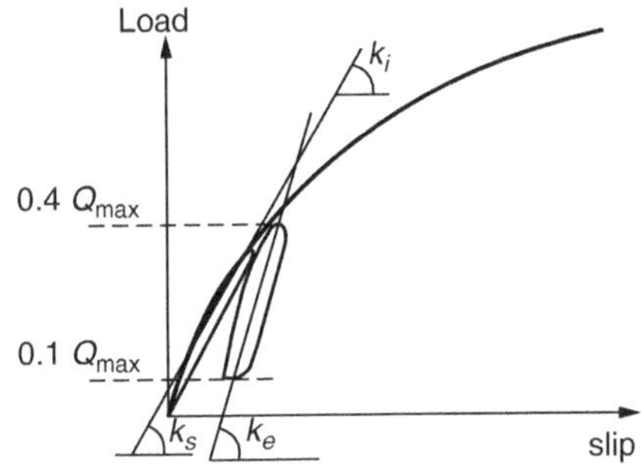

Figure 21.16 Load-slip behaviour and slip coefficients definition according to EN 26891: k_i, initial, k_s, short-term, k_e, elastic

It is evident that the ratio between the actual deflection $w(x)$ and the deflection of the 'glued' case $w_\infty(x)$ varies with x along the entire beam. This ratio equals the values of $1/\eta_q$ and $1/\eta_Q$ at mid-span for uniformly distributed load q and concentrated load Q at mid-span, respectively.

In the case with a sinusoid-like distributed load along a simply supported beam, the solution is particularly simple(Figure 21.18).

The deflection $w(x)$ along the beam is the following:

$$w(x) = \frac{M_0}{(EI)_\infty}\left(\frac{l}{\pi}\right)^2 \frac{1+\left(\dfrac{\pi}{l\delta}\right)^2}{1+\beta^2\left(\dfrac{\pi}{l\delta}\right)^2}\cos\frac{\pi x}{l} \quad (21.21)$$

Because the $w_\infty(x)$ deflection of the corresponding 'glued' beam is

$$w_\infty(x) = \frac{M_0}{(EI)_\infty}\left(\frac{l}{\pi}\right)^2 \cos\frac{\pi x}{l} \quad (21.22)$$

the ratio $1/\eta_{\sin}$ between that actual deflection and the deflection of the 'glued' case is *constant* along the entire beam:

$$1/\eta_{\sin} = \frac{1+\left(\dfrac{\pi}{l\delta}\right)^2}{1+\beta^2\left(\dfrac{\pi}{l\delta}\right)^2} \quad (21.23)$$

Therefore, this ratio can be used to define an equivalent or *effective* moment of inertia of

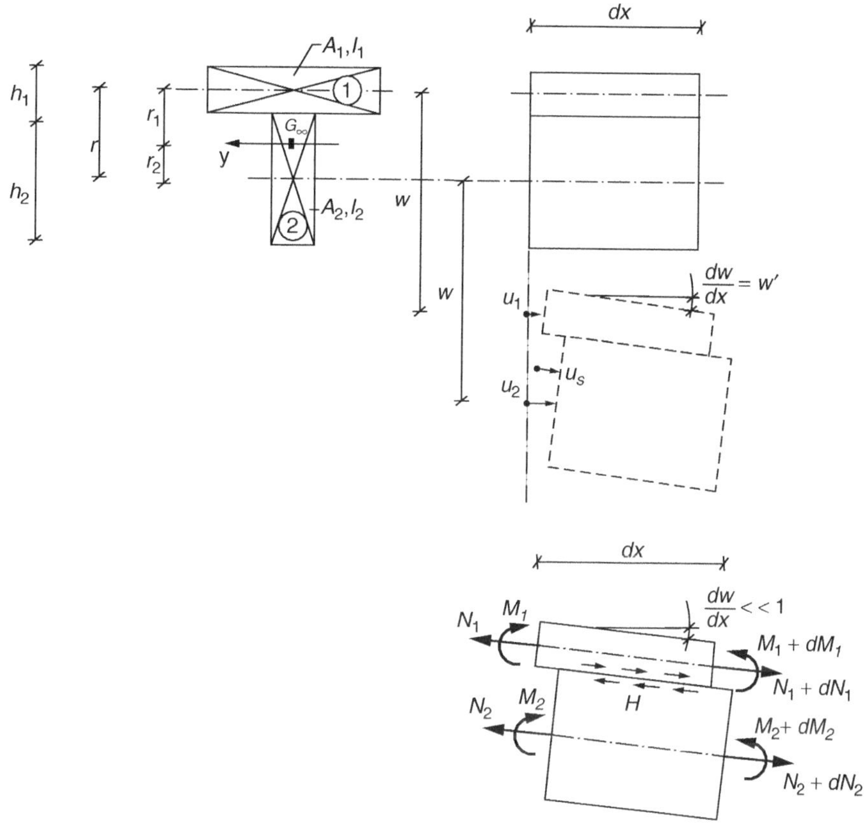

Figure 21.17 Basic picture for equilibrium/deformation equations for a two member composite section (NB: G_∞ is the cross-section centroid, as the parts were glued together). (Reproduced by permission of Danish Building and Urban Research)

the beam:

$$(EI)_{\text{eff}} = \eta_{\sin}(EI)_\infty \tag{21.24}$$

so that, from Equations 21.11 and 21.13:

$$(EI)_{\text{eff}} = (EI)_0 + \gamma[(EI)_\infty - (EI)_0] \tag{21.25}$$

with

$$\gamma = \frac{1}{1+\left(\frac{\pi}{l\delta}\right)^2} = \frac{1}{1+\frac{\pi^2(EA)_r}{kl^2}} \tag{21.26}$$

with $\gamma = 1$ for a perfect-rigid (glued) connection and $\gamma = 0$ for no connection at all. The coefficient γ is called the 'connection efficiency factor'.

The definition of an *effective (EI)* is 100% correct in the case of a simple supported beam with a sinusoidal load acting on it: the relevant calculated deformation equals the real one at each point of the beam.

One could also define an equivalent stiffness value for other load cases (i.e. uniform load or a concentrated load), i.e. equalising the deflection at mid-span and referring to $1/\eta_q$ and $1/\eta_Q$, respectively. In these cases, the deflection calculated at other points would not actually be an exact value. However, errors would be very small indeed.

Moreover, even for uniform distributed and concentrated loads, the effective value of $(EI)_{\text{eff}}$ obtained from the sinusoidal load case could be adopted: errors are bigger now than in the previous case, but they remain acceptable for engineering purposes.

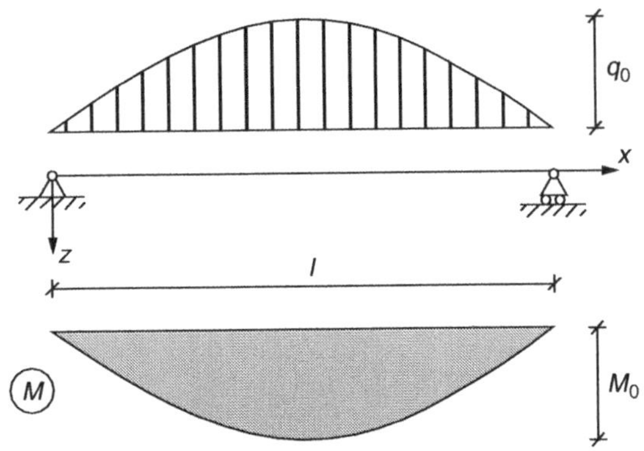

Figure 21.18 Example of a sinusoidal load distribution. (Reproduced by permission of Danish Building and Urban Research)

In fact the actual error in evaluating a mid-span deflection is very limited, as can be seen from Table 21.1, where the ratio of the deflection of beams with semi-rigid connections to the deflection of corresponding *glued* beams is given. The table shows that this ratio, for two different beams – 2.5 m and 10 m span, respectively, T-type timber/concrete composite cross-section – is almost unaffected by the load distribution. The worst case, one concentrated load in the middle (i.e. a load distribution very different from a sinusoid) gives a discrepancy of less than 3%. Therefore, it can be concluded that the assumption of an *effective* stiffness value is a reasonable one. The coefficient η is then called the 'stiffness reduction factor' of the beam.

Actually, the effective *(EI)* value depends upon the length of the beams (as does γ). In other words, $(EI)_{\text{eff}}$ is not a characteristic of the cross-section *per se*, but it depends also upon the extension of the beam, i.e. the distance between two zero-moment points. This has some effect on multi-supported beams, i.e. hyperstatic beams, because zero moment points are not known in advance. Scientifically correct solutions do exist for this problem, but just a simple, even rough, trial and error approach can be used, as outlined below.

Table 21.1 Ratio between mid-span deflection for a timber-concrete T beam with deformable connections (values calculated exactly) and the deflection of the same beam but with perfect rigid connections

	Short span T beam	Long span T beam
Concentrated load at mid-span	1.9313	1.3492
Concentrated loads at middle third	1.9060	1.3266
Uniform load	1.9039	1.3258
Sinusoidal load	1.9021	1.3190

Start by considering the beam with a constant generic *(EI)* value all along the beam. This leads to the identification of zero-moment points in a very easy way. Then, apply the calculated $(EI)_{\text{eff}}$ to the respective lengths of the single parts. Then carry the structural analysis on the entire beam with different stiffness values along the beam. A simple commercial computer program can do this very easily. Find new zero-moment points and repeat the procedure. With up to three repetitions a stable solution is usually obtained. Of course, the hypothesis of a linear behaviour of the beam must be retained, but this is usually a well accepted assumption (see Section 21.6).

Another well accepted assumption is the following: if fastener spacing s is not uniform along the length of the beam but varies according to the shear force, an effective, constant, s_{ef} is assumed:

$$s_{\text{ef}} = 0.75 s_{\text{e}} + 0.25 s_{\text{m}} \quad \text{with} \quad s_{\text{e}} < s_{\text{m}} < 4 s_{\text{e}} \tag{21.27}$$

where s_{e} is the spacing at the ends of the beam and s_{m} is the spacing in the middle of the beam.

When the value and the concept of an effective stiffness has been accepted, then it is easy to calculate deflections, and stresses on sections (Larsen and Riberholt, 1983).

21.3.2 General Formulae

Extending the exercise to 'I' and 'Box' cross-section beams, it is possible to obtain expressions for effective bending stiffness and stresses. The formulas reported below are also valid for not double symmetric cross-sections and, by default, also for 'T' shaped sections where $h_3 = 0$ (Figure 21.19, formula 33).

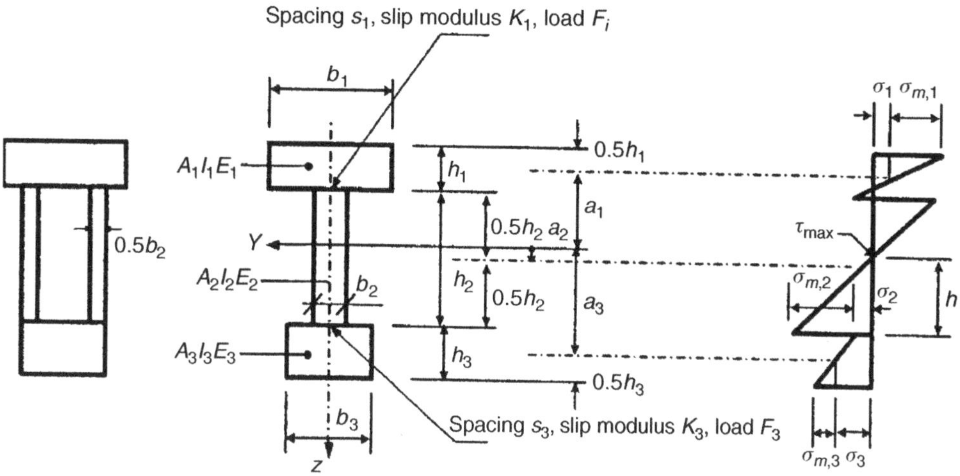

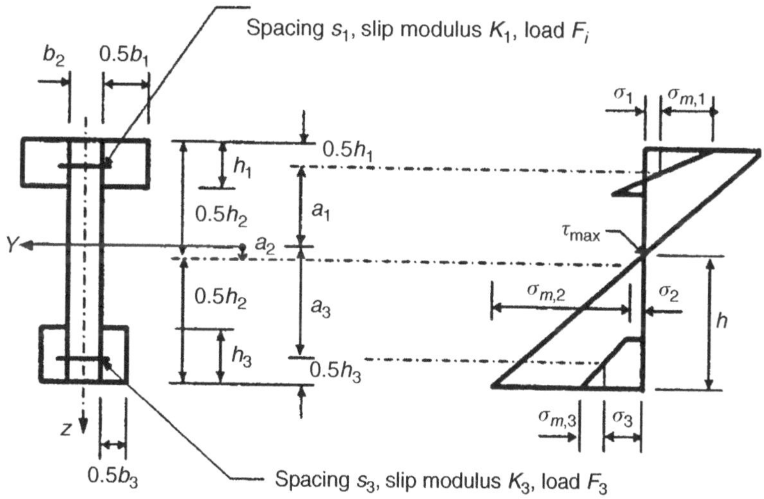

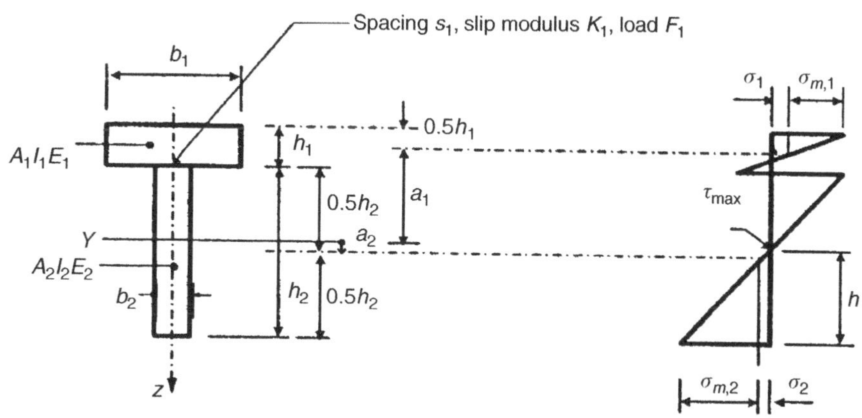

Figure 21.19 Cross-section (left) and distribution of bending stresses (right) for box, I and T shaped composite beams. (Reproduced by permission of Danish Building and Urban Research)

To give general applicability to the formulas below, more γ_i values are shown, each one affecting one single part of the section, instead of just one single γ value.

The only consequence of this assumption, which is purely formal, is the following: in Figure 21.19 the y-axis does not cross the symmetry axis at the centroid of the corresponding glued cross-section (as in Figure 21.17), but at the centre of axial rigidities of the cross-section, single parts are reduced by γ_i factors. This means that modifying the connection stiffness at this point will change its position.

The effective bending stiffness will be assumed as

$$(EI)_{\text{eff}} = \sum_{l=1}^{3} (E_i I_i + \gamma_i E_i A_i a_i^2) \quad (21.28)$$

where

$$A_i = b_i h_i \quad (21.29)$$

$$I_i = \frac{b_i h_i^2}{12} \quad (21.30)$$

$$\gamma_2 = 1 \quad (21.31)$$

$$\gamma_i = \frac{1}{1 + \dfrac{\pi^2 E_i A_i}{k_i l^2}} \quad \text{for } i = 1 \text{ and } i = 3 \quad (21.32)$$

$$a_2 = \frac{\gamma_1 E_1 A_1 \dfrac{(h_1 + h_2)}{2} - \gamma_3 E_3 A_3 \dfrac{(h_2 + h_3)}{2}}{\sum_{l=1}^{3} \gamma_i E_i A_i} \quad (21.33)$$

Normal stresses are

$$\sigma_i = \frac{\gamma_i E_i a_i M}{(EI)_{\text{eff}}} \quad (21.34)$$

$$\sigma_{m,i} = \frac{0.5 E_i h_i M}{(EI)_{\text{eff}}} \quad (21.35)$$

The maximum shear stresses in the web member due to the shear V are

$$\tau_{2,\max} = \frac{\gamma_3 E_3 A_3 a_3 + 0.5 E_2 b_2 h_2}{b_2 (EI)_{\text{eff}}} V \quad (21.36)$$

The load on a fastener is ($i = 1$ and $i = 3$)

$$Q_i = \frac{\gamma_i E_i A_i a_i s_i}{(EI)_{\text{eff}}} V \quad (21.37)$$

21.4 SPECIAL BEAMS

21.4.1 Beams with Thin Web

For I beams with thin webs of plywood, particle boards or fibre boards, the shear deformation of the web must be taken into account. This is done simply by 'adding' the shear deformation of the web to the slip deformation of the upper and lower connections:

$$\gamma_i = \frac{1}{1 + \dfrac{\pi^2 E_i A_i}{l^2}\left(\dfrac{1}{k_i} + \dfrac{h_2}{2Gb_2}\right)}$$

where G is the tangent modulus of elasticity of the web.

21.4.2 Beams with Discontinuous Connection

Beams built with internal packs (Figure 21.20), external battens (Figure 21.21) or diagonals (Figure 21.22) can be analysed as a composite structure where the timber flanges are basically seen as two parts joined via a 'deformable link', which can be represented as a web with zero area with an equivalent shear stiffness.

Beams with Packs

A beam can be analysed utilising formulae (21.28) to (21.37) by simply putting A_2 and I_2 to zero, but with an equivalent unit slip modulus:

$$k_1 = \frac{l_1}{nK_1} + \frac{(h_1 + h_2)^2 l_1^2}{48 E_1 I_1} \quad (21.38)$$

where K_1 is the slip modulus of one of the n nails as shown in Figure 21.20 (Larsen and Riberholt, 1983). Due to the assumed symmetry about the Y-axis, $K_3 = K_1$.

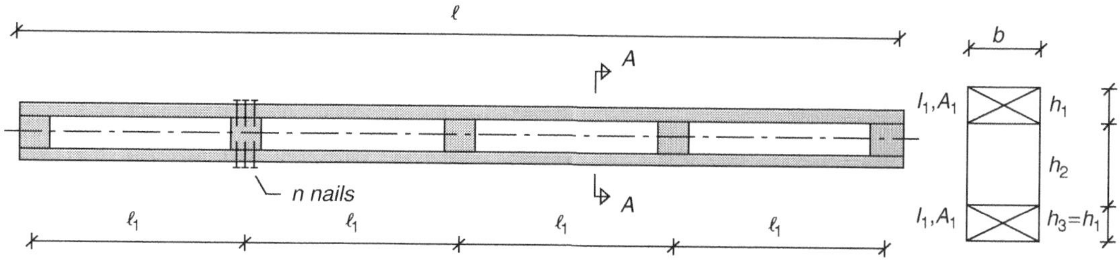

Figure 21.20 Composite beam with nailed wooden packs (Reproduced by permission of Danish Building and Urban Research)

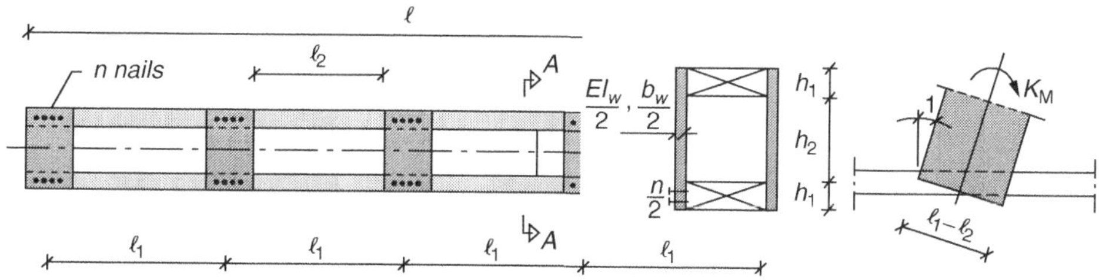

Figure 21.21 Beam with nailed battens (drafted by J. Heergard and O. Nesdam. Reproduced by permission of Danish Building and Urban Research)

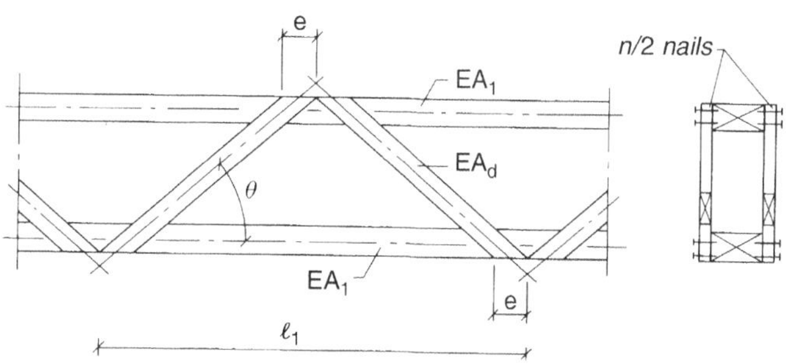

Figure 21.22 Lattice beam. Diagonals can be made of wood or steel. (Reproduced by permission of Danish Building and Urban Research)

Beams with Battens

This kind of beam, also known as a Vierendeel beam, has a greater deformation than the packed one due to the batten rotation (K_m is the rotational stiffness in Figure 21.22) and their flexural and shear deformation (G modulus appears in formula (21.39)). Analysis can be made as for a three member beam with an internal member with no area or stiffness: thickness and other geometric parameters of the web appear only in formula (21.39) to quantify the equivalent slip modulus (Larsen and Riberholt, 1983):

$$k_1 = \frac{l_1}{nK_1} + (h_1 + h_2)^2 l_1 \left[\frac{l_1}{48E_1I_1} + \frac{h_1 + h_2}{24(EI)_w} + \frac{1}{4K_m}\right] + \frac{h_2 l_1}{2Gb_w(l_1 - l_2)} \tag{21.39}$$

where

$$I_w = \frac{1}{12} b_w (l_1 - l_2)^3 \qquad (21.40)$$

and $k_3 = k_1$

Lattice Beams

For lattice beams the same approach can be followed. With reference to Figure 21.22, the equivalent unit length slip modulus is given by formula (21.41), where K_d is the stiffness of the connection between the diagonal and one chord, i.e. $K_d = nK_1$ if n nails of K_1 slip modulus each are present. It is assumed that $k_3 = k_1$:

$$k_1 = \frac{l_1}{2\cos^2\theta}\left[\frac{l_1}{4E_d A_d \cos\theta} + \frac{1}{K_d} + \frac{l_1 e^2}{6E_1 I_1}\left(1 - \frac{2e}{l_1}\right)^2 \sin^2\theta\right] \qquad (21.41)$$

21.5 BUILT-UP COLUMNS

Once the concept of an effective stiffness has been accepted, the verification of composite timber columns can be performed as for solid timber cross-section columns, just using the expressions determining the effective stiffness and thus the effective slenderness ratio.

21.5.1 Full-length Jointed Columns

The effective slenderness ratio is taken as

$$\lambda_{\text{ef}} = l\sqrt{\frac{E_{\text{mean}} A_{\text{tot}}}{(EI)_{\text{eff}}}} \qquad (21.42)$$

where E_{mean} is the average E value of the connected parts, l is the actual length of the column supposed to be simply supported at its ends, and $(EI)_{\text{eff}}$ calculated as in Section 21.3.

For the T cross-section (bottom of Figure 21.19) and the I cross-section (top of Figure 21.19), the load-carrying capacity for deflection in the Y-direction is found as the sum of the load-carrying capacities of the individual members, i.e. the stiffening effect that the members might have on each other is not taken into account.

The load on fasteners can be calculated according to formula (21.36), where V is supposed to have the following values:

$$V = \begin{cases} F_c/(120\,k_c), & \lambda_{\text{ef}} \le 30 \\ F_c\lambda_{\text{ef}}/(3600\,k_c), & 30 < \lambda_{\text{ef}} \le 60 \\ F_c/(60\,k_c), & 60 < \lambda_{\text{ef}} \end{cases} \qquad (21.43)$$

where F_c is the axial force supposed to act at the geometric centre of gravity of the column, and k_c is the column factor determined as for a corresponding column with rigid joints between the cross-section members.

In cases where small moments resulting from, for example, a column's own weight acting apart from the axial load, the usual interaction formulae can be used for the stresses determined above.

21.5.2 Spaced Columns with Nailed Packs or Battens

In spaced mechanically jointed columns (Figure 21.23), the basic concept for verification remains the evaluation of an effective slenderness. Nevertheless, to take into account the interaction between the instability of the entire column and the instability of single shafts between two consecutive packs or battens, the resulting formula is a sort of average square root of the slenderness of the entire column and the slenderness of single shafts. Therefore, the formula for the effective slenderness is necessarily very different from the previous one for full-length jointed columns:

$$\lambda_{\text{ef}} = \sqrt{\lambda^2 + \mu_{pb}\frac{m}{2}\lambda_1^2} \qquad (21.44)$$

where

$$\lambda = l\sqrt{\frac{A_{\text{tot}}}{I_{\text{tot}}}}$$

is the slenderness ratio for a solid column (glued) with the same cross-section and an l free length,

$$\lambda_1 = l_1\sqrt{\frac{A}{I}}$$

is the slenderness ratio for the shafts, with $A =$ area of one shaft, $A_{\text{tot}} = mA =$ total area, m being

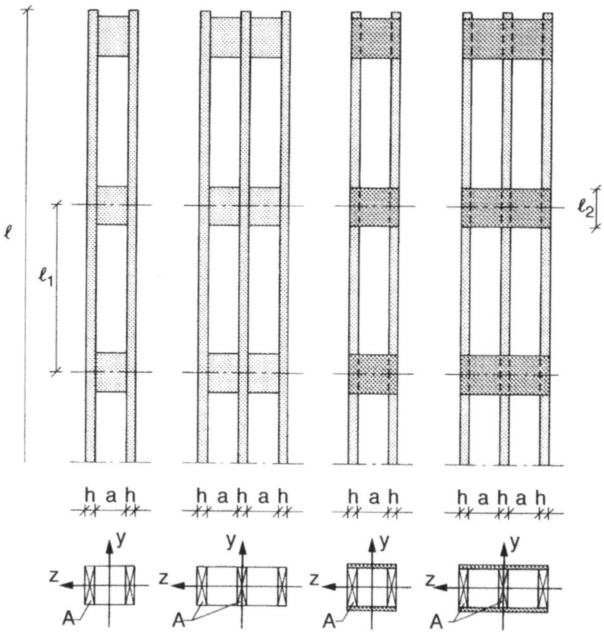

Figure 21.23 Spaced columns (drafted by J. Heergard and O. Nesdam. Reproduced by permission of Danish Building and Urban Research)

the number of shafts, I = minor moment of inertia of one shaft, I_{tot} = total moment of inertia about the Y-axis as for a solid column, l_1 = the distance between midpoints of packs or battens, and μ_{pb} = a factor given in Table 21.2.

- The structure is assumed to be symmetric about the Y- and Z-axes of the cross-section, and the columns are solely subjected to axial loads.
- The cross-section is composed of two, three or four identical shafts, and the number of free fields is at least three, i.e. the shafts are at least jointed at the ends and in the third points.

Table 21.2 μ_{pb} factor for spaced columns

	Packs		Battens (nailed)
	nailed	bolted, with connectors	
Long-term loading	4	3.5	6
Short-term loading	3	2.5	4.5

- The free distance between the shafts is not greater than five times the lamella thickness ($a/h \leq 5$).
- The length of the packs should satisfy the condition $l_2 \geq 1.5$.
- For nailed joints, in each section there should be at least four nails or two bolts with metal plate connectors. For nailed joints at the ends, there should be at least four nails in a row in the longitudinal direction of the column.
- Battens should be made of structural plywood, and their length should satisfy the condition $l_2/a \geq 2$.

The joints, packs and battens are designed for a shear force V. The load on the fasteners and battens or packs can be calculated as stated in Figure 21.24 assuming for V the same as in formula (21.43) and for T:

$$T = \frac{Vl_1}{a_1} \tag{21.45}$$

where $a_1 = a + h$.

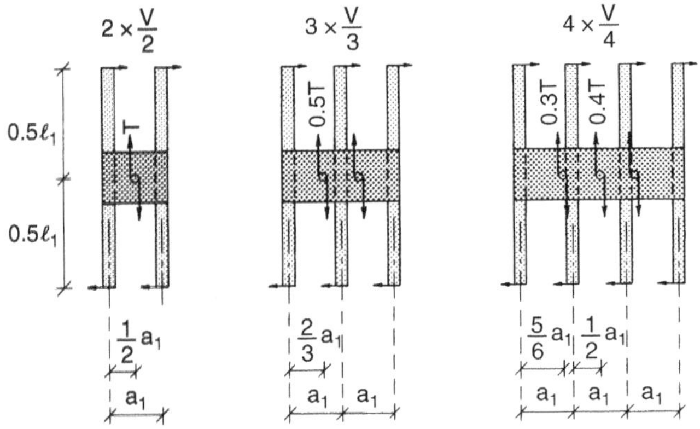

Figure 21.24 Shear force distribution and loads on battens or packs. (Reproduced by permission of Danish Building and Urban Research)

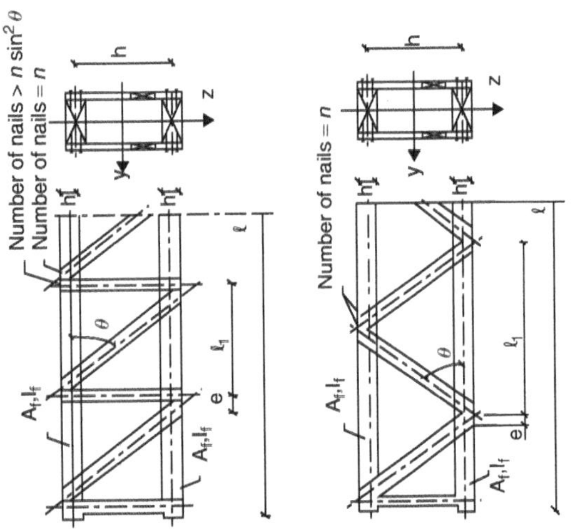

Figure 21.25 Lattice columns, *N*-truss (left), *V*-truss (right). (Reproduced by permission of Danish Building and Urban Research)

21.5.3 Lattice Columns with Nailed Joints

Also, in this case the basic point are the expressions to determine an effective moment of inertia, and thus the effective slenderness ratio, whereupon the critical column stress is determined as for a column of solid timber with the same slenderness ratio.

- The structure is assumed to be symmetric about the Y- and Z-axes of the cross-section.
- However, the lattice of the two sides are allowed to be staggered along the length $l_1/2$.
- There should be at least three fields, i.e. $l > 3l_1$.
- The column should be designed for a shear force V as given in formula (21.43), i.e. the diagonals and joints should be designed for $V/\sin\theta$.
- In nailed structures, at least four nails per shear should be used in each diagonal in each nodal point. At the end, bracings should be used.
- The slenderness ratio for the individual flange corresponding to the length l_1 must not exceed 60, i.e. $\lambda_{\text{eff}} \leq 60$.
- It assumed that no local buckling is occurring in the flanges corresponding to the column length l_1.

The load-carrying capacity corresponding to a deflection in the Y-direction is equal to the sum of the load-carrying capacity of the flanges for deflection in this direction.

For deflection in the Z-direction, the load carrying capacity is calculated as for a corresponding column of solid timber, but instead of the geometrical slenderness ratio of the column:

$$\lambda_{\text{tot}} = l\sqrt{\frac{2A_f}{I_{\text{tot}}}} \qquad (21.46)$$

the effective slenderness ratio is used:

$$\lambda_{\text{eff}} = \lambda_{\text{tot}}\sqrt{1+\mu_{VN}} \qquad (21.47)$$

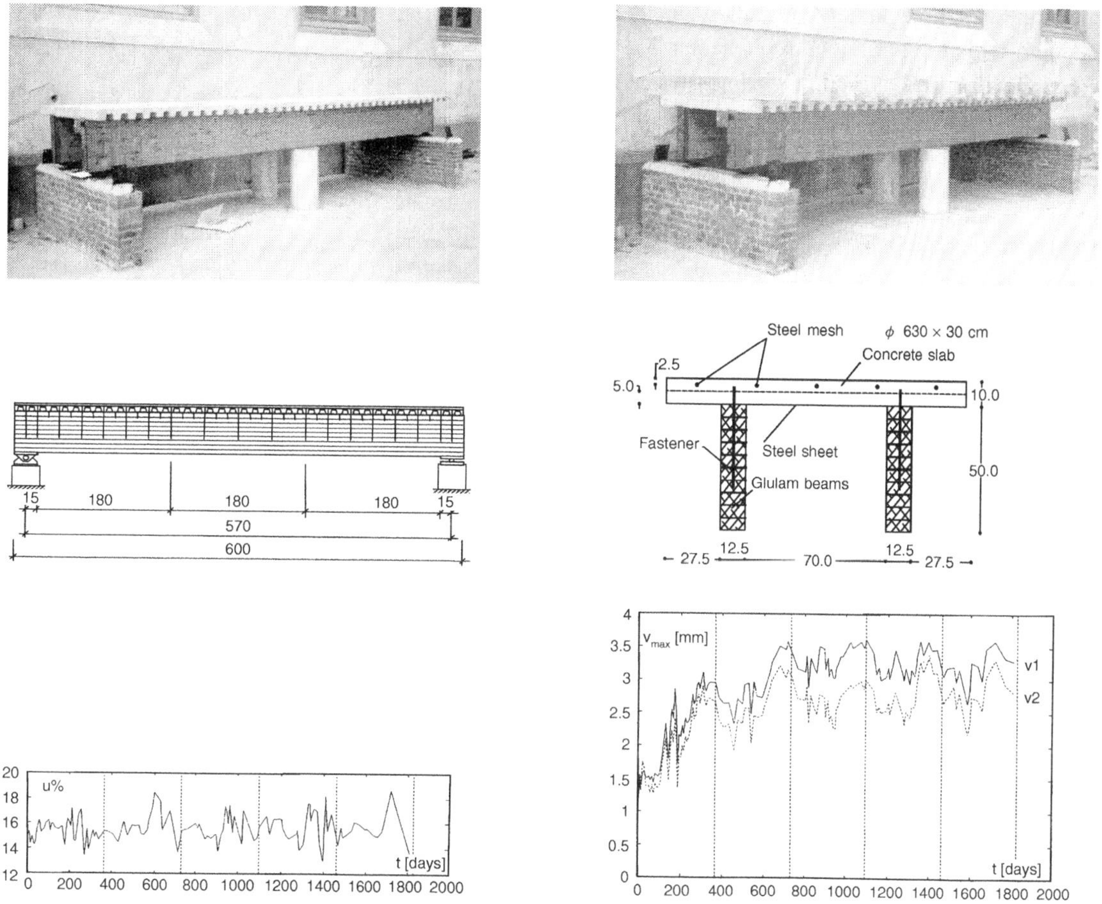

Figure 21.26 Outdoor timber-concrete composite beams loaded with quasi-permanent design load and with moisture content in the wood varying between 13% and 18% in five years time shows stabilised deflection after two years

where μ_{VN} assumes the values reported in the following formulae:

For a nailed V-truss:

$$\mu_V = 25\frac{hE_{\text{mean}}A_f}{l^2nK_u\sin 2\theta} \tag{21.48}$$

For a nailed N-truss:

$$\mu_N = 50\frac{hE_{\text{mean}}A_f}{l^2nK_u\sin 2\theta} \tag{21.49}$$

where n is the number of nails in a diagonal and other notations are defined in Figure 21.25. If a diagonal consists of two or more pieces, n is the sum of the nails (not the number of nails per shear plane), and K_u is the slip modulus of one nail for the ultimate limit states.

21.6 DESIGN

21.6.1 Serviceability Limit State – Long-term Behaviour

For serviceability limit states, the single fastener slip coefficient K_i is assumed to correspond to $K_{i,s}$ (Figure 21.16), with linear-elastic connection behaviour.

When calculating long-term performance, a simplified assumption is made to take into account creep phenomena by conventionally adopting reduced stiffness values of the single parts. This can be done, for example, by dividing the modulus of elasticity by a factor $(1 + k_{\text{def}})$, where k_{def} is a parameter that takes into account the increase of deflection with time under constant load, due to creep and moisture variations see Chapter 13.

If a specific load is acting during time under its nominal value, the relevant k_{def} has to be multiplied by an appropriate reduction coefficient, i.e. the ratio between the quasi-permanent value and the nominal value of the variable action. In this way, not only long-term deflection, but also stress re-distribution across the cross-section, can be evaluated with a sufficient approximation.

The use of conventionally-reduced E moduli to predict long-term deflection of composite beams may appear a very crude approach. Actually, one of the most extreme cases is composite timber-concrete beams, where the two materials behave very differently: they have different rheological behaviour, including very different reaction to relative air humidity and air temperature changes. Theoretical work is still underway, though experimental evidence obtained under very severe external testing conditions proves that deformation tends to stabilise after some years under quasi-permanent design load, see Figure 21.26. Therefore, correct final deformation values can be predicted by adopting the above-mentioned calculation procedure by choosing appropriate k_{def} values.

21.6.2 Ultimate State – Short-term Behaviour

At ultimate limit state calculations, for the sake of simplicity, a linear behaviour of connections is also assumed, but with a reduced stiffness (secant), i.e. $K_{i,u} = 2/3K_{i,s}$, to take into account the bigger deformability of connections entering the plastic deformation field. Such an assumption works reasonably well unless connection strength and stiffness are under-estimated for some reason. In such a case, in fact, a simply linear elastic behaviour of the beam will be achieved (Figure 21.27), which means that the ductile behaviour of joints does not necessarily mean ductile behaviour of the composite beam.

21.7 CONCLUSIONS

Composite timber structures are an economic construction that is easy to build and easy to calculate, when the basic parameters are known. They provide good architectural appearance

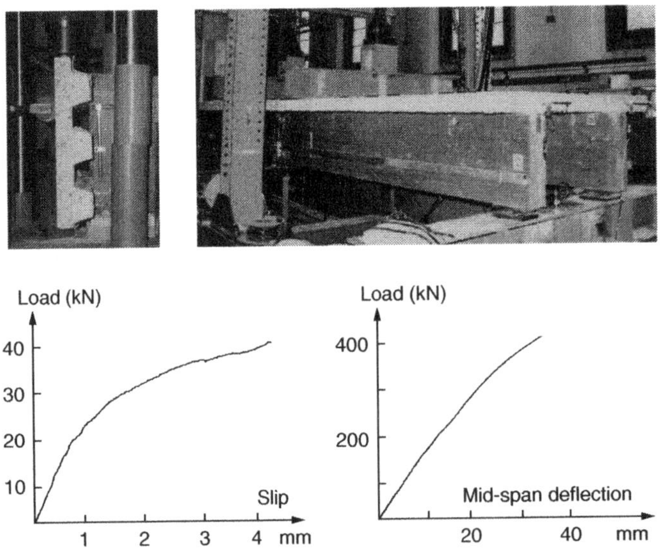

Figure 21.27 The same beams as Figure 21.26 after long-term test outdoors have been loaded until failure. The diagram on the left shows the load-slip behaviour of basic connections under shear loading; the diagram on the right shows the actual load-deflection behaviour of the beams

coupled with good mechanical short- and long-term performance. Therefore, they have not been superceded by glulam techniques, and do present great possibilities in today's construction market both for new and renovated buildings.

REFERENCES

Mohler K. (1956) Über das Tragverhalten von Biegeträgern und Druckstäben mit zusammengesteztem Querschnitt und nachgiebigen Verbindungsmitteln. Habilitation, Technische Universität Karlsruhe, Germany.

Larsen H.J. and Riberholt H. (1983) *Trækonstruktioner – Beregning*. SBI-Anvisning 135 – Statens Byggeforskningsinstitut, Hørsholm, Denmark.

Kreutzinger H. (1994) Mechanically jointed beams and columns. Lecture B11, *STEP-Timber Engineering 1*, Centrum Hout, Almere, The Netherlands.

Index

Note: Figures and Tables are indicated by *italic page numbers*

Printed and bound by CPI Group (UK) Ltd, Croydon, CR0 4YY
18/09/2026
14964994-0001